BIRKHÄUSER

Operator Theory: Advances and Applications

Vol. 200

Founded in 1979 by Israel Gohberg

A State Space Approach to Canonical Factorization with Applications

Harm Bart
Israel Gohberg
Marinus A. Kaashoek
André C.M. Ran

L | Linear
O | Operators
&
L | Linear
S | Systems

Birkhäuser

Authors:

Harm Bart
Econometrisch Instituut
Erasumus Universiteit Rotterdam
Postbus 1738
3000 DR Rotterdam
The Netherlands
e-mail: bart@ese.eur.nl

Israel Gohberg (Z"L)

Marinus A. Kaashoek, André C. M. Ran
Department of Mathematics, FEW
Vrije Universiteit
De Boelelaan 1081A
1081 HV Amsterdam
The Netherlands
e-mail: m.a.kaashoek@few.vu.nl
acm.ran@few.vu.nl

2010 Mathematics Subject Classification: Primary 46C20; 47A48, 47A56, 47A68, 93B36; secondary: 42A85, 82D75

Library of Congress Control Number: 2010923703

Bibliographic information published by Die Deutsche Bibliothek.
Die Deutsche Bibliothek lists this publication in the Deutsche Nationalbibliografie; detailed bibliographic data is available in the Internet at http://dnb.ddb.de

ISBN 978-3-7643-8752-5

P.O. Box 133, CH-4010 Basel, Switzerland
Part of Springer Science+Business Media
Printed on acid-free paper produced from chlorine-free pulp. TCF∞
Printed in Germany

ISBN 978-3-7643-8752-5
e-ISBN 978-3-7643-8753-2

9 8 7 6 5 4 3 2 1

www.birkhauser.ch

Contents

Preface

The present book deals with canonical factorization problems for different classes of matrix and operator functions. Such problems appear in various areas of mathematics and its applications. The functions we consider have in common that they appear in the state space form or can be represented in such a form. The main results are all expressed in terms of the matrices or operators appearing in the state space representation. This includes necessary and sufficient conditions for canonical factorizations to exist and explicit formulas for the corresponding factors. Also, in the applications the entries in the state space representation play a crucial role.

The theory developed in the book is based on a geometric approach which has its origins in different fields. One of the initial steps can be found in mathematical systems theory and electrical network theory, where a cascade decomposition of an input-output system or a network is related to a factorization of the associated transfer function.

Canonical factorization has a long and interesting history which starts in the theory of convolution equations. Solving Wiener-Hopf integral equations is closely related to canonical factorization. The problem of canonical factorization also appears in other branches of applied analysis and in mathematical systems theory, in H_∞-control theory in particular.

The first book devoted to the state space factorization theory was published in 1979 as the monograph "Minimal factorization of matrix and operator functions," Operator Theory: Advances and Applications **1**, Birkhäuser Verlag, written by the first three authors. Some of the factorization results published in the 1979 book appeared there in print for the first time.

The present book is the second book written by the four of us in which the state space factorization method is systematically used and developed further. In the earlier book [20], published in 2008, the emphasis is on non-canonical factorizations and degree 1 factorizations, in particular. In the present book we concentrate on canonical factorizations. Together both books present a rich and far reaching update of the 1979 monograph [11].

In the present book the emphasis is on canonical factorization and symmetric factorization with applications to different classes of convolution equations. For the latter we have in mind the transport equation, singular integral equations, equations with symbols analytic in a strip, and equations involving factorization of non-proper rational matrix functions. A large part of the book will deal with factorization of matrix functions satisfying various symmetries. A main theme will be the effect of these symmetries on factorization and how the symmetries can be used in effective ways to get state space formulas for the factors. Applications to H_∞-control theory, which have been developed in the 1980s and 1990s, will also be included. The text is largely self-contained, and will be of interest to experts and students in mathematics, sciences and engineering.

The authors gratefully acknowledge a visitor fellowship for the second author from the Netherlands Organization for Scientific Research (NWO), and the financial support from the School of Economics of the Erasmus University at Rotterdam, from the School of Mathematical Sciences of Tel-Aviv University and the Nathan and Lily Silver Family Foundation, and from the Mathematics Department of the Vrije Universiteit at Amsterdam. These funds allowed us to meet and to work together on the book for different extended periods of time in Amsterdam and Tel-Aviv.

The authors *Amsterdam – Rotterdam – Tel-Aviv, Summer 2009*

Postscript

On Monday October 12, 2009, Israel Gohberg, the second author of this book, passed away at the age of 81. At that time the preparation of the book was in a final phase and only some minor work had to be done. Israel Gohberg was one of the initiators using state space methods in solving problems appearing in various branches of mathematical analysis and its applications. His fundamental insights and inspiring leadership have been driving forces in our joint work.

Chapter 0

Introduction

This monograph presents a unified approach for solving canonical factorization problems for different classes of matrix and operator functions. The notion of canonical factorization originates from the theory of convolution equations. For instance, canonical factorization, provided it exists, allows one to invert Wiener-Hopf, Toeplitz and singular integral operators, and when the factors are known one can also build explicitly the inverses of these operators. The problem of canonical factorization also appears in various branches of applied analysis, in linear transport theory, in interpolation theory, in mathematical systems theory, in particular, in H_∞-control theory.

The various matrix and operator functions that are considered in this book have in common that they appear in a natural way as functions of the form

$$W(\lambda) = D + C(\lambda I - A)^{-1}B \tag{1}$$

or (after a suitable transformation) can be represented in this form. In the above formula λ is a complex variable, and A, B, C, and D are matrices or linear operators acting between appropriate Banach or Hilbert spaces, which in this book often will be finite dimensional. When the underlying spaces are all finite dimensional, A, B, C, and D can be viewed as matrices and the function W is a rational matrix function which is analytic at infinity. From mathematical systems theory it is known that, conversely, any rational matrix function which is analytic at infinity admits a representation of the above form. In systems theory the right hand side of (1) is called a *state space realization* of the function W, and one refers to the space in which A is acting as the *state space.*

The method of factorization employed in this book uses realizations as in (1), and for this reason it is referred to as the *state space method.* It allows one to deal with factorization from a geometric point of view. This state space factorization approach has its origins in different fields, for instance, in the theory of non-selfadjoint operators [27], [141], in mathematical systems theory and electrical network theory [23], [95], [94], and in the factorization theory of matrix polynomials [67], [131]. In all three areas a state space representation of the function to be factored is used, and the factors are also expressed in state space form.

The first book to deal with factorization problems in a systematic way using the state space approach is the monograph [11] of the first three authors. This monograph appeared in 1979, very soon after the first main results were obtained. In fact, some of the factorization results were published in [11] for the first time.

The present book is the second book written by the four of us in which the state space factorization method is systematically used and developed further. In our first book [20], published in 2008, the emphasis is on non-canonical factorizations and degree 1 factorizations, in particular. In the present book we concentrate on canonical factorizations. As a result the overlap between the main parts of the two books is minor. Together both books present a rich and far reaching update of the 1979 monograph [11].

In the present book special attention is paid to various factorizations with additional symmetries such as spectral factorization, inner-outer factorization, and J-spectral factorization. The latter require elements of the theory of spaces with an indefinite metric. Factorizations with symmetries appear in a natural way in H_∞-control problems and the related Nehari approximation problem. In fact, the latter problems are the main topic of the final part of the book. We also deal with applications to problems in the theory of algebraic Riccati equations, to inversion problems for Wiener-Hopf, Toeplitz and singular integral operators, and to Riemann-Hilbert problems. The linear transport equation from mathematical physics is another important area of application in this book. It requires infinite dimensional realizations of a special type.

We have made an effort to make the text reasonably self-contained. For that reason we included some known material about realizations, minimal factorizations of rational matrix functions, angular operators, and the theory of matrices in indefinite inner product spaces. In the final part we also briefly review elements of control theory of linear systems.

Not counting the present introduction, the book consists of 20 chapters grouped into 7 parts. We shall now give a short description of the contents of the book.

Part I. The first part has a preparatory character. In the first chapter we review the role of canonical factorization in inverting Wiener-Hopf integral operators and block Toeplitz operators. Also the role of this factorization in solving singular integral equations is described. The second chapter presents in detail the elements of the state space method that are used in this book.

Part II. This part starts with the canonical factorization theorem for rational matrix functions in state space form. This theorem is then used to invert explicitly Wiener-Hopf, Toeplitz and singular integral operators with a rational matrix symbol, with the inverses being presented explicitly in state space formulas. For rational matrix symbols the solution to the homogeneous Riemann-Hilbert boundary value problem is also given in state space form. In the first chapter of this part we consider proper rational matrix functions, that is, rational matrix functions that are analytic at infinity. The case of non-proper rational symbols is treated in the second chapter of this part. In this case the realization (1) is replaced by

$$W(\lambda) = I + C(\lambda G - A)^{-1}B, \tag{2}$$

where I is an identity matrix, G and A are square matrices, and B and C are matrices of appropriate sizes. A square rational matrix function, proper or not, always admits such a realization. We develop this realization result, and prove a canonical factorization theorem for the realization (2). As an application we solve the homogeneous Riemann-Hilbert boundary value problem for an arbitrary rational matrix symbol.

Part III. In this part we carry out a program analogous to that of the second part, but now for certain classes of non-rational matrix and operator functions. For instance, for matrix functions analytic on a strip but not at infinity we develop a realization theory, prove a canonical factorization theorem in state space form, and develop its applications to Wiener-Hopf integral equations. A new feature is that the problems involved require us to employ realizations with an unbounded main operator A and deal with curves cutting through the spectrum of this main operator. In this part it is also shown that, after an appropriate modification, the state space method can be used to solve the integro-differential equation appearing in linear transport theory, which forces us to use realizations of operator-valued functions. In the final chapter of this part we make an excursion into non-canonical Wiener-Hopf factorization for analytic operator-valued functions on a curve, and identify the so-called factorization indices in state space terms.

Part IV. The fourth part deals with factorization of rational matrix functions that have Hermitian values on the imaginary axis, the real line or the unit circle. In the analysis of such functions, minimal realizations play an important role. These are realizations of which the order of the state matrix in (1) is a small possible. Also the so-called state space similarity theorem, which tells us that a minimal realization is unique up to a basis transformation in the state space, enters into the analysis. These facts are reviewed in the first chapter of this part. In this first chapter, using the notion of local minimality, also the concept of a pseudo-canonical factorization relative to a curve is introduced and studied for rational matrix functions with singularities on the given curve. The effect on minimal realizations of the function having Hermitian values on the imaginary axis, the real line or the unit circle is described in the second chapter of this part. This then leads to the construction of special canonical and pseudo-canonical factorizations with additional relations between the factors. Included are spectral factorization for positive definite rational matrix functions and pseudo-spectral factorization for nonnegative rational matrix functions. In the final chapter we present (without proofs) some background material on matrices in indefinite inner product spaces, and review the main results from this area that are used in this book.

Part V. In this part the canonical factorization theorem is presented in a different way using the notion of an angular subspace and Riccati equations. In this case one has to look for angular subspaces that are also spectral subspaces, and the solutions of the Riccati equation must have additional spectral properties. These results, which have a preliminary character, are presented in the first chapter of this part. In the second chapter we introduce the symmetric algebraic Riccati equation, and describe spectral factorization as well as pseudo-spectral factorization in terms of Hermitian solutions of such a Riccati equation. In the final chapter of this part we continue the study of rational matrix functions that take Hermitian values on certain curves. The emphasis will be on rational matrix functions that have Hermitian values for which the inertia is independent of the point on the curve. Such functions may still admit a symmetric canonical factorization, provided we allow for a constant Hermitian invertible matrix in the middle. Such a factorization is commonly known as a J-spectral factorization. Necessary and sufficient conditions for its existence are given, first in terms of invariant subspaces and then in terms of solutions of a corresponding symmetric algebraic Riccati equation. We also study the question when a function which admits a left J-spectral factorization admits a right J-spectral factorization too.

Part VI. In this part we study rational matrix functions that are unitary or of the form identity matrix plus contractions, and rational matrix functions that have a positive real part. Because of the state space similarity theorem, these additional symmetries can be restated in terms of special properties of the minimal realizations of the rational matrix functions considered. These reformulations involve an algebraic Riccati equation. The results are known in systems theory as the bounded real lemma and the positive real lemma, respectively. They allow us to solve related canonical and pseudo-canonical factorization problems in state space form. In the final chapter of this part realizations are used to analyze rational matrix functions of which the values on the imaginary axis are J-unitary matrices. Solutions to various factorization problems are given. Special attention is paid to factorization of J-unitary rational matrix functions into J-unitary factors. In this chapter we also discuss problems of embedding a contractive rational matrix function into a unitary rational matrix function of larger size.

Part VII. In this part the state space theory of J-spectral factorization, developed in the final chapter of the fifth part, is used to solve H_∞ problems. The first chapter of this part contains the solution of the Nehari interpolation problem for rational matrix interpolants. The second chapter presents a short review of elements of control theory that play an important role in the third (and final) chapter of this part. This final chapter is about H_∞-control. Here we use the J-spectral factorization theory to obtain the solutions of some of the main problems in this area, namely the standard problem, the one-sided problem, and the full model matching problem.

As the description of the contents given above shows, the emphasis in the book is mainly on rational matrix functions and finite dimensional realizations. An exception is Part III. The latter part deals with non-rational matrix functions and operator-valued functions, and it uses realizations that have an infinite dimensional state space. Other exceptions are Chapter 2 in Part I and Chapter 12 in Part V. For the material in the other chapters of the book, in particular, in Parts IV–VII, often extensions to an infinite dimensional setting exist; they require appropriate modifications. See, e.g., the books [5], [35], [42], [73], and the references therein.

A few remarks about terminology and notation

At the end of this book, after the bibliography, the reader will find a List of Symbols and an Index. The latter contains in alphabetical order the various terms that are used in this book with references to the pages where they are introduced. In addition, we would like to mention the following.

In the sequel, whenever convenient, a $p \times q$ matrix with complex entries will be identified with the (linear) operator from $\mathbb{C}^q$ into $\mathbb{C}^p$ defined by the canonical action of the matrix on the standard orthogonal basis of $\mathbb{C}^q$. Conversely, a linear operator from $\mathbb{C}^q$ into $\mathbb{C}^p$ is identified with its $p \times q$ matrix representation with respect to the standard orthogonal bases of $\mathbb{C}^q$ and $\mathbb{C}^p$.

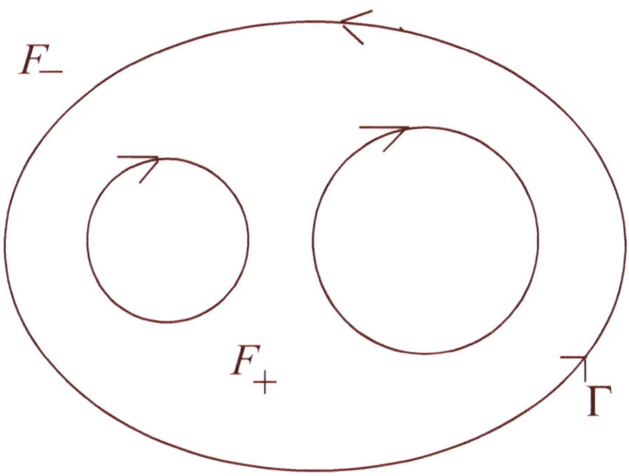

Throughout the word "operator" refers to a bounded linear transformation acting between Banach or Hilbert spaces (finite or infinite dimensional). We assume the reader to be familiar with Sections I.1 and I.2 in [51] which contain the standard spectral theory of operators, including the notion of a Riesz projection and the corresponding functional calculus (see, also Chapter V in [144]). In particular, we shall often use the notions of a Cauchy domain and Cauchy contour which are defined as follows. A *Cauchy domain* is an open set in the complex plane $\mathbb{C}$ consisting of a finite number of components such that its boundary is composed of a finite number of simple closed non-intersecting rectifiable curves. A *Cauchy contour* Γ is the positively oriented boundary of a bounded Cauchy domain. We write F_+ for the interior domain of Γ, and F_- for the exterior domain, i.e., the complement of the closure $\overline{F_+}$ of F_+ in the Riemann sphere $\mathbb{C}_\infty = \mathbb{C} \cup \{\infty\}$. The picture on the previous page illustrates this notion. We shall also work with the extended real line and the extended imaginary axis as contours on the Riemann sphere $\mathbb{C}_\infty$. For the real line the orientation will be from left to right and for the imaginary axis from bottom to top. Thus for the extended real line the interior domain is the open upper half plane, which will be denoted by $\mathbb{C}_+$; for the extended imaginary axis it is the open left half plane, which is denoted by $\mathbb{C}_{\text{left}}$.

We shall also freely use the Lesbesgue integral and related L_p spaces (see, e.g., Appendix 2 in [53]). Functions which are equal almost everywhere (shorthand: a.e.) are often identified, sometimes without explicitly mentioning this.

Finally, when dealing with inner-outer factorization, we shall always assume that the outer factor is invertible outer (see Section 17.6). In the outer-co-inner factorizations considered in this book, the outer factor will be assumed to be invertible outer as well.

Part I
Convolution equations, canonical factorization and the state space method

This part has a preparatory character. It consists of two chapters. In the first chapter we review the role of canonical factorization in inverting Wiener-Hopf integral operators and block Toeplitz operators. The role of this factorization in solving singular integral equations is described as well. The second chapter presents in detail the basic elements of the state space method that are used throughout this book. The central notion is that of a realization of a matrix or operator function. Three important operations on realizations are studied.

Chapter 1

The role of canonical factorization in solving convolution equations

This chapter has a preparatory character. We review (without giving proofs) the role of canonical factorization in inverting systems of convolution equations. The chapter consists of three sections. Section 1.1 deals with Wiener-Hopf integral equations, Section 1.2 with block Toeplitz equations, and Section 1.3 with singular integral equations.

1.1 Wiener-Hopf integral equations and factorization

In this section we outline the factorization method of [61] to solve systems of Wiener-Hopf integral equations. Such a system may be written as a single vector-valued *Wiener-Hopf equation*

$$\phi(t) - \int_0^\infty k(t-s)\phi(s)\,ds = f(t), \qquad t \geq 0. \tag{1.1}$$

Here ϕ and f are m-dimensional vector functions and $k \in L_1^{m\times m}(-\infty,\infty)$, that is, the kernel function k is an $m \times m$ matrix function whose entries are in $L_1(-\infty,\infty)$. We assume that the given vector function f has its component functions in the Lebesgue space $L_p[0,\infty)$, and we express this property by writing $f \in L_p^m[0,\infty)$. Throughout this section p will be fixed and $1 \leq p < \infty$. The problem we shall consider is to find a solution ϕ of equation (1.1) that also belongs to the space $L_p^m[0,\infty)$.

The usual method (see [61]) for solving equation (1.1) is as follows. First assume that (1.1) has a solution ϕ in $L_p^m[0,\infty)$. Extend ϕ and f to the full real line by putting

$$\phi(t) = 0, \qquad f(t) = -\int_0^\infty k(t-s)\phi(s)\,ds, \quad t < 0.$$

Then $\phi, f \in L_p^m(-\infty,\infty)$ and the full line convolution equation

$$\phi(t) - \int_{-\infty}^\infty k(t-s)\phi(s)\,ds = f(t), \qquad -\infty < t < \infty$$

is satisfied. By applying the Fourier transformation and leaving the part of f that is given in the right-hand side, one gets

$$W(\lambda)\Phi_+(\lambda) - F_-(\lambda) = F_+(\lambda), \qquad \lambda \in \mathbb{R}, \tag{1.2}$$

where

$$W(\lambda) = I_m - \int_{-\infty}^\infty e^{i\lambda t}k(t)\,dt, \qquad F_+(\lambda) = \int_0^\infty e^{i\lambda t}f(t)\,dt, \tag{1.3}$$

$$\Phi_+(\lambda) = \int_0^\infty e^{i\lambda t}\phi(t)\,dt, \qquad F_-(\lambda) = \int_{-\infty}^0 e^{i\lambda t}f(t)\,dt. \tag{1.4}$$

Here I_m is the $m \times m$ identity matrix. Note that the functions K and F_+ are given, but the functions Φ_+ and F_- have to be found. In fact in this way the problem to solve (1.1) is reduced to that of finding two functions Φ_+ and F_- such that (1.2) holds, while furthermore Φ_+ and F_- must be as in (1.4) with $\phi \in L_p^m[0,\infty)$ and $f \in L_p^m(-\infty, 0]$.

To find Φ_+ and F_- of the desired form such that (1.2) holds, one factorizes the $m \times m$ matrix function W appearing in (1.2). This function is called the *symbol* of the integral equation (1.1). Note that W is continuous on the real line, and by the Riemann-Lebesgue lemma $\lim_{\lambda\in\mathbb{R},\,\lambda\to\infty} W(\lambda)$ exists and is equal to I_m.

Assume that the symbol admits a factorization of the following form:

$$W(\lambda) = \big(I_m + G_-(\lambda)\big)\big(I_m + G_+(\lambda)\big), \qquad \lambda \in \mathbb{R}, \tag{1.5}$$

where

$$G_+(\lambda) = \int_0^\infty e^{i\lambda t}g_+(t)\,dt, \qquad G_-(\lambda) = \int_{-\infty}^0 e^{i\lambda t}g_-(t)\,dt,$$

with $g_+ \in L_1^{m\times m}[0,\infty)$ and $g_- \in L_1^{m\times m}(-\infty, 0]$ while, in addition, the determinants

$$\det\big(I_m + G_+(\lambda)\big), \qquad \det\big(I_m + G_-(\lambda)\big)$$

do not vanish in the closed upper and lower half plane, respectively. We shall refer to the factorization (1.5) as a *right canonical factorization* of W *with respect to the real line*. Under the conditions stated above the functions $\big(I_m+G_+(\lambda)\big)^{-1}$ and $\big(I_m+G_-(\lambda)\big)^{-1}$ admit representations as Fourier transforms:

$$\big(I_m+G_+(\lambda)\big)^{-1} = I_m+\int_0^\infty e^{i\lambda t}\gamma_+(t)\,dt, \tag{1.6}$$

$$\big(I_m+G_-(\lambda)\big)^{-1} = I_m+\int_{-\infty}^0 e^{i\lambda t}\gamma_-(t)\,dt, \tag{1.7}$$

with $\gamma_+\in L_1^{m\times m}[0,\infty)$ and $\gamma_-\in L_1^{m\times m}(-\infty,0]$. Using the factorization (1.5) and omitting the variable λ, equation (1.2) can be rewritten as

$$(I_m+G_+)\Phi_+-(I_m+G_-)^{-1}F_- = (I_m+G_-)^{-1}F_+. \tag{1.8}$$

Let $\mathcal{P}$ be the projection acting on the Fourier transforms of $L_p^m(-\infty,\infty)$-functions according to the following rule:

$$\mathcal{P}\left(\int_{-\infty}^\infty e^{i\lambda t}h(t)\,dt\right)=\int_0^\infty e^{i\lambda t}h(t)\,dt.$$

Applying $\mathcal{P}$ to (1.8) one gets

$$(I_m+G_+)\Phi_+=\mathcal{P}\big((I_m+G_-)^{-1}F_+\big),$$

and hence

$$\Phi_+=(I_m+G_+)^{-1}\mathcal{P}\big((I_m+G_-)^{-1}F_+\big), \tag{1.9}$$

which is the formula for the solution of equation (1.2). To obtain the solution ϕ of the original equation (1.1), i.e., to obtain the inverse Fourier transform of Φ_+, one can employ the formulas (1.6) and (1.7). In fact

$$\phi(t)=f(t)+\int_0^\infty \gamma(t,s)f(s)\,ds, \qquad t\geq 0,$$

where the $m\times m$ matrix function $\gamma(t,s)$ is given by

$$\gamma(t,s) = \gamma_+(t-s)+\gamma_-(t-s)+\int_0^{\min(t,s)}\gamma_+(t-r)\gamma_-(r-s)\,dr.$$

We conclude the description of this factorization method by mentioning that the equation (1.1) has a unique solution in $L_p^m[0,\infty)$ for each f in $L_p^m[0,\infty)$ if and only if its symbol admits a factorization as in (1.5). For details, see [50], [61].

Let T be the *Wiener-Hopf integral operator* on $L_p^m[0,\infty)$ associated with equation (1.1), that is, T is the operator on $L_p^m[0,\infty)$ given by

$$(T\phi)(t) = \phi(t)-\int_0^\infty k(t-s)\phi(s)\,ds, \qquad t\geq 0.$$

The function W in the left-hand side of (1.3) is also referred to as the *symbol* of T. Obviously the operator T is invertible if and only if the equation (1.1) has a unique solution in $L_p^m[0,\infty)$ for each f in $L_p^m[0,\infty)$. Thus the results reviewed above can be summarized as follows.

Theorem 1.1. *Let T be the Wiener-Hopf integral operator on $L_p^m[0,\infty)$ with symbol W. Then T is invertible if and only if W admits a right canonical factorization with respect to the real line. Furthermore, if (1.5) is such a factorization of W, then the inverse of T is the integral operator given by*

$$(T^{-1}f)(t) = f(t) + \int_0^\infty \gamma(t,s)f(s)\,ds, \qquad t \geq 0,$$

where the kernel function γ is defined by

$$\gamma(t,s) = \begin{cases} \gamma_+(t-s) + \displaystyle\int_0^s \gamma_+(t-r)\gamma_-(r-s)\,dr, & 0 \leq s < t, \\ \gamma_-(t-s) + \displaystyle\int_0^t \gamma_+(t-r)\gamma_-(r-s)\,dr, & 0 \leq t < s \end{cases} \tag{1.10}$$

with γ_- and γ_+ as in (1.6) and (1.7), respectively.

To illustrate the method, let us consider a special choice for the right-hand side f (cf., [61]). Take

$$f(t) = e^{-iqt}x_0, \tag{1.11}$$

where x_0 is a fixed vector in $\mathbb{C}^m$ and q is a complex number with $\Im q < 0$. Then

$$F_+(\lambda) = \int_0^\infty e^{i(\lambda-q)t}x_0\,dt = \frac{i}{\lambda-q}x_0, \qquad \Im\lambda \geq 0.$$

Now observe that

$$\frac{i}{\lambda-q}\Big(\big(I_m + G_-(\lambda)\big)^{-1} - \big(I_m + G_-(q)\big)^{-1}\Big)x_0$$

is the Fourier transform of an $L_p^m(-\infty,0]$-function and hence it vanishes when the projection $\mathcal{P}$ is applied. It follows that in the present case the formula for Φ_+ may be written as

$$\Phi_+(\lambda) = \frac{i}{\lambda-q}\big(I_m + G_+(\lambda)\big)^{-1}\big(I_m + G_-(q)\big)^{-1}x_0.$$

Recall that the solution ϕ is the inverse Fourier transform of Φ_+. So we have

$$\phi(t) = e^{-iqt}\left(I_m + \int_0^t e^{iqs}\gamma_+(s)\,ds\right)\big(I_m + G_-(q)\big)^{-1}x_0. \tag{1.12}$$

1.2 Block Toeplitz equations and factorization

In this section we consider the discrete analogue of a Wiener-Hopf integral equation, that is, a *block Toeplitz equation* . So we consider an equation of the type

$$\sum_{k=0}^{\infty} a_{j-k}\xi_k = \eta_j, \qquad j = 0, 1, 2, \dots . \tag{1.13}$$

Throughout we assume that the coefficients a_j are given complex $m \times m$ matrices satisfying

$$\sum_{j=-\infty}^{\infty} \|a_j\| < \infty, \tag{1.14}$$

and $\eta = (\eta_j)_{j=0}^{\infty}$ is a given vector from $\ell_p^m = \ell_p(\mathbb{C}^m)$. The problem is to find $\xi = (\xi_k)_{k=0}^{\infty} \in \ell_p^m$ such that (1.13) is satisfied. We shall restrict ourselves to the case $1 \le p \le 2$; the final results however are valid for $2 < p \le \infty$ as well.

Assume $\xi \in \ell_p^m$ is a solution of (1.13). Then one can write (1.13) in the form

$$\sum_{k=-\infty}^{\infty} a_{j-k}\xi_k = \eta_j, \qquad j = 0, \pm 1, \pm 2, \dots , \tag{1.15}$$

where $\xi_k = 0$ for $k < 0$ and η_j is defined by (1.15) for $j < 0$. Multiplying both sides of (1.15) by λ^j with $|\lambda| = 1$ and summing over j, one gets

$$a(\lambda)\xi_+(\lambda) - \eta_-(\lambda) = \eta_+(\lambda), \qquad |\lambda| = 1, \tag{1.16}$$

where

$$a(\lambda) = \sum_{j=-\infty}^{\infty} \lambda^j a_j, \qquad \eta_+(\lambda) = \sum_{j=0}^{\infty} \lambda^j \eta_j, \tag{1.17}$$

$$\xi_+(\lambda) = \sum_{j=0}^{\infty} \lambda^j \xi_j, \qquad \eta_-(\lambda) = \sum_{j=-\infty}^{-1} \lambda^j \eta_j.$$

In this way the problem to solve (1.13) is reduced to that of finding two sequences ξ_+ and η_- such that (1.16) holds, while moreover, ξ_+ and η_- must be as in (1.2) with $(\xi_j)_{j=0}^{\infty}$ and $(\eta_{-j-1})_{j=0}^{\infty}$ from ℓ_p^m.

The usual way (cf., [61] or the book [40]) of solving (1.16) is again by factorizing the *symbol* $a(\lambda)$ of the given block Toeplitz equation. Assume that $a(\lambda)$ admits a *right canonical factorization with respect to the unit circle* . By definition this means that $a(\lambda)$ can be written as

$$\begin{aligned} a(\lambda) &= h_-(\lambda)h_+(\lambda), \qquad |\lambda| = 1, \\ h_+(\lambda) &= \sum_{j=0}^{\infty} \lambda^j h_j^+, \qquad h_-(\lambda) = \sum_{j=-\infty}^{0} \lambda^j h_j^-, \end{aligned} \tag{1.18}$$

where $(h_j^+)_{j=0}^\infty$ and $(h_{-j}^-)_{j=0}^\infty$ belong to the space $\ell_1^{m\times m}$ of all absolutely convergent sequences of complex $m\times m$ matrices, $\det h_+(\lambda)\neq 0$ for $|\lambda|\leq 1$ and $\det h_-(\lambda)\neq 0$ for $|\lambda|\geq 1$ (including $\lambda=\infty$). Then h_+^{-1} and h_-^{-1} also admit a representation of the form

$$h_+^{-1}(\lambda)=\sum_{j=0}^{\infty}\lambda^j\gamma_j^+,\qquad h_-^{-1}(\lambda)=\sum_{j=-\infty}^{0}\lambda^j\gamma_j^-, \tag{1.19}$$

with $(\gamma_j^+)_{j=0}^\infty$ and $(\gamma_{-j}^-)_{j=0}^\infty$ from $\ell_1^{m\times m}$. Defining the projection $\mathcal{P}$ by

$$\mathcal{P}\Big(\sum_{j=-\infty}^{\infty}\lambda^j b_j\Big)=\sum_{j=0}^{\infty}\lambda^j b_j,$$

one gets from (1.16) and (1.18)

$$\xi_+=h_+^{-1}\mathcal{P}(h_-^{-1}\eta_+). \tag{1.20}$$

Here, for convenience, the variable λ is omitted. The solution of the original equation (1.13) can now be written as

$$\xi_k=\sum_{s=0}^{\infty}\gamma_{ks}\eta_s,\qquad k=0,1,\dots, \tag{1.21}$$

where

$$\gamma_{ks}=\begin{cases}\displaystyle\sum_{r=0}^{s}\gamma_{k-r}^+\gamma_{r-s}^-, & s\leq k,\\ \displaystyle\sum_{r=0}^{k}\gamma_{k-r}^+\gamma_{r-s}^-, & s\geq k.\end{cases}$$

Note that for $s=k$ both sums in the above formula define the same matrix.

The assumption that $a(\lambda)$ admits a right canonical factorization as in (1.18) is equivalent to the requirement that for each $\eta=(\eta_j)_{j=0}^\infty$ in ℓ_p^m the equation (1.13) has a unique solution $\xi=(\xi_k)_{k=0}^\infty$ in ℓ_p^m. For details we refer to [61], [40].

Let T be the block Toeplitz operator on ℓ_p^m associated with the Toeplitz equation (1.13), that is, T is the operator on ℓ_p^m given by

$$T\xi=\eta\quad\Longleftrightarrow\quad\sum_{k=0}^{\infty}a_{j-k}\xi_k=\eta_j,\quad j=0,1,2,\dots.$$

The function a appearing in the left-hand side of (1.17) is also referred to as the *symbol* of T. Obviously T is invertible if and only if for each $\eta=(\eta_j)_{j=0}^\infty$ in ℓ_p^m the equation (1.13) has a unique solution $\xi=(\xi_k)_{k=0}^\infty$ in ℓ_p^m. This allows us to summarize the results reviewed above as follows.

Theorem 1.2. *Let T be the block Toeplitz operator on ℓ_p^m with symbol $a(\lambda)$ satisfying (1.14). Then T is invertible if and only $a(\lambda)$ admits a right canonical factorization with respect to the unit circle. Furthermore, if (1.18) is such a factorization of the function $a(\lambda)$, then the inverse of T is given by*

$$T^{-1} = \begin{bmatrix} \gamma_{11} & \gamma_{12} & \gamma_{13} & \cdots \\ \gamma_{21} & \gamma_{22} & \gamma_{23} & \cdots \\ \gamma_{31} & \gamma_{32} & \gamma_{33} & \cdots \\ \vdots & \vdots & \vdots & \ddots \end{bmatrix},$$

where the matrices γ_{ks} are defined by

$$\gamma_{ks} = \begin{cases} \sum_{r=0}^{s} \gamma_{k-r}^{+}\gamma_{r-s}^{-}, & s \le k, \\ \sum_{r=0}^{k} \gamma_{k-r}^{+}\gamma_{r-s}^{-}, & s \ge k, \end{cases} \tag{1.22}$$

with γ_j^+ and γ_j^- being determined by (1.19).

By way of illustration, we consider the special case when

$$\eta_j = q^j \eta_0, \qquad j = 0, 1, \ldots . \tag{1.23}$$

Here η_0 is a fixed vector in $\mathbb{C}^m$ and q is a complex number with $|q| < 1$. Then clearly

$$\eta_+(\lambda) = \frac{1}{1 - \lambda q}\eta_0, \qquad |\lambda| \le 1,$$

and one checks without difficulty that formula (1.21) becomes

$$\xi_k = q^k \sum_{s=0}^{k} q^{-s}\gamma_s^{+} h_-^{-1}(q^{-1})\eta_0, \qquad k = 0, 1, \ldots . \tag{1.24}$$

This is the analogue of formula (1.12) in the previous section.

1.3 Singular integral equations and factorization

In this section we review the factorization method that is used to solve systems of singular integral equations [48]. Consider the *singular integral equation*

$$a(t)\phi(t) + b(t)\frac{1}{\pi i}\int_\Gamma \frac{\phi(\tau)}{\tau - t}\, d\tau = f(t), \qquad t \in \Gamma, \tag{1.25}$$

with integration taken over a Cauchy contour Γ. (For the definition of the latter notion see the final paragraphs of Chapter 0 dealing with terminology and notation.) We write F_+ for the interior domain of Γ, and F_- for the exterior domain (i.e., the complement of $\overline{F}_+$ in the Riemann sphere $\mathbb{C}\cup\{\infty\}$). The functions a and b in (1.25) are given continuous $m \times m$ matrix functions defined on Γ, and f is a given function from $L_p^m(\Gamma)$, p fixed, $1 < p < \infty$. As usual in the theory of singular integral equations, it is assumed that the interior domain F_+ of Γ is connected and contains 0; the exterior domain F_- of Γ contains ∞. The problem is to find $\phi \in L_p^m(\Gamma)$ such that(1.25) is satisfied.

For ϕ a rational function without poles on Γ we put

$$(S\phi)(t) = \frac{1}{\pi i}\int_\Gamma \frac{\phi(\tau)}{\tau - t}\, d\tau = f(t), \qquad t \in \Gamma, \tag{1.26}$$

where the integral is taken in the sense of the Cauchy principal value. The operator S defined in this way can be extended by continuity to a bounded linear operator, again denoted by S, on all of $L_p^m(\Gamma)$. Equation (1.25) can now be written as

$$aI\phi + bS\phi = f, \tag{1.27}$$

where I is the identity operator on $L_p^m(\Gamma)$. In other words, the study of the equation (1.25) reduces to that of the operator $aI + bS$. Here a and b are viewed as multiplication operators. Equation (1.25) has a unique solution $\phi \in L_p^m(\Gamma)$ for each choice of $f \in L_p^m(\Gamma)$ if and only if the operator $aI + bS$ is invertible as an operator on $L_p^m(\Gamma)$. In the remainder of this section we shall discuss a necessary and sufficient condition for this to happen, and we shall give formulas for the inverse $(aI + bS)^{-1}$.

The operator S enjoys the property $S^2 = I$. Hence the operators

$$P_\Gamma = \frac{1}{2}(I + S), \qquad Q_\Gamma = \frac{1}{2}(I - S)$$

are complementary projections on $L_p^m(\Gamma)$. The image of P_Γ consists of all functions in $L_p^m(\Gamma)$ that admit an analytic continuation into F_+. Similarly, the image of Q_Γ is the set of all functions in $L_p^m(\Gamma)$ that admit an analytic continuation into F_- vanishing at ∞. Putting $c = a + b$ and $d = a - b$, one can write the equation (1.27) in the form $cP_\Gamma\phi + dQ_\Gamma\phi = f$.

The following is known (see [62] for the case when the coefficients a and b are scalar functions and [48] for the matrix-valued case). The operator $aI + bS = cP_\Gamma + dQ_\Gamma$ is invertible if and only if the matrices $c(\lambda)$ and $d(\lambda)$ are invertible for each $\lambda \in \Gamma$ and the function w given by $w(\lambda) = d(\lambda)^{-1}c(\lambda)$ admits a *right canonical factorization with respect to* Γ . By this we mean a factorization

$$w(\lambda) = w_-(\lambda)w_+(\lambda), \qquad \lambda \in \Gamma, \tag{1.28}$$

where w_- and w_+ are $m\times m$ matrix functions, analytic and taking invertible values on an open neighborhood of $\overline{F}_-$ and $\overline{F}_+$, respectively. With the help of (1.28), the operator $aI+bS = cP_\Gamma+dQ_\Gamma$ can be rewritten as $aI+bS = dw_-(w_+P_\Gamma+w_-^{-1}Q_\Gamma)$, and its inverse is given by

$$\begin{aligned}(aI+bS)^{-1} &= (w_+^{-1}P_\Gamma + w_-Q_\Gamma)w_-^{-1}d^{-1}\\ &= w_+^{-1}P_\Gamma w_-^{-1}d^{-1} + w_-Q_\Gamma w_-^{-1}d^{-1}.\end{aligned} \tag{1.29}$$

Replacing P_Γ and Q_Γ by $\frac{1}{2}(I+S)$ and $\frac{1}{2}(I-S)$, respectively, one gets

$$\begin{aligned}(aI+bS)^{-1} &= \frac{1}{2}(c^{-1}+d^{-1})I + \frac{1}{2}(w_+^{-1}-w_-)Sw_-^{-1}d^{-1}\\ &= \frac{1}{2}[(a+b)^{-1}+(a-b)^{-1}]I + \frac{1}{2}(w_+^{-1}-w_-)Sw_-^{-1}(a-b)^{-1}\\ &= (a+b)^{-1}a(a-b)^{-1}I + \frac{1}{2}(w_+^{-1}-w_-)Sw_-^{-1}(a-b)^{-1}.\end{aligned}$$

Summarizing we get the following theorem.

Theorem 1.3. *The singular integral operator $T=aI+bS$ on $L_p^m(\Gamma)$ is invertible if and only if the matrices $a(\lambda)+b(\lambda)$ and $a(\lambda)-b(\lambda)$ are invertible for each $\lambda\in\Gamma$ and the function w given by*

$$w(\lambda) = \big(a(\lambda)+b(\lambda)\big)^{-1}\big(a(\lambda)+b(\lambda)\big)$$

admits a right canonical factorization with respect to Γ. Furthermore, if (1.28) *is such a factorization of w, then the inverse of T is given by*

$$T^{-1} = (a+b)^{-1}a(a-b)^{-1}I + \frac{1}{2}(w_+^{-1}-w_-)Sw_-^{-1}(a-b)^{-1}. \tag{1.30}$$

Thus, as before for Wiener-Hopf and block Toeplitz operators, canonical factorization is a useful method for inverting singular integral operators too.

Notes

The material in this chapter is standard, and can be found in much more detail and greater generality in various monographs and papers, for instance, see the books [29] and [50]. A first introduction to the theory of Wiener-Hopf integral equations and the theory of (block) Toeplitz operators can be found in Chapters XII and XIII of [51] and Chapters XXIII–XXV of [52], respectively. More information can be found in the monographs [37], [62], [63], [64] and [24]. For an extensive review (with many additional references) of the factorization theory of matrix functions with respect to a curve and its applications to inversion of singular integral operators of different types, including Wiener-Hopf and block Toeplitz operators, the reader is referred to the recent survey paper [59].

Chapter 2

The state space method and factorization

This chapter describes in detail the elements of the state space method that are used throughout this book. The central notion is that of a realization of a matrix or operator function. The chapter consists of six sections. Section 2.1 presents preliminaries on realization, including the relevant definitions and the connection with systems theory. In the next two sections the realization problem is discussed. First for rational matrix functions in Section 2.2, and then for analytic operator functions in a possibly infinite dimensional setting in Section 2.3. The last three sections are devoted to the main operations on realizations that are needed in this book: inversion (Section 2.4), taking products (Section 2.5), and factorization (Section 2.6).

2.1 Preliminaries on realization

Let W be a rational matrix function which is also *proper*, that is, W has no pole at infinity. As is well-known such a function can always be represented (see the next section for an explicit construction) in the form

$$W(\lambda) = D + C(\lambda I - A)^{-1}B. \tag{2.1}$$

Here λ is a complex variable, A is a square matrix, I is the identity matrix of the same size as A, and B and C are matrices of appropriate sizes. Since A, B, C and D are matrices, it is immediate from Cramer's rule that the right-hand side of (2.1) is also a proper rational matrix function. We shall understand the equality in (2.1) as an equality between rational matrix functions, and we shall refer to (2.1) as a *matrix-valued realization* of W. Sometimes we simply say that the quadruple of matrices (A, B, C, D) is a *realization* of W. A rational matrix function has many different realizations. Of particular interest are those matrix-valued realizations of W of which the order of the matrix A is as small as possible. These realizations are called *minimal*; we shall describe their properties in Chapter 8.

For operator-valued functions W, expressions of the type (2.1) are important too but have to be considered with some care. Let W be an $\mathcal{L}(U,Y)$-valued function on a subset Ω of $\mathbb{C}$. Here U and Y are possibly infinite dimensional complex Banach spaces. We say that W admits a *realization on* Ω whenever W can be written as

$$W(\lambda) = D + C(\lambda I_X - A)^{-1}B, \qquad \lambda \in \Omega. \tag{2.2}$$

Here A is a bounded linear operator on a complex Banach space X such that Ω is a subset of $\rho(A)$, the *resolvent set* of A. Furthermore, I_X is the identity operator on X, and

$$B \in \mathcal{L}(U,X), \qquad C \in \mathcal{L}(X,Y), \qquad D \in \mathcal{L}(U,Y),$$

that is $B : U \to X$, $C : X \to Y$, and $D : U \to Y$, are bounded linear operators. The fact that $\Omega \subset \rho(A)$ implies that the right-hand side of (2.2) is a well-defined bounded linear operator which maps U into Y for each $\lambda \in \Omega$. Also, $W(\lambda)$ is a bounded linear operator mapping U into Y for each $\lambda \in \Omega$. Note that (2.2) requires these operators to be equal for each $\lambda \in \Omega$. When Ω is open, an obvious necessary condition for W to admit a realization on Ω is that W be analytic on Ω. When Ω is a punctured open neighborhood of ∞, then (2.2) implies $\lim_{\lambda\to\infty} W(\lambda) = D$ and so W is proper.

Often the identity matrix I in (2.1) and the identity operator I_X in (2.2) will be suppressed, and we simply write $(\lambda - A)^{-1}$ in place of $(\lambda I - A)^{-1}$ or $(\lambda I_X - A)^{-1}$.

When X and Y are both finite dimensional, then the realization (2.2) is called *finite dimensional*. In that case $W(\lambda)$, A, B, C and D can be identified in the usual way with matrices.

In the next two sections we shall address the realization problem, i.e., the question under what conditions a given matrix or operator function admits a realization. First however, we sketch a connection with systems theory which reflects itself in some terminology to be introduced at the end of the present section.

A system Σ can be considered as a physical object which produces an output in response to an input. Schematically:

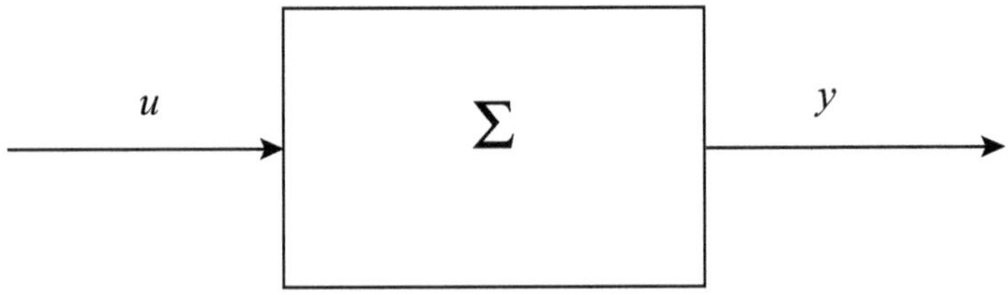

where u denotes the input and y denotes the output. Mathematically, the input u and the output y are vector-valued functions of a parameter t. The input can be chosen freely (at least in principle), but the output is uniquely determined by the choice of the input. The relationship between the input and the output can be quite complicated. Here we consider the simplest model which means that the relationship in question is described by a causal linear time invariant system, i.e., a system of differential equations of the type

$$\Sigma \left\{ \begin{array}{rcl} x'(t) & = & Ax(t) + Bu(t), \\ y(t) & = & Cx(t) + Du(t), \qquad t \geq 0, \\ x(0) & = & 0, \end{array} \right. \tag{2.3}$$

where A, B, C and D are matrices of appropriate sizes, A and D square. Application of the Laplace transform (under appropriate conditions on the input and output functions) changes (2.3) into

$$\left\{ \begin{array}{rcl} \lambda\widehat{x}(s) & = & A\widehat{x}(\lambda) + B\widehat{u}(\lambda), \\ \widehat{y}(\lambda) & = & C\widehat{x}(\lambda) + D\widehat{u}(\lambda), \end{array} \right.$$

and from these expressions one can solve $\widehat{y}(\lambda)$ in terms of $\widehat{u}(\lambda)$, resulting in

$$\widehat{y}(\lambda) = \big(D + C(\lambda - A)^{-1}B\big)\widehat{u}(\lambda).$$

So in what is called the frequency domain, the input-output behavior of (2.3) is determined by the function $D + C(\lambda - A)^{-1}B$, which is called the *transfer function* of the system (2.3). Note that this function appears in the realized form.

The connection with systems theory indicated above is reflected in the terminology which is customarily used in dealing with realizations. Returning to (2.2), the space X is usually called the *state space* of the realization, and the operator A is referred to as its *state space operator* or *main operator*. Further we call B the *input operator*, C the *output operator*, and D the *external operator* of (2.2). The realization is called *strictly proper* when $D = 0$ and *biproper* if D is an invertible operator. In the latter case, the operator $A - BD^{-1}C$ is well-defined. It is referred to by the term the *associate state space operator* or *associate main operator* and (by slight abuse of notation as $A^\times$ does not depend only on A) denoted by $A^\times$. This operator will play a crucial role in the inversion and factorization results to be discussed later on. In the situation where $U = Y$ and D is the identity operator, we say that (2.2) is a *unital realization*. The associate main operator then has the form $A^\times = A - BC$. In the case of a matrix-valued realization, the terms *state space matrix*, *main matrix*, *input matrix*, *output matrix*, *external matrix*, *associate state space matrix*, and *associate main matrix* will be used.

Other elements of systems theory involving stability properties, feedback and stabilization, will be reviewed in Chapter 19. These will be of central importance in Chapter 20 (the final chapter of the book) which is concerned with H_∞-control.

2.2 Realization of rational matrix functions

In this section we construct a matrix-valued realization for a given proper rational (possibly non-square) matrix function.

Theorem 2.1. *Every proper rational matrix function has a matrix-valued realization. Moreover, the realization can be chosen in such a way that the set of eigenvalues of the main matrix coincides with the set of poles of W.*

Proof. Let W be a proper rational $r \times m$ matrix function, and let w_{ij} be the (i,j)-entry of W. Since W is rational, we have

$$w_{ij}(\lambda) = \frac{p_{ij}(\lambda)}{q_{ij}(\lambda)}, \qquad i = 1, \ldots, r, \ j = 1, \ldots, n,$$

where p_{ij} and q_{ij} are scalar polynomials. The polynomials q_{ij} are non-zero and can be taken to be monic. Without loss of generality we may assume that the polynomials p_{ij} and q_{ij} have no common zero. Taking the least common multiple of the polynomials q_{ij}, we obtain a monic polynomial q.

Define Ω_W to be the set of all complex λ for which $q(\lambda) \neq 0$. Notice that $\mathbb{C} \setminus \Omega_W$ is precisely the set of all points in $\mathbb{C}$ where W has a pole. One checks without difficulty that W has a representation of the form

$$W(\lambda) = W(\infty) + \frac{1}{q(\lambda)} H(\lambda), \qquad \lambda \in \Omega_W,$$

where H is an $r \times m$ matrix polynomial. Since W is proper, this matrix polynomial is either identically equal to zero or it has degree strictly smaller than k, the degree of the scalar polynomial q. Write

$$q(\lambda) = \lambda^k + \sum_{j=0}^{k-1} \lambda^j q_j, \qquad H(\lambda) = \sum_{j=0}^{k-1} \lambda^j H_j,$$

and, with I_r the $r \times r$ identity matrix,

$$A = \begin{bmatrix} 0 & 0 & \ldots & 0 & -q_0 I_r \\ I & 0 & \ldots & 0 & -q_1 I_r \\ & \ddots & & & \vdots \\ 0 & \ldots & & I & -q_{k-1} I_r \end{bmatrix}, \quad B = \begin{bmatrix} H_0 \\ H_1 \\ \vdots \\ H_{k-1} \end{bmatrix}, \quad C = \begin{bmatrix} 0 & \ldots & 0 & I_r \end{bmatrix}.$$

Then the resolvent set $\rho(A)$ of A coincides with Ω_W, the subset of $\mathbb{C}$ on which q takes non-zero values. For $\lambda \in \rho(A)$, define $C_1(\lambda), \ldots, C_k(\lambda)$ by

$$\begin{bmatrix} C_1(\lambda) & C_2(\lambda) & \ldots & C_k(\lambda) \end{bmatrix} = C(\lambda - A)^{-1}.$$

From the special form of the matrix A (second companion type) we see that

$$C_{j+1}(\lambda) = \lambda C_j(\lambda), \qquad j = 0, \dots, k-1,$$

and $C_1(\lambda) = q(\lambda)^{-1}I$. Hence

$$C(\lambda - A)^{-1}B = \sum_{j=0}^{k-1} C_{j+1}(\lambda)H_j = \frac{1}{q(\lambda)}H(\lambda).$$

It follows that $W(\lambda) = W(\infty) + C(\lambda - A)^{-1}B$ for each $\lambda \in \Omega_W = \rho(A)$. Thus W has a matrix-valued realization such that the set of eigenvalues of the main matrix A is equal to $\mathbb{C} \setminus \Omega_W$. In other words, the set of eigenvalues of A coincides with the set of poles of W, as desired. □

Let W be a proper rational matrix function. Elaborating on Theorem 2.1 and its proof, we note that W does not admit any realization involving a main matrix A whose spectrum $\sigma(A)$ is strictly smaller than $\mathbb{C} \setminus \Omega_W$, the set of poles of W. Indeed, we would then have a realization of W on an open subset of $\mathbb{C}$ strictly larger than Ω_W and such a subset would contain a pole of W, contradicting the fact that W has to be analytic on it. It is not difficult to construct realizations of W having a main matrix A with spectrum strictly larger than $\mathbb{C} \setminus \Omega_W$ and where certain eigenvalues of A (namely those belonging to Ω_W) do not correspond with poles of W. So the realization constructed in the proof of Theorem 2.1 enjoys a certain minimality property. However, it does this only in a weak sense. This one sees, for instance, by looking at the pole orders. If μ is a pole of W, its order as a pole of W is generally strictly smaller than the order of μ as a pole of the resolvent $(\lambda - A)^{-1}$. With the proper notion of minimality to be introduced in Section 8.1, this anomaly disappears so that the two pole orders are the same. The key point is that the state space dimension (which is equal to rk) of the realization of the proof of Theorem 2.1 is generally not the least possible.

2.3 Realization of analytic operator functions

In this section we consider the realization problem for possibly non-rational operator functions. First we consider operator functions that are analytic on a bounded Cauchy domain in $\mathbb{C}$. Recall from Chapter 0 that the boundary of such a Cauchy domain consists of a finite number of simple closed non-intersecting rectifiable curves.

Theorem 2.2. *Let Ω be a bounded Cauchy domain, and let W be an operator function with values in $\mathcal{L}(U, Y)$, where U and Y are complex Banach spaces. Suppose W is analytic on Ω and continuous on the closure of Ω. Then, given a bounded linear operator $D : U \to Y$, there exists a realization for W on Ω having D as its external operator. In particular, if $U = Y$, then W admits a unital realization on Ω.*

Proof. Let Γ be the positively oriented boundary of Ω (so that Ω is the interior domain of Γ). With Γ and U we associate the space $C(\Gamma;U)$ of all U-valued continuous functions on Γ endowed with the supremum norm. This will become the state space of the realization to be constructed.

Write B for the canonical embedding of U into $C(\Gamma;U)$, so $(Bu)(z) = u$ for each $u \in U$ and $z \in \Gamma$. Next, define $C : C(\Gamma;U) \to Y$ by setting

$$Cf = \frac{1}{2\pi i}\int_\Gamma \big(D - W(z)\big)f(z)\,dz, \qquad f \in C(\Gamma;U).$$

Here D is the given operator from U into Y. Finally, the operator A from $C(\Gamma;U)$ into $C(\Gamma;U)$ is the multiplication operator given by

$$(Af)(z) = zf(z), \qquad f \in C(\Gamma;U),\ z \in \Gamma.$$

All these operators are linear and bounded. We claim that

$$W(\lambda) = D + C(\lambda - A)^{-1}B, \qquad \lambda \in \Omega \subset \rho(A).$$

Take $\lambda \in \Omega$. Then $\lambda - A$ is invertible with inverse given by

$$\big((\lambda - A)^{-1}g\big)(z) = \frac{1}{\lambda - z}\,g(z), \qquad g \in C(\Gamma;U),\ z \in \Gamma.$$

It follows that

$$\big((\lambda - A)^{-1}Bu\big)(z) = \frac{1}{\lambda - z}\,u, \qquad u \in U,\ z \in \Gamma,$$

and hence

$$C(\lambda - A)^{-1}Bu = \frac{1}{2\pi i}\int_\Gamma \frac{1}{\lambda - z}\big(D - W(z)\big)u\,dz, \qquad u \in U.$$

By the Cauchy integral formula, the right-hand side of this identity is $W(\lambda)u - Du$, and the desired result is immediate. □

Theorem 2.2 remains true when the conditions on Ω and W are replaced by the simpler hypotheses that Ω is any bounded open set in $\mathbb{C}$ and W is just analytic on Ω. In that case the space $C(\Gamma;U)$ must be replaced by an appropriate Banach space defined in terms of the behavior of W near the boundary of Ω. For details, cf., [113]; see also the next theorem.

Theorem 2.3. *Let $\Omega \subset \mathbb{C}$ be an open punctured neighborhood of ∞ in the Riemann sphere $\mathbb{C}_\infty$, let U and Y be complex Banach spaces, and let $W : \Omega \to \mathcal{L}(U,Y)$ be analytic and proper. Then W admits a realization on Ω with external operator $D = \lim_{\lambda\to\infty} W(\lambda)$.*

Proof. First assume Ω is the full complex plane. Then, by Liouville's theorem, the function W has the constant value $D = \lim_{\lambda\to\infty} W(\lambda)$. Now take for the state space X the zero space $\{0\}$, and the desired realization for W on $\mathbb{C}$ is obtained trivially.

Next, consider the more interesting case where Ω is different from $\mathbb{C}$. For notational reasons we will assume that $0 \notin \Omega$. The general case can be reduced to this situation by a simple translation.

Define X to be the space of all Y-valued functions, analytic on $\Omega \cup \{\infty\}$, such that

$$\|f\|_\bullet = \sup_{z\in\Omega\cup\{\infty\}} \frac{\|f(z)\|}{\max(1, \|W(z)\|)} < \infty.$$

Taking $\|\cdot\|_\bullet$ for the norm, X is a Banach space. Introduce $B : U \to X$ by

$$(Bu)(z) = \begin{cases} z\big(W(z)u - W(\infty)u\big), & z \in \Omega, \\ \lim_{z\to\infty} z\big(W(z)u - W(\infty)u\big), & z = \infty. \end{cases}$$

Further, let $C : X \to Y$ be given by $Cf = f(\infty)$. Finally, define $A : X \to X$ by

$$(Af)(z) = \begin{cases} z\big(f(z) - f(\infty)\big), & z \in \Omega, \\ \lim_{z\to\infty} z\big(f(z) - f(\infty)\big), & z = \infty. \end{cases}$$

All these operators are linear and bounded. We claim that

$$W(\lambda) = W(\infty) + C(\lambda - A)^{-1}B, \qquad \lambda \in \Omega \subset \rho(A).$$

Take $\lambda \in \Omega$. For $g \in X$, put

$$h(z) = \begin{cases} \dfrac{zg(\lambda) - \lambda g(z)}{z - \lambda}, & z \in \Omega,\ z \neq \lambda, \\ g(\lambda) - \lambda g'(\lambda), & z = \lambda, \\ g(\lambda), & z = \infty, \end{cases}$$

where g' stands for the derivative of g. Then $h \in X$, and by direct computation one sees that $\big((\lambda - A)h\big)(z) = \lambda g(z)$, $z \in \Omega \cup \{\infty\}$. Now λ is non-zero (since Ω does not contain the origin), and it follows that $\lambda - A$ is surjective. But $\lambda - A$ is injective too. Indeed, if $f \in X$ and $Af = \lambda f$, then

$$f(z) = \frac{z}{z - \lambda} f(\infty), \qquad z \in \Omega,\ z \neq \lambda,$$

which, on account of the definition of the norm $\|\cdot\|_\bullet$ on X, implies $f(\infty) = 0$ (cf., the behavior of f when $z \to \lambda$), hence $f = 0$. It follows that $\lambda \in \rho(A)$ and

$(\lambda - A)^{-1}g = \lambda^{-1}h$. We now apply this result to $g = Bu$ with $u \in U$. With this g, we have $h(\infty) = (Bu)(\lambda) = \lambda\big(W(\lambda)u - W(\infty)\big)u$, and so

$$\big((\lambda - A)^{-1}Bu\big)(\infty) = \lambda^{-1}h(\infty) = \big(W(\lambda)u - W(\infty)\big)u.$$

In other words $C(\lambda - A)^{-1}Bu = \big(W(\lambda)u - W(\infty)\big)u$. As $u \in U$ was taken arbitrarily, we get $W(\lambda) = W(\infty) + C(\lambda - A)^{-1}B$ for each $\lambda \in \Omega$. □

2.4 Inversion

We begin with some heuristics. Consider the realization

$$W(\lambda) = D + C(\lambda - A)^{-1}B, \qquad \lambda \in \rho(A), \tag{2.4}$$

and view W as the transfer function of the linear time invariant system

$$\Sigma \begin{cases} x'(t) &= Ax(t) + Bu(t), \\ y(t) &= Cx(t) + Du(t), \qquad t \geq 0, \\ x(0) &= 0. \end{cases}$$

Assuming that we are in the biproper situation where D is invertible, we can solve u in terms of x and y:

$$u(t) = -D^{-1}Cx(t) + D^{-1}y(t), \qquad t \geq 0.$$

Inserting this into Σ yields

$$\Sigma^{\times} \begin{cases} x'(t) &= A^{\times}x(t) + BD^{-1}y(t), \\ u(t) &= -D^{-1}Cx(t) + D^{-1}y(t), \qquad t \geq 0, \\ x(0) &= 0. \end{cases}$$

Here $A^{\times} = A - BD^{-1}C$ is the associate main operator of the given realization as introduced in the last paragraph of Section 2.1. The linear time invariant systems Σ and $\Sigma^{\times}$ can be seen as each other's inverse. The transfer function of Σ is given by (2.4), the transfer function of $\Sigma^{\times}$ by

$$W^{\times}(\lambda) = D^{-1} - D^{-1}C(\lambda - A^{\times})^{-1}BD^{-1}, \qquad \lambda \in \rho(A^{\times}).$$

So it is to be expected that W and $W^{\times}$ are related by inversion. We shall now make this precise.

Theorem 2.4. *Consider the biproper realization*

$$W(\lambda) = D + C(\lambda - A)^{-1}B, \qquad \lambda \in \rho(A).$$

Put $A^\times = A - BD^{-1}C$, *and take* $\lambda \in \rho(A)$. *Then* $W(\lambda)$ *is invertible if and only if* λ *belongs to* $\rho(A^\times)$. *In that case, for* $\lambda \in \rho(A) \cap \rho(A^\times)$, *the following identities hold:*

$$\begin{aligned} W(\lambda)^{-1} &= D^{-1} - D^{-1}C(\lambda - A^\times)^{-1}BD^{-1}, \\ (\lambda - A^\times)^{-1} &= (\lambda - A)^{-1} - (\lambda - A)^{-1}BW(\lambda)^{-1}C(\lambda - A)^{-1}. \end{aligned}$$

Moreover, again for $\lambda \in \rho(A) \cap \rho(A^\times)$, *we have*

$$\begin{aligned} W(\lambda)D^{-1}C(\lambda - A^\times)^{-1} &= C(\lambda - A)^{-1}, \\ (\lambda - A^\times)^{-1}BD^{-1}W(\lambda) &= (\lambda - A)^{-1}B. \end{aligned}$$

Proof. For $\lambda \in \rho(A^\times)$, put $W^\times(\lambda) = D^{-1} - D^{-1}C(\lambda - A^\times)^{-1}BD^{-1}$. Then, when $\lambda \in \rho(A) \cap \rho(A^\times)$, one has

$$\begin{aligned} W(\lambda)W^\times(\lambda) &= \left(D + C(\lambda - A)^{-1}B\right)\left(D^{-1} - D^{-1}C(\lambda - A^\times)^{-1}BD^{-1}\right) \\ &= I_Y + C(\lambda - A)^{-1}BD^{-1} - C(\lambda - A^\times)^{-1}BD^{-1} + \\ &\qquad -C(\lambda - A)^{-1}BD^{-1}C(\lambda - A^\times)^{-1}BD^{-1}. \end{aligned}$$

Now use that $BD^{-1}C = A - A^\times = (\lambda - A^\times) - (\lambda - A)$. It follows that $W(\lambda)W^\times(\lambda) = I_Y$. Analogously one has $W^\times(\lambda)W(\lambda) = I_U$. The expression for $(\lambda - A^\times)^{-1}$ as well as the last two identities in the theorem are obtained in a similar way. □

Instead of the previous proof one can also give an argument using Schur complements of the operator matrix

$$\begin{bmatrix} A - \lambda I & B \\ C & I \end{bmatrix}.$$

For details, see the second proof of Theorem 2.1 in [20] or Sections 2 and 4 in [19].

2.5 Products

Again we begin with some heuristical remarks. This time we start with two linear time invariant systems

$$\Sigma_1 \left\{ \begin{aligned} x_1'(t) &= A_1x_1(t) + B_1u_1(t), \\ y_1(t) &= C_1x_1(t) + D_1u_1(t), \qquad t \geq 0, \\ x_1(0) &= 0, \end{aligned} \right.$$

$$\Sigma_2 \begin{cases} x_2'(t) &= A_2x_2(t) + B_2u_2(t), \\ y_2(t) &= C_2x_2(t) + D_2u_2(t), \qquad t \geq 0, \\ x_2(0) &= 0, \end{cases}$$

and we assume that the output y_2 of Σ_2 can be and is used as the input $u_1 = y_2$ for Σ_1, resulting in the cascade synthesis Σ of the systems Σ_1 and Σ_2. The input for Σ is $u = u_2$ and the output (modulo $u_1 = y_2$) is $y = y_1$. The equations governing the relationship between u and y then are

$$\begin{cases} x_1'(t) &= A_1x_1(t) + B_1C_2x_2(t) + B_1D_2u(t), \\ x_2'(t) &= A_2x_2(t) + B_2u(t), \\ y(t) &= C_1x_1(t) + D_1C_2x_2(t) + D_1D_2u(t), \qquad t \geq 0, \\ x_1(0) &= 0, \\ x_2(0) &= 0, \end{cases}$$

and this is a linear time invariant system which can be rewritten as

$$\Sigma : \begin{cases} \begin{bmatrix} x_1 \\ x_2 \end{bmatrix}' &= \begin{bmatrix} A_1 & B_1C_2 \\ 0 & A_2 \end{bmatrix} \begin{bmatrix} x_1 \\ x_2 \end{bmatrix} + \begin{bmatrix} B_1D_2 \\ B_2 \end{bmatrix} u, \\ y &= \begin{bmatrix} C_1 & D_1C_2 \end{bmatrix} \begin{bmatrix} x_1 \\ x_2 \end{bmatrix} + D_1D_2u, \\ \begin{bmatrix} x_1 \\ x_2 \end{bmatrix}(0) &= \begin{bmatrix} 0 \\ 0 \end{bmatrix}. \end{cases}$$

The transfer functions of Σ_1 and Σ_2 are

$$W_1(\lambda) = D_1 + C_1(\lambda - A_1)^{-1}B_1, \qquad \lambda \in \rho(A_1), \tag{2.5}$$

$$W_2(\lambda) = D_2 + C_2(\lambda - A_2)^{-1}B_2, \qquad \lambda \in \rho(A_2), \tag{2.6}$$

respectively, and the transfer function of Σ is the product W_1W_2 of W_1 and W_2, in other words

$$W(\lambda) = W_1(\lambda)W_2(\lambda).$$

So our considerations lead to a product formula for realizations. Here are the details.

First we specify the spaces associated with the realizations (2.5) and (2.6), and the actions of the operators involved:

$$A_1 : X_1 \to X_1, \qquad B_1 : U_1 \to X_1, \qquad C_1 : X_1 \to Y_1, \qquad D_1 : U_1 \to Y_1,$$

$$A_2 : X_2 \to X_2, \qquad B_2 : U_2 \to X_2, \qquad C_2 : X_2 \to Y_2, \qquad D_2 : U_2 \to Y_2.$$

Now assume $Y_1 = U_2$. Put $U = U_1$, $Y = Y_2$, and introduce

$$\begin{aligned} A &= \begin{bmatrix} A_1 & B_1C_2 \\ 0 & A_2 \end{bmatrix} : X_1 \dot{+} X_2 \to X_1 \dot{+} X_2, \\ B &= \begin{bmatrix} B_1 D_2 \\ B_2 \end{bmatrix} : Y \to X_1 \dot{+} X_2, \\ C &= \begin{bmatrix} C_1 & D_1C_2 \end{bmatrix} : X_1 \dot{+} X_2 \to Y, \\ D &= D_1 D_2 : U \to Y. \end{aligned}$$

Then the following result holds true.

Theorem 2.5. *Let W_1 and W_2 be given by the realizations* (2.5) *and* (2.6), *respectively. Then, with A, B, C and D as above,*

$$W_1(\lambda)W_2(\lambda) = D + C(\lambda - A)^{-1}B, \qquad \lambda \in \rho(A_1) \cap \rho(A_2) \subset \rho(A).$$

Proof. Take $\lambda \in \rho(A_1) \cap \rho(A_2)$. Then $\lambda \in \rho(A)$. Indeed, $\lambda - A$ is invertible with inverse given by

$$(\lambda - A)^{-1} = \begin{bmatrix} (\lambda - A_1)^{-1} & H(\lambda) \\ 0 & (\lambda - A_2)^{-1} \end{bmatrix} : X_1 \dot{+} X_2 \to X_1 \dot{+} X_2,$$

where $H(\lambda) = -(\lambda - A_1)^{-1} B_1 C_2 (\lambda - A_2)^{-1}$. Employing this, and the expressions for B, C and D given prior to the theorem, $D + C(\lambda - A)^{-1}B$ is seen to be equal to

$$\begin{aligned} & D_1 D_2 + \begin{bmatrix} C_1 & D_1C_2 \end{bmatrix} \begin{bmatrix} (\lambda - A_1)^{-1} & H(\lambda) \\ 0 & (\lambda - A_2)^{-1} \end{bmatrix} \begin{bmatrix} B_1 D_2 \\ B_2 \end{bmatrix} \\ &= D_1 D_2 + \begin{bmatrix} C_1 (\lambda - A_1)^{-1} & C_1 H(\lambda) + D_1 C_2 (\lambda - A_2)^{-1} \end{bmatrix} \begin{bmatrix} B_1 D_2 \\ B_2 \end{bmatrix} \\ &= \left(D_1 + C_1(\lambda - A_1)^{-1}B_1\right)\left(D_2 + C_2(\lambda - A_2)^{-1}B_2\right). \end{aligned}$$

Thus $D + C(\lambda - A)^{-1}B = W_1(\lambda)W_2(\lambda)$, as desired. □

The product $W_1(\lambda)W_2(\lambda)$ is defined for $\lambda \in \rho(A_1) \cap \rho(A_2)$, a punctured neighborhood of ∞ in $\mathbb{C} \cup \{\infty\}$. On the other hand $D + C(\lambda - A)^{-1}B$ is defined for $\lambda \in \rho(A)$. As we have seen above $\rho(A_1) \cap \rho(A_2) \subset \rho(A)$. In general, this inclusion is strict. Equality occurs, for instance, when the spectra $\sigma(A_1)$ and $\sigma(A_2)$ of the operators A_1 and A_2 are disjoint. Another case where one has the equality $\rho(A) = \rho(A_1) \cap \rho(A_1)$ is when $\rho(A)$ is connected. In particular, the equality in question is valid when W_1 and W_2 are rational matrix functions, and (2.5) and (2.6) are matrix-valued realizations.

The realization of Theorem 2.5 is called the *product* of the realizations (2.5) and (2.6), in that order.

The counterpart of taking products is factorization. In the next section this topic will be discussed for functions given by a biproper realization. We close the present section with a remark preparing for this discussion.

The main operator A in the product realization is given in the form of a 2×2 upper triangular operator matrix:

$$A = \begin{bmatrix} A_1 & B_1C_2 \\ 0 & A_2 \end{bmatrix} : X_1 \dot{+} X_2 \to X_1 \dot{+} X_2.$$

Analogously, assuming the external operators to be invertible, the associate main operator $A^\times = A - BD^{-1}C$ is of 2×2 lower triangular type:

$$A^\times = \begin{bmatrix} A_1^\times & 0 \\ B_2D^{-1}C_1 & A_2^\times \end{bmatrix} : X_1 \dot{+} X_2 \to X_1 \dot{+} X_2$$

where $A_1^\times = A_1 - B_1D_1^{-1}C_1$ and $A_2^\times = A_2 - B_2D_2^{-1}C_2$ are the associate main operators of (2.5) and (2.6), respectively. Note that $M = X_1 \dot{+} \{0\}$ is an invariant subspace for A, that $M^\times = \{0\} \dot{+} X_2$ is an invariant subspace for $A^\times$, and that M and $M^\times$ match in the sense that the state space of the product realization is the direct sum of M and $M^\times$. This state of affairs turns out to be a key point in the discussion of factorization we now turn to.

2.6 Factorization

The theorems in this section will serve as a basis for the more involved factorization results to be given in the sequel. Subspaces of Banach spaces are always assumed to be closed, otherwise we use the term linear manifold. For simplicity (and without loss of generality) we assume the external spaces U and Y to be equal.

Theorem 2.6. *Consider the biproper realization*

$$W(\lambda) = D + C(\lambda I_X - A)^{-1}B, \qquad \lambda \in \rho(A), \tag{2.7}$$

and let $A^\times = A - BD^{-1}C$ be its associate main operator. Let M and $M^\times$ be invariant subspaces for A and $A^\times$, respectively, and suppose

$$X = M \dot{+} M^\times. \tag{2.8}$$

Assume $D = D_1D_2$, where D_1 and D_2 are invertible operators on Y, and write

$$
\begin{aligned}
A &= \begin{bmatrix} A_1 & A_+ \\ 0 & A_2 \end{bmatrix} : M\dot{+}M^\times \to M\dot{+}M^\times, \\
B &= \begin{bmatrix} B_1 \\ B_2 \end{bmatrix} : Y \to M\dot{+}M^\times, \\
C &= \begin{bmatrix} C_1 & C_2 \end{bmatrix} : M\dot{+}M^\times \to Y.
\end{aligned}
$$

Introduce the functions W_1 and W_2 via the biproper realizations

$$W_1(\lambda) = D_1 + C_1(\lambda I_M - A_1)^{-1}B_1D_2^{-1}, \qquad \lambda \in \rho(A_1), \tag{2.9}$$

$$W_2(\lambda) = D_2 + D_1^{-1}C_2(\lambda I_{M^\times} - A_2)^{-1}B_2, \qquad \lambda \in \rho(A_2). \tag{2.10}$$

Then W admits the factorization

$$W(\lambda) = W_1(\lambda)W_2(\lambda), \qquad \lambda \in \rho(A_1) \cap \rho(A_2) \subset \rho(A).$$

The function W is defined and analytic on $\rho(A)$, while the factors W_1 and W_2 are defined and analytic on the sets $\rho(A_1)$ and $\rho(A_2)$, respectively. In particular, the factors may be defined and analytic on domains where the left-hand side is not. This will turn out to be relevant in applications (cf., the remarks made at the end of this section).

Proof. Identifying X and $M\dot{+}M^\times$ in the usual manner, the product of the realizations (2.9) and (2.10) is precisely the realization (2.7). The desired result now follows from Theorem 2.5. □

We shall refer to (2.8) as the *matching condition*, and when this condition is satisfied we refer to $M, M^\times$ as a *pair of matching subspaces*. A pair of matching subspaces $M, M^\times$ satisfying

$$A[M] \subset M, \qquad A^\times[M^\times] \subset M^\times$$

will be called a *supporting pair of subspaces* for the realization (2.7). Matching pairs of subspaces correspond to projections. So Theorem 2.6 has a counterpart in terms of projections. We say that a projection $\Pi : X \to X$ is a *supporting projection* for the realization (2.7) if

$$A[\operatorname{Ker} \Pi] \subset \operatorname{Ker} \Pi, \qquad A^\times[\operatorname{Im} \Pi] \subset \operatorname{Im} \Pi.$$

Here $\operatorname{Ker} T$ stands for the null space of an operator or matrix T, and $\operatorname{Im} T$ for its range.

Theorem 2.7. *Let Π be a supporting projection for the biproper realization*

$$W(\lambda) = D + C(\lambda I_X - A)^{-1}B, \qquad \lambda \in \rho(A).$$

Assume $D = D_1 D_2$, where D_1 and D_2 are invertible operators on Y, and introduce the functions W_1 and W_2 via the biproper realizations

$$\begin{aligned} W_1(\lambda) &= D_1 + C(\lambda I_X - A)^{-1}(I_X - \Pi)BD_2^{-1}, \qquad && \lambda \in \rho(A), \\ W_2(\lambda) &= D_2 + D_1^{-1}C\Pi(\lambda I_X - A)^{-1}B, && \lambda \in \rho(A). \end{aligned}$$

Then $W(\lambda) = W_1(\lambda)W_2(\lambda)$ for all $\lambda \in \rho(A)$.

This factorization holds on the resolvent set $\rho(A)$ of A. However, in many cases (relevant for applications), the factors in the right-hand side have an analytic extension to larger domains (see Theorem 2.6; cf., also the remarks made at the end of this section).

Proof. The fact that Π is a supporting projection for the given biproper realization means nothing else than that the identities $\Pi A = \Pi A \Pi$ and $A^\times \Pi = \Pi A^\times \Pi$ are satisfied. Hence $(I - \Pi)(A - A^\times)\Pi = A\Pi - \Pi A$. Now take $\lambda \in \rho(A)$. Then

$$\begin{aligned} W_1(\lambda)W_2(\lambda) &= D + C(\lambda - A)^{-1}(I - \Pi)B + C\Pi(\lambda - A)^{-1}B \\ &\qquad + C(\lambda - A)^{-1}(I - \Pi)BD^{-1}C\Pi(\lambda - A)^{-1}B \\ &= D + C(\lambda - A)^{-1}(I - \Pi)B + C\Pi(\lambda - A)^{-1}B \\ &\qquad + C(\lambda - A)^{-1}(I - \Pi)(A - A^\times)\Pi(\lambda - A)^{-1}B \\ &= D + C(\lambda - A)^{-1}(I - \Pi)B + C\Pi(\lambda - A)^{-1}B \\ &\qquad + C(\lambda - A)^{-1}(A\Pi - \Pi A)(\lambda - A)^{-1}B \\ &= D + C(\lambda - A)^{-1}(I - \Pi)B + C\Pi(\lambda - A)^{-1}B \\ &\qquad + C(\lambda - A)^{-1}\big(\Pi(\lambda - A) - (\lambda - A)\Pi\big)(\lambda - A)^{-1}B \\ &= D + C(\lambda - A)^{-1}(I - \Pi)B + C\Pi(\lambda - A)^{-1}B \\ &\qquad + C(\lambda - A)^{-1}\Pi B - C\Pi(\lambda - A)^{-1}B, \\ &= D + C(\lambda - A)^{-1}B = W(\lambda), \end{aligned}$$

as desired. □

The material presented above contains two factorization results: Theorems 2.6 and 2.7. These theorems contain not only different expressions for the factors, these factors also have different domains. For rational matrix functions and matrix-valued realizations, the differences are not essential. In the case of an infinite dimensional state space one has to be more careful, the reason being that

$\rho(A_1) \cap \rho(A_2)$ can then be a proper subset of $\rho(A)$. For an exhaustive discussion of the issues involved, see Section 2.5 in [20].

We shall meet the differences referred to above when the factorization results are applied, as will be done later on, for solving Wiener-Hopf, Toeplitz or singular integral equations. In that context, it is also necessary to have information on the sets where the factors take invertible values and to have expressions for the inverses. In other words, it is necessary to have a good understanding of the relationship between Theorems 2.6 and 2.7 on the one hand, and the inversion result Theorem 2.4 on the other. The point here is that, by taking inverses, the factorizations of the function $W(\lambda)$ given in Theorems 2.6 and 2.7 directly induce factorizations of the point-wise inverse W^{-1} of W, that is the function given by $W^{-1}(\lambda) = W(\lambda)^{-1}$, while on the other hand factorizations of W^{-1} can also be obtained by applying Theorems 2.6 and 2.7 to the realization

$$W^{-1}(\lambda) = D^{-1} - D^{-1}C(\lambda - A^{\times})^{-1}BD^{-1}. \tag{2.11}$$

Note here that if $M, M^{\times}$ is a supporting pair of subspaces for the realization (2.7), then $M^{\times}, M$ is a supporting pair of subspaces for the realization (2.11), and, analogously, if Π is a supporting projection for (2.7), then $I - \Pi$ is a supporting projection for (2.11). The analysis in [20], Section 2.5 also clarifies these matters; the upshot is that the two approaches lead to essentially the same result.

Notes

The notion of a realization originates from the Kalman theory of linear time-invariant systems [95]. The literature on the subject is rich; see, e.g., the text books [94], [33]. In a somewhat different form the notion of realization also appears in the theory of characteristic operator functions [27], [141]. The realization problem has many different faces, depending on the class of matrix or operator functions one is dealing with. The material of the first two sections is standard. Theorem 2.1 is a variation on Theorem 4.20 in [10]. Other constructions of matrix-valued realizations, including realizations with smallest possible state space dimension, can be found in text books; see, e.g., [94], [33] or [85] and references given there. The realization theorems for analytic operator functions in Section 2.3 originate from [57]. The operations of inversion and taking products are standard in systems theory. Theorem 2.11 has a natural Schur complement interpretation; see Section 2.2 in [20] and the paper [19]. The factorization theorem in the final section originates from [21]; see also the first chapter of [11]. For a brief description of the history of the factorization principle presented here, we refer to the Editorial introduction in [54].

Part II
Convolution equations with rational matrix symbols

The canonical factorization theorem for rational matrix functions in state space form is the first result presented and proved in this part. This theorem is then used to invert explicitly Wiener-Hopf, Toeplitz and singular integral operators with a rational matrix symbol, with the inverses being presented explicitly in state space formulas. For rational matrix symbols the solution to the homogeneous Riemann-Hilbert boundary value problem is also given in state space form.

This part consists of two chapters. In the first chapter (Chapter 3) we consider proper rational matrix functions, that is, rational matrix functions that are analytic at infinity. The case of non-proper rational symbols is treated in the second chapter (Chapter 4). This requires a different type of realization. This modified realization result is developed and a corresponding canonical factorization theorem is proved. As an application the homogeneous Riemann-Hilbert boundary value problem is solved for an arbitrary rational matrix symbol.

Chapter 3

Explicit solutions using realizations

As we have seen in Chapter 1, canonical factorization serves as a tool to solve Wiener-Hopf integral equations, their discrete analogues, and the more general singular integral equations. In this chapter the state space factorization method developed in Chapter 2 is used to solve the problem of canonical factorization (necessary and sufficient conditions for its existence) and to derive explicit formulas for its factors. This is done in Section 3.1 for rational matrix functions and later in Section 7.1 for operator-valued transfer functions that are analytic on an open neighborhood of a curve. The results are applied to invert Wiener-Hopf integral equations with a rational matrix symbol (Section 3.2), block Toeplitz operators (Section 3.3) and singular integral equations (Section 3.4). The methods developed in this chapter also allow us to deal with the Riemann-Hilbert boundary value problem. This is done in the final section which also contains material on the homogeneous Wiener-Hopf equation.

3.1 Canonical factorization of rational matrix functions in state space form

In this section and the next one we shall consider the factorization theorems of Section 2.6 for the special case when the two factors satisfy additional spectral conditions. Recall from Chapter 0 that a Cauchy contour is the positively oriented boundary of a bounded Cauchy domain in $\mathbb{C}$ and that such a contour consists of a finite number of simple closed non-intersecting rectifiable curves. We say that a Cauchy contour Γ *splits the spectrum* $\sigma(S)$ of a bounded linear operator S if $\Gamma \cap \sigma(S) = \emptyset$. In that case $\sigma(S)$ decomposes into two disjoint compact sets σ_+ and σ_- such that σ_+ is in the interior domain of Γ and σ_- is in the exterior domain of Γ. If Γ splits the spectrum of S, then we have a *Riesz projection*, also called *spectral projection*, associated with S and Γ, namely

$$P(S;\Gamma) = \frac{1}{2\pi i}\int_\Gamma (\lambda - S)^{-1}\,d\lambda.$$

The subspace $N = \operatorname{Im} P(S;\Gamma)$ will be called the *spectral subspace* for S corresponding to the contour Γ (or to the spectral set σ_+).

Lemma 3.1. *Let Y_1 and Y_2 be complex Banach spaces, and consider the operator*

$$S = \begin{bmatrix} S_{11} & S_{12} \\ 0 & S_{22} \end{bmatrix} : Y_1 \dot{+} Y_2 \to Y_1 \dot{+} Y_2. \tag{3.1}$$

Let Π be any projection of $Y = Y_1 \dot{+} Y_2$ such that $\operatorname{Ker}\Pi = Y_1$. Then the compression $\Pi S|_{\operatorname{Im}\Pi} : \operatorname{Im}\Pi \to \operatorname{Im}\Pi$ and $S_{22} : Y_2 \to Y_2$ are similar. Furthermore, Y_1 is a spectral subspace for S if and only if $\sigma(S_{11}) \cap \sigma(S_{22}) = \emptyset$, and in that case $\sigma(S) = \sigma(S_{11}) \cup \sigma(S_{22})$ while, in addition,

$$Y_1 = \operatorname{Im} P(S;\Gamma) = \operatorname{Im}\left(\frac{1}{2\pi i}\int_\Gamma (\lambda I - S)^{-1}\,d\lambda\right), \tag{3.2}$$

where Γ is a Cauchy contour around $\sigma(S_{11})$ separating $\sigma(S_{11})$ from $\sigma(S_{22})$.

Proof. Let P be the projection of $Y = Y_1 \dot{+} Y_2$ along Y_1 onto Y_2. As $\operatorname{Ker} P = \operatorname{Ker}\Pi$, we have $P = P\Pi$ and the map $E = P|_{\operatorname{Im}\Pi} : \operatorname{Im}\Pi \to Y_2$ is an invertible operator. Write S_0 for the compression $\Pi S|_{\operatorname{Im}\Pi} : \operatorname{Im}\Pi \to \operatorname{Im}\Pi$ of S to $\operatorname{Im}\Pi$, and take $x = \Pi y$. Then $ES_0x = P\Pi S\Pi y = PS\Pi y = PSP\Pi y = S_{22}Ex$, and hence S_0 and S_{22} are similar.

Now suppose $\sigma(S_{11}) \cap \sigma(S_{22}) = \emptyset$. Then $\rho(S_{11}) \cup \rho(S_{22}) = \mathbb{C}$ and hence

$$\rho(S) = \Big(\rho(S) \cap \rho(S_{11})\Big) \bigcup \Big(\rho(S) \cap \rho(S_{22})\Big).$$

The upper triangular form of S in (3.1) ensues

$$\rho(S) \cap \rho(S_{11}) = \rho(S) \cap \rho(S_{22}) = \rho(S_{11}) \cap \rho(S_{22})$$

and it follows that $\rho(S_{11}) \cup \rho(S_{22}) = \rho(S)$, an identity which can be rewritten as $\sigma(S) = \sigma(S_{11}) \cup \sigma(S_{22})$.

Still under the assumption that $\sigma(S_{11}) \cap \sigma(S_{22}) = \emptyset$, let Γ be a Cauchy contour Γ around $\sigma(S_{11})$ separating $\sigma(S_{11})$ from $\sigma(S_{22})$. Then Γ splits the spectrum of S. In fact, if $\lambda \in \Gamma$, then both $\lambda - S_{11}$ and $\lambda - S_{22}$ are invertible and

$$(\lambda - S)^{-1} = \begin{bmatrix} (\lambda - S_{11})^{-1} & (\lambda - S_{11})^{-1}S_{12}(\lambda - S_{22})^{-1} \\ 0 & (\lambda - S_{22})^{-1} \end{bmatrix}$$

which leads to an expression of the type

$$P(S;\Gamma) = \begin{bmatrix} I & * \\ 0 & 0 \end{bmatrix}$$

for the Riesz projection associated with S and Γ. In particular, it is clear that $Y_1 = \operatorname{Im} P(S;\Gamma)$. So Y_1 is a spectral subspace for S and (3.2) holds.

Next assume that $Y_1 = \operatorname{Im} Q$, where Q is a Riesz projection for S. Put $\Pi = I - Q$, and let S_0 be the restriction of S to $\operatorname{Im} \Pi$. Then $\sigma(S_{11}) \cap \sigma(S_0) = \emptyset$. By the first part of the proof, the operators S_0 and S_{22} are similar. So $\sigma(S_0) = \sigma(S_{22})$, and hence we have shown that $\sigma(S_{11}) \cap \sigma(S_{22}) = \emptyset$. □

Let Γ be a Cauchy contour. As before (see the one but last paragraph in Chapter 0) we denote by F_+ and F_- the interior and exterior domain of Γ, respectively. Note that $\infty \in F_-$. Let W be a rational $m \times m$ matrix function, with $W(\infty) = I$, analytic on an open neighborhood of Γ, whose values on Γ are invertible matrices. By a *right canonical factorization* of W with respect to Γ we mean a factorization

$$W(\lambda) = W_-(\lambda) W_+(\lambda), \qquad \lambda \in \Gamma, \tag{3.3}$$

where W_- and W_+ are rational $m \times m$ matrix functions, analytic and taking invertible values on (an open neighborhood of) $\overline{F}_-$ and $\overline{F}_+$, respectively. If in (3.3) the factors W_- and W_+ are interchanged, we speak of a *left canonical factorization*.

Theorem 3.2. *Let Γ be a Cauchy contour and let W be a rational $m \times m$ matrix function, Suppose W admits the realization $W(\lambda) = I_m + C(\lambda I_n - A)^{-1}B$ such that the main matrix A has no eigenvalues on Γ. Then W admits a right canonical factorization with respect to Γ if and only if the following two conditions are satisfied:*

(i) *$A^\times = A - BC$ has no eigenvalues on Γ,*

(ii) $\mathbb{C}^n = \operatorname{Im} P(A;\Gamma) \dotplus \operatorname{Ker} P(A^\times;\Gamma)$.

In that case, a right canonical factorization of W is given by

$$W(\lambda) = W_-(\lambda) W_+(\lambda), \qquad \lambda \in \Gamma,$$

where the factors and their inverses can be written as

$$\begin{aligned} W_-(\lambda) &= I_m + C(\lambda I_n - A)^{-1}(I - \Pi)B, \\ W_+(\lambda) &= I_m + C\Pi(\lambda I_n - A)^{-1}B, \\ W_-^{-1}(\lambda) &= I_m - C(I - \Pi)(\lambda I_n - A^\times)^{-1}B, \\ W_+^{-1}(\lambda) &= I_m - C(\lambda I_n - A^\times)^{-1}\Pi B. \end{aligned}$$

Here Π is the projection of $\mathbb{C}^n$ along $\operatorname{Im} P(A;\Gamma)$ onto $\operatorname{Ker} P(A^\times;\Gamma)$.

For left canonical factorizations an analogous theorem holds. In the result in question, (ii) is replaced by $\mathbb{C}^n = \operatorname{Ker} P(A;\Gamma) \dot{+} \operatorname{Im} P(A^\times;\Gamma)$.

The expressions for the functions W_- and W_+ suggest that these functions are defined on the resolvent set $\rho(A)$ of A. Similarly, W_-^{-1} and W_+^{-1} seem to have $\rho(A^\times)$ as their domain. At first sight this is at variance with the requirements for Wiener-Hopf factorization. We will address this point in the proof.

Proof. From the definition given above it is clear that a necessary condition in order that W admits a right canonical factorization with respect to Γ is that W takes invertible values on Γ. By Theorem 2.4 this necessary condition is met if and only if (i) holds true.

Assume that (i) is satisfied. The spectral projections $P(A;\Gamma)$ and $P(A^\times;\Gamma)$ are then well-defined. The image $X_- = \operatorname{Im} P(A;\Gamma)$ of $P(A;\Gamma)$ and the null space $X_+ = \operatorname{Ker} P(A^\times;\Gamma)$ of $P(A^\times;\Gamma)$ are invariant for A and $A^\times$, respectively. Suppose (ii) is fulfilled too, and write

$$A = \begin{bmatrix} A_- & A_0 \\ 0 & A_+ \end{bmatrix}, \qquad B = \begin{bmatrix} B_- \\ B_+ \end{bmatrix}, \qquad C = \begin{bmatrix} C_- & C_+ \end{bmatrix}$$

for the matrix presentations of A, B and C with respect to the decomposition $\mathbb{C}^n = X_- \dot{+} X_+$. With

$$\begin{aligned} W_-(\lambda) &= I_{X_-} + C_-(\lambda - A_-)^{-1}B_-, \qquad & \lambda \in \rho(A_-), \\ W_+(\lambda) &= I_{X_+} + C_+(\lambda - A_+)^{-1}B_+, \qquad & \lambda \in \rho(A_+), \end{aligned}$$

we have (from Theorem 2.6) the factorization

$$W(\lambda) = W_-(\lambda)W_+(\lambda), \qquad \lambda \in \rho(A_-) \cap \rho(A_+) \subset \rho(A).$$

As X_- is a spectral subspace for A, we can apply Lemma 3.1 to show that $\sigma(A_-)$ and $\sigma(A_+)$ are disjoint. But then $\rho(A) = \rho(A_-) \cap \rho(A_+)$ and it follows that

$$W(\lambda) = W_-(\lambda)W_+(\lambda), \qquad \lambda \in \rho(A_-) \cap \rho(A_+) = \rho(A). \tag{3.4}$$

Applying Lemma 3.1 once again we see that

$$\sigma(A_-) = \sigma(A) \cap F_+, \qquad \sigma(A_+) = \sigma(A) \cap F_-, \tag{3.5}$$

where F_+ and F_- are the interior and exterior domain of Γ, respectively. In a similar way one proves that

$$\sigma(A_-^\times) = \sigma(A^\times) \cap F_+, \qquad \sigma(A_+^\times) = \sigma(A^\times) \cap F_-. \tag{3.6}$$

Using the first parts of (3.5) and (3.6), it now follows that W_- is analytic and has invertible values on an open neighborhood of $\overline{F}_-$. Analogously, employing the second parts of (3.5) and (3.6), one gets that W_+ is analytic and has invertible values on an open neighborhood of $\overline{F}_+$. Thus (3.4) is a right canonical Wiener-Hopf factorization with respect to Γ.

The projection Π of $\mathbb{C}^n$ along $\operatorname{Im} P(A;\Gamma)$ onto $\operatorname{Ker} P(A^\times;\Gamma)$ is a supporting projection for the given realization of W. Also $I_n - \Pi$ is a supporting projection for the realization $W(\lambda)^{-1} = I_m - C(\lambda I_n - A^\times)^{-1}B$ of W^{-1}. With this in mind, one checks without difficulty that W_-, W_+, W_-^{-1} and W_+^{-1} can also be written as in the theorem. For an exhaustive discussion of the intricacies concerning inversion, factorization, and the combination of these operations (in fact: the relationship between Theorems 2.6, 2.7 and 2.4), see Section 2.5 in [20]. Note, however, that in the present case there is no ambiguity because we are working here with rational matrix functions.

Next, suppose that $W(\lambda) = W_-(\lambda)W_+(\lambda)$ is a right canonical factorization with respect to Γ. We only have to show that $\mathbb{C}^n = \operatorname{Im} P(A;\Gamma) \dotplus \operatorname{Ker} P(A^\times;\Gamma)$. We first prove that $\operatorname{Im} P(A;\Gamma) \cap \operatorname{Ker} P(A^\times;\Gamma) = \{0\}$. Without loss of generality it may be assumed that the values of W_- and W_+ at infinity are equal to I_m.

Suppose $x \in \operatorname{Im} P(A;\Gamma) \cap \operatorname{Ker} P(A^\times;\Gamma)$, and consider $(\lambda - A)^{-1}x$. This function is analytic on an open neighborhood of $\overline{F}_-$. On the other hand the function $(\lambda - A^\times)^{-1}x$ is analytic on an open neighborhood of $\overline{F}_+$. For λ in the intersection $\rho(A) \cap \rho(A^\times)$, we have

$$\begin{aligned} W(\lambda)C(\lambda - A^\times)^{-1} &= C(\lambda - A^\times)^{-1} + C(\lambda - A)^{-1}BC(\lambda - A^\times)^{-1} \\ &= C(\lambda - A^\times)^{-1} + C(\lambda - A)^{-1}(A - A^\times)(\lambda - A^\times)^{-1} \\ &= C(\lambda - A)^{-1}, \end{aligned}$$

and it follows that $W_+(\lambda)C(\lambda - A^\times)^{-1} = W_-(\lambda)^{-1}C(\lambda - A)^{-1}$. The analyticity properties of the factors W_-, W_+ and their inverses now imply that the function $W_+(\lambda)C(\lambda - A^\times)^{-1}x = W(\lambda)^{-1}C(\lambda - A)^{-1}x$ is analytic on the Riemann sphere $\mathbb{C}_\infty$. By Liouville's theorem it must be constant. As it takes the value zero at infinity, it is identically zero. Hence both $C(\lambda - A^\times)^{-1}x$ and $C(\lambda - A)^{-1}x$ vanish. Next use the identity

$$(\lambda - A^\times)^{-1}BC(\lambda - A)^{-1} = (\lambda - A)^{-1} - (\lambda - A^\times)^{-1}$$

to obtain $(\lambda - A^\times)^{-1}x = (\lambda - A)^{-1}x$. But then this function is analytic on the Riemann sphere too. Using Liouville's theorem again, we see that it must be identically zero. Thus $x = 0$.

Observe that up to this point in the proof we have not used the finite dimensionality of the state space. It will play a role in the next paragraph.

We now finish the proof by a duality argument. Let Γ^* be the adjoint curve of Γ, i.e., the curve obtained from Γ by complex conjugation. Also introduce the functions V, V_+ and V_- by putting $V(\lambda) = W(\bar{\lambda})^*$, $V_-(\lambda) = W_-(\bar{\lambda})^*$ and $V_+(\lambda) = W_+(\bar{\lambda})^*$. Clearly V has the realization $V(\lambda) = I + B^*(\lambda - A^*)^{-1}C^*$ and $V(\lambda) = V_+(\lambda)V_-(\lambda)$ is a left canonical factorization. Arguing as above, we may conclude that $\operatorname{Ker} P(A^*, \Gamma^*) \cap \operatorname{Im} P\big((A^\times)^*, \Gamma^*\big) = 0$. It follows that

$\operatorname{Ker} P(A^*, \Gamma^*) + \operatorname{Im} P\big((A^\times)^*, \Gamma^*\big) = \mathbb{C}^n$. In the first instance, this equality holds for the closure of $\operatorname{Ker} P(A^*, \Gamma^*) + \operatorname{Im} P\big((A^\times)^*, \Gamma^*\big)$, but in $\mathbb{C}^n$ all linear manifolds are closed. □

With minor modifications we could have worked in Theorem 3.2 with two curves, one splitting the spectrum of A and the other splitting the spectrum of $A^\times$ (cf., [100]). Finally, let us mention that Theorem 3.2 remains true if the Cauchy contour Γ is replaced by the extended real line $\mathbb{R}_\infty$, i.e., the closure of the real line in the Riemann sphere $\mathbb{C}_\infty$. In that case F_+ is the open upper half plane and F_- is the open lower half plane. For details, see Theorem 4.5 at the end of Section 4.3 below which, by the way, deals with the situation where W is a not necessarily proper rational matrix function.

3.2 Wiener-Hopf integral operators

In this section the general factorization result proved in the preceding sections is used to provide explicit formulas for solutions of finite systems of the Wiener-Hopf equation

$$\phi(t) - \int_0^\infty k(t-s)\phi(s)\,ds = f(t), \qquad t \geq 0, \tag{3.7}$$

where ϕ and f are m-dimensional vector functions and $k \in L_1^{m\times m}(-\infty,\infty)$, i.e., the kernel function k is an $m \times m$ matrix function of which the entries are in $L_1(-\infty,\infty)$. We assume that the given vector function f has its component functions in $L_p[0,\infty)$, and we express this property by writing $f \in L_p^m[0,\infty)$. Throughout this section, p will be fixed and $1 \leq p < \infty$. The problem we shall consider is to find a solution ϕ for equation (3.7) that also belongs to the space $L_p^m[0,\infty)$. As was explained in Section 1.1 the equation (3.7) has a unique solution in $L_p^m[0,\infty)$ for each f in $L_p^m[0,\infty)$ if and only if its symbol $I_m - K(\lambda)$ admits a factorization as in (1.5).

Our aim is to apply the factorization theory developed in the previous sections to get the canonical factorization (1.5). Therefore, in the sequel we assume that the symbol is a rational $m \times m$ matrix function. As $K(\lambda)$ is the Fourier transform of an $L_1^{m\times m}(-\infty,\infty)$–function, the symbol is continuous on the real line. In particular, $I_m - K(\lambda)$ has no poles on the real line. Furthermore, by the Riemann-Lebesgue lemma,

$$\lim_{\lambda\in\mathbb{R},\ |\lambda|\to\infty} K(\lambda) = 0,$$

which implies that the symbol $I_m - K(\lambda)$ has the value I_n at ∞. The fact that $I_m - K(\lambda)$ is rational is equivalent to the requirement that the kernel function k is in the linear space spanned by all functions of the form

$$h(t) = \begin{cases} p(t)e^{i\alpha t}, & t > 0, \\ q(t)e^{i\beta t}, & t < 0, \end{cases}$$

where $p(t)$ and $q(t)$ are matrix polynomials in t with coefficients in $\mathbb{C}^{m\times m}$, and α and β are complex numbers with $\Im\alpha > 0$ and $\Im\beta < 0$.

From Section 2.2 we know that the matrix function $I_m - K(\lambda)$ admits a realization

$$I_m - K(\lambda) = I_m + C(\lambda I_n - A)^{-1}B$$

such that the main matrix A has no real eigenvalues. In the next theorem we express the solvability of equation (3.7) in terms of such a realization and give explicit formulas for its solutions in the same terms.

Theorem 3.3. *Let $I_m - K(\lambda) = I_m + C(\lambda I_n - A)^{-1}B$ be a realization for the symbol of equation* (3.7), *and suppose A has no real eigenvalues. In order that* (3.7) *has a unique solution ϕ in $L_p^m[0,\infty)$ for each f in $L_p^m[0,\infty)$, the following two conditions are necessary and sufficient:*

(i) *$A^\times = A - BC$ has no real eigenvalues;*

(ii) *$\mathbb{C}^n = M \dot{+} M^\times$, where M is the spectral subspace of A corresponding to the eigenvalues of A in the upper half plane, and $M^\times$ is the spectral subspace of $A^\times$ corresponding to the eigenvalues of $A^\times$ in the lower half plane.*

Assume conditions (i) *and* (ii) *hold true, and let Π be the projection of $\mathbb{C}^n$ along M onto $M^\times$. Then $I_m - K(\lambda)$ admits a right canonical factorization with respect to the real line that has the form*

$$I_m - K(\lambda) = \big(I_m + G_-(\lambda)\big)\big(I_m + G_+(\lambda)\big), \qquad \lambda \in \mathbb{R},$$

where the factors and their inverses can be written as

$$\begin{aligned} I_m + G_+(\lambda) &= I_m + C\Pi(\lambda I_n - A)^{-1}B, \\ I_m + G_-(\lambda) &= I_m + C(\lambda I_n - A)^{-1}(I_n - \Pi)B, \\ \big(I_m + G_+(\lambda)\big)^{-1} &= I_m - C(\lambda I_n - A^\times)^{-1}\Pi B, \\ \big(I_m + G_-(\lambda)\big)^{-1} &= I_m - C(I_n - \Pi)(\lambda I_n - A^\times)^{-1}B. \end{aligned}$$

The functions γ_+ and γ_- in (1.6) *and* (1.7) *are given by*

$$\begin{aligned} \gamma_+(t) &= +iCe^{-itA^\times}\Pi B, \qquad t > 0, \\ \gamma_-(t) &= -iC(I_n - \Pi)e^{-itA^\times t}B, \qquad t < 0. \end{aligned}$$

Finally, the solution ϕ to (3.7) *can be written as*

$$\phi(t) = f(t) + \int_0^\infty \gamma(t,s)f(s)\,ds,$$

where

$$\gamma(t,s) = \begin{cases} +iCe^{-itA^\times}\Pi e^{isA^\times}B, & s < t, \\ -iCe^{-itA^\times}(I_n - \Pi)e^{isA^\times}B, & s > t. \end{cases}$$

Proof. We have already mentioned that equation (3.7) has a unique solution in $L_p^m[0,\infty)$ for each f in $L_p^m[0,\infty)$ if and only if the symbol $I_m - K(\lambda)$ admits a right canonical factorization as in (1.5). So to prove the necessity and sufficiency of the conditions (i) and (ii), it suffices to show that the conditions (i) and (ii) together are equivalent to the statement that $I_m - K(\lambda)$ admits a right canonical factorization as in (1.5). We first observe that condition (i) is equivalent to the requirement that $I_m - K(\lambda)$ is invertible for all $\lambda \in \mathbb{R}$ (see Theorem 2.4). But then we can apply Theorem 3.2 in combination with the remark made at the end of Section 3.1 to prove the first part of the theorem.

Next assume that conditions (i) and (ii) hold true. Applying Theorem 3.2 once again, we get the desired formulas for $I_m + G_+(\lambda)$, $I_m + G_-(\lambda)$ and their inverses. The formulas for γ_+ and γ_- are now obtained by noticing that

$$\begin{aligned}\int_0^\infty e^{i\lambda t}e^{-itA^\times}\Pi\,dt &= i(\lambda - A^\times)^{-1}\Pi, \qquad \lambda\in\rho(A^\times),\ \Im\lambda\geq 0,\\ \int_{-\infty}^0 e^{i\lambda t}(I-\Pi)e^{-itA^\times}\,dt &= -i(I-\Pi)(\lambda - A^\times)^{-1}, \qquad \lambda\in\rho(A^\times),\ \Im\lambda\leq 0,\end{aligned}$$

where $I = I_n$. The proof of the latter identity uses (the first conclusion in) Lemma 3.1.

It remains to prove the final formula for $\gamma(t,s)$. We use (1.10), and compute first that

$$\gamma_+(t-r)\gamma_-(r-s) = Ce^{-i(t-r)A^\times}\Pi BC(I-\Pi)e^{-i(r-s)A^\times}B.$$

Now $\operatorname{Ker}\Pi = M$ is A-invariant and $\operatorname{Im}\Pi = M^\times$ is $A^\times$-invariant. Thus $\Pi A(I-\Pi) = 0$ and $(I-\Pi)A^\times\Pi = 0$, and it follows that $\Pi BC(I-\Pi) = \Pi(A-A^\times)(I-\Pi) = \Pi A^\times - A^\times\Pi$. But then

$$\begin{aligned}\gamma_+(t-r)\gamma_-(r-s) &= Ce^{-i(t-r)A^\times}(A^\times\Pi - \Pi A^\times)e^{-i(r-s)A^\times}B\\ &= -i\frac{d}{dr}Ce^{-i(t-r)A^\times}\Pi e^{-i(r-s)A^\times}B.\end{aligned}$$

Inserting this in (1.30) we obtain for $s<t$ that

$$\begin{aligned}\gamma(t,s) &= iCe^{-i(t-s)A^\times}\Pi B - \int_0^s i\frac{d}{dr}Ce^{-i(t-r)A^\times}\Pi e^{-i(r-s)A^\times}B\,dr\\ &= iCe^{-i(t-s)A^\times}\Pi B - Ce^{-i(t-r)A^\times}\Pi e^{-i(r-s)A^\times}B\big|_{r=0}^{s}\\ &= iCe^{-itA^\times}\Pi e^{isA^\times}B,\end{aligned}$$

while for $s > t$ we get

$$\begin{aligned}\gamma(t,s) &= -iC(I-\Pi)e^{-i(t-s)A^\times}B + \int_0^t i\frac{d}{dr}Ce^{-i(t-r)A^\times}\Pi e^{-i(r-s)A^\times}B\,dr \\ &= -iC(I-\Pi)e^{-i(t-s)A^\times}B - Ce^{-i(t-r)A^\times}\Pi e^{-i(r-s)A^\times}B|_{r=0}^{t} \\ &= -iCe^{-itA^\times}(I-\Pi)e^{isA^\times}B.\end{aligned}$$

This completes the proof. □

Corollary 3.4. *Let $I_m - K(\lambda) = I_m + C(\lambda I_n - A)^{-1}B$ be a realization for the symbol of equation (3.7). Assume that A and $A^\times = A - BC$ have no spectrum on the real line, and that*

$$\mathbb{C}^n = \operatorname{Im} P \dot{+} \operatorname{Ker} P^\times, \tag{3.8}$$

where P and $P^\times$ are the Riesz projections of A and $A^\times$, respectively, corresponding to the spectra in the upper half plane. Fix $x \in \operatorname{Ker} P$, and let the right-hand side of (3.7) be given by $f(t) = Ce^{-itA}x$, $t \geq 0$. Then the unique solution ϕ in $L_p^m[0,\infty)$ of equation (3.7) is given by

$$\phi(t) = Ce^{-itA^\times}\Pi x, \qquad t \geq 0.$$

Here Π is the projection of $\mathbb{C}^n$ onto $\operatorname{Ker} P^\times$ *along* $\operatorname{Im} P$.

Proof. Since $x \in \operatorname{Ker} P$, the vector $e^{-itA}x$ is exponentially decaying in norm when $t \to \infty$, and thus the function f belongs to $L_p^m[0,\infty)$. Furthermore, the conditions (i) and (ii) in Theorem 3.3 are fulfilled, and hence equation (3.7) has a unique solution $\phi \in L_p^m[0,\infty)$. Moreover from Theorem 3.3 we know that ϕ is given by

$$\begin{aligned}\phi(t) = f(t) + iCe^{-itA^\times}\Big(\int_0^t \Pi e^{isA^\times}BCe^{-isA}x\,ds\Big) \\ -iCe^{-itA^\times}\Big(\int_t^\infty (I-\Pi)e^{isA^\times}BCe^{-isA}x\,ds\Big).\end{aligned}$$

Now use that

$$e^{isA^\times}BCe^{-isA} = ie^{isA^\times}(iA^\times - iA)e^{-isA} = i\frac{d}{ds}e^{isA^\times}e^{-isA}.$$

It follows that

$$\begin{aligned}\phi(t) = f(t) - Ce^{-itA^\times}\Big(\Pi e^{isA^\times}e^{-isA}x|_0^t\Big) \\ +Ce^{-itA^\times}\Big((I-\Pi)e^{isA^\times}e^{-isA}x|_t^\infty\Big).\end{aligned}$$

Since $(I - \Pi) = (I - \Pi)P^\times$, the function $(I - \Pi)e^{isA^\times} = (I - \Pi)P^\times e^{isA^\times}$ is exponentially decaying for $s \to \infty$. As we have seen, the same holds true for $e^{-isA}x$. Thus

$$\begin{aligned}\phi(t) &= f(t) - Ce^{-itA^\times}\Pi e^{itA^\times}e^{-itA}x + Ce^{-itA^\times}\Pi x \\ &\quad -Ce^{-itA^\times}(I-\Pi)e^{itA^\times}e^{-itA}x \\ &= f(t) + Ce^{-itA^\times}\Pi x - Ce^{-itA}x \\ &= Ce^{-itA^\times}\Pi x,\end{aligned}$$

which completes the proof. □

Finally, let us return to the special situation where the function$f$ is given by formula (1.11), and assume that the conditions (i) and (ii) in Theorem 3.3 are satisfied. Then the solution ϕ admits the representation

$$\phi(t) = e^{-iqt}\{I_m + i\int_0^t Ce^{i(q-A^\times)s}\Pi B\,ds\} \tag{3.9}$$

$$\cdot\{I_m - C(I-\Pi)(q-A^\times)^{-1}B\}x_0; \tag{3.10}$$

see formula (1.12).

3.3 Block Toeplitz operators

In the previous section the factorization theory was applied to finite systems of Wiener-Hopf integral equations. In this section we carry out a similar program for their discrete analogues, block Toeplitz equations (cf., Section 1.2). So we consider an equation of the type

$$\sum_{k=0}^{\infty} a_{j-k}\xi_k = \eta_j, \qquad j = 0, 1, 2, \ldots . \tag{3.11}$$

Throughout we assume that the coefficients a_j are given complex $m \times m$ matrices satisfying

$$\sum_{j=-\infty}^{\infty} \|a_j\| < \infty,$$

and $\eta = (\eta_j)_{j=0}^{\infty}$ is a given vector from $\ell_p^m = \ell_p(\mathbb{C}^m)$. The problem is to find $\xi = (\xi_k)_{k=0}^{\infty} \in \ell_p^m$ such that (3.11) is satisfied.

As before, we shall apply our factorization theory. For that reason we assume that the symbol $a(\lambda) = \sum_{j=-\infty}^{\infty} \lambda^j a_j$ is a rational $m \times m$ matrix function whose value at ∞ is I_m. Note that $a(\lambda)$ has no poles on the unit circle. Therefore the conditions on $a(\lambda)$ are equivalent to the following assumptions:

(j) the sequence $(a_j - \delta_{j0}I_m)_{j=0}^{\infty}$ is a linear combination of sequences of the form

$$\left(\alpha^j j^r D\right)_{j=0}^{\infty},$$

where $|\alpha| < 1$, r is a nonnegative integer and D is a complex $m \times m$ matrix;

(jj) the sequence $(a_{-j})_{j=1}^{\infty}$ is a linear combination of sequences of the form

$$\left(\beta^{-j} j^s E\right)_{j=1}^{\infty}, \qquad \left(\delta_{jk} F\right)_{j=1}^{\infty},$$

where $|\beta| > 1$, s and k are nonnegative integers and E and F are complex $m \times m$ matrices.

From Section 2.2 we know that the matrix function $a(\lambda)$ admits a realization

$$a(\lambda) = I_m + C(\lambda I_n - A)^{-1}B \tag{3.12}$$

such that the main matrix A has no eigenvalues on the unit circle. The next theorem is the analogue of Theorem 3.3.

Theorem 3.5. *Let* (3.12) *be a realization for the symbol* $a(\lambda)$ *of the equation* (3.11), *and suppose* A *has no eigenvalues on the unit circle. Then* (3.11) *has a unique solution* $\xi = (\xi_k)_{k=0}^{\infty}$ *in* ℓ_p^m *for each* $\eta = (\eta_j)_{j=0}^{\infty}$ *in* ℓ_p^m *if and only if the following two conditions are satisfied:*

(i) $A^{\times} = A - BC$ *has no eigenvalues on the unit circle,*

(ii) $\mathbb{C}^n = M \dot{+} M^{\times}$, *where* M *is the spectral subspace of* A *corresponding to the eigenvalues of* A *inside the unit circle, and* $M^{\times}$ *is the spectral subspace of* $A^{\times}$ *corresponding to the eigenvalues of* $A^{\times}$ *outside the unit circle.*

Assume conditions (i) *and* (ii) *are satisfied, and let* Π *be the projection of* $\mathbb{C}^n$ *along* M *onto* $M^{\times}$. *Then the function* $a(\lambda)$ *admits a right canonical factorization with respect to the unit circle that has the form*

$$a(\lambda) = h_-(\lambda)h_+(\lambda), \qquad |\lambda| = 1,$$

where the factors and their inverses can be written as

$$\begin{aligned}
h_+(\lambda) &= I_m + C\Pi(\lambda I_n - A)^{-1}B,\\
h_-(\lambda) &= I_m + C(\lambda I_n - A)^{-1}(I_n - \Pi)B,\\
h_+^{-1}(\lambda) &= I_m - C(\lambda I_n - A^{\times})^{-1}\Pi B,\\
h_-^{-1}(\lambda) &= I_m - C(I_n - \Pi)(\lambda I_n - A^{\times})^{-1}B.
\end{aligned}$$

The sequences $(\gamma_j^+)_{j=0}^{\infty}$ *and* $(\gamma_{-j}^-)_{j=0}^{\infty}$ *in* (1.19) *are given by*

$$\begin{aligned}
\gamma_0^+ &= I_m + C(A^{\times})^{-1}\Pi B,\\
\gamma_j^+ &= C(A^{\times})^{-(j+1)}\Pi B, \qquad j = 1, 2, \ldots,\\
\gamma_0^- &= I_m,\\
\gamma_j^- &= -C(I_n - \Pi)(A^{\times})^{-(j+1)}B, \qquad j = -1, -2, \ldots.
\end{aligned}$$

Finally, the solution ξ to (3.11) *can be written as $\xi_k = \sum_{s=0}^{\infty} \gamma_{ks}\eta_s$ where*

$$\gamma_{ks} = \begin{cases} C(A^\times)^{-(k+1)}\Pi(A^\times)^s B, & s < k, \\ I_m + C(A^\times)^{-(s+1)}\Pi(A^\times)^s B, & s = k, \\ -C(A^\times)^{-(k+1)}(I_n - \Pi)(A^\times)^s B, & s > k. \end{cases}$$

Proof. The proof of Theorem 3.5 is similar to that of Theorem 3.3. Here we only derive the final formula for γ_{ks}.

With respect to the formulas for γ_j^+, we note that $\operatorname{Im}\Pi$ is $A^\times$-invariant and the restriction of $A^\times$ to $\operatorname{Im}\Pi$ is invertible. So, with slight abuse of notation as far as inverses of $A^\times$ are involved,

$$\begin{aligned} h_+(\lambda)^{-1} &= I_m - C(\lambda - A^\times)^{-1}\Pi B \\ &= I_m + C\big(I - \lambda(A^\times)^{-1}\big)^{-1}(A^\times)^{-1}\Pi B \\ &= I_m + \sum_{j=0}^{\infty} \lambda^j C(A^\times)^{-(j+1)}\Pi B. \end{aligned}$$

Now compare coefficients with $h_+(\lambda)^{-1} = \sum_{j=0}^{\infty} \lambda^j \gamma_j^+$. Similarly, the formulas for γ_j^- are obtained by comparing

$$\begin{aligned} h_-(\lambda)^{-1} &= I_m - C(I - \Pi)(\lambda - A^\times)^{-1}B \\ &= I_m - C(I - \Pi)\sum_{j=1}^{\infty} \frac{1}{\lambda^j}(A^\times)^{j-1}B \\ &= I_m - \sum_{j=-\infty}^{-1} \lambda^j C(I - \Pi)(A^\times)^{-(j+1)}B \end{aligned}$$

with $h_-(\lambda)^{-1} = \sum_{j=-\infty}^{0} \lambda^j \gamma_j^-$. Here $I = I_n$.

To obtain the formulas for γ_{ks} we employ (1.22). For $s < k$ we must find

$$\gamma_{ks} = \gamma_{k-s}^+\gamma_0^- + \sum_{r=0}^{s-1} \gamma_{k-r}^+\gamma_{r-s}^-,$$

while for $s > k$ we need to calculate

$$\gamma_{ks} = \gamma_0^+\gamma_{k-s}^- + \sum_{r=0}^{k-1} \gamma_{k-r}^+\gamma_{r-s}^-.$$

Again by slight abuse of notation

$$\begin{aligned}\gamma_{k-r}^{+}\gamma_{r-s}^{-} &= -C(A^{\times})^{-(k-r+1)}\Pi BC(I-\Pi)(A^{\times})^{-(r-s+1)}B\\ &= -C(A^{\times})^{-(k-r+1)}(A^{\times}\Pi-\Pi A^{\times})(A^{\times})^{-(r-s+1)}B\\ &= -C(A^{\times})^{-(k-r)}\Pi(A^{\times})^{-(r-s+1)}B+\\ &\qquad +C(A^{\times})^{-(k-r+1)}\Pi(A^{\times})^{-(r-s)}B.\end{aligned}$$

Observe that if we replace r by $r+1$ in the last one of the latter two terms we get the first one. So the summation in the formula for γ_{ks} is telescoping and collapses into just a few terms. We proceed as follows.

For $s<k$ we get

$$\gamma_{ks}=\gamma_{k-s}^{+}\gamma_0^{-}-C(A^{\times})^{-(k-s+1)}\Pi B+C(A^{\times})^{-(k+1)}\Pi(A^{\times})^{s}B.$$

Since $\gamma_0^{-}=I$ and $\gamma_{k-s}^{+}=C(A^{\times})^{-(k-s+1)}\Pi B$, this results in

$$\gamma_{ks}=C(A^{\times})^{-(k+1)}\Pi(A^{\times})^{s}B.$$

For $s>k$ the computation is a little more involved as $\gamma_0^{+}=I_n+C(A^{\times})^{-1}\Pi B$. Using that $\Pi BC(I-\Pi)=A^{\times}\Pi-\Pi A^{\times}$, it goes this way:

$$\begin{aligned}\gamma_{ks} &= -\big(I+C(A^{\times})^{-1}\Pi B\big)C\big(I-\Pi)(A^{\times})^{-(k-s+1)}B\\ &\qquad +C(A^{\times})^{-(k+1)}\Pi(A^{\times})^{s}B-C(A^{\times})^{-1}\Pi(A^{\times})^{-(k-s)}B\\ &= -C\big(I-\Pi)(A^{\times})^{-(k-s+1)}B\\ &\qquad +C(A^{\times})^{-1}(\Pi A^{\times}-A^{\times}\Pi)(A^{\times})^{-(k-s+1)}B\\ &\qquad +C(A^{\times})^{-(k+1)}\Pi(A^{\times})^{s}B-C(A^{\times})^{-1}\Pi(A^{\times})^{-(k-s)}B\\ &= C(A^{\times})^{-(k+1)}\Pi(A^{\times})^{s}B-C(A^{\times})^{-(k-s+1)}B\\ &= -C(A^{\times})^{-(k+1)}(I-\Pi)(A^{\times})^{s}B.\end{aligned}$$

It remains to consider the case $k=s$. Then we have

$$\gamma_{ss}=\gamma_0^{+}\gamma_0^{-}+\sum_{r=0}^{k-1}\gamma_{s-r}^{+}\gamma_{r-s}^{-}.$$

Following the line of argument as in the case $s<k$ this yields

$$\begin{aligned}\gamma_{ss} &= I_m+C(A^{\times})^{-1}\Pi B-C(A^{\times})^{-1}\Pi B+C(A^{\times})^{-(k+1)}\Pi(A^{\times})^{k}B\\ &= I_m+C(A^{\times})^{-(k+1)}\Pi(A^{\times})^{k}B,\end{aligned}$$

which completes the proof. □

The main step in the factorization method for solving the equation (3.11) is to construct a right canonical factorization of the symbol $a(\lambda)$ with respect to the unit circle. In Theorem 3.5 we obtained explicit formulas for the case when $a(\lambda)$ is rational and has the value I_n at ∞. The latter condition is not essential. Indeed, by a suitable Möbius transformation one can transform the symbol $\alpha(\lambda)$ into a function which is invertible at infinity (see Section 3.6). Next one makes the Wiener-Hopf factorization of the transformed symbol with respect to the image of the unit circle under the Möbius transformation. Here one can use the same formulas as in Theorem 3.5. Finally, using the inverse Möbius transformation, one can obtain explicit formulas for the factorization with respect to the unit circle, and hence also for the solution of equation (3.11).

3.4 Singular integral equations

In this section we apply Theorem 3.2 to solve the singular integral equation from Section 1.3:

$$a(t)\phi(t) + b(t)\frac{1}{\pi i}\int_\Gamma \frac{\phi(\tau)}{\tau - t}\,d\tau = f(t), \qquad t \in \Gamma, \tag{3.13}$$

where Γ is a Cauchy contour. The problem is to find $\phi \in L_p^m(\Gamma)$ such that (3.13) is satisfied. Recall that (3.13) can be rewritten in the form $aI\phi + bS\phi = f$, where S is the singular integral operator as in (1.26). Put $c = a + b$ and $d = a - b$. Then we know from Section 1.3 that the operator $aI + bS$ is invertible if and only if $c(\lambda)$ and $d(\lambda)$ are invertible for all $\lambda \in \Gamma$ and the function $w(\lambda) = d(\lambda)^{-1}c(\lambda)$ admits a right canonical factorization with respect to Γ. The next theorem deals with the case when $w(\lambda)$ is rational and has the value I_m at ∞.

Theorem 3.6. *Suppose* $\det\big(a(\lambda) + b(\lambda)\big)$ *and* $\det\big(a(\lambda) - b(\lambda)\big)$ *do not vanish on* Γ, *and assume* $w(\lambda) = \big(a(\lambda) - b(\lambda)\big)^{-1}\big(a(\lambda) + b(\lambda)\big)$ *is a rational function which has the value* I_m *at infinity. Let*

$$w(\lambda) = I_m + C(\lambda I_n - A)^{-1}B$$

be a realization for w. *Suppose* A *and* $A^\times = A - BC$ *have no spectrum on* Γ. *Then* $aI + bS$ *is invertible if and only if* $\mathbb{C}^n = M \dot{+} M^\times$, *where* M *is the spectral subspace corresponding to the eigenvalues of* A *inside* Γ, *and* $M^\times$ *is the spectral subspace corresponding to the eigenvalues of* $A^\times$ *outside* Γ. *In that case the functions* w_+, w_+^{-1}, w_- *and* w_-^{-1} *appearing in the expressions for* $(aI + bS)^{-1}$ *given in Section* 1.3 *can be specified as follows:*

$$\begin{aligned}
w_+(\lambda) &= I_m + C\Pi(\lambda I_n - A)^{-1}B,\\
w_-(\lambda) &= I_m + C(\lambda I_n - A)^{-1}(I_n - \Pi)B,\\
w_+^{-1}(\lambda) &= I_m - C(\lambda I_n - A^\times)^{-1}\Pi B,\\
w_-^{-1}(\lambda) &= I_m - C(I_n - \Pi)(\lambda I_n - A^\times)^{-1}B.
\end{aligned}$$

Here Π *is the projection of* $\mathbb{C}^n$ *along* M *onto* $M^\times$ *and* $I = I_n$ *is the identity operator on* $\mathbb{C}^n$.

By way of illustration, we consider the special case when

$$f(t) = \frac{1}{t-\alpha}\Big(a(t) - b(t)\Big)\eta,$$

where α is a complex number outside Γ and $\eta \in \mathbb{C}^m$. Put

$$g(t) = \frac{1}{t-\alpha}\eta.$$

Then one can write $f = dg$, where as before $d = a - b$. Hence $w_-^{-1}d^{-1} = w_-^{-1}g$. Observe now that the function

$$\frac{1}{t-\alpha}\Big(w_-^{-1}(t) - w_-^{-1}(\alpha)\Big)\eta$$

is analytic outside Γ and vanishes at ∞. So when we apply P_Γ to it, we get zero. It follows that

$$(P_\Gamma w_-^{-1}g)(t) = \frac{1}{t-\alpha}w_-^{-1}(\alpha)\eta.$$

But then

$$(Q_\Gamma w_-^{-1}g)(t) = \frac{1}{t-\alpha}\Big(w_-^{-1}(t) - w_-^{-1}(\alpha)\Big)\eta,$$

and hence

$$\Big((aI+bS)^{-1}f\Big)(t) = \frac{1}{t-\alpha}w_+^{-1}(t)w_-^{-1}(\alpha)\eta + \frac{1}{t-\alpha}\Big(I_m - w_-(t)w_-^{-1}(\alpha)\Big)\eta.$$

In the situation of Theorem 3.6, the right-hand side of this equality becomes

$$\frac{1}{t-\alpha}\eta - \frac{1}{t-\alpha}C\Big((t-A^\times)^{-1}\Pi + (t-A)^{-1}(I-\Pi)\Big)B$$
$$\cdot\Big(I_m - C(I-\Pi)(\alpha - A^\times)^{-1}B\Big)\eta.$$

The case when $w(\lambda)$ is rational, but does not have the value I_m at ∞, can be treated by applying a suitable Möbius transformation. The argument is similar to that indicated at the end of Section 3.3.

3.5 The Riemann-Hilbert boundary value problem

In this section we consider the (homogeneous) *Riemann-Hilbert boundary value problem* (on the real line):

$$W(\lambda)\Phi_+(\lambda) = \Phi_-(\lambda), \qquad -\infty < \lambda < +\infty. \tag{3.14}$$

The precise formulation of this problem is as follows. Let W be a given $m \times m$ matrix function, with entries that are integrable on the real line. The problem is to describe all pairs Φ_+, Φ_- of $\mathbb{C}^m$-valued functions such that (3.14) is satisfied while, in addition, Φ_+ and Φ_- are the Fourier transforms of integrable $\mathbb{C}^m$-valued functions with support in $[0, \infty)$ and $(-\infty, 0]$, respectively. For such a pair of functions Φ_+, Φ_- we have that Φ_+ is continuous on the closed upper half plane, analytic in the open upper half plane and vanishes at infinity, the same being true for Φ_- with the understanding that the upper half plane is replaced by the lower.

The functions W that we shall deal with are rational $m \times m$ matrix functions with the value I_m at infinity and given in the form of a realization.

Theorem 3.7. *Let W be a rational $m \times m$ matrix function, and suppose W admits the realization $W(\lambda) = I_m + C(\lambda I_n - A)^{-1}B$. Suppose further that both A and $A^\times = A - BC$ have no eigenvalues on the real line. Let M be the spectral subspace of A corresponding to the eigenvalues of A in the upper half plane, and let $M^\times$ be the spectral subspace of $A^\times$ corresponding to the eigenvalues of $A^\times$ in the lower half plane. Then the pair of functions Φ_+, Φ_- is a solution of the Riemann-Hilbert boundary value problem* (3.14) *if and only if there exists $x \in M \cap M^\times$ such that*

$$\Phi_+(\lambda) = C(\lambda I_n - A^\times)^{-1}x, \qquad \Phi_-(\lambda) = C(\lambda I_n - A)^{-1}x. \tag{3.15}$$

Moreover, the vector x in (3.15) *is uniquely determined by the pair Φ_+, Φ_-.*

Proof. Take $x \in M \cap M^\times$ and define Φ_+ and Φ_- by (3.15). From Theorem 2.4 we know that $W(\lambda)C(\lambda - A^\times)^{-1} = C(\lambda - A)^{-1}$. It follows that (3.14) is satisfied. Here the specific choice of x does not even play a role. Put $\phi_+(t) = -iCe^{-itA^\times}x$, $t \geq 0$. Since $x \in M^\times$, the function ϕ_+ is integrable on $[0, \infty)$. Similarly, as $x \in M$, the function ϕ_- given by $\phi_-(t) = iCe^{-itA}x$, $t \leq 0$ is integrable on $(-\infty, 0]$. A straightforward computation shows that

$$\Phi_+(\lambda) = \int_0^\infty e^{i\lambda t}\phi_+(t)dt, \qquad \Phi_-(\lambda) = \int_{-\infty}^0 e^{i\lambda t}\phi_-(t)dt \tag{3.16}$$

and the proof of the "if part" of the theorem is complete.

The proof of the "only if part" is somewhat more involved. Let Φ_+, Φ_- be a solution of (3.14) given in the form (3.16) with integrable ϕ_+ and ϕ_-. It will be convenient to extend ϕ_+ and ϕ_- to integrable functions on the full real line by stipulating that they vanish on $[-\infty, 0)$ and $[0, \infty)$, respectively. For $\lambda \in \mathbb{R}$ put $\rho(\lambda) = (\lambda - A)^{-1}B\Phi_+(\lambda)$. Note that $(\lambda - A)^{-1}$ appears as a Fourier transform of a matrix function with entries from $L_1(\mathbb{R})$. In fact

$$(\lambda - A)^{-1} = \int_{-\infty}^\infty e^{i\lambda t}\ell(t)dt, \qquad \lambda \in \mathbb{R},$$

where

$$\ell(t) = \begin{cases} ie^{-itA}P, & t < 0, \\ -ie^{-itA}(I_n - P), & t < 0. \end{cases}$$

Using inverse Fourier transforms and the fact that the support of ϕ_+ is contained in $[0,\infty)$, we have

$$\rho(\lambda) = \int_{-\infty}^{\infty} e^{i\lambda t}\Big(\int_0^\infty \ell(t-s)B\phi_+(s)\,ds\Big)\,dt, \qquad \lambda\in\mathbb{R}.$$

Introduce

$$\gamma_-(t) = \int_0^\infty \ell(t-s)B\phi_+(s)\,ds, \quad (t<0), \qquad \gamma_-(t)=0 \quad (t>0),$$

$$\gamma_+(t) = \int_0^\infty \ell(t-s)B\phi_+(s)\,ds, \quad (t>0), \qquad \gamma_+(t)=0 \quad (t<0),$$

and for each $\lambda\in\mathbb{R}$ set

$$\rho_+(\lambda) = \int_{-\infty}^{\infty} e^{i\lambda t}\gamma_+(t)\,dt = \int_0^\infty e^{i\lambda t}\gamma_+(t)\,dt,$$

$$\rho_-(\lambda) = \int_{-\infty}^{\infty} e^{i\lambda t}\gamma_-(t)\,dt = \int_{-\infty}^0 e^{i\lambda t}\gamma_-(t)\,dt.$$

Obviously, $\rho(\lambda)=\rho_-(\lambda)+\rho_+(\lambda)$ for each $\lambda\in\mathbb{R}$. From (3.14) and the definition of ρ it follows that

$$\Phi_+(\lambda)+C\rho_+(\lambda) \;=\; \Phi_-(\lambda)-C\rho_-(\lambda), \qquad \lambda\in\mathbb{R}. \tag{3.17}$$

The left-hand side of (3.17) is continuous on the closed upper half plane, analytic in the open upper half plane and vanishes at infinity. The same is true for the right-hand side of (3.17) provided the upper half plane is replaced by the lower half plane. But then we can apply Liouville's theorem to show that both sides of (3.17) are identically zero. Hence

$$\Phi_-(\lambda) = C\rho_-(\lambda) = \int_{-\infty}^0 e^{i\lambda t}C\gamma_-(t)\,dt, \qquad \Im\lambda\le 0, \tag{3.18}$$

$$\Phi_+(\lambda) = -C\rho_+(\lambda) = -\int_0^\infty e^{i\lambda t}C\gamma_+(t)\,dt, \qquad \Im\lambda\ge 0. \tag{3.19}$$

For $t<0$ we have

$$\begin{aligned}\gamma_-(t) &= \int_0^\infty \ell(t-s)B\phi_+(s)\,ds\\ &= ie^{-itA}\int_0^\infty e^{isA}PB\phi_+(s)\,ds = ie^{-itA}x,\end{aligned}$$

where $x = \int_0^\infty e^{isA}PB\phi_+(s)\,ds$. Clearly $x\in\operatorname{Im}P$, and we conclude that

$$\rho_-(\lambda) = \int_{-\infty}^0 e^{i\lambda t}\big(ie^{-itA}x\big)\,dt = (\lambda-A)^{-1}x, \qquad \Im\lambda\le 0. \tag{3.20}$$

Next, fix $\lambda \in \mathbb{R}$. Since $(\lambda - A)\rho(\lambda) = B\Phi_+(\lambda)$ and $(\lambda - A)\rho_-(\lambda) = x$, we can use the first part of (3.19) to show that

$$(\lambda - A)\rho_+(\lambda) + x = (\lambda - A)\rho(\lambda) = B\Phi_+(\lambda) = -BC\rho_+(\lambda).$$

Recall that $A^\times = A - BC$. It follows that

$$\rho_+(\lambda) = -(\lambda - A^\times)^{-1}x, \qquad \lambda \in \mathbb{R}. \tag{3.21}$$

The left-hand side of (3.21) is continuous on the closed upper half plane and analytic in the open upper half plane. Thus (3.21) implies that $P^\times x = 0$, where $P^\times$ is the spectral projection of $A^\times$ corresponding to the eigenvalues in the upper half plane. Since $\operatorname{Im} P = M$ and $\operatorname{Ker} P^\times = M^\times$, we see that $x \in M \cap M^\times$. From (3.19) and (3.21) it follows that the first identity in (3.15) holds. Similarly, (3.18) and (3.20) yield the second identity in (3.15).

It remains to prove the unicity of x. Take $u \in M \cap M^\times$, and assume that $C(\lambda - A)^{-1}u = 0$. It suffices to show that $u = 0$. To do this, recall (see Theorem 2.4) that

$$(\lambda - A^\times)^{-1} = (\lambda - A)^{-1} - (\lambda - A)^{-1}BW(\lambda)^{-1}C(\lambda - A)^{-1}, \qquad \lambda \in \mathbb{R}.$$

Thus the assumption $C(\lambda - A)^{-1}u = 0$ yields

$$(\lambda - A^\times)^{-1}u = (\lambda - A)^{-1}u, \qquad \lambda \in \mathbb{R}. \tag{3.22}$$

The fact that $u \in M^\times$ implies that $(\lambda - A^\times)^{-1}u$ is analytic on $\Im\lambda \geq 0$. On the other hand, $u \in M$ gives that $(\lambda - A)^{-1}$ is analytic on $\Im\lambda \leq 0$. Since both $(\lambda - A^\times)^{-1}u$ and $(\lambda - A)^{-1}u$ vanish at infinity, Liouville's theorem implies that $(\lambda - A)^{-1}u$ is identically zero on $\mathbb{R}$, hence $u = 0$. □

There is an intimate connection between the Riemann-Hilbert boundary value problem (on the real line) and the homogeneous Wiener-Hopf integral equation. This is already clear from the material presented in Section 1.1 by specializing to the situation where $f = 0$. The fact is further underlined by the above proof of Theorem 3.7. Indeed, notice that (3.19) implies that $\phi_+ = -C\gamma_+$, and hence we see from the definition of γ_+ that

$$\phi_+(t) - \int_0^\infty k(t-s)\phi_+(s)\,ds = 0, \qquad t > 0,$$

where $k(t) = -C\ell(t)B$, and hence $\widehat{k}(\lambda) = -C(\lambda I_n - A)^{-1}B$. Thus ϕ_+ is the solution of the homogeneous Wiener-Hopf integral equation with symbol given by $I_m + C(\lambda I_n - A)^{-1}B$. A more detailed (but straightforward) analysis gives the following result, the formulation of which is in line with Theorem 3.3.

Theorem 3.8. *Let $I_m - K(\lambda) = I_m + C(\lambda I_n - A)^{-1}B$ be a realization for the symbol of the homogeneous Wiener-Hopf equation*

$$\phi(t) - \int_0^\infty k(t-s)\phi(s)ds = 0, \qquad t \geq 0, \tag{3.23}$$

and let $A^\times = A - BC$. Assume that both A and $A^\times$ have no real eigenvalues, in other words,

$$\det\left(I_m - K(\lambda)\right) \neq 0, \qquad -\infty < \lambda < +\infty.$$

Let M be the spectral subspace of A corresponding to the eigenvalues of A in the upper half plane, and let $M^\times$ be the spectral subspace of $A^\times$ corresponding to the eigenvalues of $A^\times$ in the lower half plane. Then ϕ is a solution of (3.23) if and only if there exists $x \in M \cap M^\times$ such that

$$\phi(t) = Ce^{-itA^\times}x, \qquad t \geq 0. \tag{3.24}$$

Moreover, the vector x in (3.24) is uniquely determined by ϕ.

Formula (3.24) has to be understood in the sense of equality in the solution space $L_1^m[0,\infty)$ (or, more generally, $L_p^m[0,\infty)$ with $1 \leq p < \infty$; cf., Section 1.1 and Theorem 3.3).

As a direct consequence of Theorem 3.8, one sees that the dimension of the null space of the Wiener-Hopf integral operator T defined by the left-hand side of (3.23) is equal to $\dim(M \cap M^\times)$. It can also be proved that the codimension of its range is equal to $\operatorname{codim}(M + M^\times)$. In fact, under the conditions of Theorem 3.8, the operator T is a Fredholm operator (see Section XI.1 in [51] for the definition of this notion), and its Fredholm index, which is defined as the difference of the codimension of its range and the dimension of its null space, is equal to

$$\begin{aligned}
\operatorname{ind} T &= \operatorname{codim}(M + M^\times) - \dim(M \cap M^\times) \\
&= \dim \frac{\mathbb{C}^n}{M + M^\times} - \dim(M \cap M^\times) \\
&= \dim \frac{\mathbb{C}^n}{M^\times} - \dim \frac{M + M^\times}{M^\times} - \dim(M \cap M^\times) \\
&= \dim \frac{\mathbb{C}^n}{M^\times} - \dim \frac{M}{M \cap M^\times} - \dim(M \cap M^\times) \\
&= \dim \frac{\mathbb{C}^n}{M^\times} - \dim M \\
&= \operatorname{rank} P^\times - \operatorname{rank} P.
\end{aligned}$$

Here P and $P^\times$ are the spectral projections corresponding to the eigenvalues in the upper half plane of A and $A^\times$, respectively. (In the step from the third to the fourth equality in the above calculation we used Lemma 2 in [89].) More detailed information about the null space and range of the Wiener-Hopf integral operator T can be obtained in this way (see, e.g., Theorem XIII.8.1 in [51]). We shall return to this theme, in a more general context, in Chapter 7, where it will be shown that the factorization indices in a non-canonical Wiener-Hopf factorization can be expressed in terms of the spaces M and $M^{\times}$, and related subspaces defined in terms of these spaces and the matrices appearing in the realization of the symbol.

Notes

The first section of this chapter originates from Section 1.2 in [11]. The basic facts about Cauchy domains (see also the final paragraphs of Chapter 0), Riesz projections and spectral subspaces, used in this first section, can be found in Sections I.1 – I.3 of [51]. The material in Sections 3.2, 3.3 and 3.4 goes back to Chapter 4 in [11]. For Section 3.5 we refer to [12]. We shall return to canonical factorization in a more general setting in Chapters 5 and 7; see Theorems 5.14 and 7.1. Other state space methods for solving convolution equations, also based on matrix-valued realizations but not employing factorization, are developed in [12] and [13].

Chapter 4

Factorization of non-proper rational matrix functions

In this chapter we treat the problem of factorizing a non-proper rational matrix function. The realization used in the earlier chapters is replaced by

$$W(\lambda) = I + C(\lambda G - A)^{-1}B. \tag{4.1}$$

Here $I = I_m$ is the $m \times m$ identity matrix, A and G are square matrices of order n say, and the matrices C and B are of sizes $m \times n$ and $n \times m$, respectively. Any rational $m \times m$ matrix function W, proper or non-proper, admits such a representation. The representation (4.1) allows us to extend the results obtained in the previous chapter to arbitrary rational matrix functions. As an application we treat the problem to invert a singular integral operator with a rational matrix symbol.

This chapter consists of five sections. In Section 4.1 we review the spectral theory of matrix pencils. Section 4.2 presents the realization theorem for non-proper rational matrix functions referred to in the previous paragraph. The corresponding canonical factorization theorem is given in Section 4.3. The final two sections deal with applications to inverting singular integral operators (Section 4.4) and solving Riemann-Hilbert problems (Section 4.5).

4.1 Preliminaries about matrix pencils

Let A and G be complex $n \times n$ matrices. The linear matrix-valued function $\lambda G - A$, where λ is a complex variable, is called a (*linear matrix*) *pencil*. We say that the pencil $\lambda G - A$ is *regular on* Ω or Ω-*regular* if $\lambda G - A$ is invertible for each $\lambda \in \Omega$. Here Ω is a subset of $\mathbb{C}$.

From now on Γ will be a Cauchy contour. Its interior domain is denoted by F_+ and its exterior domain by F_-. We shall assume that $\infty \in F_-$. Pencils that are Γ-regular admit block matrix partitionings that are comparable to spectral decompositions of a single matrix. This fact is summarized by the following theorem, the proof of which can be found in [140] (see also Section IV.1 of [51]).

Theorem 4.1. *Let $\lambda G - A$ be a Γ-regular pencil, and let the matrices P and Q be defined by*

$$P = \frac{1}{2\pi i}\int_\Gamma G(\lambda G - A)^{-1} d\lambda, \qquad Q = \frac{1}{2\pi i}\int_\Gamma (\lambda G - A)^{-1} G d\lambda. \tag{4.2}$$

Then P and Q are projections such that

(i) *$PA = AQ$ and $PG = GQ$,*

(ii) *$(\lambda G - A)^{-1}P = Q(\lambda G - A)^{-1}$ on Γ and this function has an analytic continuation on F_- which vanishes at ∞,*

(iii) *$(\lambda G - A)^{-1}(I - P) = (I - Q)(\lambda G - A)^{-1}$ on Γ and this function has an analytic continuation on F_+.*

The properties (i)–(iii) in the above proposition determine P and Q uniquely, that is, if P and Q are projections such that (i)–(iii) hold, then P and Q are given by the integral formulas in (4.2).

For a better understanding of the above result, let us write A and G as block matrices relative to the decompositions of $\mathbb{C}^m$ induced by the projections P and Q. Condition (i) in Theorem 4.1 implies that A and G have block diagonal representations:

$$A = \begin{bmatrix} A_1 & 0 \\ 0 & A_2 \end{bmatrix} : \operatorname{Im} Q \dot{+} \operatorname{Ker} Q \to \operatorname{Im} P \dot{+} \operatorname{Ker} P,$$

$$G = \begin{bmatrix} G_1 & 0 \\ 0 & G_2 \end{bmatrix} : \operatorname{Im} Q \dot{+} \operatorname{Ker} Q \to \operatorname{Im} P \dot{+} \operatorname{Ker} P.$$

Property (ii) is equivalent to saying that the pencil $\lambda G_1 - A_1$ is regular on F_- and G_1 is invertible; property (iii) amounts to regularity of the pencil $\lambda G_2 - A_2$ on F_+.

In the particular case when G is the identity matrix I, the two projections P and Q coincide, and P is just the spectral (or Riesz) projection of A corresponding to the eigenvalues in F_+. The latter means (see Section 3.1 or Section I.2 in [51]) that P is a projection commuting with A, the eigenvalues of $A|_{\operatorname{Im} P}$ are in F_+ and the eigenvalues of $A|_{\operatorname{Ker} P}$ are in F_-. In that case, $\operatorname{Im} P$ is the spectral subspace of A corresponding to the eigenvalues of A in F_+, and $\operatorname{Ker} P$ is the spectral subspace of A corresponding to the eigenvalues of A in F_-.

4.2 Realization of a non-proper rational matrix function

In this section we derive the representation (4.1), and present some useful identities related to (4.1).

Theorem 4.2. *Let W be a rational $m \times m$ matrix function, and let Ω be the subset of $\mathbb{C}$ on which W is analytic. Then, given an $m \times m$ matrix D, the function W admits a representation*

$$W(\lambda) = D + C(\lambda G - A)^{-1}B, \qquad \lambda \in \Omega, \tag{4.3}$$

where $\lambda G - A$ is an Ω-regular $m \times m$ matrix pencil, and B and C are matrices of sizes $n \times m$ and $m \times n$, respectively.

The set Ω is the complement in $\mathbb{C}$ of the set of finite poles of W (i.e., the poles of W in $\mathbb{C}$). In later applications, D will be taken to be I_m, the $m \times m$ identity matrix.

Proof. Let us first remark that W admits a decomposition

$$W(\lambda) = K(\lambda) + L(\lambda), \qquad \lambda \in \Omega, \tag{4.4}$$

where L is an $m \times m$ matrix polynomial and K is a proper rational $m \times m$ matrix such that the subset of $\mathbb{C}$ on which K is analytic coincides with Ω. Such a decomposition is not unique. In fact, given (4.4) we can obtain another decomposition of F with the same properties by adding a constant matrix to K and subtracting the same matrix from L. This, however, is all the freedom one has. In other words the decomposition (4.4) will be unique if we fix the value of K at infinity.

From now on we shall assume that $K(\infty) = D$. The results obtained in Section 2.2 then imply that K admits a realization

$$K(\lambda) = D + C_K(\lambda - A_K)^{-1}B_K, \qquad \lambda \in \Omega,$$

where A_K, B_K and C_K are matrices of appropriate sizes and the resolvent set of the (square) matrix A_K coincides with Ω. The latter can be reformulated by saying that the eigenvalues of A_K are just the finite poles of W.

Proceeding with the second term in the right-hand side of the identity (4.4), we write $L(\lambda) = L_0 + \lambda L_1 + \cdots + \lambda^q L_q$, and introduce

$$G_L = \begin{bmatrix} 0 & I_m & & \\ & 0 & \ddots & \\ & & \ddots & I_m \\ & & & 0 \end{bmatrix}, \quad B_L = \begin{bmatrix} L_0 \\ L_1 \\ \vdots \\ L_q \end{bmatrix}, \quad C_L = \begin{bmatrix} -I_m & 0 & \cdots & 0 \end{bmatrix},$$

where the blanks in G_L indicate zero entries. The matrix G_L is square of size $l = m(q+1)$. Also G_L is nilpotent (of order $q+1$), and hence $I_l - \lambda G_L$ is invertible for each λ in $\mathbb{C}$. The first row in the block matrix representation of $(I_l - \lambda G_L)^{-1}$ is equal to $[\, I_m \ \lambda I_m \ \ldots \ \lambda^q I_m]$ and it follows that $L(\lambda) = C_L(\lambda G_L - I_l)^{-1} B_L$ on all of the (finite) complex plane.

By combining the representation results for K and L we see that W can be written in the form (4.3) with

$$A = \begin{bmatrix} A_K & 0 \\ 0 & I_l \end{bmatrix}, \quad B = \begin{bmatrix} B_K \\ B_L \end{bmatrix}, \quad C = \begin{bmatrix} C_K & C_L \end{bmatrix}, \quad G = \begin{bmatrix} I & 0 \\ 0 & G_L \end{bmatrix}.$$

Here I is the identity matrix of the same size as A_K. The fact that G_L is nilpotent, implies that the matrix $\lambda G - A$ is invertible if and only if λ is an eigenvalue of A_K, that is if and only if λ is a finite pole of W. □

The following proposition, which describes some elementary operations on a rational matrix function in terms of a given realization, is the natural analogue of Theorem 2.4 for realizations of the form (4.1).

Theorem 4.3. *Let $W(\lambda) = I + C(\lambda G - A)^{-1}B$, and put $A^\times = A - BC$. Then $W(\lambda)$ is invertible if and only if $\lambda G - A^\times$ is invertible, and in that case the following identities hold:*

$$W(\lambda)^{-1} = I - C(\lambda G - A^\times)^{-1}B, \tag{4.5}$$

$$W(\lambda)C(\lambda G - A^\times)^{-1} = C(\lambda G - A)^{-1}, \tag{4.6}$$

$$(\lambda G - A^\times)^{-1}BW(\lambda) = (\lambda G - A)^{-1}B, \tag{4.7}$$

$$(\lambda G - A^\times)^{-1} = (\lambda G - A)^{-1} - (\lambda G - A)^{-1}BW(\lambda)^{-1}C(\lambda G - A)^{-1}. \tag{4.8}$$

Proof. Fix $\lambda \in \mathbb{C}$ such that $\lambda G - A$ is invertible. Then

$$\begin{aligned} \det W(\lambda) &= \det\big(I + C(\lambda G - A)^{-1}B\big) = \det\big(I + (\lambda G - A)^{-1}BC\big) \\ &= \det\big((\lambda G - A)^{-1}\big)\det(\lambda G - A + BC) \\ &= \frac{\det(\lambda G - A^\times)}{\det(\lambda G - A)}. \end{aligned}$$

It follows that $W(\lambda)$ is invertible if and only if $\lambda G - A^\times$ is invertible. Also, in that case, a straightforward computation yields

$$
\begin{aligned}
W(\lambda)C(\lambda G - A^\times)^{-1} - C(\lambda G - A^\times)^{-1}
&= C(\lambda G - A)^{-1}BC(\lambda G - A^\times)^{-1}\\
&= C(\lambda G - A)^{-1}(A - A^\times)(\lambda G - A^\times)^{-1}\\
&= C(\lambda G - A)^{-1}\big((\lambda G - A^\times) - (\lambda G - A)\big)(\lambda G - A^\times)^{-1}\\
&= C(\lambda G - A)^{-1} - C(\lambda G - A^\times)^{-1}.
\end{aligned}
$$

Since $W(\lambda)$ is invertible, this proves (4.6). The identity (4.7) is proved in a similar way. Using (4.6) a straightforward computation shows that

$$W(\lambda)\big(I - C(\lambda G - A^\times)^{-1}B\big) = I,$$

and hence (4.5) holds. Finally, (4.8) follows by applying (4.6) and again using the identity $BC = (\lambda G - A^\times) - (\lambda G - A)$. □

Instead of the above argument one can also use an analogue of the second proof of Theorem 2.1 in [20], which uses Schur complements arguments (cf., the remark made in the final paragraph of Section 2.4).

4.3 Explicit canonical factorization

In this section we show how the realization (4.1) can be used to construct a canonical factorization of an arbitrary rational matrix function. Necessary and sufficient conditions for the existence of such a factorization and formulas for the factors are stated explicitly in terms of the data appearing in the realization. The next theorem, a counterpart of Theorem 3.2 for non-proper rational matrix functions, is the main result.

Theorem 4.4. *Let W be a rational $m \times m$ matrix function without poles on the curve Γ, and let W be given by the Γ-regular realization*

$$W(\lambda) = I + C(\lambda G - A)^{-1}B, \qquad \lambda \in \Gamma. \tag{4.9}$$

Put $A^\times = A - BC$. Then W admits a right canonical factorization with respect to Γ if and only if the following two conditions are satisfied:

(i) *the pencil $\lambda G - A^\times$ is Γ-regular,*

(ii) $\mathbb{C}^n = \operatorname{Im} P \dot{+} \operatorname{Ker} P^\times$ *and* $\mathbb{C}^n = \operatorname{Im} Q \dot{+} \operatorname{Ker} Q^\times$.

Here n is the order of the matrices G and A, and

$$P = \frac{1}{2\pi i}\int_\Gamma G(\lambda G - A)^{-1}d\lambda, \qquad P^\times = \frac{1}{2\pi i}\int_\Gamma G(\lambda G - A^\times)^{-1}d\lambda,$$

$$Q = \frac{1}{2\pi i}\int_\Gamma (\lambda G - A)^{-1} G d\lambda, \qquad Q^\times = \frac{1}{2\pi i}\int_\Gamma (\lambda G - A^\times)^{-1} G d\lambda.$$

If the conditions (i) and (ii) are satisfied, a right canonical factorization with respect to Γ *is given by*

$$W(\lambda) = W_-(\lambda) W_+(\lambda), \qquad \lambda \in \Gamma,$$

where the factors and their inverses can be written as

$$W_-(\lambda) = I + C(\lambda G - A)^{-1}(I - \Delta)B, \tag{4.10}$$

$$W_+(\lambda) = I + C\Lambda(\lambda G - A)^{-1}B, \tag{4.11}$$

$$W_-^{-1}(\lambda) = I - C(I - \Lambda)(\lambda G - A^\times)^{-1}B, \tag{4.12}$$

$$W_+^{-1}(\lambda) = I - C(\lambda G - A^\times)^{-1}\Delta B. \tag{4.13}$$

Here Δ *is the projection along* $\operatorname{Im} P$ *onto* $\operatorname{Ker} P^\times$, *and* Λ *is the projection of* $\mathbb{C}^n$ *along* $\operatorname{Im} Q$ *onto* $\operatorname{Ker} Q^\times$. *Finally, the first equality in* (ii) *implies the second and conversely.*

Proof. We split the proof into four parts. The first part concerns the condition (i). In the second part we prove that the first equality in (ii) implies the second and conversely. In the third part we use (i) and (ii) to derive the canonical factorization and the formulas for its factors. The final part concerns the necessity of the condition (ii).

Part 1. From the definition given in Section 3.1 it is clear that a necessary condition in order that W admits a right canonical factorization with respect to Γ is that W takes invertible values on Γ. By Theorem 4.3 this necessary condition is fulfilled if and only if (i) holds true. In what follows we shall assume that (i) is satisfied.

Part 2. In this part we prove the last statement of the theorem. Consider the operators

$$P^\times|_{\operatorname{Im} P} : \operatorname{Im} P \to \operatorname{Im} P^\times, \qquad Q^\times|_{\operatorname{Im} Q} : \operatorname{Im} Q \to \operatorname{Im} Q^\times. \tag{4.14}$$

The first equality in (ii) is equivalent to the invertibility of the first operator in (4.14). To see this, note that $\operatorname{Ker}(P^\times|_{\operatorname{Im} P}) = \operatorname{Ker} P^\times \cap \operatorname{Im} P$, and thus $P^\times|_{\operatorname{Im} P}$ is injective if and only if $\operatorname{Ker} P^\times \cap \operatorname{Im} P = \{0\}$. Next, observe that for each $y \in \operatorname{Im} P$ we have $y = (I - P^\times)y + P^\times|_{\operatorname{Im} P} y \in \operatorname{Ker} P^\times + \operatorname{Im}(P^\times|_{\operatorname{Im} P})$. Thus $\operatorname{Ker} P^\times + \operatorname{Im} P \subset \operatorname{Ker} P^\times + \operatorname{Im}(P^\times|_{\operatorname{Im} P})$. The reverse inclusion is also true. Indeed, for $z \in \operatorname{Im} P$ we have $P^\times z = (P^\times z - z) + z \in \operatorname{Ker} P^\times + \operatorname{Im} P$. It follows that $\operatorname{Ker} P^\times + \operatorname{Im}(P^\times|_{\operatorname{Im} P}) = \operatorname{Ker} P^\times + \operatorname{Im} P$, and hence $P^\times|_{\operatorname{Im} P}$ considered as an operator into $\operatorname{Im} P^\times$ is surjective if and only if $\mathbb{C}^n = \operatorname{Ker} P^\times + \operatorname{Im} P$. Thus, as claimed, the first identity in (ii) amounts to the same as the invertibility of the first operator in (4.14). Similarly, the second equality in (ii) is equivalent to the invertibility of the second operator in (4.14). Notice that

$$GQ = PG, \qquad GQ^\times = P^\times G, \tag{4.15}$$

which is clear from the definitions of the projections Q, P and $Q^\times$, $P^\times$. Furthermore, from the material presented in Section 4.1, applied to $\lambda G - A$ as well as to $\lambda G - A^\times$, we see that G maps $\operatorname{Im} Q$ and $\operatorname{Im} Q^\times$ in a one-one manner onto $\operatorname{Im} P$ and $\operatorname{Im} P^\times$, respectively. Thus the operators $E = G|_{\operatorname{Im} Q} : \operatorname{Im} Q \to \operatorname{Im} P$ and $E^\times = G|_{\operatorname{Im} Q^\times} : \operatorname{Im} Q^\times \to \operatorname{Im} P^\times$ are invertible and, in addition,

$$E^\times (Q^\times|_{\operatorname{Im} Q}) = (P^\times|_{\operatorname{Im} P})E.$$

So the operators in (4.14) are equivalent, and hence the first operator in (4.14) is invertible if and only if the same is true for the second operator in (4.14). This proves that the first equality in (ii) implies the second and vice versa.

Part 3. Next assume that (i) and the direct sum decompositions in (ii) hold true. Our aim is to obtain a canonical factorization of W. Write A, G, B, C as well as $A^\times = A - BC$ in block form relative to the decompositions in (ii):

$$A = \begin{bmatrix} A_{11} & A_{12} \\ 0 & A_{22} \end{bmatrix} : \operatorname{Im} Q \dotplus \operatorname{Ker} Q^\times \to \operatorname{Im} P \dotplus \operatorname{Ker} P^\times, \tag{4.16}$$

$$G = \begin{bmatrix} G_{11} & 0 \\ 0 & G_{22} \end{bmatrix} : \operatorname{Im} Q \dotplus \operatorname{Ker} Q^\times \to \operatorname{Im} P \dotplus \operatorname{Ker} P^\times, \tag{4.17}$$

$$B = \begin{bmatrix} B_1 \\ B_2 \end{bmatrix} : \mathbb{C}^n \to \operatorname{Im} P \dotplus \operatorname{Ker} P^\times, \tag{4.18}$$

$$C = \begin{bmatrix} C_1 & C_2 \end{bmatrix} : \operatorname{Im} Q \dotplus \operatorname{Ker} Q^\times \to \mathbb{C}^n, \tag{4.19}$$

$$A^\times = \begin{bmatrix} A_{11}^\times & 0 \\ A_{21}^\times & A_{22}^\times \end{bmatrix} : \operatorname{Im} Q \dotplus \operatorname{Ker} Q^\times \to \operatorname{Im} P \dotplus \operatorname{Ker} P^\times. \tag{4.20}$$

From Theorem 4.1, applied to $\lambda G - A$ as well as to $\lambda G - A^\times$, we know that

$$AQ = PA, \qquad A^\times Q^\times = P^\times A^\times. \tag{4.21}$$

The first identity in (4.21) implies that A maps $\operatorname{Im} Q$ into $\operatorname{Im} P$. This explains the zero entry in the left lower corner of the block matrix for A. From (4.15) we conclude that G has the desired block diagonal form. From the second identity in (4.21) it follows that $A^\times$ maps $\operatorname{Ker} Q^\times$ into $\operatorname{Ker} P^\times$, which justifies the zero in the right upper corner of the block matrix for $A^\times$. Taking into account the identity $A^\times = A - BC$ gives

$$A_{12} = B_1 C_2, \qquad A_{21}^\times = -B_2 C_1, \tag{4.22}$$

$$A_{11}^\times = A_{11} - B_1 C_1, \qquad A_{22}^\times = A_{22} - B_2 C_2. \tag{4.23}$$

Define the matrix functions W_- and W_+ by (4.10) and (4.11), respectively. Using the block matrix representations of A, G, B, and C we may rewrite W_- and W_+ in the form

$$W_-(\lambda) = I + C_1(\lambda G_1 - A_{11})^{-1}B_1, \qquad \lambda \in \Gamma, \tag{4.24}$$

$$W_+(\lambda) = I + C_2(\lambda G_2 - A_{22})^{-1}B_2, \qquad \lambda \in \Gamma. \tag{4.25}$$

From the block matrix representation of A and the first identity in (4.22) we see that

$$\begin{aligned} W_-(\lambda)W_+(\lambda) &= I + \begin{bmatrix} C_1 & C_2 \end{bmatrix} \begin{bmatrix} \lambda G_1 - A_{11} & -B_1C_2 \\ 0 & \lambda G_2 - A_{22} \end{bmatrix}^{-1} \begin{bmatrix} B_1 \\ B_2 \end{bmatrix} \\ &= I + C(\lambda G - A)^{-1}B \\ &= W(\lambda), \end{aligned}$$

which gives the factorization $W = W_-W_+$.

Next, we check the analytic properties of the factors. Obviously, W_- and W_+ have no poles on Γ. Note that

$$\lambda G_1 - A_{11} = (\lambda G - A)|_{\operatorname{Im} Q} : \operatorname{Im} Q \to \operatorname{Im} P.$$

Thus we know from Section 4.1 that $(\lambda G_1 - A_{11})^{-1}$ has an analytic extension on F_- which vanishes at infinity. So W_- is continuous on $F_- \cup \Gamma$ and analytic on F_- (including infinity). To see that a similar statement holds true for W_+ on F_+, we first note that the linear maps

$$\begin{aligned} J &= (I - Q)|_{\operatorname{Ker} Q^\times} : \operatorname{Ker} Q^\times \to \operatorname{Ker} Q, \\ H &= (I - P)|_{\operatorname{Ker} P^\times} : \operatorname{Ker} P^\times \to \operatorname{Ker} P, \end{aligned}$$

are invertible. In fact, $J^{-1} = \Lambda|_{\operatorname{Ker} Q}$ and $H^{-1} = \Delta|_{\operatorname{Ker} P}$, where Λ is the projection along $\operatorname{Im} Q$ onto $\operatorname{Ker} Q^\times$, and Δ is the projection along $\operatorname{Im} P$ onto $\operatorname{Ker} P^\times$. Next, take $x \in \operatorname{Ker} Q^\times$. Then

$$(\lambda G_2 - A_{22})x = \Delta(\lambda G - A)x = \Delta(\lambda G - A)(I - Q)x = \Delta(\lambda G - A)Jx,$$

which shows that $H(\lambda G_2 - A_{22}) = \big((\lambda G - A)|_{\operatorname{Ker} Q}\big)J$. But then we can use Theorem 4.1 and the invertibility of the operators H and J to show that the function $(\lambda G_2 - A_2)^{-1}$ has an analytic extension on F_+. Hence W_+ is continuous on $F_+ \cup \Gamma$ and analytic on F_+.

From the factorization $W(\lambda) = W_-(\lambda)W_+(\lambda)$ for $\lambda \in \Gamma$ it follows that $W_-(\lambda)$ and $W_+(\lambda)$ are both invertible for each $\lambda \in \Gamma$. So we can apply Theorem 4.3 to show that

$$W_-^{-1}(\lambda) = I - C_1(\lambda G_1 - A_{11}^\times)^{-1}B_1, \tag{4.26}$$

$$W_+^{-1}(\lambda) = I - C_2(\lambda G_2 - A_{22}^\times)^{-1}B_2. \tag{4.27}$$

Here we use the two identities in (4.23). Using the block matrix representations of A, G, B and C given above, it is clear that (4.26) and (4.27) yield the formulas (4.12) and (4.13), respectively.

We proceed by checking the analyticity properties of the functions W_-^{-1} and W_+^{-1}. First note that

$$\lambda G_2 - A_{22}^\times = (\lambda G - A^\times)|_{\operatorname{Ker} Q^\times} : \operatorname{Ker} Q^\times \to \operatorname{Ker} P^\times.$$

Thus by applying Theorem 4.1 with $\lambda G - A^\times$ in place of $\lambda G - A$ we see that the function $(\lambda G_2 - A_{22}^\times)^{-1}$ has an analytic extension on F_+. It follows that the function W_+^{-1} is continuous on $F_+ \cup \Gamma$ and analytic on F_+. To prove the analogous result for W_-^{-1} with respect to F_- we use that

$$H^\times(\lambda G_1 - A_{11}^\times) = \big((\lambda G - A^\times)|_{\operatorname{Im} Q^\times}\big) J^\times,$$

where $J^\times = Q^\times|_{\operatorname{Im} Q} : \operatorname{Im} Q \to \operatorname{Im} Q^\times$ and $H^\times = P^\times|_{\operatorname{Im} P} : \operatorname{Im} P \to \operatorname{Im} P^\times$ are invertible linear maps of which the inverses are given by

$$(J^\times)^{-1} = (I - \Lambda)|_{\operatorname{Im} Q^\times}, \qquad (H^\times)^{-1} = (I - \Delta)|_{\operatorname{Im} P^\times}.$$

Since $\big((\lambda G - A^\times)|_{\operatorname{Im} Q^\times}\big)^{-1}$ is analytic on F_- by virtue of Theorem 4.1 applied to $\lambda G - A^\times$, we conclude that the same holds true for $(\lambda G_1 - A_{11}^\times)^{-1}$. Hence the function $W_-(\lambda)^{-1}$ is continuous on $F_- \cup \Gamma$ and analytic on F_-. Thus we have proved that $W = W_- W_+$ is a right canonical factorization with respect to the curve Γ.

Part 4. In this part we prove the necessity of the equalities in (ii). So in what follows we assume that $W = W_- W_+$ is a canonical factorization of W with respect to Γ. Take $x \in \operatorname{Im} P \cap \operatorname{Ker} P^\times$ and, for $\lambda \in \Gamma$, put

$$\varphi_-(\lambda) = C(\lambda G - A)^{-1}x, \qquad \varphi_+(\lambda) = C(\lambda G - A^\times)^{-1}x.$$

Since $x \in \operatorname{Im} P$, the first identity in (4.21) allows us to rewrite φ_- as

$$\varphi_-(\lambda) = (C|_{\operatorname{Im} Q})\big((\lambda G - A)|_{\operatorname{Im} Q}\big)^{-1}x,$$

and hence Theorem 4.1(ii) implies that φ_- has an analytic continuation on F_- which vanishes at infinity. Similarly, since

$$\varphi_+(\lambda) = (C|_{\operatorname{Ker} Q^\times})\big((\lambda G - A^\times)^{-1}|_{\operatorname{Ker} Q^\times}\big)^{-1}x,$$

we conclude from Theorem 4.1(iii) applied to $\lambda G - A^\times$ that φ_+ has an analytic continuation on F_+. Note that $W(\lambda)^{-1}\varphi_-(\lambda) = \varphi_+(\lambda)$ for each $\lambda \in \Gamma$, because of formula (4.6) in Theorem 4.3. It follows that

$$W_-(\lambda)^{-1}\varphi_-(\lambda) = W_+(\lambda)\varphi_+(\lambda), \qquad \lambda \in \Gamma.$$

Now use the analyticity properties of the factors W_- and W_+. We conclude that $W_-^{-1}\varphi_-$ has an analytic continuation on F_- which vanishes at infinity, and $W_+\varphi_+$ has an analytic continuation on F_+. Liouville's theorem implies that both functions are identically zero. It follows that $\varphi_-(\lambda) = 0$ for each $\lambda \in \Gamma$. But then we can apply formula (4.8) to show that

$$(\lambda G - A^\times)^{-1}x = (\lambda G - A)^{-1}x, \qquad \lambda \in \Gamma.$$

Now, repeat part of the above reasoning. Note that $(\lambda G - A)^{-1}x$ has an analytic continuation on F_- which vanishes at infinity, and $(\lambda G - A^\times)^{-1}x$ has an analytic continuation on F_+. Again using Liouville's theorem we conclude that both matrix functions $(\lambda G - A)^{-1}x$ and $(\lambda G - A^\times)^{-1}x$ are identically zero on Γ. This yields $x = 0$.

We proved that $\operatorname{Im} P \cap \operatorname{Ker} P^\times = \{0\}$. Recall that G maps $\operatorname{Im} Q$ in a one-one manner onto $\operatorname{Im} P$. Thus (4.15) shows that G maps $\operatorname{Im} Q \cap \operatorname{Ker} Q^\times$ in a one-one manner into $\operatorname{Im} P \cap \operatorname{Ker} P^\times$. Hence $\operatorname{Im} Q \cap \operatorname{Ker} Q^\times = \{0\}$ too.

Next we show that $\operatorname{Im} Q + \operatorname{Ker} Q^\times = \mathbb{C}^n$. Take $y \in \mathbb{C}^n$ such that y is orthogonal to $\operatorname{Im} Q + \operatorname{Ker} Q^\times$. Let y^* be the row vector of which the j-th entry is equal to the complex conjugate of the j-th entry of y $(j = 1, \ldots, m)$. For $\lambda \in \Gamma$, put

$$\psi_-(\lambda) = y^*(\lambda G - A^\times)^{-1}B, \qquad \psi_+(\lambda) = y^*(\lambda G - A)^{-1}B.$$

Since $y^*(I - Q)^\times = 0$, Theorem 4.1 shows that $\psi_-(\lambda) = y^*(\lambda G - A^\times)^{-1}P^\times B$, and thus ψ_- has an analytic continuation on F_- which vanishes at infinity. Similarly, $y^*Q = 0$ implies that ψ_+ has an analytic continuation on F_+. Now, use the canonical factorization $W = W_-W_+$ and (4.7) to show that

$$\psi_+(\lambda)W_+(\lambda)^{-1} = \psi_-(\lambda)W_-(\lambda), \qquad \lambda \in \Gamma.$$

But then, as before, we can use Liouville's theorem to show that both sides of the identity are equal to zero. It follows that $\psi_+(\lambda) = 0$ for each $\lambda \in \Gamma$, and we can use formula (4.8) to show that

$$y^*(\lambda G - A^\times)^{-1} = y^*(\lambda G - A)^{-1}, \qquad \lambda \in \Gamma.$$

Recall that y^*Q and $y^*(I - Q^\times)$ are both zero. Thus Theorem 4.1 implies that $y^*(\lambda G - A^\times)^{-1}$ has an analytic continuation on F_- which vanishes at infinity, and the function $y^*(\lambda G - A)^{-1}$ has an analytic continuation on F_+. So, by Liouville's theorem, $y^*(\lambda G - A)^{-1} = 0$ on Γ, and thus $y = 0$. This gives $\operatorname{Im} Q + \operatorname{Ker} Q^\times = \mathbb{C}^n$. Combining this with with what we saw in the preceding paragraph, we obtain $\operatorname{Im} Q \dot{+} \operatorname{Ker} Q^\times = \mathbb{C}^n$. But then the result of Part 2 yields the direct sum decomposition $\operatorname{Im} P \dot{+} \operatorname{Ker} P^\times = \mathbb{C}^n$, and (ii) is proved. □

The fact that in Theorem 4.4 the curve Γ is bounded is not essential. We only use that Γ is a closed curve on the Riemann sphere $\mathbb{C}_\infty$ and that W has no poles on Γ. Thus Γ may pass through infinity. For instance, let us replace Γ by the extended real line $\mathbb{R}_\infty$ which passes through infinity. By the results of Section 2.2, the condition that the $m \times m$ rational matrix function W has no poles on $\mathbb{R} \cup \{\infty\}$ implies that W can be represented in the form

$$W(\lambda) = D + C(\lambda - A)^{-1}B, \qquad \lambda \in \mathbb{R}, \tag{4.28}$$

where A is a square matrix with no real eigenvalues. The condition that W takes invertible values on $\mathbb{R} \cup \{\infty\}$ now amounts to the requirement that D is invertible and the matrix $A - BD^{-1}C$ has no real eigenvalues. Also, in that case,

$$W^{-1}(\lambda) = D^{-1} - D^{-1}C(\lambda - A^{\times})^{-1}BD^{-1}, \qquad \lambda \in \mathbb{R},$$

where $A^{\times} = A - BD^{-1}C$. With these minor modifications the proof of Theorem 4.4 also applies to realizations of the form (4.28), and yields the following theorem.

Theorem 4.5. *Let W be a rational $m \times m$ matrix function without poles on the real line, and let W be given by the realization*

$$W(\lambda) = D + C(\lambda I_n - A)^{-1}B, \qquad \lambda \in \mathbb{R}, \tag{4.29}$$

where A is an $n \times n$ matrix with no real eigenvalues. Then W admits a right canonical factorization with respect to $\mathbb{R} \cup \{\infty\}$ if and only if the following conditions are satisfied:

(i) *D is invertible and $A^{\times} = A - BD^{-1}C$ has no real eigenvalues,*

(ii) $\mathbb{C}^n = M \dot{+} M^{\times}$.

Here n is the order of the matrix A, the space M is the spectral subspace of A corresponding to its eigenvalues in the upper half plane, and $M^{\times}$ is the spectral subspace of $A^{\times}$ corresponding to its eigenvalues in the lower half plane. Furthermore, if the conditions (i) *and* (ii) *are fulfilled, then a right canonical factorization with respect to $\mathbb{R} \cup \{\infty\}$ is given by*

$$W(\lambda) = W_-(\lambda)W_+(\lambda), \qquad \lambda \in \Gamma,$$

where the factors and their inverses can be written as

$$\begin{aligned}
W_-(\lambda) &= D + C(\lambda I_n - A)^{-1}(I - \Pi)B,\\
W_+(\lambda) &= I + D^{-1}C\Pi(\lambda I_n - A)^{-1}B,\\
W_-^{-1}(\lambda) &= D^{-1} - D^{-1}C(I - \Pi)(\lambda I_n - A^{\times})^{-1}BD^{-1},\\
W_+^{-1}(\lambda) &= I - D^{-1}C(\lambda I_n - A^{\times})^{-1}\Pi B.
\end{aligned}$$

Here Π is the projection of $\mathbb{C}^n$ along M onto $M^{\times}$.

Since there is no a priori assumption on the invertibility of (the external) operator D, Theorem 4.5 is a slight extension of Theorem 3.2 dealing with matrix functions too. The results can be generalized to the case of operator functions (cf., Section 7.1 below).

4.4 Inversion of singular operators with a rational matrix symbol

In this section we apply the results of the previous sections to solve the problem of inverting the singular integral equation

$$a(t)\varphi(t) + b(t)\frac{1}{\pi i}\int_\Gamma \frac{\varphi(\tau)}{\tau - t}\,d\tau = g(t), \qquad t \in \Gamma. \tag{4.30}$$

Throughout we assume that a and b are rational $m \times m$ matrix functions which do not have poles on the Cauchy contour Γ. We shall analyze equation (4.30) under the additional condition that the difference $a(\lambda) - b(\lambda)$ is invertible for each $\lambda \in \Gamma$. Since we are interested in invertibility, the latter condition is not an essential restriction (cf., Theorem 1.3).

The fact that the matrix $a(\lambda) - b(\lambda)$ is invertible for $\lambda \in \Gamma$ allows us to introduce the operator $T = M_W P_\Gamma + Q_\Gamma$ which we consider on $L_2^m(\Gamma)$. Here

$$W(\lambda) = \big(a(\lambda) - b(\lambda)\big)^{-1}\big(a(\lambda) + b(\lambda)\big),$$

and M_W is the operator of multiplication by W on $L_2^m(\Gamma)$, that is, for $\varphi \in L_2^m(\Gamma)$ we have $(M_W\varphi)(t) = W(t)\varphi(t)$ for almost all $t \in \Gamma$. Furthermore, P_Γ and Q_Γ are the orthogonal projections on $L_2^m(\Gamma)$ associated with the singular integral operator introduced in Section 1.3. Thus, for $\varphi \in L_2^m(\Gamma)$,

$$(P_\Gamma\varphi)(t) = \frac{1}{2}\varphi(t) + \frac{1}{2\pi i}\int_\Gamma \frac{\varphi(\tau)}{\tau - t}\,d\tau, \tag{4.31}$$

$$(Q_\Gamma\varphi)(t) = \frac{1}{2}\varphi(t) - \frac{1}{2\pi i}\int_\Gamma \frac{\varphi(\tau)}{\tau - t}\,d\tau, \tag{4.32}$$

for almost all $t \in \Gamma$. The image of P_Γ consists of all functions in $L_2^m(\Gamma)$ that admit an analytic continuation into F_+. Similarly, the image of Q_Γ is the subspace of all functions in $L_2^m(\Gamma)$ that admit an analytic continuation into F_- and vanish at infinity. Note that equation (4.30) is equivalent to

$$(M_W P_\Gamma + Q_\Gamma)\varphi = \widetilde{g}, \quad \text{where} \quad \widetilde{g}(\lambda) = \big(a(\lambda) - b(\lambda)\big)^{-1} g(\lambda).$$

Since W is a rational $m \times m$ matrix function without poles on Γ, we know from Theorem 4.2 that W admits a Γ-regular realization

$$W(\lambda) = I + C(\lambda G - A)^{-1}B, \qquad \lambda \in \Gamma. \tag{4.33}$$

The main result of this section provides an explicit inversion formula for the operator $M_W P_\Gamma + Q_\Gamma$ in terms of the realization (4.33).

Theorem 4.6. *Let the rational $m \times m$ matrix function W be given by the Γ-regular realization (4.33), and put $A^\times = A - BC$. Then $M_W P_\Gamma + Q_\Gamma$ is an invertible operator on $L_2^m(\Gamma)$ if and only if the following two conditions are satisfied:*

(1) *the pencil* $\lambda G - A^\times$ *is* Γ*-regular,*

(2) $\mathbb{C}^n = \operatorname{Im} P \dot{+} \operatorname{Ker} P^\times$,

where n *is the order of the matrices* A *and* G, *and*

$$P = \frac{1}{2\pi i}\int_\Gamma G(\lambda G - A)^{-1} d\lambda, \qquad P^\times = \frac{1}{2\pi i}\int_\Gamma G(\lambda G - A^\times)^{-1} d\lambda. \tag{4.34}$$

In that case

$$\begin{aligned}((M_W P_\Gamma + Q_\Gamma)^{-1} g)(\lambda) &= g(\lambda) - C(\lambda G - A^\times)^{-1} B (P_\Gamma g)(\lambda) \\ &\quad + \Big(C(\lambda G - A^\times)^{-1} - C(\lambda G - A^\times)^{-1} \Big)(I - \Pi) \\ &\quad \cdot \Big(\frac{1}{2\pi i}\int_\Gamma P^\times G(\zeta G - A^\times)^{-1} B g(\zeta) d\zeta \Big), \quad \lambda \in \Gamma.\end{aligned}$$

Here Π *is the projection of* $\mathbb{C}^n$ *along* $\operatorname{Im} P$ *onto* $\operatorname{Ker} P^\times$.

With suitable changes, the theorem remains true when P and $P^\times$ are replaced by the projections Q and $Q^\times$ (also) appearing in Theorem 4.4.

Proof. From the general theory of singular integral equations reviewed in Section 1.3 we know that the operator $M_W P_\Gamma + Q_\Gamma$ is invertible if and only if W admits a right canonical factorization with respect to Γ. Since W is given by (4.33), the latter is the case if and only if conditions (i) and (ii) in Theorem 4.4 are fulfilled. By the final statement in Theorem 4.4, conditions (i) and (ii) in Theorem 4.4 are equivalent to conditions (1) and (2) in the present theorem. Thus we have proved that $M_W P_\Gamma + Q_\Gamma$ is invertible if and only if (1) and (2) are satisfied.

To get the formula for the inverse of $M_W P_\Gamma + Q_\Gamma$ we again use the general theory of singular integral equations, the inversion formula (1.29) in particular. Let $W = W_- W_+$ be a right canonical factorization of W with respect to Γ. For $g \in L_2^m(\Gamma)$ we then have, suppressing the variable λ,

$$(M_W P_\Gamma + Q_\Gamma)^{-1} g = W_+^{-1}\Big(P_\Gamma(W_-^{-1} g)\Big) + W_-\Big(Q_\Gamma(W_-^{-1} g)\Big).$$

Taking into account the form of P_Γ and Q_Γ in (4.31) and (4.32), this identity can be rewritten as

$$\begin{aligned}((M_W P_\Gamma + Q_\Gamma)^{-1} g)(\lambda) &= \frac{1}{2} g(\lambda) + \frac{1}{2} W(\lambda)^{-1} g(\lambda) \\ &\quad + \frac{1}{2\pi i}\int_\Gamma \frac{1}{\tau - \lambda}\Big(W_+(\lambda)^{-1} - W_-(\lambda)\Big) W_-(\tau)^{-1} g(\tau)\, d\tau, \quad \lambda \in \Gamma.\end{aligned} \tag{4.35}$$

Next, we use the formulas for W_+, W_- and their inverses given in Theorem 4.4. This yields

$$\begin{aligned}\Big(W_+(\lambda)^{-1} - W_-(\lambda)\Big)&W_-(\tau)^{-1}\\ &= -C(\lambda G - A^\times)^{-1}\Delta B - C(\lambda G - A)^{-1}(I-\Delta)B \\ &\quad + \lambda G - A^\times)^{-1}\Delta BC(I-\Lambda)(\tau G - A^\times)^{-1}B \\ &\quad + C(\lambda G - A)^{-1}(I-\Delta)BC(I-\Lambda)(\tau G - A^\times)^{-1}B.\end{aligned} \tag{4.36}$$

Here Δ and Λ are the projections defined in Theorem 4.4. Using these definitions, and the partitionings of A, G, and $A^\times$ in (4.16), (4.17) and (4.20), respectively, we obtain

$$\Delta A(I-\Lambda) = 0, \qquad (I-\Delta)A^\times\Lambda = 0, \qquad \Delta G = G\Lambda.$$

Since $BC = A - A^\times$, it follows that

$$\begin{aligned}\Delta BC(I-\Lambda) &= A^\times\Lambda - \Delta A^\times \\ &= (A^\times - \lambda G)\Lambda - \Delta(A^\times - \tau G) - (\tau - \lambda)\Delta G,\end{aligned}$$

and

$$\begin{aligned}(I-\Delta)BC(I-\Lambda) &= A(I-\Lambda) - (I-\Delta)A^\times \\ &= (A - \lambda G)(I-\Lambda) - (I-\Delta)(A^\times - \tau G) - (\tau-\lambda)(I-\Delta)G.\end{aligned}$$

Inserting these expressions into (4.36) gives

$$\begin{aligned}\Big(W_+(\lambda)^{-1} - W_-(\lambda)\Big)W_-(\tau)^{-1} &= -C(\tau G - A^\times)^{-1}B \\ &\quad -(\tau-\lambda)C(\lambda G - A^\times)^{-1}\Delta G(\tau G - A^\times)^{-1}B \\ &\quad -(\tau-\lambda)C(\lambda G - A)^{-1}(I-\Delta)G(\tau G - A^\times)^{-1}B.\end{aligned}$$

Next we use that $(\tau-\lambda)C(\lambda G - A^\times)^{-1}G(\tau G - A^\times)^{-1}B$ can be written as

$$C(\lambda G - A^\times)^{-1}\big((\tau G - A^\times) - (\lambda G - A^\times)\big)(\tau G - A^\times)^{-1}B$$

which in turn is equal to $C(\lambda G - A^\times)^{-1}B - C(\tau G - A^\times)^{-1}B$, and this leads to

$$\begin{aligned}\Big(W_+(\lambda)^{-1} - W_-(\lambda)\Big)W_-(\tau)^{-1} &= -C(\lambda G - A^\times)^{-1}B \\ &\quad +(\tau-\lambda)\Big(C(\lambda G - A^\times)^{-1} - C(\lambda G - A)^{-1}\Big) \\ &\quad \cdot(I-\Delta)G(\tau G - A^\times)^{-1}B.\end{aligned} \tag{4.37}$$

Using (4.37) and (4.5) in (4.35) we obtain

$$
\begin{aligned}
((M_W P_\Gamma + Q_\Gamma)^{-1} g)(\lambda) &= g(\lambda) - \frac{1}{2} C(\lambda G - A^\times)^{-1} B g(\lambda) \\
&\quad - C(\lambda G - A^\times)^{-1} B \Big(\frac{1}{2\pi i} \int_\Gamma \frac{1}{\tau - \lambda} g(\tau)\, d\tau \Big) \\
&\quad + \big(C(\lambda G - A^\times)^{-1} - C(\lambda G - A)^{-1} \big)(I - \Delta) \\
&\qquad \cdot \Big(\frac{1}{2\pi i} \int_\Gamma G(\tau G - A^\times)^{-1} B g(\tau)\, d\tau \Big), \qquad \lambda \in \Gamma.
\end{aligned}
$$

Finally, note that $\Delta = \Pi$ and $(I - \Pi)P^\times = I - \Pi$. Since P_Γ is given by (4.31), we see that we have derived the desired expression for the inverse of the operator $M_W P_\Gamma + Q_\Gamma$. □

4.5 The Riemann-Hilbert boundary value problem revisited (1)

In this section we treat the (homogeneous) Riemann-Hilbert boundary value problem for non-proper rational matrix functions. As before Γ is a Cauchy contour. As usual, the interior domain of Γ is denoted by F_+, and its exterior domain, which contains the point infinity, by F_-. Throughout W is a rational $m \times m$ matrix function which does not have poles on Γ.

We say that a pair of $\mathbb{C}^m$-valued functions Φ_+, Φ_- is a *solution of the Riemann-Hilbert boundary problem* of W with respect to Γ if Φ_+ and Φ_- are continuous on $F_+ \cup \Gamma$ and $F_- \cup \Gamma$, respectively, Φ_+ and Φ_- are analytic in F_+ and F_-, respectively, Φ_- vanishes at infinity, and

$$W(\lambda)\Phi_+(\lambda) = \Phi_-(\lambda), \qquad \lambda \in \Gamma. \tag{4.38}$$

Since W is assumed to be a rational $m \times m$ matrix function which has no poles on Γ, we may assume that W is given by a Γ-regular realization

$$W(\lambda) = I + C(\lambda G - A)^{-1} B, \qquad \lambda \in \Gamma. \tag{4.39}$$

We shall also assume that W takes invertible values on Γ. This additional condition is equivalent to the requirement that the pencil $\lambda G - A^\times$ is Γ-regular. The following theorem is the natural analogue of Theorem 3.7.

Theorem 4.7. *Let W be given by* (4.39), *and assume that the pencil $\lambda G - A^\times$ is a Γ-regular. Put*

$$P = \frac{1}{2\pi i} \int_\Gamma G(\lambda G - A)^{-1}\, d\lambda, \qquad P^\times = \frac{1}{2\pi i} \int_\Gamma G(\lambda G - A^\times)^{-1}\, d\lambda.$$

Then the pair of functions Φ_+ *and* Φ_- *is a solution of the Riemann-Hilbert boundary value problem of* W *with respect to* Γ *if and only if there exists* x *belonging to* $\operatorname{Im} P \cap \operatorname{Ker} P^\times$ *such that*

$$\Phi_+(\lambda) = C(\lambda G - A^\times)^{-1}x, \qquad \Phi_-(\lambda) = C(\lambda G - A)^{-1}x. \tag{4.40}$$

Moreover the vector x *in* (4.40) *is uniquely determined by* Φ_+, Φ_-

With the appropriate modifications, the theorem remains true when P and $P^\times$ are replaced by the projections Q and $Q^\times$ (also) appearing in Theorem 4.4.

Proof. Take $x \in \operatorname{Im} P \cap \operatorname{Ker} P^\times$, and define Φ_+ and Φ_- by (4.40). Formula (4.6) implies that (4.38) is satisfied. Since $x = Px$, Theorem 4.1 (ii) shows that Φ_- is continuous on $F_- \cup \Gamma$, analytic in F_-, and vanishes at infinity. Similarly, using $x = (I - P^\times)x$, Theorem 4.1 (iii), applied to $\lambda G - A^\times$, yields that Φ_+ is continuous on $F_+ \cup \Gamma$ and analytic on F_+. Thus the functions Φ_+ and Φ_- have the desired properties, and the pair Φ_+, Φ_- is a solution.

To prove the converse, assume that the pair Φ_+, Φ_- is a solution of the Riemann-Hilbert problem for W with respect to Γ. For $\lambda \in \Gamma$, introduce $\rho(\lambda) = (\lambda G - A)^{-1}B\Phi_+(\lambda)$. The $n \times m$ matrix function ρ is continuous on Γ, thus it makes sense to put

$$\begin{aligned} \rho_+(\lambda) &= \frac{1}{2}\rho(\lambda) + \frac{1}{2\pi i}\int_\Gamma \frac{\rho(\tau)}{\tau - \lambda}\,d\tau, \qquad \lambda \in \Gamma, \\ \rho_-(\lambda) &= \frac{1}{2}\rho(\lambda) - \frac{1}{2\pi i}\int_\Gamma \frac{\rho(\tau)}{\tau - \lambda}\,d\tau, \qquad \lambda \in \Gamma; \end{aligned}$$

cf., the expressions (4.31) and (4.32). The function ρ_+ is continuous on $F_+ \cup \Gamma$ and analytic in F_+, and ρ_- has the same properties with F_- in place of F_+. Moreover, ρ_- vanishes at infinity.

We first show that

$$\Phi_+(\lambda) = -C\rho_+(\lambda), \qquad \lambda \in F_+ \cup \Gamma, \tag{4.41}$$

$$\Phi_-(\lambda) = C\rho_-(\lambda), \qquad \lambda \in F_- \cup \Gamma. \tag{4.42}$$

Since the pair Φ_+, Φ_- satisfies (4.38), we have

$$\Phi_-(\lambda) = \Phi_+(\lambda) + C(\lambda G - A)^{-1}B\Phi_+(\lambda) = \Phi_+(\lambda) + C\rho(\lambda), \qquad \lambda \in \Gamma.$$

But $\rho(\lambda) = \rho_-(\lambda) + \rho_+(\lambda)$ on Γ, and therefore

$$\Phi_-(\lambda) - C\rho_-(\lambda) = \Phi_+(\lambda) + C\rho_+(\lambda), \qquad \lambda \in \Gamma. \tag{4.43}$$

The right-hand side of (4.43) is continuous on $F_+ \cup \Gamma$ and analytic in F_+. On the other hand, the left-hand side of (4.43) is continuous on $F_- \cup \Gamma$, analytic in

F_- and vanishes at infinity. Thus, by Liouville's theorem, both sides of (4.43) are identically zero on Γ, and the identities (4.41) and (4.42) hold.

Next, we compute the function ρ_-. From the definition of $\rho(\lambda)$ we see that $(\lambda G - A)\rho(\lambda) = B\Phi_+(\lambda)$ for $\lambda \in \Gamma$. Since Φ_+ is continuous on $F_+ \cup \Gamma$ and analytic in F_+, we conclude that for each $\lambda \in \Gamma$,

$$\begin{aligned}
\frac{1}{2}(\lambda G - A)\rho(\lambda) &= \frac{1}{2\pi i}\int_\Gamma \frac{1}{\tau - \lambda}(\tau G - A)\rho(\tau)\,d\tau \\
&= \frac{1}{2\pi i}\int_\Gamma \frac{1}{\tau - \lambda}\Big((\lambda G - A) + (\tau - \lambda)G\Big)\rho(\tau)\,d\tau \\
&= (\lambda G - A)\Big(\frac{1}{2\pi i}\int_\Gamma \frac{\rho(\tau)}{\tau - \lambda}\,d\tau\Big) + x,
\end{aligned}$$

where

$$x = \frac{1}{2\pi i}\int_\Gamma G\rho(\tau)\,d\tau.$$

Using the definition of ρ_-, the above calculation shows that

$$\rho_-(\lambda) = (\lambda G - A)^{-1}x, \qquad \lambda \in \Gamma. \tag{4.44}$$

To compute ρ_+, recall that $\rho(\lambda) = \rho_-(\lambda) + \rho_+(\lambda)$ on Γ. This, together with (4.41) and (4.42), yields

$$\begin{aligned}
(\lambda G - A)\rho_+(\lambda) &= (\lambda G - A)\rho(\lambda) - (\lambda G - A)\rho_-(\lambda) \\
&= B\Phi_+(\lambda) - x = -BC\rho_+(\lambda) - x, \qquad \lambda \in \Gamma.
\end{aligned}$$

Since $A^\times = A - BC$, we obtain

$$\rho_+(\lambda) = -(\lambda G - A^\times)^{-1}x, \qquad \lambda \in \Gamma. \tag{4.45}$$

From (4.44) and the fact that ρ_- is continuous on $F_- \cup \Gamma$, analytic in F_-, and vanishes at infinity, we conclude that $x = Px$. Similarly, we obtain from (4.45) that $x = (I - P^\times)x$. Thus $x \in \operatorname{Im} P \cap \operatorname{Ker} P^\times$. Formulas (4.41), (4.42), (4.44) and (4.45) now show that the functions Φ_+ and Φ_- have the desired representation (4.40).

It remains to prove the uniqueness of the vector x in (4.40). To do this assume that $u \in \operatorname{Im} P \cap \operatorname{Ker} P^\times$, and let $C(\lambda G - A)^{-1}u$ be identically zero on Γ. It suffices to show that $u = 0$. For this purpose we use the identity (4.8). Applying this identity to the vector u, we see that

$$(\lambda G - A^\times)^{-1}u = (\lambda G - A)^{-1}u, \qquad \lambda \in \Gamma. \tag{4.46}$$

Since $u \in \operatorname{Ker} P^\times$, the left-hand side of (4.46) has an analytic continuation on F_+; see Theorem 4.1 (iii). Similarly, $u \in \operatorname{Im} P$ implies that the right side of (4.46) has an analytic continuation on F_- which vanishes at infinity; see Theorem 4.1 (ii). But then we can apply Liouville's theorem to show that these functions are identically zero on Γ, which yields $u = 0$. □

Notes

The extension of the Riesz spectral theory for operators to operator pencils, which is described in Section 4.1, is due to Stummel [140]; the results can also be found in Section IV.1 of [51]. Section 4.2 combines the classical realization theory for proper rational matrix functions with that of matrix polynomials; for the latter, see [65]. The main source for the material in Sections 4.2 and 4.3 is the paper [55]; Section 4.4 is based on [56]. Section 4.5 seems to be new. For realizations of the form considered in this chapter, non-canonical Wiener-Hopf factorization has been studied in [151]. Instead of (4.3) other realizations of W can be used; see for instance [79], where (4.3) is replaced by the realization

$$W(\lambda) = D + (\lambda - \alpha)C(\lambda G - A)^{-1}B$$

which can also be used for non-square matrix functions.

Part III
Equations with non-rational symbols

In this part we carry out a program analogous to that of the second part, but now for certain classes of non-rational matrix and operator functions. Included are matrix functions analytic in a strip but not at infinity, an operator function appearing in linear transport theory, and operator functions analytic on a given curve.

There are three chapters. The main topic of the first chapter (Chapter 5) is a canonical factorization theorem for matrix functions analytic in a strip but not necessarily at infinity. Its applications to different classes of Wiener-Hopf equations are included too. The realizations of such matrix functions require that we consider systems with an infinite dimensional state space and with a state operator that is unbounded and exponentially dichotomous. Thus the theory of strongly continuous semigroups plays an important role in this material. Chapter 6 is entirely dedicated to the solution of an integro-differential equation from mathematical physics describing stationary migration of particles in a medium. To illustrate the approach, the special case of a finite number of scattering directions is considered first. This restriction makes it possible to reduce the problem to a canonical factorization problem for rational matrix functions. The general situation features an infinite dimensional separable Hilbert space as state space. The final chapter (Chapter 7) deals with canonical factorization and non-canonical Wiener-Hopf factorization for operator-valued functions that are analytic on a given curve. In this chapter the so-called factorization indices are described in state space terms.

Chapter 5

Factorization of matrix functions analytic in a strip

This chapter deals with $m \times m$ matrix-valued functions of the form

$$W(\lambda) = I - \int_{-\infty}^{\infty} e^{i\lambda t} k(t)\, dt, \tag{5.1}$$

where k is an $m \times m$ matrix-valued function with the property that for some $\omega < 0$ the entries of $e^{-\omega|t|}k(t)$ are Lebesgue integrable on the real line. In other words, k is of the form

$$k(t) = e^{\omega|t|}h(t) \text{ with } h \in L_1^{m\times m}(\mathbb{R}). \tag{5.2}$$

It follows that the function W is analytic in the strip $|\Im\lambda| < \tau$, where $\tau = -\omega$. This strip contains the real line. The aim is to extend the canonical factorization theorem of Chapter 5 to functions of the type (5.1).

In general, the function W in (5.1) is not a rational matrix function, and hence one cannot expect a representation of W in the form

$$W(\lambda) = I + C(\lambda - A)^{-1}B \tag{5.3}$$

with A, B, C matrices. Also a realization with A, B and C bounded linear operators will not work. Indeed, in that case the function W would be analytic at infinity, however in general it is not. Thus to get a representation of the type (5.3) one has to allow for unbounded linear operators. In fact, we shall have to allow for A and C to be unbounded while B can be taken to be bounded.

This chapter consists of nine sections. In Sections 5.1 and 5.2 we present preliminary material on exponentially dichotomous operators and associated bisemigroups. These exponentially dichotomous operators appear as state operators in the realization triples defined in Section 5.3. In Section 5.4 we construct realization triples for $m \times m$ matrix-valued functions W of the form (5.1) with k as in

(5.2), and in Section 5.5 we use the realization triples to invert such a matrix function W. It turns out that inversion is only possible when the associate operator $A^\times = A - BC$ is exponentially dichotomous too. The inversion formula of Section 5.5 is used in Section 5.6 to derive an explicit formula for the kernel function of the inverse of a full line convolution integral operator when the symbol W is given by (5.1) and (5.2). This section also contains some preliminary material about Hankel operators. The final three sections concern applications. Sections 5.7 and 5.8 deal with inversion of a Wiener-Hopf integral equation with a kernel function k of the form (5.2) and with canonical factorization of the corresponding symbol. In Section 5.9 we revisit the Riemann-Hilbert boundary value problem.

5.1 Exponentially dichotomous operators and bisemigroups

We begin with some preliminaries about strongly continuous semigroups of operators (also called C_0-semigroups). Free use will be made of the standard theory of these semigroups as explained, for instance, in Chapter XIX of [51]. Besides ordinary C_0-semigroups defined on the positive half line $[0,\infty)$, henceforth to be called *right semigroups*, we shall also consider semigroups defined on the negative half line $(-\infty, 0]$. The latter will be called *left semigroups*. Notice that $T(t)$ is a left semigroup if and only if $T(-t)$ is a right semigroup.

Let $T(t)$ be a strongly continuous right or left semigroup. As is well-known, there exist constants M and ω such that

$$\|T(t)\| \le Me^{\omega|t|}, \qquad t \in J.$$

Here J is the half line $[0,\infty)$ or $(-\infty,0]$ according to $T(t)$ being a right or a left semigroup. If the above inequality is satisfied for a given real number ω and some positive constant M, we say that $T(t)$ is of *exponential type* ω. Semigroups of negative exponential type will be called *exponentially decaying*.

Next we introduce the concept of an exponentially dichotomous operator. Let X be a complex Banach space, and let A be a (possibly unbounded) linear operator with domain $\mathcal{D}(A)$ in X and with values in X, in short $A(X \to X)$. Further, let $P : X \to X$ be a (bounded linear) projection of X commuting with A. The latter means that P maps $\mathcal{D}(A)$ into itself and $PAx = APx$ for each $x \in \mathcal{D}(A)$. Put $X_- = \operatorname{Im} P$ and $X_+ = \operatorname{Ker} P$. Then

$$X = X_- \dot{+} X_+, \tag{5.4}$$

and this decomposition reduces A, that is,

$$\mathcal{D}(A) = [\mathcal{D}(A) \cap X_-] \dot{+} [\mathcal{D}(A) \cap X_+], \tag{5.5}$$

with A mapping $[\mathcal{D}(A) \cap X_-]$ into X_- and $[\mathcal{D}(A) \cap X_+]$ into X_+. So with respect to the decompositions (5.4) and (5.5), the operator A has the matrix representation

$$A = \begin{bmatrix} A_- & 0 \\ 0 & A_+ \end{bmatrix}. \tag{5.6}$$

Here $A_-(X_- \to X_-)$ is the restriction of A to X_-, and $A_+(X_+ \to X_+)$ is the restriction of A to X_+. In particular, the domain $\mathcal{D}(A_-)$ of A_- is $\mathcal{D}(A) \cap X_-$ and the domain $\mathcal{D}(A_+)$ of A_-+ is $\mathcal{D}(A) \cap X_+$. Thus (5.5) can be rewritten as $\mathcal{D}(A) = \mathcal{D}(A_-) \dot{+} \mathcal{D}(A_+)$.

The operator A is said to be *exponentially dichotomous* if the operators A_- and A_+ in (5.6) are generators of exponentially decaying strongly continuous left and right semigroups, respectively. In that case the projection P, which will turn out to be unique (see Proposition 5.1 below), is called the *separating projection* for A. We say that A is of *exponential type* ω (< 0) if this is true for the semigroups generated by A_- and A_+.

Suppose, for the moment, that $A : X \to X$ is a bounded linear operator. Then A is exponentially dichotomous if and only if the spectrum $\sigma(A)$ of A does not meet the imaginary axis. In that situation the separating projection for A is simply the Riesz projection corresponding to the part of $\sigma(A)$ lying in the open right half plane $\Re\lambda > 0$.

Next, observe that generators of exponentially decaying strongly continuous semigroups belong to the class of exponentially dichotomous operators, the left semigroup case corresponding to the separating projection being the identity operator and the right semigroup case corresponding to the separating projection being the zero operator on X.

Returning to the general case, we note that the operators A_- and A_+ in the definition of an exponentially dichotomous operator are closed and densely defined. Hence the same is true for their direct sum A. Furthermore, if A is of (negative) exponential type ω, then, by the Hille-Yosida-Phillips theorem (see, e.g., Theorem XIX.2.3 in [51]), the spectrum $\sigma(A_-)$ of A_- is contained in the closed half plane $\Re\lambda \geq -\omega$, whereas $\sigma(A_+)$ is a subset of $\Re\lambda \leq \omega$. In particular, the strip $|\Re\lambda| < -\omega$ is contained in $\rho(A)$, the resolvent set of A. This justifies the use of the term "separating projection" for P.

It is convenient to adopt the following notation and terminology. Suppose $A(X \to X)$ is an exponentially dichotomous operator with separating projection P, and let A_- and A_+ be as above. Thus A_- and A_+ are the restrictions of A to $X_- = \operatorname{Im} P$ and $X_+ = \operatorname{Ker} P$, respectively. With A we associate a function $E(\cdot; A)$ with domain $\mathbb{R} \setminus \{0\}$ and with values in $\mathcal{L}(X)$, the space of all bounded operators on X. The definition is as follows: for $x \in X$,

$$E(t; A)x = \begin{cases} -e^{tA_-}Px, & t < 0, \\ e^{tA_+}(I - P)x, & t > 0, \end{cases} \tag{5.7}$$

where, following standard conventions, e^{tA_-} denotes the value at $t(< 0)$ of the semigroup generated by A_- and e^{tA_+} denotes the value at $t(> 0)$ of the semigroup generated by A_+. We call $E(\cdot\,;A)$ the *bisemigroup generated by* A. The operator A will be referred to as the *bigenerator* of $E(\cdot\,;A)$.

For each $x \in X$ the function $E(t;A)x$ is continuous on $\mathbb{R}\setminus\{0\}$, and

$$\lim_{t\uparrow 0} E(t;A)x = -Px, \qquad \lim_{t\downarrow 0} E(t;A)x = (I-P)x. \tag{5.8}$$

We conclude that $E(\cdot\,;A)$ is an exponentially decaying operator function which is strongly continuous on the real line, except at the origin where it has (at worst) a jump discontinuity. For $x \in \mathcal{D}(A) = \mathcal{D}(A_-)\dot{+}\mathcal{D}(A_+)$, the function $E(t;A)x$ is even differentiable on $\mathbb{R}\setminus\{0\}$. In fact, we have

$$\frac{d}{dt}E(t;A)x \;=\; AE(t;A)x \;=\; E(t;A)Ax, \qquad t \neq 0.$$

Obviously the derivative of $E(\cdot\,;A)x$ is continuous on $\mathbb{R}\setminus\{0\}$, exponentially decaying (in both directions) and has (at worst) a jump discontinuity at the origin. From (5.7) it is clear that

$$E(t,A)P = PE(t,A) = E(t;A), \qquad t < 0,$$

$$E(t,A)(I-P) = (I-P)E(t,A) = E(t;A), \qquad t > 0.$$

Also the following semigroup properties hold:

$$\begin{aligned} E(t+s,A) &= -E(t;A)E(s;A), && t,s < 0,\\ E(t+s,A) &= E(t,A)E(s;A), && t,s > 0. \end{aligned}$$

One of the reasons for the different signs to appear in the definition of $E(t;A)$ is that in this way the following identity holds:

$$(\lambda - A)^{-1}x = \int_{-\infty}^{\infty} e^{-\lambda t}E(t;A)x\,dt, \qquad x \in X,\ |\Re\lambda| < -\omega. \tag{5.9}$$

Here ω is a negative constant such that A is of exponential type ω. The proof of (5.9) is based on standard semigroup theory (see, e.g., Theorem XIX.2.2 in [51]).

With the help of (5.8) and (5.9) we now can prove the uniqueness of the separating projection.

Proposition 5.1. *Let $A(X \to X)$ be an exponentially dichotomous operator. Then A has precisely one separating projection.*

Proof. Let P be a separating projection for A, and let $E(\cdot\,;A)$ be the associate bisemigroup. A priori $E(\cdot\,;A)$ depends not only on A but also on P. However, (5.9) and the fact that $E(\cdot\,;A)$ is strongly continuous on $\mathbb{R}\setminus\{0\}$ imply that $E(\cdot\,;A)$ is uniquely determined by A. On the other hand the first identity in (5.8) shows that P is uniquely determined by $E(\cdot\,;A)$. So along with $E(\cdot\,;A)$ the separating projection is uniquely determined by A. □

From (5.9) it follows that on a strip around the imaginary axis, the resolvent $(\lambda - A)^{-1}$ of A is the pointwise two-sided Laplace transform of an exponentially decaying operator function which is strongly continuous on $\mathbb{R} \setminus \{0\}$ and has (at worst) a jump discontinuity at zero. The following theorem shows that this property characterizes exponentially dichotomous operators.

Theorem 5.2. *Let $A(X \to X)$ be a densely defined closed linear operator on the complex Banach space X. Then A is exponentially dichotomous if and only if the imaginary axis is contained in the resolvent set of A and*

$$(\lambda - A)^{-1}x = \int_{-\infty}^{\infty} e^{-\lambda t} E(t)x \, dt, \qquad x \in X, \ \Re\lambda = 0, \tag{5.10}$$

where $E : \mathbb{R} \setminus \{0\} \to \mathcal{L}(X)$ is exponentially decaying and strongly continuous, and E has (at worst) a jump discontinuity at zero. In that case the function E is the bisemigroup generated by A.

The above theorem will play an important role in Section 5.5. For the sake of completeness its proof is given in the next section. The reader who is ready to accept Theorem 5.2 may proceed directly to Section 5.3.

5.2 Spectral splitting and proof of Theorem 5.2

In this section we prove Theorem 5.2. The proof will be based on the spectral splitting results proved in Section XV.3 of [51], which originate from [16]. It will be convenient first to prove the following result which is the semigroup version of Theorem 5.2.

Theorem 5.3. *Let $S(X \to X)$ be a densely defined closed linear operator on the complex Banach space X. Then S is the infinitesimal generator of a strongly continuous right semigroup of negative exponential type if and only if the imaginary axis is contained in the resolvent set of S and*

$$(\lambda - S)^{-1}x = \int_{0}^{\infty} e^{-\lambda t} E(t)x \, dt, \qquad x \in X, \ \Re\lambda = 0, \tag{5.11}$$

where $E : [0, \infty) \to \mathcal{L}(X)$ is exponentially decaying and strongly continuous. In that case the function E is the right semigroup generated by S.

Proof. The "only if part" of Theorem 5.2 is immediate from standard semigroup theory. To prove the "if part" let ω be a negative real number and L a positive constant such that

$$\|E(t)\| \leq Le^{\omega t}, \qquad t \geq 0. \tag{5.12}$$

For $\Re\lambda > \omega$ and $x \in X$, put

$$R(\lambda)x = \int_{0}^{\infty} e^{-\lambda t} E(t)x \, dt. \tag{5.13}$$

Then $R(\lambda)$ is a well-defined bounded linear operator on X with norm not exceeding L. The function R is pointwise analytic on $\Re\lambda > \omega$, and hence it is analytic on $\Re\lambda > \omega$. We shall prove that $\Re\lambda > \omega$ implies that $\lambda \in \rho(S)$ and $R(\lambda) = (\lambda - S)^{-1}$.

Let $T = S^{-1}$ be the (bounded) inverse of S. For $0 \neq \lambda \in \rho(S)$, one has $\lambda^{-1} \in \rho(T)$ and $(\lambda - S)^{-1} = -\lambda^{-1}T(\lambda^{-1} - T)^{-1}$. Take λ on the imaginary axis, $\lambda \neq 0$. Combining (5.11) and (5.13) we get

$$R(\lambda) = (\lambda - S)^{-1} = -\lambda^{-1}T(\lambda^{-1} - T)^{-1},$$

and hence $R(\lambda) = \big(\lambda R(\lambda) - I\big)T$. But then the unicity theorem for analytic functions gives that these identities hold on all of $\Re\lambda > \omega$. A simple computation now shows that $R(\lambda) = (\lambda - S)^{-1}$ for each λ with $\Re\lambda > \omega$.

We have seen that the open half plane $\Re\lambda > \omega$ is contained in $\rho(S)$ and

$$(\lambda - S)^{-1}x = \int_0^\infty e^{-\lambda t}E(t)x\,dt, \qquad x \in X,\ \Re\lambda > \omega. \tag{5.14}$$

Differentiating the left-and right-hand side of (5.14) for the variable λ, one finds

$$(\lambda - S)^{-n}x = \frac{(-1)^n}{(n-1)!}\int_0^\infty t^{n-1}e^{-\lambda t}E(t)x\,dt, \qquad x \in X,\ \Re\lambda > \omega. \tag{5.15}$$

Here n is an arbitrary positive integer. Taking $\lambda > \omega$ and combining (5.12) and (5.15) we get the estimate

$$\|(\lambda - S)^{-n}x\| \leq \frac{L}{(n-1)!}\left(\int_0^\infty t^{n-1}e^{-(\lambda-\omega)t}\,dt\right)\|x\|.$$

Observe that

$$\int_0^\infty t^{n-1}e^{-(\lambda-\omega)t}\,dt = \frac{1}{(\lambda-\omega)^n}\int_0^\infty s^{n-1}e^{-s}\,ds = \frac{(n-1)!}{(\lambda-\omega)^n}.$$

Thus $\|(\lambda - S)^{-n}\| \leq L(\lambda - \omega)^{-n}$ for real $\lambda > \omega$ and $n = 1, 2, \ldots$. The Hille-Yosida-Phillips theorem ([51], page 419) now guarantees that S is the generator of a strongly continuous right semigroup $T(t)$ of exponential type $\omega < 0$. But then (5.11) holds with $E(t)$ replaced by $T(t)$. As the operator-valued functions E) and T are both strongly continuous, they must coincide, and the proof is complete. □

Proof of Theorem 5.2. We split the proof into three parts. Throughout ω is a negative real number and L is a positive constant such that

$$\|E(t)\| \leq Le^{\omega|t|}, \qquad 0 \neq t \in \mathbb{R}. \tag{5.16}$$

Part 1. In this part we show that $(\lambda - A)^{-1}$ is well-defined and uniformly bounded on each closed strip $|\Re\lambda| \leq h$ where $0 < h < -\omega$. To do this, let us consider the following expressions:

$$\begin{aligned}\Psi_+(\lambda)x &= \int_0^\infty e^{-\lambda t}E(t)x\,dt, \qquad \Re\lambda > \omega,\\ \Psi_-(\lambda)x &= \int_{-\infty}^0 e^{-\lambda t}E(t)x\,dt, \qquad \Re\lambda < -\omega.\end{aligned}$$

Here $x \in X$. Clearly $\Psi_+(\lambda)$ is a well-defined bounded linear operator on X which depends analytically on λ on the open half plane $\Re\lambda > \omega$, and an analogous statement holds of course for $\Psi_-(\lambda)$. Note that $\Psi_-(\lambda) + \Psi_+(\lambda)$ is analytic on the strip $|\Re\lambda| < -\omega$ and coincides on the imaginary axis with $(\lambda - A)^{-1}$. Thus $|\Re\lambda| < -\omega$ implies $\lambda \in \rho(A)$ and $(\lambda - A)^{-1} = \Psi_-(\lambda) + \Psi_+(\lambda)$, i.e.,

$$(\lambda - A)^{-1}x = \int_{-\infty}^\infty e^{-\lambda t}E(t)x\,dt, \qquad x \in X,\ |\Re\lambda| < -\omega. \tag{5.17}$$

A detailed argument can be given along the lines indicated in the second paragraph of the proof of Theorem 5.3.

From (5.16) one easily deduces that

$$\begin{aligned}\|\Psi_+(\lambda)\| &\leq \frac{L}{\Re\lambda - \omega}, \qquad \Re\lambda > \omega,\\ \|\Psi_-(\lambda)\| &\leq \frac{-L}{\Re\lambda + \omega}, \qquad \Re\lambda < -\omega.\end{aligned}$$

On the strip $|\Re\lambda| < -\omega$, the norm of $(\lambda - A)^{-1} = \Psi_-(\lambda) + \Psi_+(\lambda)$ can now be estimated as follows:

$$\|(\lambda - A)^{-1}\| \leq \frac{-2L\omega}{\omega^2 - (\Re\lambda)^2}, \qquad |\Re\lambda| < -\omega. \tag{5.18}$$

In particular $(\lambda - A)^{-1}$ is uniformly bounded on each closed strip $|\Re\lambda| \leq h$ with $0 < h < -\omega$.

Part 2. Fix $0 < h < -\omega$. From what has been proved in the previous part, we know that

$$\sup_{|\Re\lambda| \leq h} \|(\lambda - A)^{-1}\| < \infty. \tag{5.19}$$

This allows us to use the spectral theory developed in Section XV.3 of [51]. First we introduce the operators

$$\begin{aligned}Q_- &= \frac{1}{2\pi i}\int_{-\alpha - i\infty}^{-\alpha + i\infty} \lambda^{-2}(\lambda - A)^{-1}\,d\lambda,\\ Q_+ &= \frac{-1}{2\pi i}\int_{\alpha - i\infty}^{\alpha + i\infty} \lambda^{-2}(\lambda - A)^{-1}\,d\lambda.\end{aligned}$$

Here $0 < \alpha < h$, and hence (5.19) implies that Q_- and Q_+ are well-defined bounded linear operators on X. It can be proved that these operators do not depend on the particular choice of α; nevertheless, in what follows we keep α fixed. (Notice that in Section XV.3 of [51] the operators Q_- and Q_+ are denoted by S_- and S_+, respectively.) We define

$$M_- = \overline{\operatorname{Im} Q_-}, \qquad M_+ = \overline{\operatorname{Im} Q_+}.$$

Put $T = A^{-1}$. Then T is a bounded linear operator on X commuting with $(\lambda - A)^{-1}$ for each λ in the strip $|\Re\lambda| \leq h$. It follows that T commutes with Q_- and Q_+. Since T is bounded, this implies that $TM_- \subset M_-$ and $TM_+ \subset M_+$. We also know that $\operatorname{Im} T = \mathcal{D}(A)$, and thus TM_- and TM_+ belong to $\mathcal{D}(A)$. This allows us to define operators $A_-(M_- \to M_-)$ and $A_+(M_+ \to M_+)$ by setting

$$\mathcal{D}(A_-) = TM_-, \qquad A_-x = Ax, \qquad x \in \mathcal{D}(A_-),$$

$$\mathcal{D}(A_+) = TM_+, \qquad A_+x = Ax, \qquad x \in \mathcal{D}(A_+).$$

In other words,

$$A_- = \left(T|_{M_-}\right)^{-1}, \qquad A_+ = \left(T|_{M_+}\right)^{-1}.$$

The first part of Lemma XV.3.3 in [51] shows that A_- and A_+ are closed and densely defined linear operators, and their spectra satisfy the inclusion relations

$$\sigma(A_-) \subset \{\lambda \in \mathbb{C} \mid \Re\lambda \leq -h\},$$

$$\sigma(A_+) \subset \{\lambda \in \mathbb{C} \mid \Re\lambda \geq h\}.$$

We shall now prove that

$$(\lambda - A_-)^{-1}x = \int_0^\infty e^{-\lambda t}E(t)x\,dt, \qquad x \in M_-,\ Re\lambda > -h, \tag{5.20}$$

$$(\lambda - A_+)^{-1}x = \int_{-\infty}^0 e^{-\lambda t}E(t)x\,dt, \qquad x \in M_+,\ Re\lambda < h. \tag{5.21}$$

Following Section XV.3, page 330, of [51], we introduce two auxiliary sets N_- and N_+. By definition N_- is the set of all vectors $x \in X$ for which there exists an X-valued function φ_x^-, bounded and analytic on $\Re\lambda > -h$, which takes its values in $\mathcal{D}(A)$ and satisfies

$$(\lambda - A)\varphi_x^-(\lambda) = x, \qquad \Re\lambda > -h.$$

Roughly speaking, N_- consists of all vectors $x \in X$ such that $(\lambda - A)^{-1}x$ has a bounded analytic continuation to the open half plane $\Re\lambda > -h$. The function φ_x^- (assuming it exists) is uniquely determined by x. Analogously, we let N_+ be the set of all vectors $x \in X$ for which there exists an X-valued function φ_x^+, bounded and analytic on $\Re\lambda < h$, which takes its values in $\mathcal{D}(A)$ and satisfies

$$(\lambda - A)\varphi_x^+(\lambda) = x, \qquad \Re\lambda < h.$$

Also φ_x^+ is unique, provided it exists. Obviously, the sets N_- and N_+ are (possibly non-closed) linear manifolds of X.

The second part of Lemma XV.3.3 in [51] states that $\mathcal{D}(A_-^2) \subset N_-$ and $\mathcal{D}(A_+^2) \subset N_+$. Now, fix $x \in \mathcal{D}(A_-^2)$. Then $x \in N_-$, and hence $(\lambda - A)^{-1}x$ extends to a bounded analytic function on $\Re\lambda > -h$. Notice that $\Psi_+(\lambda)$ is also bounded and analytic on $\Re\lambda > -h$. Recall that $\Psi_-(\lambda)$ is equal to $(\lambda - A)^{-1} - \Psi_+(\lambda)$ for each λ in the strip $|\Re\lambda| < \omega$. It follows that $\Psi_-(\lambda)x$ extends to a bounded analytic function on $\Re\lambda > -h$. On the other hand $\Psi_-(\lambda)x$ is analytic on $\Re\lambda < -\omega$ and bounded on $\Re\lambda \leq h$. Hence $\Psi_-(\lambda)x$ determines a bounded entire function. From the estimate given for $\|\Psi_-(\lambda)\|$ in the previous part, it is clear that

$$\lim_{\lambda\in\mathbb{R},\, \lambda\to-\infty} \Psi_-(\lambda)x \;=\; 0.$$

But then we can use Liouville's theorem to show that $\Psi_-(\lambda)x$ vanishes identically. We conclude that

$$\int_{-\infty}^{0} e^{-\lambda t}E(t)x\,dt = 0, \qquad \Re\lambda < -\omega.$$

Since $E(t)x$ is continuous on $-\infty < t < 0$, it follows that $E(t)x = 0$ for all negative real numbers t.

Now recall that $T|_{M_-}$ is one-to-one and that $A_- = \left(T|_{M_-}\right)^{-1}$ is densely defined. This implies that

$$\mathcal{D}(A_-^2) = \operatorname{Im}\left(T|_{M_-}\right)^2,$$

and that $\mathcal{D}(A_-^2)$ is dense in M_-. Thus the result of the previous paragraph shows that $E(t)$ vanishes on M_- for $-\infty < t < 0$. For $x \in M_-$ and $|\Re\lambda| < h$, we have

$$(\lambda - A)^{-1}x = -(I - \lambda T)^{-1}Tx = -\left(I - \lambda(T|_{M_-})\right)^{-1}(T|_{M_-})x = (\lambda - A_-)^{-1}x.$$

Hence, for $x \in M_-$ and $|\Re\lambda| < h$,

$$(\lambda - A_-)^{-1}x = (\lambda - A)^{-1}x = \int_{-\infty}^{\infty} e^{-\lambda t}E(t)x\,dt = \int_0^{\infty} e^{-\lambda t}E(t)x\,dt.$$

By analytic continuation this proves (5.20). Formula (5.21) is proved in a similar manner.

Part 3. In this part we complete the proof. First we show that for $t > 0$ the operator $E(t)$ maps M_- into M_-. To see this, take $x \in M_-$, and let f be a continuous linear functional on X annihilating M_-. Then $f\big((\lambda - A_-)^{-1}x\big) = 0$ for $\Re\lambda > -h$, and thus (5.20) yields

$$\int_0^\infty e^{-\lambda t} f\big(E(t)x\big)\,dt = 0, \qquad \Re\lambda > -h.$$

This implies that $f\big(E(t)x\big) = 0$ for $t > 0$, and so, by the Hahn-Banach theorem, $E(t)x \in M_-$ for $t > 0$. Thus $E(t)M_- \subset M_-$ for $t > 0$.

The result of the previous paragraph enables us to define an operator-valued function $E_- : (0, \infty) \to \mathcal{L}(M_-)$ by stipulating that $E_-(t) = E(t)|_{M_-}$. Our assumptions on the behavior of E near the origin (together with the Banach-Steinhaus theorem) imply that E_- can be extended to a strongly continuous function, defined on $0 \le t < \infty$, by putting

$$E_-(0)x = \lim_{t\downarrow 0} E(t)x, \qquad x \in M_-.$$

The identity (5.20) can now be written as

$$(\lambda - A_-)^{-1}x = \int_0^\infty e^{-\lambda t} E_-(t)\,dt, \qquad x \in M_-,\ \Re\lambda > -h.$$

Since $A_-(M_- \to M_-)$ is closed and densely defined, it follows from Theorem 5.3 that E_- is a strongly continuous right semigroup, and that A_- is its infinitesimal generator.

In the same way one proves that $E(t)M_+ \subset M_+$ for $t < 0$, and we define $E_+ : (-\infty, 0] \to \mathcal{L}(M_+)$ by setting

$$E_+(t) = -E(t)|_{M_+}, \qquad E_+(0)x = \lim_{t\uparrow 0} -E(t)x, \qquad x \in M_+.$$

Then the analogue of Theorem 5.3 for left semigroups shows that E_+ is a strongly continuous left semigroup which has $A_+(M_+ \to M_+)$ as its generator.

Next, consider the operator P on X defined by

$$Px = \lim_{t\uparrow 0} -E(t)x, \qquad x \in X.$$

By the Banach-Steinhaus theorem, P is a bounded linear operator on X. For $t < 0$ we have that $E(t)$ vanishes on M_-, and so $Px = 0$ for each $x \in M_-$. For $x \in M_+$ and $t < 0$ we have $E(t)x = -E_+(t)x$, and thus $Px = x$. These properties of P imply that

$$M_- \cap M_+ = \{0\} \text{ and } M_- + M_+ \text{ is closed.} \tag{5.22}$$

The first part of (5.22) is obvious. To prove the second part, let $x_1, x_2, \ldots$ be a sequence in M_-, let $y_1, y_2, \ldots$ be a sequence in M_+, and assume that $x_n + y_n \to z$ for $n \to \infty$. It suffices to show that $z \in M_- + M_+$. Since P is continuous on X and P is zero on M_-, we have

$$Pz = \lim_{n\to\infty} P(x_n + y_n) = \lim_{n\to\infty} Py_n.$$

But $Py_n = y_n \in M_+$ and M_+ is closed. Thus $Pz \in M_+$. Moreover, $y_n = Py_n$ converges to Pz if $n \to \infty$. Thus $x_n = (x_n+y_n)-y_n$ converges to $z-Pz$ if $n \to \infty$. Also, M_- is closed. We conclude that $z - Pz \in M_-$, and hence $z = z - Pz + Pz$ belongs to $M_- + M_+$. So $M_- + M_+$ is closed.

Finally, the first part of the proof of Theorem XV.3.1 in [51] shows that $M_- + M_+$ is dense in X. We conclude that $X = M_- \dot{+} M_+$, and that P is the projection of X along M_- onto M_+. Recall that $\mathcal{D}(A) = \operatorname{Im} T$, where $T = A^{-1}$. It follows that

$$\mathcal{D}(A) = TX = T(M_- \dot{+} M_+) = TM_- \dot{+} TM_+ = \mathcal{D}(A_-) \dot{+} \mathcal{D}(A_+).$$

Hence P maps $\mathcal{D}(A)$ into itself, and P commutes with A. Thus relative to the decompositions

$$X = M_- \dot{+} M_+, \qquad \mathcal{D}(A) = \mathcal{D}(A_-) \dot{+} \mathcal{D}(A_+),$$

the operator A admits the partitioning

$$A = \begin{bmatrix} A_- & 0 \\ 0 & A_+ \end{bmatrix}.$$

Therefore A is an exponentially dichotomous operator, P is the separating projection for A, and $E(\cdot) = E(\cdot\,; A)$. □

5.3 Realization triples

In this section we introduce the realizations that will be used to obtain representations of the type (5.3). We begin with some additional notation.

By $\mathbf{D}_1^m(\mathbb{R})$ we denote the linear submanifold of $L_1^m(\mathbb{R}) = L_1(\mathbb{R}, \mathbb{C}^m)$ consisting of all $f \in L_1^m(\mathbb{R})$ for which there exists $g \in L_1^m(\mathbb{R})$ such that

$$f(t) = \begin{cases} \displaystyle\int_{-\infty}^{t} g(s)\, ds, & \text{a.e. on } (-\infty, 0), \\ \displaystyle\int_{t}^{\infty} g(s)\, ds, & \text{a.e. on } (0, \infty). \end{cases} \tag{5.23}$$

If $f \in \mathbf{D}_1^m(\mathbb{R})$, then there is only one $g \in L_1^m(\mathbb{R})$ such that (5.23) holds. This g is called the *derivative* of f and is denoted by f'. From (5.23) it follows that $f(0+) = \lim_{t\downarrow 0} f(t)$ and $f(0-) = \lim_{t\uparrow 0} f(t)$ exist; in fact,

$$f(0+) = \int_0^{\infty} g(s)\, ds, \qquad f(0-) = \int_{-\infty}^{0} g(s)\, ds.$$

Let ω be a negative constant. A triple $\Theta = (A, B, C)$ of operators is called a *realization triple of exponential type* ω if the following conditions are satisfied:

(C1) $-iA$ is an exponentially dichotomous operator of exponential type ω with domain $\mathcal{D}(A)$ and range in a Banach space X;

(C2) $B : \mathbb{C}^m \to X$ is a linear operator;

(C3) C is a possibly unbounded operator with domain $\mathcal{D}(C)$ in X and range in $\mathbb{C}^m$ such that $\mathcal{D}(A) \subset \mathcal{D}(C)$ and C is A-bounded;

(C4) there exists a linear operator Λ_Θ from X into $L_1^m(\mathbb{R})$ such that

(i) $\displaystyle\sup_{\|x\|\leq 1} \int_{-\infty}^{\infty} e^{-\omega|t|}\|(\Lambda_\Theta x)(t)\|\,dt < \infty,$

(ii) for every $x \in \mathcal{D}(A)$ we have $(\Lambda_\Theta x)(t) = iCE(t; -iA)x$, $t \in \mathbb{R}$, and the function $\Lambda_\Theta x$ belongs to $\mathbf{D}_1^m(\mathbb{R})$.

In (ii), the function $E(t; -iA)$ is the bisemigroup generated by $-iA$. Note that B, being a linear operator from $\mathbb{C}^m$ into X, is automatically bounded. Observe also that (i) implies that Λ_Θ is bounded and maps X into $L_{1,\omega}^m(\mathbb{R})$ where

$$L_{1,\omega}^m(\mathbb{R}) = \{f \in L_1^m(\mathbb{R}) \mid e^{-\omega|\cdot|}f(\cdot) \in L_1^m(\mathbb{R})\}. \tag{5.24}$$

Taking into account (ii) and the fact that $\mathcal{D}(A)$ is dense in X, one sees that Λ_Θ is uniquely determined. Since ω is negative, $L_{1,\omega}^m(\mathbb{R})$ given by (5.24) is a linear manifold in $\mathbf{D}_1^m(\mathbb{R})$.

The space X is called the *state space* and the space $\mathbb{C}^m$ the *input/output space* of the triple. We shall refer to A as the *main operator* of the triple.

Suppose Θ is a realization triple of exponential type ω and $\omega \leq \omega_1 < 0$. Then Θ is a realization triple of exponential type ω_1 too. To see this, note that (i) and (ii) are fulfilled with ω replaced by ω_1. When the actual value of ω is not relevant, we simply call Θ a realization triple. Thus $\Theta = (A, B, C)$ is a *realization triple* if Θ is a realization triple of exponential type ω for some $\omega < 0$. The operator Λ_Θ does not depend on the value of ω, and the same is true with regard to the separating projection for $-iA$. This projection will be denoted by P_Θ, although it is defined in terms of A alone.

The case when C is a bounded linear operator from X into $\mathbb{C}^m$ is of special interest. In that case C is obviously A-bounded, and (C4) is fulfilled with $\Lambda_\Theta x = iCE(\cdot, -iA)x$ for each $x \in X$. Thus when C is bounded, then conditions (C3) and (C4) are automatically satisfied.

Let $\Theta = (A, B, C)$ be a realization triple with state space X. Notice that item (i) in (C4) implies that $\Lambda_\Theta : X \to L_1^m(\mathbb{R})$ is a bounded linear operator. Since (ii) prescribes Λ_Θ on $\mathcal{D}(A)$, the boundedness of Λ_Θ and the density of $\mathcal{D}(A)$ in X imply that Λ_Θ is uniquely determined by the operators A and C. The operator Λ_Θ plays the role of the observability operator in systems theory. For its dual analogue (the controllability operator) we refer to the following proposition.

Proposition 5.4. *Suppose $\Theta = (A, B, C)$ is a realization triple of exponential type $\omega < 0$, and let $\Gamma_\Theta : L_1^m(\mathbb{R}) \to X$ be defined by*

$$\Gamma_\Theta \varphi = \int_{-\infty}^{\infty} E(-t; -iA)B\varphi(t)\,dt, \qquad \varphi \in L_1^m(\mathbb{R}). \tag{5.25}$$

Then Γ_Θ is a bounded linear operator, and Γ_Θ maps $\mathbf{D}_1^m(\mathbb{R})$ into $\mathcal{D}(A)$.

Proof. The operator function $E(\cdot\,; -iA)$ is strongly continuous. Now recall the following well-known fact: if a sequence of operators converges in the strong operator topology, then the convergence is uniform on compact subsets of the underlying space. Because of the finite dimensionality of $\mathbb{C}^m$, the operator B is of finite rank, hence compact. It follows that the function $E(\cdot\,; -iA)B$ is continuous on $\mathbb{R} \setminus \{0\}$ with a possible jump at the origin where continuity is taken with respect the operator norm. It follows that the integral in (5.25) is well-defined for each $\varphi \in L_1^m(\mathbb{R})$, and that Γ_Θ is a bounded linear operator.

Now fix $\varphi \in \mathbf{D}_1^m(\mathbb{R})$. For simplicity we restrict ourselves to the case when φ vanishes almost everywhere on $(-\infty, 0)$. By our assumption on φ there exists $\psi \in L_1^m(\mathbb{R})$ such that

$$\varphi(t) = -\int_t^{\infty} \psi(s)\,ds, \qquad t > 0.$$

But then

$$\begin{aligned}
\Gamma_\Theta \varphi &= -\int_0^{\infty} E(-t; -iA)B\Big(\int_t^{\infty} \psi(s)\,ds\Big)\,dt \\
&= -\int_0^{\infty} \Big(\int_t^{\infty} E(-t; -iA)B\psi(s)\,ds\Big)\,dt \\
&= -\int_0^{\infty} \Big(\int_0^{s} E(-t; -iA)B\psi(s)\,dt\Big)\,ds.
\end{aligned}$$

The last equality follows by applying Fubini's theorem. Since A is exponentially dichotomous, zero belongs to the resolvent set of A. So it makes sense to consider the operator $iE(-t; -iA)A^{-1}B$. This function is differentiable on $[0, \infty)$, and its derivative is the continuous operator-valued function $-E(-t; -iA)B$. Here differentiation and continuity are taken with respect to the operator norm which we can use because of the compactness of B. Thus

$$-\int_0^{s} E(-t; -iA)B\,dt = iE(-s; -iA)A^{-1}B - iP_\Theta A^{-1}B,$$

where P_Θ is the separating projection of $-iA$. Hence

$$\begin{aligned}
\Gamma_\Theta \varphi &= \int_0^{\infty} \big(iE(-s; -iA)A^{-1}B - iP_\Theta A^{-1}B\big)\psi(s)\,ds \\
&= A^{-1}\Big(\int_0^{\infty} \big(iE(-s; -iA)B - iP_\Theta B\big)\psi(s)\,ds\Big).
\end{aligned}$$

This shows that $\Gamma_\Theta\varphi$ belongs to $\operatorname{Im} A^{-1} = \mathcal{D}(A)$. □

Let $\Theta = (A, B, C)$ be a realization triple of exponential type $\omega < 0$ and having input/output space $\mathbb{C}^m$. With Θ we associate two $m \times m$ matrix functions. These functions will be denoted by k_Θ and W_Θ, and they are called the *kernel function* associated with Θ and the *transfer function* of Θ, respectively. The first of these is defined as follows. For every u in $\mathbb{C}^m$, we have that $\Lambda_\Theta Bu$ belongs to $L^m_{1,\omega}(\mathbb{R})$. Thus the expression

$$k_\Theta(.)u = \big(\Lambda_\Theta Bu\big)(.), \qquad u \in \mathbb{C}^m, \tag{5.26}$$

determines a unique element k_Θ of $L^{m\times m}_{1,\omega}(\mathbb{R})$, that is each column of k_Θ belongs to $L^m_{1,\omega}(\mathbb{R})$. In fact $k_\Theta(.)u \in L^m_{1,\omega}(\mathbb{R}) \subset \mathbf{D}^m_1(\mathbb{R}) \subset L^m_1(\mathbb{R})$ for each $u \in \mathbb{C}^m$.

Next let us turn to W_Θ. This function is given by

$$W_\Theta(\lambda) = I + C(\lambda - A)^{-1}B, \qquad |\Im\lambda| < -\omega. \tag{5.27}$$

To see that W_Θ is well-defined, fix λ in the resolvent set $\rho(A)$ of A. Since the operator $(\lambda - A)^{-1}$ maps X into the domain $\mathcal{D}(A)$ of A, and $\mathcal{D}(A)$ is contained in the domain of C, the product $C(\lambda - A)^{-1}$ is well-defined. Hence $C(\lambda - A)^{-1}B$ is a well-defined linear transformation on $\mathbb{C}^n$. The fact that $-iA$ is an exponentially dichotomous operator of exponential type ω implies that $|\Re\lambda| < -\omega$ is contained in $\rho(-iA)$, and thus $|\Im\lambda| < -\omega$ is contained in $\rho(A)$. We conclude that W_Θ is a well-defined analytic $m \times m$ matrix function on $|\Im\lambda| < -\omega$.

The next proposition explains the relation between the two functions W_Θ and k_Θ.

Proposition 5.5. *Suppose $\Theta = (A, B, C)$ is a realization triple of exponential type $\omega < 0$. Then*

$$W_\Theta(\lambda) = I - \int_{-\infty}^{\infty} e^{i\lambda t} k_\Theta(t)\, dt, \qquad |\Im\lambda| < -\omega. \tag{5.28}$$

Proof. It suffices to show that for $x \in X$ and $|\Im\lambda| < -\omega$ we have

$$C(\lambda - A)^{-1}x = -\int_{-\infty}^{\infty} e^{i\lambda t}(\Lambda_\Theta x)(t)\, dt, \tag{5.29}$$

that is, $-C(\lambda - A)^{-1}x$ is equal to the Fourier transform $\widehat{(\Lambda_\Theta x)}(\lambda)$ of $\Lambda_\Theta x$. In what follows λ is fixed subject to $|\Im\lambda| < -\omega$.

We already know that $C(\lambda - A)^{-1}$ is a well-defined map from X into $\mathbb{C}^m$. Obviously, this map is linear. To show that it is also bounded, take $x \in X$. Using the fact that C is A-bounded, there exists a constant M such that

$$\|C(\lambda - A)^{-1}x\| \le M\big(\|(\lambda - A)^{-1}x\| + \|A(\lambda - A)^{-1}x\|\big).$$

Now $A(\lambda - A)^{-1}x = -x + \lambda(\lambda - A)^{-1}x$. Thus

$$\|C(\lambda - A)^{-1}x\| \le M\big(\|(\lambda - A)^{-1}\| + 1 + |\lambda|\|(\lambda - A)^{-1}\|\big)\|x\|.$$

It follows that $C(\lambda - A)^{-1}$ is a bounded linear operator from X into $\mathbb{C}^m$.

Now consider the map $x \mapsto \widehat{(\Lambda_\Theta x)}(\lambda)$ from X into $\mathbb{C}^m$. This map is linear and bounded too. Linearity is obvious. Boundedness follows from the estimate

$$\|\widehat{(\Lambda_\Theta x)}(\lambda)\| \leq \int_{-\infty}^{\infty} e^{-\omega|t|}\|\Lambda_\Theta x(t)\|\, dt,$$

together with condition (i) in the definition of a realization triple.

We have now shown that, for λ fixed, both sides of (5.29) are continuous in x. Hence it suffices to prove (5.29) for $x \in \mathcal{D}(A)$ because of $\overline{\mathcal{D}(A)} = X$.

Take $x \in \mathcal{D}(A)$, and put $y = Ax$. Since $-iA$ is an exponentially dichotomous operator of exponential type ω, we use (5.9) for $-iA$ in place of A and $-i\lambda$ in place of λ to show that

$$(\lambda - A)^{-1}y = -i\int_{-\infty}^{\infty} e^{i\lambda t}E(t;-iA)y\, dt. \tag{5.30}$$

Recall that CA^{-1} is a bounded linear operator. It follows that

$$\begin{aligned} C(\lambda - A)^{-1}x &= CA^{-1}(\lambda - A)^{-1}y \\ &= -i\int_{-\infty}^{\infty} e^{i\lambda t}CA^{-1}E(t;-iA)y\, dt \\ &= -i\int_{-\infty}^{\infty} e^{i\lambda t}CE(t;-iA)x\, dt \\ &= -i\int_{-\infty}^{\infty} e^{i\lambda t}(\Lambda_\Theta x)(t)\, dt, \end{aligned}$$

the latter equality holding by virtue of condition (ii) in the definition of a realization triple. Thus (5.29) is proved. □

From (5.29) it follows that $C(\lambda - A)^{-1}$ is analytic on $|\Im\lambda| < -\omega$. This result can also be proved directly using that C is A-bounded. In fact, employing the C-boundedness of A one can show that the function $\lambda \mapsto C(\lambda - A)^{-1}$ is analytic on the resolvent set $\rho(A)$.

5.4 Construction of realization triples

In this section we construct a representation of the form (5.3) for the $m \times m$ matrix-valued function W in (5.1) with the kernel function k being given by (5.2). The following theorem is the main result.

Theorem 5.6. *An $m \times m$ matrix function W is the transfer function of a realization triple if and only if W is of the form*

$$W(\lambda) = I - \int_{-\infty}^{\infty} e^{i\lambda t}k(t)\, dt, \tag{5.31}$$

where k is an $m \times m$ matrix function with the property that there exist $\omega < 0$ and $h \in L_1^{m\times m}(\mathbb{R})$ such that

$$k(t) = e^{\omega|t|}h(t). \tag{5.32}$$

If W is given by (5.31) and (5.32) for some $\omega < 0$ and $h \in L_1^{m\times m}(\mathbb{R})$, then $W = W_\Theta$ with $\Theta = (A, B, C)$ constructed in the following way: the state space X of Θ is $L_1^m(\mathbb{R})$, the input/output space is $\mathbb{C}^m$,

$$\mathcal{D}(A) = \mathcal{D}(C) = \mathbf{D}_1^m(\mathbb{R}),$$

$$(Af)(t) = \begin{cases} -i\omega f(t) + if'(t), & \text{a.e. on } -\infty < t < 0, \\ i\omega f(t) + if'(t), & \text{a.e. on } 0 < t < \infty, \end{cases}$$

$$(By)(t) = e^{-\omega|t|}k(t)y, \qquad \text{a.e. on } \mathbb{R},$$

$$Cf = i\int_{-\infty}^{\infty} f'(s)\,ds.$$

Proof. Let Θ be a realization triple, and let $W = W_\Theta$ be its transfer function. Then, by Proposition 5.5 in the preceding section, (5.31) holds with $k = k_\Theta$. Using the fact that the second operator in a realization triple is bounded, we see from (i) in the definition of a realization triple that

$$\sup_{\|y\|\leq 1}\int_{-\infty}^{\infty} e^{-\omega|t|}\|k_\Theta(t)y\|\,dt < \infty,$$

for some $\omega < 0$. Hence $k = k_\Theta$ satisfies (5.32). This proves the "if part" of the theorem.

Next, let W be given by (5.31) and (5.32) for some $\omega < 0$ and $h \in L_1^{m\times m}(\mathbb{R})$, and let $\Theta = (A, B, C)$ be the triple of operators defined in the second part of the theorem. We need to show that this triple is a realization triple and that $W = W_\Theta$.

As is well-known (cf., [51], page 420), the backward translation semigroup on $L_1^m[0,\infty)$ is strongly continuous. The infinitesimal generator of this semigroup has $\mathbf{D}_1^m[0,\infty)$ as its domain and its action amounts to taking the derivative. Here $\mathbf{D}_1^m[0,\infty)$ is the linear manifold consisting of all functions $f \in \mathbf{D}_1^m(\mathbb{R})$ with the property that $f(t) = 0$ for $t < 0$, and hence the derivative f' is well-defined for each $f \in \mathbf{D}_1^m[0,\infty)$. Using this, one sees that $-iA$ an exponentially dichotomous operator of exponential type ω and that the bisemigroup associated with $-iA$ acts as follows: for $t < 0$,

$$\big(E(t; -iA)f\big)(s) = \begin{cases} -e^{-\omega t}f(t+s), & \text{a.e. on } -\infty < s < 0, \\ 0, & \text{a.e. on } 0 < s < \infty, \end{cases}$$

and for $t > 0$,

$$\big(E(t; -iA)f\big)(s) = \begin{cases} 0, & \text{a.e. on } -\infty < s < 0, \\ e^{\omega t}f(t+s), & \text{a.e. on } 0 < s < \infty. \end{cases}$$

The separating projection for $-iA$ is the projection of the state space $X = L_1^m(\mathbb{R})$ onto $L_1^m(-\infty, 0]$ along $L_1^m[0, \infty)$.

Condition (5.32) on k implies that the operator B from $\mathbb{C}^m$ into $L_1^m(\mathbb{R})$ is bounded. From the definition of C and A we see that

$$\|Cf\| \leq -\omega\|f\| + \|Af\|, \qquad f \in \mathcal{D}(A).$$

Thus C is A-bounded.

Define $\Lambda : X \to L_1^m(\mathbb{R})$ by

$$(\Lambda f)(t) = e^{\omega|t|} f(t), \qquad \text{a.e. on } \mathbb{R}. \tag{5.33}$$

Then Λ satisfies the conditions (i) and (ii) in the definition of a realization triple with Λ in place of Λ_Θ. For (i) this is obvious. To check the first part of (ii), one uses the above description of the bisemigroup $E(t; -iA)$ and the definition of C. As to the second part of (ii), observe that $f \in \mathbf{D}_1^m(\mathbb{R})$ and $\omega < 0$ imply that the function $e^{\omega t} f(t)$ belongs to $\mathbf{D}_1^m(\mathbb{R})$ too.

We have now proved that $\Theta = (A, B, C)$ is a realization triple We claim that the kernel function k_Θ associated with Θ coincides with k. Indeed, for $y \in \mathbb{C}^m$ the following identities hold almost everywhere on $\mathbb{R}$:

$$k_\Theta(t)y \;=\; (\Lambda By)(t) \;=\; (e^{\omega|t|}By)(t) \;=\; k(t)y.$$

Since $\mathbb{C}^m$ has a finite basis, it follows that $k_\Theta(t) = k(t)$ almost everywhere on $\mathbb{R}$. In other words, k_Θ and k coincide as elements of $L_1^{m\times m}(\mathbb{R})$. □

5.5 Inverting matrix functions analytic in a strip

Let $\Theta = (A, B, C)$ be a realization triple with state space X. In this section we shall employ the operator $A^\times (X \to X)$. Here is the definition: the domain of $A^\times$ is equal to the domain of A, and its action is defined by $A^\times = A - BC$. We call $A^\times$ the *associate main operator* of the triple Θ. As one may expect from Section 2.4, the operator $A^\times$ plays an important role in inverting $W_\Theta(\lambda)$. In fact, we have the following theorem.

Theorem 5.7. *Let the $m \times m$ matrix function W be given by*

$$W(\lambda) = I + C(\lambda - A)^{-1}B,$$

with $\Theta = (A, B, C)$ being a realization triple. Let $A^\times$ be the associate main operator of Θ. Then $W(\lambda)$ is invertible for each $\lambda \in \mathbb{R}$ if and only if the spectrum of $A^\times$ does not intersect the real line. In that case $(A^\times, B, -C)$ is a realization triple, and

$$W(\lambda)^{-1} = I - C(\lambda - A^\times)^{-1}B, \qquad \lambda \in \mathbb{R}, \tag{5.34}$$

$$W(\lambda)C(\lambda - A^\times)^{-1} = C(\lambda - A)^{-1}, \qquad \lambda \in \mathbb{R}, \tag{5.35}$$

$$(\lambda - A^\times)^{-1}BW(\lambda) = (\lambda - A)^{-1}B, \qquad \lambda \in \mathbb{R}, \tag{5.36}$$

$$(\lambda - A^\times)^{-1} = (\lambda - A)^{-1} - (\lambda - A)^{-1}BW(\lambda)^{-1}C(\lambda - A)^{-1}, \qquad \lambda \in \mathbb{R}. \tag{5.37}$$

Proof. We split the proof into four parts. In the first part we show that $W(\lambda)$ is invertible for each $\lambda \in \mathbb{R}$ if and only if the spectrum of $A^\times$ does not intersect the real line, and we derive the expressions (5.34) – (5.37). The remaining three parts are concerned with the statement that $(A^\times, B, -C)$ is a realization triple.

Part 1. Suppose $A^\times$ has no spectrum on the real line. This condition means that for each real λ the linear operator $\lambda - A^\times$ maps $\mathcal{D}(A^\times) = \mathcal{D}(A)$ in a one-one way onto X, and hence the linear operator $I - C(\lambda - A^\times)^{-1}B$ acting on $\mathbb{C}^m$ is well-defined. We claim that it is the inverse of $W(\lambda)$. To see this we first prove (5.35). From $BCx = (A - A^\times)x$ for each $x \in \mathcal{D}(A)$, it follows that

$$BC(\lambda - A)^{-1} = (A - A^\times)(\lambda - A)^{-1}.$$

Using the latter identity and fixing $\lambda \in \mathbb{R}$, we obtain the equality (5.35) from the following calculation:

$$\begin{aligned} W(\lambda)C(\lambda - A^\times)^{-1} &= C(\lambda - A^\times)^{-1} + C(\lambda - A)^{-1}BC(\lambda - A)^{-1} \\ &= C(\lambda - A^\times)^{-1} + C(\lambda - A)^{-1}(A - A^\times)(\lambda - A)^{-1} \\ &= C(\lambda - A^\times)^{-1} + C(\lambda - A)^{-1}\big((A - \lambda) + (\lambda - A^\times)\big)(\lambda - A)^{-1} \\ &= C(\lambda - A)^{-1}. \end{aligned}$$

From (5.35) we obtain that

$$W(\lambda)\big(I - C(\lambda - A^\times)^{-1}B\big) = W(\lambda) - C(\lambda - A)^{-1} = I, \qquad \lambda \in \mathbb{R}.$$

Hence $W(\lambda)$ is invertible for each $\lambda \in \mathbb{R}$.

Next assume $W(\lambda)$ is invertible for each $\lambda \in \mathbb{R}$. We claim that $A^\times$ has no spectrum on the real line and that (5.37) holds. To prove this, fix $\lambda \in \mathbb{R}$ and let $R(\lambda)$ be the operator on X defined by the right-hand side of (5.37). Since $(\lambda - A^\times)(\lambda - A)^{-1} = I + BC(\lambda - A)^{-1}$, we have

$$\begin{aligned} (\lambda - A^\times)R(\lambda) &= I + BC(\lambda - A)^{-1} \\ &\qquad + \big(-I - BC(\lambda - A)^{-1}\big)BW(\lambda)^{-1}C(\lambda - A)^{-1} \\ &= I + BC(\lambda - A)^{-1} \\ &\qquad + B\big(-I - C(\lambda - A)^{-1}B\big)W(\lambda)^{-1}C(\lambda - A)^{-1} \\ &= I + BC(\lambda - A)^{-1} - BC(\lambda - A)^{-1}, \end{aligned}$$

and so $(\lambda - A^\times)R(\lambda) = I$. Thus to prove (5.37) it remains to show that $\lambda - A^\times$ is one-to-one.

Let $x \in \mathcal{D}(A^\times) = \mathcal{D}(A)$ and suppose $(\lambda - A^\times)x = 0$. Since $A^\times x = Ax - BCx$, we have $(\lambda - A)^{-1}BCx = -x$, and hence

$$W(\lambda)Cx = Cx + C(\lambda - A)^{-1}BCx = Cx - Cx = 0.$$

By assumption $W(\lambda)$ is invertible. Therefore $Cx = 0$ and, consequently, $(\lambda - A)x = (\lambda - A^\times)x = 0$. Now use the fact that A has no spectrum on the real line. It follows that $x = 0$, and hence $\lambda - A^\times$ is one-to-one.

Note that in passing we established (5.34), (5.35) and (5.37). The argument for (5.36) is analogous to that for (5.35).

In the remaining three parts it is assumed that $A^\times$ has no spectrum on the real line, or equivalently, that $W(\lambda)$ is invertible for each $\lambda \in \mathbb{R}$.

Part 2. We show that $A^\times$ is closed and that C is $A^\times$-bounded. Applying (5.37) with $\lambda = 0$ we see that

$$(A^\times)^{-1} = A^{-1} + A^{-1}BW(0)^{-1}CA^{-1}. \tag{5.38}$$

Since C is A-bounded, the operator CA^{-1} is bounded. Thus in the right-hand side of (5.38) the operators B, A^{-1} and CA^{-1} are all bounded. It follows that $(A^\times)^{-1}$ is bounded too. Hence $A^\times$ is a closed operator. Recall that the operators A^{-1} and $(A^\times)^{-1}$ map X into $\mathcal{D}(A) = \mathcal{D}(A^\times)$. Since the latter space is contained in $\mathcal{D}(C)$, we can apply C to both sides of (5.38). This yields

$$C(A^\times)^{-1} = CA^{-1} + CA^{-1}BW(0)^{-1}CA^{-1}.$$

But CA^{-1} is bounded. Hence $C(A^\times)^{-1}$ is bounded, which implies that C is $A^\times$-bounded.

Part 3. In this part we show that $-iA^\times$ is exponentially dichotomous. To do this we apply Theorem 5.2. First some preparations. Recall that

$$W(\lambda) = I - \int_{-\infty}^{\infty} e^{i\lambda t} k_\Theta(t)\, dt, \qquad \lambda \in \mathbb{R},$$

with k_Θ belonging to the space $e^{\omega|\cdot|}L_1^{m\times m}(\mathbb{R})$. By the matrix-valued version of Wiener's theorem (see, e.g., [52], page 830), the fact that $W(\lambda)$ is invertible for each $\lambda \in \mathbb{R}$ implies that

$$W(\lambda)^{-1} = I - \int_{-\infty}^{\infty} e^{i\lambda t} k^\times(t)\, dt, \qquad \lambda \in \mathbb{R}, \tag{5.39}$$

for some $k^\times \in L_1^{m\times m}(\mathbb{R})$. In fact (see [47], Section 18), taking $|\omega|$ smaller if necessary we may assume that $k^\times$ also belongs to $e^{\omega|\cdot|}L_1^{m\times m}(\mathbb{R})$. Next note that for each $x \in X$ and each $y \in \mathbb{C}^m$,

$$\begin{aligned}
C(\lambda - A)^{-1}x &= -i\int_{-\infty}^{\infty} e^{i\lambda t}(\Lambda_\Theta x)(t)\, dt, && \lambda \in \mathbb{R},\\
(\lambda - A)^{-1}x &= -i\int_{-\infty}^{\infty} e^{i\lambda t}E(t; -iA)x\, dt, && \lambda \in \mathbb{R},\\
(\lambda - A)^{-1}By &= -i\int_{-\infty}^{\infty} e^{i\lambda t}E(t; -iA)By\, dt, && \lambda \in \mathbb{R};
\end{aligned}$$

cf., (5.29) and (5.30). Using these formulas in (5.37), and taking inverse Fourier transforms, we see that

$$(\lambda - A^\times)^{-1}x = -i\int_{-\infty}^{\infty} e^{i\lambda t}\big(E(t;-iA) + E_1(t) + E_2(t)\big)x\,dt, \qquad \lambda\in\mathbb{R},$$

where for each $x\in X$ we have

$$E_1(t)x = i\int_{-\infty}^{\infty} E(t-s;-iA)B(\Lambda_\Theta x)(s)\,ds, \tag{5.40}$$

$$E_2(t)x = -i\int_{-\infty}^{\infty} E(t-s;-iA)B\left(\int_{-\infty}^{\infty} k^\times(s-r)(\Lambda_\Theta x)(r)\,dr\right)ds. \tag{5.41}$$

Recall that the function $E(\cdot\,;-iA)B$ is exponentially decaying, that $k^\times$ belongs to $e^{\omega|\cdot|}L_1^{m\times m}(\mathbb{R})$, and that for each $x\in X$ the function $\Lambda_\Theta x$ belongs to $e^{\omega|\cdot|}L_1^m(\mathbb{R})$. These facts imply that E_1 and E_2 are exponentially decaying too. Moreover, a routine argument shows that these functions are strongly continuous, that is, for each $x\in X$ the functions $E_1(\cdot)x$ and $E_2(\cdot)x$ are continuous in the norm of X. We conclude that the function $E(\cdot\,;-iA)+E_1(\cdot)+E_2(\cdot)$ is exponential decaying, strongly continuous on $\mathbb{R}\setminus\{0\}$, and that at zero it has (at worst) a jump discontinuity. But then we can apply Theorem 5.2 with A replaced by $-iA^\times$ and λ replaced by $-i\lambda$ to show that $-iA^\times$ is exponentially dichotomous. Furthermore, the bisemigroup generated by $-iA^\times$ is given by

$$E(\cdot\,;-iA^\times) = E(\cdot\,;-iA) + E_1(\cdot) + E_2(\cdot), \tag{5.42}$$

where $E(\cdot\,;-iA)$ is the bisemigroup generated by $-iA$, and the functions $E_1(\cdot)$ and $E_2(\cdot)$ are given by (5.40) and (5.41), respectively.

Part 4. In this part we complete the proof and show $\Theta^\times = (A^\times, B, -C)$ is a realization triple. The negative constant ω having been taken sufficiently close to zero, one has that Θ is of exponential type ω and $k^\times$ belongs to $e^{\omega|\cdot|}L_1^{m\times m}(\mathbb{R})$. A standard reasoning now shows that the convolution product $k^\times * (\Lambda_\Theta x)$, given by

$$\big(k^\times * (\Lambda_\Theta x)\big)(t) = \int_{-\infty}^{\infty} k^\times(t-s)(\Lambda_\Theta x)(s)\,ds, \qquad \text{a.e. on } \mathbb{R},$$

determines a bounded linear operator from X into $L_1^m(\mathbb{R})$ such that

$$\sup_{\|x\|\le 1}\int_{-\infty}^{\infty} e^{-\omega|t|}\|k^\times * (\Lambda_\Theta x)(t)\|\,dt < \infty.$$

But then the expression

$$\Lambda^\times x = -\Lambda_\Theta x + \big(k^\times * (\Lambda_\Theta x)\big) \tag{5.43}$$

defines a bounded linear operator $\Lambda^\times : X\to L_1^m(\mathbb{R})$ for which condition (i) in the definition of a realization triple (Section 5.3), with Λ_Θ replaced by $\Lambda^\times$, is satisfied.

Next, take $x \in X$, and consider the Fourier transform of $\Lambda^\times x$. Using formula (5.43) we see that for each $\lambda \in \mathbb{R}$ we have

$$\begin{aligned}\widehat{(\Lambda^\times x)}\lambda &= -\widehat{(\Lambda_\Theta x)}(\lambda) + \widehat{k^\times}(\lambda)\widehat{(\Lambda_\Theta x)}(\lambda)\\ &= \big(I - \widehat{k^\times}(\lambda)\big)\widehat{(\Lambda_\Theta x)}(\lambda)\\ &= -W(\lambda)^{-1}C(\lambda - A)^{-1}x.\end{aligned}$$

In this calculation the final equality results from (5.29) and (5.39). Next, using (5.35) we see that

$$\widehat{(\Lambda^\times x)}(\lambda) = C(\lambda - A^\times)^{-1}x, \qquad \lambda \in \mathbb{R}. \tag{5.44}$$

Note that this equality actually holds in a strip $|\Im\lambda| < -\omega$ containing the real line.

Now take $x \in \mathcal{D}(A^\times) = \mathcal{D}(A)$, and put $z = A^\times x$. Then $C(\lambda - A^\times)^{-1}x = C(A^\times)^{-1}(\lambda - A^\times)^{-1}z$, and the operator $C(A^\times)^{-1}$ is bounded by the result of the second part of the proof. Since $-iA^\times$ is exponentially dichotomous, by the third part of the proof, we can use formula (5.9), with A replaced by $-iA^\times$ and by $-i\lambda$, to show that

$$\begin{aligned}\widehat{(\Lambda^\times x)}(\lambda) &= -iC(A^\times)^{-1}\int_{-\infty}^{\infty} e^{i\lambda t}E(t;-iA^\times)z\,dt\\ &= -i\int_{-\infty}^{\infty} e^{i\lambda t}C(A^\times)^{-1}E(t;-iA^\times)z\,dt\\ &= -i\int_{-\infty}^{\infty} e^{i\lambda t}CE(t;-iA^\times)x\,dt, \qquad \lambda \in \mathbb{R}.\end{aligned}$$

Thus we have proved that $(\Lambda^\times x)(t) = -iCE(t;-iA^\times)x$ almost everywhere on $\mathbb{R}$. It remains to show that $\Lambda^\times x \in \mathbf{D}_1^m(\mathbb{R})$.

In view of the properties of Λ_Θ and the identity (5.43), it suffices to show that $k^\times * (\Lambda_\Theta x)$ belongs to $\mathbf{D}_1^m(\mathbb{R})$. Since $\Lambda_\Theta x = \mathbf{D}_1^m(\mathbb{R})$, we can consider its derivative g, that is the function given by

$$(\Lambda_\Theta x)(t) = \begin{cases} \displaystyle\int_{-\infty}^{t} g(s)\,ds, & \text{a.e. on } (-\infty, 0),\\[2ex] \displaystyle -\int_{t}^{\infty} g(s)\,ds, & \text{a.e. on } (0, -\infty).\end{cases}$$

Now use that

$$(k^\times * f)' = k^\times * f' + k^\times(\cdot)\big(f(0+) - f(0-)\big), \qquad f \in \mathbf{D}_1^m(\mathbb{R}).$$

If follows that

$$(k^\times * (\Lambda_\Theta x))(t) = \begin{cases} \displaystyle\int_{-\infty}^{t} h(s)\,ds, & \text{a.e. on } (-\infty, 0), \\ \displaystyle -\int_{t}^{\infty} h(s)\,ds < & \text{a.e. on } (0, -\infty), \end{cases}$$

where $h \in L_1^m(\mathbb{R})$ is given by

$$h = k^\times * g \;-\; k^\times(\cdot)\left(\int_{-\infty}^{\infty} g(s)\,ds\right).$$

This proves that $k^\times * (\Lambda_\Theta x) \in \mathbf{D}_1^m(\mathbb{R})$. Thus, with $\Lambda_{\Theta^\times} = \Lambda^\times$, we see that condition (C4) in the definition of a realization triple is satisfied. □

5.6 Inverting full line convolution operators

Let L be the convolution integral operator on $L_1^m(\mathbb{R})$ defined by

$$(Lf)(t) = \int_{-\infty}^{\infty} k(t-s)f(s)\,ds, \qquad \text{a.e. on } \mathbb{R}. \tag{5.45}$$

Here k is a kernel function of the form (5.2). As is well-known (see, e.g., Theorem XII.1.4 in [51]), the operator $I - L$ is invertible if and only if its symbol W, which is the $m \times m$ matrix function defined by (5.1), has the property that $W(\lambda)$ is invertible for each $\lambda \in \mathbb{R}$. Moreover we then have $(I - L)^{-1} = I - L^\times$, where $L^\times$ is the convolution integral operator on $L_1^m(\mathbb{R})$ given by

$$(L^\times g)(t) = \int_{-\infty}^{\infty} k^\times(t-s)f(s)\,ds, \qquad \text{a.e. on } \mathbb{R}, \tag{5.46}$$

the kernel function $k^\times$ of which is the inverse Fourier transform of $\big(I - W(\lambda)\big)^{-1}$, that is

$$\widehat{k^\times}(\lambda) = I - \big(I - \widehat{k}(\lambda)\big)^{-1}, \qquad \lambda \in \mathbb{R}. \tag{5.47}$$

Since the kernel function k in (5.45) is of the form (5.2), we know that $k = k_\Theta$ for some realization triple $\Theta = (A, B, C)$, and hence the symbol W of $I - L$ is the transfer function of this triple Θ. But then we can use the result of the previous section to restate the inversion theorem for $I - L$ in terms of Θ, and to give an explicit formula for $k^\times$ in terms of the operators A, B and C. The details are as follows.

Theorem 5.8. *Let L be the convolution integral operator on $L_1^m(\mathbb{R})$ given by* (5.45), *and assume that $k = k_\Theta$, where k_Θ is the kernel function associated with the realization triple $\Theta = (A, B, C)$. Then $I - L$ is invertible if and only if the spectrum* *of $A^\times$ does not intersect the real line. In that case $\Theta^\times = (-iA^\times, B, -C)$ is a realization triple and $(I-L)^{-1} = I - L^\times$, where $L^\times$ is given by (5.46) with*

$$k^\times(t)y = (\Lambda_{\Theta^\times} By)(t), \qquad \text{a.e. on } \mathbb{R},\ y \in \mathbb{C}^m. \tag{5.48}$$

Proof. Since $k = k_\Theta$, we have $W = W_\Theta$. Recall that $I - L$ is invertible if and only if $W(\lambda)$ is invertible for each $\lambda \in \mathbb{R}$. Thus Theorem 5.7 shows that $I - L$ is invertible if and only if the spectrum of $A^\times$ does not intersect the real line. Moreover, in that case $\Theta^\times = (-iA^\times, B, -C)$ is a realization triple and

$$W(\lambda)^{-1} = I - C(\lambda - A^\times)^{-1}B, \qquad \lambda \in \mathbb{R}.$$

Next we use Proposition 5.5 with $\Theta^\times$ in place of Θ. This yields

$$\widehat{k^\times}(\lambda) = I - \big(I - \widehat{k}(\lambda)\big)^{-1} = I - W(\lambda)^{-1} = \int_{-\infty}^{\infty} e^{i\lambda t} k_{\Theta^\times}(t)\, dt, \qquad \lambda \in \mathbb{R}.$$

Formula (5.48) now follows by applying (5.26) to $\Theta^\times$ in place of Θ. □

It is interesting to write $I - L$ as a 2×2 operator matrix relative to the decomposition $L_1^m(\mathbb{R}) = L_1^m[0,\infty) \dot{+} L_1^m(-\infty, 0]$. In particular, we will be interested in the first row and first column of this matrix. We have

$$I - L = \begin{bmatrix} I - K & L_+ \\ L_- & * \end{bmatrix},$$

with

$$\begin{aligned}
(K\varphi)(t) &= \int_0^\infty k(t-s)\varphi(s)\, ds, && \text{a.e. on } [0,\infty),\\
(L_-\varphi)(t) &= -\int_0^\infty k(t-s)\varphi(s)\, ds, && \text{a.e. on } (-\infty,0],\\
(L_+\psi)(t) &= -\int_{-\infty}^0 k(t-s)\psi(s)\, ds, && \text{a.e. on } [0,\infty).
\end{aligned}$$

Here φ belongs to $L_1^m[0,\infty)$ and ψ to $L_1^m(-\infty,0]$. The operator $I - K$ is called the *Wiener-Hopf operator* with kernel function k. The operators L_+ and L_- are known as Hankel operators (see, e.g., Section XII.2 in [51]). We call L_+ the *right Hankel operator* associated with k, and L_- will be referred to as the *left Hankel operator* associated with k. Notice that L_+ and L_- are uniquely determined by the restrictions of k to the half lines $[0,\infty)$ and $(-\infty,0]$, respectively. For later purpose we present the following lemma.

Lemma 5.9. *Let $\Theta = (A, B, C)$ be a realization triple, and let k_Θ be the associated kernel function. Then the right Hankel operator L_+ and left Hankel operator L_-*

associated with k_Θ are given by

$$L_+\psi = -Q\Lambda_\Theta\Gamma_\Theta\psi, \qquad \psi \in L_1^m(-\infty,0], \tag{5.49}$$

$$L_-\psi = (I-Q)\Lambda_\Theta\Gamma_\Theta\varphi, \qquad \varphi \in L_1^m[0,\infty,). \tag{5.50}$$

Here Q is the projection of $L_1^m(\mathbb{R})$ onto $L_1^m[0,\infty)$ along $L_1^m(-\infty,0]$.

Proof. We shall prove (5.49). The proof of (5.50) is similar to that of (5.49).

Let us first establish (5.49) for the case when $\operatorname{Im} B \subset \mathcal{D}(A)$. Then B can be written as $B = A^{-1}B_1$, where B_1 is a bounded linear operator from $\mathbb{C}^m$ into X. Write $C_1 = CA^{-1}$. Then $C_1 : X \to \mathbb{C}^m$ is a bounded linear operator too. For $y \in \mathbb{C}^m$, we have

$$k_\Theta(t)y = iCE(t;-iA)By = iC_1E(t;-iA)B_1y, \qquad \text{a.e. on } \mathbb{R}.$$

Since $\mathbb{C}^m$ has a finite basis, we may assume that k_Θ on all of $\mathbb{R}\setminus\{0\}$ can be represented as $k_\Theta(t) = iC_1E(t;-iA)B_1$. Take $\psi \in L_1^m(-\infty,0]$. Then $L_+\psi$ belongs to $L_1^m[0,\infty)$ and

$$(L_+\psi)(t) = -\int_{-\infty}^0 k_\Theta(t-s)\psi(s)\,ds = -\int_{-\infty}^0 iC_1E(t-s;-iA)B_1\psi(s)\,ds,$$

almost everywhere on $[0,\infty)$. Next we use the semigroup properties of the bisemigroup $E(\cdot;-iA)$ to show that

$$E(t-s;-iA) = E(t;-iA)E(-s;-iA), \qquad t>0,\ s<0.$$

It follows that, almost everywhere on $[0,\infty)$,

$$\begin{aligned}(L_+\psi)(t) &= -iC_1E(t;-iA)\int_{-\infty}^0 E(-s;-iA)B_1\psi(s)\,ds \\ &= -iCE(t;-iA)\int_{-\infty}^0 E(-s;-iA)B\psi(s)\,ds \\ &= (-Q\Lambda_\Theta\Gamma_\Theta\psi)(t),\end{aligned}$$

and (5.49) has been obtained for the case when $\operatorname{Im} B \subset \mathcal{D}(A)$.

The general situation, where $\operatorname{Im} B$ need not be contained in $\mathcal{D}(A)$, can be treated with an approximation argument based on the fact that B can be approximated (in norm) by bounded linear operators from $\mathbb{C}^m$ into X with ranges inside $\mathcal{D}(A)$. This is true because $\mathcal{D}(A)$ is dense in X and $\mathbb{C}^m$ is finite dimensional. □

5.7 Inverting Wiener-Hopf integral operators

In this section we study inversion of the Wiener-Hopf integral operator T:

$$Tf(t) = f(t) - \int_0^\infty k(t-s)f(s)\,ds, \qquad \text{a.e. on } [0,\infty). \tag{5.51}$$

It will be assumed that the $m \times m$ matrix kernel function k is the kernel function associated with some realization triple. This implies that T is a well-defined bounded linear operator on $L_1^m(\mathbb{R})$. We shall prove the following theorem.

Theorem 5.10. *Let T be the Wiener-Hopf integral operator on $L_1^m(\mathbb{R})$ given by (5.51). Assume that $k = k_\Theta$ for some realization triple $\Theta = (A, B, C)$. Then T is invertible if and only if the following two conditions are satisfied:*

(i) $\Theta^\times = (A^\times, B, -C)$ *is a realization triple,*

(ii) $X = \operatorname{Im} P_\Theta \dot{+} \operatorname{Ker} P_{\Theta^\times}$.

Here X is the state space of both Θ and $\Theta^\times$, and P_Θ and $P_{\Theta^\times}$ are the separating projections of $-iA$ and $-iA^\times$, respectively. If (i) *and* (ii) *hold, the inverse of T is given by*

$$\begin{aligned}(T^{-1}g)(t) &= g(t) - \int_0^\infty k_{\Theta^\times}(t-s)g(s)\,ds \\ &\quad - \int_0^\infty \Lambda_{\Theta^\times}(I-\Pi)E(-s,-iA^\times)Bg(s)\,ds(t), \qquad \text{a.e. on } [0,\infty).\end{aligned}$$

Here Π is the projection of X onto $\operatorname{Ker} P_{\Theta^\times}$ along $\operatorname{Im} P_\Theta$.

To facilitate the proof of Theorem 5.10 we first establish two lemmas. If Θ is a realization triple with main operator A, the separating projection of the operator $-iA$ will be denoted by P_Θ.

Lemma 5.11. *Let $\Theta = (A, B, C)$ and $\Theta^\times = (A^\times, B, -C)$ be realization triples with state space X. Then the operator*

$$J^\times : \operatorname{Im} P_\Theta \to \operatorname{Im} P_{\Theta^\times}, \qquad J^\times x = P_{\Theta^\times} x, \tag{5.52}$$

is invertible if and only if $X = \operatorname{Im} P_\Theta \dot{+} \operatorname{Ker} P_{\Theta^\times}$, and in that case

$$(J^\times)^{-1} = (I-\Pi)|_{\operatorname{Im} P_{\Theta^\times}}, \qquad \Pi = I - (J^\times)^{-1} P_{\Theta^\times}, \tag{5.53}$$

where Π is the projection of X along $\operatorname{Im} P_\Theta$ onto $\operatorname{Ker} P_{\Theta^\times}$.

Proof. Obviously $\operatorname{Ker} J^\times = \operatorname{Im} P_\Theta \cap \operatorname{Ker} P_{\Theta^\times}$. Thus $J^\times$ is one-to-one if and only if $\operatorname{Im} P_\Theta \cap \operatorname{Ker} P_{\Theta^\times} = \{0\}$. Next, assume $J^\times$ is surjective. Take $x \in X$. Then $P_{\Theta^\times} x = J^\times P_\Theta z = P_{\Theta^\times} P_\Theta z$ for some $z \in X$. This yields

$$\begin{aligned} x &= P_{\Theta^\times} x + (I - P_{\Theta^\times})x \\ &= P_{\Theta^\times} P_\Theta z + (I - P_{\Theta^\times})x \\ &= P_\Theta z + (I - P_{\Theta^\times})(x - P_\Theta z). \end{aligned}$$

Hence $x \in \operatorname{Im} P_\Theta + \operatorname{Ker} P_{\Theta^\times}$, and we conclude that $\operatorname{Im} P_\Theta + \operatorname{Ker} P_{\Theta^\times} = X$. Thus $X = \operatorname{Im} P_\Theta \dot{+} \operatorname{Ker} P_{\Theta^\times}$ provided that $J^\times$ is invertible. Moreover, the above calculations show that

$$\left(J^\times\right)^{-1} P_{\Theta^\times} x = P_\Theta z = (I-\Pi)x = (I-\Pi)P_{\Theta^\times} P_\Theta z = (I-\Pi)P_{\Theta^\times} x,$$

which proves the first identity in (5.53).

To complete the proof, assume $X = \operatorname{Im} P_\Theta \dot{+} \operatorname{Ker} P_{\Theta^\times}$. Then $J^\times$ is injective. To prove that $J^\times$ is surjective, take $y \in \operatorname{Im} P_{\Theta^\times}$. Since $P_{\Theta^\times} y = y$ and $P_{\Theta^\times}\Pi = 0$, we have

$$y = P_{\Theta^\times} y = P_{\Theta^\times}(I - \Pi)y + P_{\Theta^\times}\Pi y = P_{\Theta^\times}(I - \Pi)y.$$

Put $x = (I - \Pi)y$. Then $x \in \operatorname{Im} P_\Theta$ and $J^\times x = y$. This shows that $J^\times$ is surjective, and thus $J^\times$ is invertible. Moreover, we see that $(J^\times)^{-1}y = x = (I - \Pi)y$, which proves the second identity in (5.53). □

Lemma 5.12. *Assume that $\Theta = (A, B, C)$ and $\Theta^\times = (A^\times, B, -C)$ are realization triples, with $\mathbb{C}^m$ being the input/output space of both Θ and $\Theta^\times$. Introduce the maps*

$$K : L_1^m[0,\infty) \to L_1^m[0,\infty),$$
$$(K\varphi)(t) = \int_0^\infty k_\Theta(t-s)\varphi(s)\,ds, \quad \text{a.e. on } [0,\infty),$$

$$K^\times : L_1^m[0,\infty) \to L_1^m[0,\infty),$$
$$(K^\times\varphi)(t) = \int_0^\infty k_\Theta^\times(t-s)\varphi(s)\,ds, \quad \text{a.e. on } (-\infty,0]),$$

$$U : \operatorname{Im} P_{\Theta^\times} \to L_1^m[0,\infty), \qquad (Ux)(t) = (\Lambda_\Theta x)(t), \quad \text{a.e. on } [0,\infty),$$

$$U^\times : \operatorname{Im} P_\Theta \to L_1^m[0,\infty), \qquad (U^\times x)(t) = -(\Lambda_{\Theta^\times} x)(t), \quad \text{a.e. on } [0,\infty),$$

$$R : L_1^m[0,\infty) \to \operatorname{Im} P_\Theta, \qquad R\varphi = \int_0^\infty E(-t; -iA)B\varphi(t)\,dt,$$

$$R^\times : L_1^m[0,\infty) \to \operatorname{Im} P_{\Theta^\times}, \qquad R^\times\varphi = -\int_0^\infty E(-t; -iA^\times)B\varphi(t)\,dt,$$

$$J : \operatorname{Im} P_{\Theta^\times} \to \operatorname{Im} P_\Theta, \qquad Jx = P_\Theta x,$$

$$J^\times : \operatorname{Im} P_\Theta \to \operatorname{Im} P_{\Theta^\times}, \qquad J^\times x = P_{\Theta^\times} x.$$

Then all these operators are well-defined, linear and bounded. Moreover,

$$\begin{bmatrix} I-K & U \\ R & J \end{bmatrix} : L_1^m[0,\infty) \dot{+} \operatorname{Im} P_{\Theta^\times} \to L_1^m[0,\infty) \dot{+} \operatorname{Im} P_\Theta,$$

$$\begin{bmatrix} I-K^\times & U^\times \\ R^\times & J^\times \end{bmatrix} : L_1^m[0,\infty) \dot{+} \operatorname{Im} P_\Theta \to L_1^m[0,\infty) \dot{+} \operatorname{Im} P_{\Theta^\times},$$

are bounded linear operators, which are invertible, and

$$\begin{bmatrix} I-K & U \\ R & J \end{bmatrix}^{-1} = \begin{bmatrix} I-K^\times & U^\times \\ R^\times & J^\times \end{bmatrix}. \tag{5.54}$$

Proof. As we have seen in Section 5.6 the operators K and $K^\times$ are bounded. To see that the other operators are well-defined and bounded too it suffices to make a few observations. Let Q be the projection of $L_1^m(\mathbb{R})$ onto $L_1^m[0,\infty)$ along $L_1^m(-\infty,0]$. Then

$$U = Q\Lambda_\Theta|_{\operatorname{Im} P_{\Theta^\times}}, \qquad U^\times = -Q\Lambda_{\Theta^\times}|_{\operatorname{Im} P_\Theta},$$

and hence these two operators are well-defined and bounded. Next, viewing P_Θ and $P_{\Theta^\times}$ as operators from X onto $\operatorname{Im} P_\Theta$ and $\operatorname{Im} P_{\Theta^\times}$, respectively, we have

$$R = P_\Theta\Gamma_\Theta|_{L_1^m[0,\infty)}, \qquad R^\times = -P_{\Theta^\times}\Gamma_{\Theta^\times}|_{L_1^m[0,\infty)}.$$

From these identities and Proposition 5.4 it follows that R and $R^\times$ are also well-defined and bounded.

It remains to prove (5.54). This amounts to checking eight identities. Pairwise these identities have analogous proofs. So, actually only four identities have to be taken care of. This will be done in the remaining part of the proof which is divided into four steps.

Step 1. First we prove that $R(I-K^\times) + JR^\times = 0$. Take φ in $L_1^m[0,\infty)$. We need to show that $RK^\times\varphi = P_\Theta R^\times\varphi + R\varphi$. Whenever this is convenient, it may be assumed that φ is a continuous function with compact support in $(0,\infty)$. By applying Fubini's theorem, one gets

$$\begin{aligned} RK^\times\varphi &= \int_0^\infty \Big(\int_0^\infty E(-t;-iA)Bk_{\Theta^\times}(t-s)\varphi(s)\,ds\Big)\,dt \\ &= \int_0^\infty \Big(\int_0^\infty E(-t;-iA)Bk_{\Theta^\times}(t-s)\varphi(s)\,dt\Big)\,ds. \end{aligned}$$

For $s>0$ and $x\in X$, consider the identity

$$\begin{aligned} &\int_0^\infty E(-t;-iA)B(\Lambda_{\Theta^\times}x)(t-s)\,dt \\ &\qquad\qquad = E(-s;-iA)x - P_\Theta E(-s;-iA^\times)x. \end{aligned} \tag{5.55}$$

To prove it, we first take $x\in\mathcal{D}(A)=\mathcal{D}(A^\times)$. Then, for $t\neq 0$ and $t\neq s$,

$$\begin{aligned} &\frac{d}{dt}\big(E(-t;-iA)E(t-s;-iA^\times)x\big) \\ &\qquad = iE(-t;-iA)BCE(t-s;-iA^\times)x \\ &\qquad = iE(-t;-iA)BC(A^\times)^{-1}E(t-s;-iA^\times)A^\times x. \end{aligned}$$

Because $C(A^\times)^{-1}$ is bounded, the last expression is a continuous function of t on the intervals $[0,s]$ and $[s,\infty)$. It follows that (5.55) holds for $x \in \mathcal{D}(A)$. The validity of (5.55) for arbitrary $x \in X$ can now be obtained by a standard approximation argument based on the fact that $\mathcal{D}(A)$ is dense in X and the continuity of the operators involved. Substituting (5.55) in the expression for $RK^\times\varphi$, one immediately gets the desired identity $R(I - K^\times) + JR^\times = 0$.

Step 2. Next we show that $RU^\times + JJ^\times = I_{\operatorname{Im} P_\Theta}$. Take x in $\operatorname{Im} P_\Theta$. Then

$$RU^\times x = -\int_0^\infty E(-t; -iA)B(\Lambda_{\Theta^\times} x)(t)\,dt. \tag{5.56}$$

Apart from the minus sign, the right-hand side of (5.56) is exactly the same as the left-hand side of (5.55) for $s = 0$. It is easy to check that (5.55) also holds for $s = 0$, provided that the right-hand side is interpreted as $-P_\Theta x + P_\Theta P_{\Theta^\times} x$. Thus $RU^\times x = P_{\Theta^\times} x = x - P_\Theta P_{\Theta^\times} x$, and the desired identity $RU^\times + JJ^\times = I_{\operatorname{Im} P_\Theta}$ is proved.

Step 3. This step concerns the identity $(I - K)U^\times + UJ^\times = 0$. Take $x \in \operatorname{Im} P_\Theta$. Then $U^\times x = -Q\Lambda_{\Theta^\times} x$, where Q is the projection of $L_1^m(\mathbb{R})$ onto $L_1^m[0,\infty)$ along $L_1^m(-\infty, 0]$. Here the latter two spaces are considered as subspaces of $L_1^m(\mathbb{R})$. Observe now that $Q\Lambda_{\Theta^\times} = \Lambda_{\Theta^\times}(I - P_{\Theta^\times})x$. For $x \in \mathcal{D}(A) = \mathcal{D}(A^\times)$ this is evident, and for arbitrary x one can use an approximation argument. Hence $KU^\times x = Qh$, where $h = -k_\Theta * \big(\Lambda_{\Theta^\times}(I - P_{\Theta^\times})x\big)$, that is, h is the (full line) convolution product of $-k_\Theta$ and $\Lambda_{\Theta^\times}(I - P_{\Theta^\times})x$. Taking Fourier transforms, one gets

$$\begin{aligned}\widehat{h}(\lambda) &= C(\lambda - A)^{-1}BC(\lambda - A^\times)^{-1}(I - P_{\Theta^\times})x\\ &= C(\lambda - A)^{-1}(I - P_{\Theta^\times})x - C(\lambda - A^\times)^{-1}(I - P_{\Theta^\times})x.\end{aligned}$$

Put $g = U^\times x + UP_{\Theta^\times} x$. Since both U and $U^\times$ map into $\operatorname{Im} Q = L_1^m[0,\infty)$, we have $g = Qg$. Also $g = -\Lambda_{\Theta^\times}(I - P_{\Theta^\times})x + \Lambda_\Theta(I - P_\Theta)P_{\Theta^\times} x$, and hence

$$\widehat{g}(\lambda) = -C(\lambda - A^\times)^{-1}(I - P_{\Theta^\times})x \; - \; C(\lambda - A)^{-1}(I - P_\Theta)P_{\Theta^\times} x.$$

Since $x \in \operatorname{Im} P_\Theta$, it follows that $\widehat{h}(\lambda) - \widehat{g}(\lambda) = C(\lambda - A)^{-1}P_\Theta(I - P_{\Theta^\times})x$. So $\widehat{h}(\lambda) - \widehat{g}(\lambda)$ is the Fourier transform of $-\Lambda_\Theta P_\Theta(I - P_{\Theta^\times})x$. But then

$$h - g = -\Lambda_\Theta P_\Theta(I - P_{\Theta^\times})x = -(I - Q)\Lambda_\Theta(I - P_{\Theta^\times})x.$$

Applying Q to both sides of this identity, we get $Qh = Qg = g$. In other words, $KU^\times x = U^\times x + P_{\Theta^\times} x$ for all $x \in X$, and this is nothing else than the identity $(I - K)U^\times + UJ^\times = 0$.

Step 4. Finally, we prove $(I - K)(I - K^\times) + UR^\times = I$. Let L be the (full line) convolution integral operator associated with k_Θ, featured in Theorem 5.8. Since Θ and $\Theta^\times$ are both realization triples, the operator $I - L$ is invertible with inverse $(I - L)^{-1} = I - L^\times$, where $L^\times$ is the convolution integral operator associated with $\Theta^\times$. With respect to the decomposition $L_1^m(\mathbb{R}) = L_1^m[0,\infty) \dot{+} L_1^m(-\infty,0)$, we write $I - L$ and its inverse in the form

$$I - L = \begin{bmatrix} I-K & L_+ \\ * & * \end{bmatrix}, \qquad I - L^\times = \begin{bmatrix} I-K^\times & * \\ L_-^\times & * \end{bmatrix}. \tag{5.57}$$

Thus L_+ is the right Hankel operator associated with k_Θ, and the operator $L_-^\times$ is the left Hankel operator associated with $k_{\Theta^\times}$. But then Lemma 5.9 yields

$$L_+\psi = -Q\Lambda_\Theta\Gamma_\Theta\psi, \qquad \psi \in L_1^m(-\infty,0], \tag{5.58}$$

$$L_-^\times\varphi = (I-Q)\Lambda_{\Theta^\times}\Gamma_{\Theta^\times}\varphi, \qquad \varphi \in L_1^m[0,\infty). \tag{5.59}$$

Since $I - L^\times$ is the inverse of $I - L$, formula (5.57) shows that

$$(I-K)(I-K^\times) + L_+L_-^\times = I.$$

So, in order to get the desired identity, it suffices to show that $L_+L_-^\times = UR^\times$.

As was observed in the last paragraph of Step 2 of the present proof, (5.55) also holds for $s = 0$, that is

$$\int_0^\infty E(-t;-iA)B(\Lambda_{\Theta^\times}x)(t)\,dt = P_\Theta(I - P_{\Theta^\times})x, \qquad x \in X.$$

Analogously, one has

$$\int_{-\infty}^0 E(-t;-iA)B(\Lambda_{\Theta^\times}x)(t)\,dt = (I - P_\Theta)P_{\Theta^\times}x, \qquad x \in X.$$

Using the expressions for L_+ and $L_-^\times$ given in (5.58) and (5.59) we obtain

$$\begin{aligned} L_+L_-^\times\varphi &= -Q\Lambda_\Theta\Gamma_\Theta(I-Q)\Lambda_{\Theta^\times}\Gamma_{\Theta^\times}\varphi \\ &= -Q\Lambda_\Theta(I-P_\Theta)P_{\Theta^\times}\Gamma_{\Theta^\times}\varphi \\ &= UR_\Theta\varphi. \end{aligned}$$

Thus $(I-K)(I-K^\times) + UR^\times = I$ holds, and the lemma is proved. □

Following [13] (see also Section III.4 in [51]) we summarize the result of the preceding lemma by saying that the operators $I-K$ and $J^\times$ are *matricially coupled* with (5.54) being the *coupling relation*. The coupling relation is very useful. For instance, this relation and Corollary III.4.3 in [51] immediately yield the following result.

Corollary 5.13. *Let the operators K, $K^\times$, U, $U^\times$, R, $R^\times$, J and $J^\times$ be as in* (5.54). *Then $I-K$ is invertible if and only if $J^\times$ is invertible, and in that case*

$$(I-K)^{-1} = I - K^\times - U^\times(J^\times)^{-1}R^\times, \quad (J^\times)^{-1} = J - R(I-K)^{-1}U. \tag{5.60}$$

Proof of Theorem 5.10. We split the proof into two parts. In the first part we show that the invertibility of T implies that $\Theta^\times$ is a realization triple. In the second part we assume that $\Theta^\times$ is a realization triple and complete the proof by using Lemma 5.11 and Corollary 5.13.

Part 1. Since the kernel function k is equal to k_Θ, we know from Proposition 5.5 that the symbol of T is equal to W_Θ. Assume T is invertible. From the general theory of Wiener-Hopf operators we know that this assumption implies that $W_\Theta(\lambda)$ is invertible for all real λ. But then we can use the final part of Theorem 5.7 to show that $\Theta^\times$ is a realization triple.

Part 2. In this part we assume that $\Theta^\times$ is a realization triple. From Corollary 5.13 we know that $T = I - K$ is invertible if and only if $J^\times$ is invertible. By Lemma 5.11 the latter happens if and only if condition (ii) is satisfied. Together with the result of the first part we have now shown that T is invertible if and only if conditions (i) and (ii) are both fulfilled. Moreover, if these conditions are fulfilled we see from the first parts of formulas (5.60) and (5.53) that

$$(I - K)^{-1} = I - K^\times - U^\times(I - \Pi)R^\times,$$

where $K^\times$, $R^\times$ and $U^\times$ are as in Lemma 5.12, and Π is the projection of X along $\operatorname{Im} P_\Theta$ onto $\operatorname{Ker} P_{\Theta^\times}$. Using the definitions of the operators $K^\times$, $R^\times$ and $U^\times$ given in Lemma 5.12, the formula for T^{-1} presented in Theorem 5.10 is now clear. □

5.8 Explicit canonical factorization

In this section we use realization triples to construct a canonical factorization for an $m \times m$ matrix function W of the form (5.1) with k being given by (5.2). By Theorem 5.6 such a function is the transfer function of a realization triple $\Theta = (A, B, C)$. In what follows it is assumed that Θ is given. We present necessary and sufficient conditions for the existence of a canonical factorization in terms of the operators appearing in the realization triple, Also, supposing these conditions are fulfilled, we give formulas for the factors and their inverses in a canonical factorization of W. The main result (Theorem 5.14 below) is the natural analogue of Theorem 5.3 for the functions considered in this section. For the definition of a canonical factorization, see Section 1.1 (cf., also Section 3.1).

Theorem 5.14. *Let the $m \times m$ matrix function W be given by*

$$W(\lambda) = I + C(\lambda - A)^{-1}B,$$

with $\Theta = (A, B, C)$ a realization triple, and let $A^\times$ be the associate main operator of Θ. Then W admits a canonical factorization with respect to the real line if and only if the following two conditions are satisfied:

(i) *$\Theta^\times = (A^\times, B, -C)$ is a realization triple,*

(ii) $X = \operatorname{Im} P_\Theta \dot{+} \operatorname{Ker} P_{\Theta^\times}$.

Here X is the state space of both Θ and $\Theta^\times$, and P_Θ and $P_{\Theta^\times}$ are the separating projections of $-iA$ and $-iA^\times$, respectively. If the conditions (i) *and* (ii) *are satisfied, then the projection Π of X along* $\operatorname{Im} P_\Theta$ *onto* $\operatorname{Ker} P_{\Theta^\times}$ *maps $\mathcal{D}(A) = \mathcal{D}(A^\times)$ into itself, and a canonical factorization $W = W_-W_+$ with respect to the real line is given by*

$$W(\lambda) = W_-(\lambda)W_+(\lambda), \qquad \lambda \in \mathbb{R},$$

where the factors and their inverses can be written as

$$\begin{aligned} W_-(\lambda) &= I + C(\lambda - A)^{-1}(I - \Pi)B, \\ W_+(\lambda) &= I + C\Pi(\lambda - A)^{-1}B, \\ W_-^{-1}(\lambda) &= I - C(I - \Pi)(\lambda - A^\times)^{-1}B, \\ W_+^{-1}(\lambda) &= I - C(\lambda - A^\times)^{-1}\Pi B. \end{aligned}$$

The projection Π maps $\mathcal{D}(A) = \mathcal{D}(A^\times)$ into itself and $\mathcal{D}(A) \subset \mathcal{D}(C)$. Hence the right-hand sides of the first two of the above four expressions are well-defined on $\rho(A)$, and those of the last two are well-defined on $\rho(A^\times)$. In particular the formulas make sense for λ in a strip containing the real line. At first sight this seems to be short of the requirements for Wiener-Hopf factorization. We will come back to and resolve this point at the end of the proof.

Proof of Theorem 5.14. The proof will be divided into four parts. In the first we show that the conditions (i) and (ii) are necessary and sufficient. In the remaining three parts we assume that (i) and (ii) are satisfied.

Part 1. Let K be the Wiener-Hopf integral operator with kernel function k_Θ. Then the function W is the symbol of the operator $I - K$, and hence we know from the general theory of Wiener-Hopf integral equations that W admits a canonical factorization with respect to the real line if and only if $T = I - K$ is invertible. The first part of Theorem 5.10 implies that the latter is satisfied if and only if (i) and (ii) are fulfilled. Thus (i) and (ii) are necessary and sufficient in order that W admits a canonical factorization with respect to the real line.

In the remaining three parts of the proof we assume that conditions (i) *and* (ii) *are satisfied;* Π *will be the projection of X along* $\operatorname{Im} P_\Theta$ *onto* $\operatorname{Ker} P_{\Theta^\times}$.

Part 2. In this part we show that Π maps $\mathcal{D}(A)$ into itself. To do this we need the operator $J^\times$ defined by (5.52). Our hypotheses imply (see Lemma 5.11) that $J^\times$ is invertible and that $\Pi = I - (J^\times)^{-1}P_{\Theta^\times}$. Recall that $P_{\Theta^\times}$ maps $\mathcal{D}(A)$ into $\mathcal{D}(A) \cap \operatorname{Im} P_{\Theta^\times}$. Thus in order to prove that $\mathcal{D}(A)$ is invariant under Π it suffices to show that $(J^\times)^{-1}$ maps $\mathcal{D}(A) \cap \operatorname{Im} P_{\Theta^\times}$ into $\mathcal{D}(A)$. From the relation (5.54) and the invertibility of the operator $I - K$, it follows (see Corollary 5.13) that

$$(J^\times)^{-1} = J - R(I - K)^{-1}U,$$

where U, R and J are as in Lemma 5.12. Take $x \in \mathcal{D}(A) \cap \operatorname{Im} P_{\Theta^\times}$. Then $Ux = Q\Lambda_\Theta x \in \mathbf{D}_1^m[0,\infty)$, where Q is the projection of $L_1^m(\mathbb{R})$ onto $L_1^m[0,\infty)$ along $L_1^m(-\infty,0]$. From the general theory of Wiener-Hopf operators we know that

$$(I-K)^{-1} = (I+\Gamma_1)(I+\Gamma_2), \tag{5.61}$$

where for $j=1,2$ the operator Γ_j is the integral operator given by

$$(\Gamma_j\varphi)(t) = \int_0^\infty \gamma_j(t-s)\varphi(s)\,ds, \qquad t>0,$$

with $\gamma_j \in L_1^{m\times m}(\mathbb{R})$. In fact, γ_1 has its support in $[0,\infty)$ and γ_2 in $(-\infty,0]$; see Section 1.5. From the representation (5.61) it follows that $(I-K)^{-1}$ maps $\mathbf{D}_1^m[0,\infty)$ into itself. Note in this context that for $h \in L_1^{m\times m}(\mathbb{R})$ and $f \in \mathbf{D}_1^m(\mathbb{R})$, we have $h * f \in \mathbf{D}_1^m(\mathbb{R})$ and

$$(h*f)' = h*f' + h(\cdot)\big(f(0+)-f(0-)\big).$$

Thus $(I-K)^{-1}Ux \in \mathbf{D}_1^m(\mathbb{R})$. But then the final part of Proposition 5.4 tells us that we end up in $\mathcal{D}(A)$ by applying Γ_Θ. We conclude that $R(I-K)^{-1}U$ maps $\mathcal{D}(A) \cap \operatorname{Im} P_{\Theta^\times}$ into $\mathcal{D}(A)$. Since the separating projection P_Θ maps $\mathcal{D}(A)$ into itself, we know that J maps $\mathcal{D}(A) \cap \operatorname{Im} P_{\Theta^\times}$ also into $\mathcal{D}(A)$. Thus $(J^\times)^{-1}$ maps $\mathcal{D}(A) \cap \operatorname{Im} P_{\Theta^\times}$ into $\mathcal{D}(A)$, and hence Π maps $\mathcal{D}(A)$ into itself.

Amplifying on the above, we note that $J^\times$ maps $\mathcal{D}(A) \cap \operatorname{Im} P_\Theta$ in a one-to-one way onto $\mathcal{D}(A) \cap \operatorname{Im} P_{\Theta^\times}$. Since $J^\times$ is invertible, it suffices to show that $(J^\times)^{-1}$ maps $\mathcal{D}(A) \cap \operatorname{Im} P_{\Theta^\times}$ into $\mathcal{D}(A) \cap \operatorname{Im} P_\Theta$. We have already shown that $(J^\times)^{-1}$ maps $\mathcal{D}(A) \cap \operatorname{Im} P_{\Theta^\times}$ into $\mathcal{D}(A)$, and the inclusion $(J^\times)^{-1}\operatorname{Im} P_{\Theta^\times} \subset \operatorname{Im} P_\Theta$ is clear from the definition of $J^\times$. Thus $J^\times$ has the desired property.

Part 3. According to our hypotheses and the fact that Π maps $\mathcal{D}(A)$ into itself, we have the following direct sum decompositions:

$$X = \operatorname{Im} P_\Theta \dotplus \operatorname{Ker} P_{\Theta^\times}, \tag{5.62}$$

$$\mathcal{D}(A) = \big(\mathcal{D}(A) \cap \operatorname{Im} P_\Theta\big) \dotplus \big(\mathcal{D}(A) \cap \operatorname{Ker} P_{\Theta^\times}\big). \tag{5.63}$$

Write

$$A = \begin{bmatrix} A_1 & Z \\ 0 & A_2 \end{bmatrix}, \qquad B = \begin{bmatrix} B_1 \\ B_2 \end{bmatrix}, \qquad C = \begin{bmatrix} C_1 & C_2 \end{bmatrix} \tag{5.64}$$

for the corresponding matrix representations of A, B, and C. We now show that

$$\Theta_1 = (A_1, B_1, C_1), \qquad \Theta_1^\times = (A_1^\times, B_1, -C_1),$$

$$\Theta_2 = (A_2, B_2, C_2), \qquad \Theta_2^\times = (A_2^\times, B_2, -C_2),$$

are realization triples, and we analyze the spectral properties of their main operators. Here $A_1^\times = A_1 - B_1C_1$ and $A_2^\times = A_2 - B_2C_2$.

We start with Θ_1. Note that $A_1(\operatorname{Im} P_\Theta \to \operatorname{Im} P_\Theta)$ and $C_1(\operatorname{Im} P_\Theta \to \mathbb{C}^m)$ are the restrictions of A and C to $\mathcal{D}(A) \cap \operatorname{Im} P_\Theta$, respectively. Since P_Θ is the separating projection of Θ, this implies that Θ_1 is a realization triple. From the definition of A_1 it follows that $-iA_1$ is the infinitesimal generator of a strongly continuous left semigroup of negative exponential type. Thus the kernel function $k_1 = k_{\Theta_1}$ has its support in $(-\infty, 0]$ and

$$W_1(\lambda) = I - \widehat{k_1}(\lambda) = I + C_1(\lambda - A_1)^{-1}$$

is defined and analytic on an open half plane of the type $\operatorname{Im} \lambda < -\omega$ with ω strictly negative.

Next, we consider $\Theta_1^\times$. Let $J^\times : \operatorname{Im} P_\Theta \to \operatorname{Im} P_{\Theta^\times}$ be the operator defined by (5.52). We know that $J^\times$ is invertible, mapping $\mathcal{D}(A) \cap \operatorname{Im} P_\Theta$ onto $\mathcal{D}(A) \cap \operatorname{Im} P_{\Theta^\times}$. It is easy to check that $J^\times$ provides a similarity between the operator $A_1^\times$ and the restriction of $A^\times$ to $\mathcal{D}(A^\times) \cap \operatorname{Im} P_{\Theta^\times}$. Hence $iA_1^\times$ is the infinitesimal generator of a strongly continuous left semigroup of negative exponential type. But then Theorem 5.7 guarantees that $\Theta_1^\times$ is a realization triple. Furthermore , the kernel function $k_1^\times$ associated with $\Theta_1^\times$ has its support in $(-\infty, 0]$, and

$$W_1(\lambda)^{-1} = I - \widehat{k_1^\times}(\lambda) = I - C_1(\lambda - A_1^\times)^{-1} B_1$$

for all λ with $\operatorname{Im} \lambda < -\omega$. Here it is assumed that the negative constant ω has been taken sufficiently close to zero.

We proceed by considering Θ_2 and $\Theta_2^\times$. Obviously $\Theta_2^\times$ is a realization triple, and a similarity argument of the type presented above yields that the same is true for Θ_2. The operators $-iA_2$ and $-iA_2^\times$ are infinitesimal generators of strongly continuous right semigroups of negative exponential type. Hence the kernel functions k_2 and $k_2^\times$ associated with Θ_2 and $\Theta_2^\times$, respectively, have their support in $[0, \infty)$. Finally, taking $|\omega|$ smaller if necessary, we have that

$$W_2(\lambda) = I - \widehat{k_2}(\lambda) = I + C_2(\lambda - A_2)^{-1} B_2$$

and

$$W_2(\lambda)^{-1} = I - \widehat{k_2^\times}(\lambda) = I - C_2(\lambda - A_2^\times)^{-1} B_2$$

are defined and analytic on $\operatorname{Im} \lambda > -\omega$.

We may assume that both Θ and $\Theta^\times$ are of exponential type ω. For values of λ with $|\Im \lambda| < -\omega$ one then has

$$\begin{aligned} W(\lambda) = I + C_1(\lambda - A_1)B_1^{-1} + C_2(\lambda - A_2)^{-1} B_2 \\ + C_1(\lambda - A_1)^{-1} Z (\lambda - A_2)^{-1} B_2. \end{aligned}$$

Now $\operatorname{Ker} P_\Theta^\times$ is an invariant subspace for

$$A^\times = \begin{bmatrix} A_1^\times & Z - B_1 C_2 \\ -B_2 C_1 & A_2^\times \end{bmatrix},$$

and so $Z = B_1C_2$. Substituting this in the above expression for $W(\lambda)$, we get $W(\lambda) = W_1(\lambda)W_2(\lambda)$. Clearly this is a canonical Wiener-Hopf factorization.

The expressions obtained for the factors and their inverses are not quite the same as those given in the theorem. One verifies without difficulty, however, that for λ in the intersection of $\rho(A)$ and $\rho(A^\times)$, they amount to the same. For further information on this point we refer again (see the proof of Theorem 3.2) to Section 2.5 in [20] where the case when all three operators A, B and C are bounded is analyzed in great detail. □

Inspired by the terminology used in [20] (see also [11], Section 1.1), we introduce some additional terminology and notation. Let $\Theta = (A, B, C)$ be a realization triple with state space X, and let Π be a projection of X which maps $\mathcal{D}(A)$ into itself. We then have

$$\begin{aligned} X &= \operatorname{Ker}\Pi \dot{+} \operatorname{Im}\Pi, \\ \mathcal{D}(A) &= \big(\mathcal{D}(A) \cap \operatorname{Ker}\Pi\big) \dot{+} \big(\mathcal{D}(A) \cap \operatorname{Im}\Pi\big), \end{aligned}$$

and with respect to these decompositions the operators A, B and C have the form

$$A = \begin{bmatrix} A_{11} & A_{12} \\ A_{21} & A_{22} \end{bmatrix}, \qquad B = \begin{bmatrix} B_1 \\ B_2 \end{bmatrix}, \qquad C = \begin{bmatrix} C_1 & C_2 \end{bmatrix}.$$

The triple (A_{22}, B_2, C_2) will be called the *projection of* $\Theta = (A, B, C)$ *associated with* Π, and it is denoted by $\mathrm{pr}_\Pi(\Theta)$. Note that (A_{11}, B_1, C_1) is the projection $\mathrm{pr}_{I-\Pi}(\Theta)$ of (A, B, C) associated with the projection $I - \Pi$. A particularly interesting case for what follows is when Π is a *supporting projection* for Θ. This means that besides the Π-invariance of $\mathcal{D}(A) = \mathcal{D}(A^\times)$ the following inclusion relations are satisfied:

$$A\big[\mathcal{D}(A) \cap \operatorname{Ker}\Pi\big] \subset \operatorname{Ker}\Pi, \qquad A^\times\big[\mathcal{D}(A^\times) \cap \operatorname{Im}\Pi\big] \subset \operatorname{Im}\Pi.$$

In that situation we have $A_{12} = B_1C_2$ and $A_{21} = 0$. Also Π is a supporting projection for the realization triple $\Theta = (A, B, C)$ if and only $I - \Pi$ is a supporting projection for $\Theta^\times = (A^\times, B, -C)$. Finally, if Π is supporting for Θ, the arguments used in Part 3 of the proof of Theorem 5.14 show that $\mathrm{pr}_\Pi(\Theta)$ and $\mathrm{pr}_{I-\Pi}(\Theta)$ are again realization triples.

With this notation and terminology we have the following alternative version of Theorem 5.10.

Theorem 5.15. *Let T be the Wiener-Hopf integral operator on $L_1^m(\mathbb{R})$ given by (5.51). Assume that $k = k_\Theta$ for some realization triple $\Theta = (A, B, C)$. Then T is invertible if and only if the following two conditions are satisfied:*

(i) *$\Theta^\times = (A^\times, B, -C)$ is a realization triple,*

(ii) $X = \operatorname{Im} P_\Theta \dot{+} \operatorname{Ker} P_{\Theta^\times}$.

Here X is the state space of both Θ and $\Theta^\times$, and P_Θ and $P_{\Theta^\times}$ are the separating projections of $-iA$ and $-iA^\times$, respectively. If (i) and (ii) hold, then the projection Π of X onto $\operatorname{Ker} P_{\Theta^\times}$ along $\operatorname{Im} P_\Theta$ is a supporting projection for Θ, the complementary projection $I - \Pi$ is a supporting projection for $\Theta^\times$, and

$$(T^{-1}g)(t) = g(t) - \int_0^\infty \gamma(t,s)g(s)\,ds.$$

Here γ is given by

$$\gamma(t,s) = \begin{cases} k_+^\times(t-s) - \displaystyle\int_0^s k_+^\times(t-r)k_-^\times(r-s)\,dr, & s < t, \\ k_-^\times(t-s) - \displaystyle\int_0^t k_+^\times(t-r)k_-^\times(r-s)\,dr, & s > t, \end{cases}$$

where $k_+^\times$ and $k_-^\times$ are the kernel functions associated with the realization triples $\operatorname{pr}_\Pi(\Theta^\times)$ and $\operatorname{pr}_{I-\Pi}(\Theta^\times)$, respectively.

5.9 The Riemann-Hilbert boundary value problem revisited (2)

In this section we deal with the Riemann-Hilbert boundary value problem on the real line for matrix functions W of the form

$$W(\lambda) = I - \int_{-\infty}^\infty e^{i\lambda t}k(t)\,dt, \tag{5.65}$$

where k is an $m \times m$ matrix-valued function with the property that for some $\omega < 0$ the entries of $e^{-\omega|t|}k(t)$ are Lebesgue integrable on the real line. In this case W is analytic in a strip around the real axis. For such a function the Riemann-Hilbert problem consists of finding pairs Φ_+, Φ_- of $\mathbb{C}^m$-valued functions on the real line such that

$$W(\lambda)\Phi_+(\lambda) = \Phi_-(\lambda), \qquad -\infty < \lambda < \infty \tag{5.66}$$

while, in addition, Φ_+ and Φ_- are Fourier transforms of integrable $\mathbb{C}^m$-valued functions with support in $[0,\infty)$ and $(-\infty,0]$, respectively. These requirements imply that Φ_+ and Φ_- both vanish at infinity and that they are continuous on the closed upper and closed lower half plane, respectively.

From the special form of k in (5.65) we know that W is the transfer function of some realization triple $\Theta = (A,B,C)$. The following theorem gives the solution of the Riemann-Hilbert problem for W in terms of the operators A, B and C appearing in the triple.

Theorem 5.16. *Let W be the transfer function of realization triple $\Theta = (A,B,C)$. Assume $\Theta^\times = (A^\times, B, C)$ is a realization triple too (or, equivalently, that $W(\lambda)$*

is invertible for all $\lambda \in \mathbb{R}$). *Write* P_Θ *and* $P_{\Theta^\times}$ *for the separating projections of* $-iA$ *and* $-iA^\times$, *respectively. Then the pair of functions* Φ_+, Φ_- *is a solution of the Riemann-Hilbert boundary value problem* (5.66) *if and only if there exists* $x \in \operatorname{Im} P_\Theta \cap \operatorname{Ker} P_{\Theta^\times}$ *such that*

$$\Phi_+(\lambda) \;=\; C(\lambda - A^\times)^{-1}x = \int_0^\infty e^{i\lambda t}(\Lambda_\Theta^\times x)(t)\,dt, \tag{5.67}$$

$$\Phi_-(\lambda) \;=\; C(\lambda - A)^{-1}x \;= -\int_{-\infty}^0 e^{i\lambda t}(\Lambda_\Theta x)(t)\,dt. \tag{5.68}$$

Moreover the vector x *is uniquely determined by the functions* Φ_+, Φ_-.

Proof. Take $x \in \operatorname{Im} P_\Theta \cap \operatorname{Ker} P_{\Theta^\times}$. Condition (C4) in the definition of a realization triple implies that $(\Lambda_\Theta^\times x)(t)$ is zero almost everywhere on the half line $-\infty < t \le 0$, while $(\Lambda_\Theta x)(t)$ is zero almost everywhere on $0 \le t < \infty$. It follows that we can apply (5.29) to both Θ and $\Theta^\times$ in order to show that

$$\int_0^\infty e^{i\lambda t}(\Lambda_\Theta^\times x)(t)\,dt \;=\; \int_{-\infty}^\infty e^{i\lambda t}(\Lambda_\Theta^\times x)(t)\,dt \;=\; C(\lambda - A^\times)^{-1}x,$$

$$\int_\infty^0 e^{i\lambda t}(\Lambda_\Theta x)(t)\,dt \;=\; \int_{-\infty}^\infty e^{i\lambda t}(\Lambda_\Theta x)(t)\,dt \;=\; -C(\lambda - A)^{-1}x.$$

Thus the functions Φ_+ and Φ_- in (5.67) and (5.68) are well-defined. Furthermore, these functions are Fourier transforms of integrable $\mathbb{C}^m$-valued functions with support in $[0,\infty)$ and $(-\infty,0]$, respectively. From (5.35) we see that (5.66) is satisfied. Thus the pair Φ_+, Φ_- is a solution of the Riemann-Hilbert problem.

To prove the reverse implication, assume that the pair Φ_+, Φ_- is a solution of the Riemann-Hilbert problem (5.66). Write Φ_+ and Φ_- in the form

$$\Phi_+(\lambda) = \int_0^\infty e^{i\lambda t}\phi_+(t)\,dt, \qquad \Phi_-(\lambda) = \int_{-\infty}^0 e^{i\lambda t}\phi_-(t)\,dt,$$

where $\phi_+ \in L_1^m[0,\infty)$ and $\phi_- \in L_1^m(-\infty,0]$. Now, let k_Θ be the kernel function associated with Θ, and consider the integral operator on $L_1^m[0,\infty)$ defined by

$$(Kf)(t) = \int_0^\infty k_\Theta(t-s)f(s)\,ds, \qquad \text{a.e. on } [0,\infty).$$

Using (5.66) and the fact that $W(\lambda) = I_m - \widehat{k_\Theta}(\lambda)$, a routine argument yields that $\phi_+ - K\phi_+ = 0$. In other words, $\phi_+ \in \operatorname{Ker}(I-K)$. Next, we use the coupling relation (5.54) together with Corollary 4.3 in Section III.4 of [51]. It follows that $\phi_+ = U^\times x$ for some x in $\operatorname{Ker} J^\times$, where

$$U^\times : \operatorname{Im} P_\Theta \to L_1^m[0,\infty), \qquad (U^\times x)(t) = -(\Lambda_{\Theta^\times} x)(t), \quad \text{a.e. on } [0,\infty),$$

$$J^\times : \operatorname{Im} P_\Theta \to \operatorname{Im} P_{\Theta^\times}, \qquad J^\times x = P_{\Theta^\times} x.$$

Obviously, $\operatorname{Ker} J^\times = \operatorname{Im} P_\Theta \cap \operatorname{Ker} P_{\Theta^\times}$. Thus there exists $x \in \operatorname{Im} P_\Theta \cap \operatorname{Ker} P_{\Theta^\times}$ such that $\phi_+(t) = -(\Lambda_{\Theta^\times} x)(t)$ a.e. on the half line $0 \le t < \infty$. But then (5.67) is satisfied. By (5.66), the identity (5.29) applied to $\Theta^\times = (A^\times, B, -C)$, and (5.35) we have $\Phi_-(\lambda) = W(\lambda)\Phi_+(\lambda) = C(\lambda - A)^{-1}x$. Hence (5.68) holds too.

It remains to prove the uniqueness of the vector x in (5.67) and (5.68). Assume that x' is a second vector with the same properties as the vector x. So $x' \in \operatorname{Im} P_\Theta \cap \operatorname{Ker} P_{\Theta^\times}$ while (5.67) and (5.68) hold true with x' in place of x. Let $J^\times$ and $U^\times$ be as in the previous paragraph. Since $\operatorname{Ker} J^\times = \operatorname{Im} P_\Theta \cap \operatorname{Ker} P_{\Theta^\times}$, we have $x - x' \in \operatorname{Ker} J^\times$. Furthermore, the fact that the left-hand side of (5.67) does not depend on x nor on x' yields that $(\Lambda_\Theta^\times x)(t) = (\Lambda_\Theta^\times x')(t)$ a.e. on $[0, \infty)$. Thus $U^\times x = U^\times x'$. It follows that both $J^\times(x - x')$ and $U^\times(x - x')$ are equal to zero. If $x \neq x'$, this implies that the operator defined by the right-hand side of (5.54) is not invertible, which is impossible by Lemma 5.12. We conclude that $x = x'$, as desired. □

Notes

The material presented in this chapter is taken from the papers [16] and [15]. In [16] the reader will also find a systematic treatment of realization triples (A, B, C) with C bounded and A unbounded. The notion of an exponentially dichotomous operator, which has been introduced in [16], has proved to be quite useful in other areas. See, e.g., the papers [22] and [93]. The theory of realization triples is also used in [14] and [92]. The papers [90] and [91] present an extension of the theory of realization triples to operator-valued functions by introducing two-sided Pritchard-Salomon realizations. In particular, the factorization theory of Section 5.8 is developed further in [91].

For more information on exponentially dichotomous operators, including various perturbation theorems and a wide variety of applications, we refer to the monograph [111]. See also the notes to Chapter 6.

Chapter 6

Convolution equations and the transport equation

In this chapter the factorization theory developed in the previous chapters is applied to solve a linear transport equation. It is known that the transport equation may be transformed into a Wiener-Hopf integral equation with an operator-valued kernel function (see [40]). An equation of the latter type can be solved explicitly if a canonical factorization of its symbol is available (cf., Sections 1.1 and 3.2). In our case the symbol may be represented as a transfer function, and to make the factorization the general factorization theorem of the second chapter can be applied. This requires that one finds an appropriate pair of invariant subspaces. In the case of the transport equation the choice of the subspaces is evident, but to prove that their direct sum is the whole space takes some effort. The latter is related to a new difficulty that appears here. Namely, in this case the curve cuts through the spectra of the main operator and the associate main operator. Nevertheless, due to the special structure of the operators involved, the factorization can be made and explicit formulas are obtained.

Since our main purpose is to show how our method works, we restrict ourselves to the case when the kernel function describing the effect of the scattering is of finite rank.

In Section 6.1 we describe the transport equation that is considered in this chapter. To illustrate our approach we first study (in Section 6.2) a simplified model, namely when the scattering appears only in a finite number of directions. In Section 6.3 the vector-valued Wiener-Hopf equation associated to the transport equation is introduced. In Section 6.4 it is shown that under appropriate conditions a canonical factorization of the symbol associated with the equation can be constructed, and the matching of corresponding invariant subspaces is established in Section 6.5. In Section 6.6, the final section of the chapter, we present formulas for the solution.

6.1 The transport equation

Transport theory is a branch of mathematical physics concerned with the mathematical analysis of equations that describe the migration of particles in a medium, for instance, a flow of electrons through a metal strip or radiative transfer in a stellar atmosphere.

For the plane symmetric case, a stationary transport problem through a homogeneous medium can be modelled by an integro-differential equation of the following form:

$$\mu\frac{\partial\psi(t,\mu)}{\partial t} + \psi(t,\mu) = \int_{-1}^{1} k(\mu,\mu')\psi(t,\mu')\,d\mu', \qquad t \geq 0. \tag{6.1}$$

This equation is a balance equation. The unknown function ψ is a density function related to the expected number of particles in an infinitesimal volume element. The right-hand side describes the effect of the collisions. The variable μ is equal to $\cos\alpha$ where α is the scattering angle, and therefore $-1 \leq \mu \leq 1$. The variable t is not a time variable but a position variable, sometimes referred to as the optical depth. The kernel function k in the right-hand side of (6.1), which is called the *scattering function*, is assumed to be a real symmetric L_1-function on $[-1,1]\times[-1,1]$.

We shall consider the so-called *half range problem*, that is, we assume the medium to be semi-infinite, and hence the position variable runs over the interval $0 \leq t < \infty$. Since the density of the incoming particles is known, the values of $\psi(0,\mu)$ are known for $0 < \mu \leq 1$. Thus the above equation will be considered together with the boundary condition

$$\psi(0,\mu) = f_+(\mu), \qquad 0 < \mu \leq 1, \tag{6.2}$$

where f_+ is a given function on (0,1]. In the sequel we shall consider f_+ as a function on $[-1,1]$ by setting $f_+(\mu) = 0$ for $-1 \leq \mu < 0$, and we assume that $f_+ \in L_2[-1,1]$.

There is also a boundary condition at infinity, which appears in different forms. Here we take the condition at infinity to be

$$\lim_{t\to\infty} \psi(t,\mu)\exp\left(\frac{t}{\mu}\right) = 0, \qquad -1 \leq \mu < 0. \tag{6.3}$$

Thus the problem is to solve (6.1) under the boundary conditions (6.2) and (6.3).

In this chapter we shall assume (cf., [81], [82] and [108]) that the scattering function k is given by

$$k(\mu,\mu') = \sum_{j=0}^{n} a_j p_j(\mu)p_j(\mu'), \tag{6.4}$$

where $p_j(\mu)$ is the j-th normalized Legendre polynomial (see [53], page 26) and

$$-\infty < a_j < 1, \qquad j = 0,1,\ldots,n. \tag{6.5}$$

In particular, the integral operator defined by the right-hand side of (6.1) has finite rank.

By writing $\psi(t)(\mu) = \psi(t, \mu)$, we may consider the unknown function ψ as a vector function on $[0, \infty)$ with values in $\mathcal{H} = L_2[-1, 1]$. In this way equation (6.1) can be written as an operator differential equation:

$$T\frac{d\psi}{dt}(t) + \psi(t) = K\psi(t), \qquad t \geq 0, \tag{6.6}$$

where the derivative is taken with respect to the norm in $\mathcal{H}$. In (6.6) the operators T and K are defined by

$$(Tf)(\mu) = \mu f(\mu), \qquad Kf = \sum_{j=0}^{n} a_j \langle f, p_j \rangle p_j. \tag{6.7}$$

Because of (6.5), the operator $I - K$ is strictly positive, and hence (6.6) is equivalent to

$$(I - K)^{-1} T \frac{d\psi}{dt} = -\psi.$$

In [81], [82], [108] this equation is solved by diagonalizing the operator $(I - K)^{-1}T$.

Equation (6.1) with boundary conditions (6.2) and (6.3) can also be written as a Wiener-Hopf integral equation with an operator-valued kernel function (cf., [40]). In order to do this, let us introduce some notation. Let $\mathcal{H}_+$ and $\mathcal{H}_-$ be the subspaces of $\mathcal{H} = L_2[-1, 1]$ consisting of all functions that are zero almost everywhere on $[-1, 0]$ and $[0, 1]$, respectively. By P_+ and P_- we denote the orthogonal projections of $\mathcal{H} = L_2[-1, 1]$ onto the subspace $\mathcal{H}_+$ and $\mathcal{H}_-$, respectively. Furthermore, h will be the operator-valued function defined by

$$(h(t)f)(\mu) = \begin{cases} \dfrac{1}{\mu} \exp\left(-\dfrac{t}{\mu}\right)(P_+Kf)(\mu), & t > 0, \\[2ex] -\dfrac{1}{\mu} \exp\left(-\dfrac{t}{\mu}\right)(P_-Kf)(\mu), & t < 0, \end{cases} \tag{6.8}$$

and F is the vector-valued function given by

$$F(t)(\mu) = \begin{cases} f_+(\mu)\exp\left(-\dfrac{t}{\mu}\right), & 0 < \mu \leq 1, \\[2ex] 0, & -1 \leq \mu \leq 0. \end{cases} \tag{6.9}$$

The operator-valued function h is referred to as the *propagator function* associated with the half range problem (6.6).

Given these functions h and F, equation (6.1) with the boundary conditions (6.2) and (6.3) can be written as

$$\psi(t) - \int_0^\infty h(t-s)\psi(s)\,ds = F(t), \qquad t \geq 0. \tag{6.10}$$

To see this, multiply equation(6.1) by $\mu^{-1}\exp(t/\mu)$ and integrate over $(0,t)$ when $\mu > 0$ or over (t,∞) in case $\mu < 0$. With the help of the boundary conditions one gets in this way the integral equation (6.10). In [40] the asymptotics of solutions of equation (6.10) are found and used to describe the asymptotics of solutions of the transport equation.

6.2 The case of a finite number of scattering directions

To make the method used in this chapter more transparent we first consider the case when scattering occurs in a finite number of directions only. This assumption reduces the equation (6.1) and the boundary condition (6.2) to

$$\mu_i \frac{d\psi}{dt}(t,\mu_i) + \psi(t,\mu_i) = \sum_{j=1}^n k(\mu_i,\mu_j)\psi(t,\mu_j), \tag{6.11}$$
$$i = 1,\dots,n,\ t \geq 0,$$

where

$$\psi(0,\mu_i) = \varphi_+(\mu_i), \qquad \mu_i > 0. \tag{6.12}$$

To treat this version of the problem, introduce the $\mathbb{C}^n$-valued function

$$\psi(t) = \begin{bmatrix} \psi(t,\mu_1) \\ \vdots \\ \psi(t,\mu_n) \end{bmatrix},$$

and the matrices

$$T = \operatorname{diag}[\mu_1,\dots,\mu_n], \qquad K = [k(\mu_i,\mu_j)]_{i,j=1}^n. \tag{6.13}$$

Observe that T and K are real symmetric (hence selfadjoint) $n \times n$ matrices. Using this notation, equation (6.11) taken with the boundary condition (6.12) can be rewritten as

$$\begin{cases} T\psi'(t) + \psi(t) = K\psi(t), & 0 \leq t < \infty, \\ P_+\psi(0) = x_+, \end{cases} \tag{6.14}$$

where P_+ is the projection on $\mathbb{C}^n$ defined by

$$P_+ \begin{bmatrix} x_1 \\ \vdots \\ x_n \end{bmatrix} = \begin{bmatrix} y_1 \\ \vdots \\ y_n \end{bmatrix}, \qquad y_i = \begin{cases} 0 & \text{if } \mu_i \leq 0, \\ x_1 & \text{if } \mu_i > 0, \end{cases}$$

and x_+ is a given vector in $\operatorname{Im} P_+$. Observe that P_+ is the spectral projection of T corresponding to the positive eigenvalues of T. In what follows we assume additionally that T is invertible, which is the generic case and corresponds to the requirement that all μ_i in (6.11) are different from 0; cf., formula (6.12). We shall look for solutions ψ of (6.14) in the space $L_2^n[0,\infty)$.

The first step in solving (6.14) is based on the observation that, for invertible T, equation (6.14) is equivalent to a Wiener-Hopf integral equation with a rational matrix symbol. In fact, the following theorem holds.

Theorem 6.1. *Suppose T in* (6.13) *is invertible and let $\psi \in L_2^n[0,\infty)$. Then ψ is a solution of equation* (6.14) *if and only if ψ is a solution of the Wiener-Hopf integral equation with a special right-hand side, namely*

$$\psi(t) - \int_0^\infty h(t-s)K\psi(s)\,ds = e^{-tT^{-1}}x_+, \qquad t \geq 0, \tag{6.15}$$

where h is the propagator function associated with problem (6.14), *that is,*

$$h(t) = \begin{cases} T^{-1}e^{-tT^{-1}}P_+, & t > 0, \\ -T^{-1}e^{-tT^{-1}}P_-, & t < 0. \end{cases} \tag{6.16}$$

Here $P_- = I - P_+$.

Proof. Assume ψ is a solution of (6.14). Applying T^{-1} to the first identity in (6.14), and solving the resulting equation by using variation of constants, yields

$$\psi(t) = e^{-tT^{-1}}\psi(0) + e^{-tT^{-1}}\int_0^t e^{sT^{-1}}T^{-1}K\psi(s)ds, \qquad t \geq 0. \tag{6.17}$$

Next, apply $e^{tT^{-1}}P_-$ to both sides of (6.17) and use that $e^{tT^{-1}}$ and P_- commute. Since $e^{tT^{-1}}P_-$ is exponentially decaying on $[0,\infty)$, the function $e^{tT^{-1}}P_-K\psi(t)$ is integrable on $[0,\infty)$, and thus

$$\lim_{t\to\infty} e^{tT^{-1}}P_-\psi(t) = P_-\psi(0) + \int_0^\infty e^{sT^{-1}}P_-T^{-1}K\psi(s)ds. \tag{6.18}$$

Again using that the function $e^{tT^{-1}}P_-\psi(t)$ is integrable on $[0,\infty)$, we see that the left-hand side of (6.18) has to be equal to zero, which proves that

$$P_-\psi(0) = -\int_0^\infty e^{sT^{-1}}P_-T^{-1}K\psi(s)ds. \tag{6.19}$$

Now, replace $\psi(0)$ in (6.17) by $P_+\psi(0) + P_-\psi(0)$, use the boundary condition in

(6.14), and apply (6.19). This gives

$$\begin{aligned}\psi(t) &= e^{-tT^{-1}}x_+ - \int_t^\infty e^{-(t-s)T^{-1}}P_-T^{-1}K\psi(s)ds \\ &\qquad + \int_0^t e^{-(t-s)T^{-1}}T^{-1}K\psi(s)ds \\ &= e^{-tT^{-1}}x_+ + \int_0^\infty h(t-s)K\psi(s)ds, \qquad t \geq 0.\end{aligned}$$

Thus ψ is a solution of (6.15).

To prove the converse statement, assume that ψ is a solution of (6.15). Thus

$$\begin{aligned}\psi(t) &= e^{-tT^{-1}}x_+ + e^{-tT^{-1}}\int_0^t e^{sT^{-1}}P_+T^{-1}K\psi(s)ds \\ &\quad -e^{-tT^{-1}}\int_t^\infty e^{sT^{-1}}P_-T^{-1}K\psi(s)ds, \qquad t \geq 0.\end{aligned} \tag{6.20}$$

It follows that ψ is absolutely continuous on each compact subinterval of $[0,\infty)$, and hence the integrands in the right-hand side of (6.20) are continuous functions of the variable s. But then ψ is differentiable on $(0,\infty)$, and we see that

$$\begin{aligned}\psi'(t) &= -T^{-1}\psi(t) + P_+T^{-1}K\psi(t) + P_-T^{-1}K\psi(t) \\ &= -T^{-1}\psi(t) + T^{-1}K\psi(t), \qquad t \geq 0,\end{aligned}$$

and hence ψ satisfies the first equation in (6.14). From (6.20) it also follows that

$$\psi(0) = x_+ - \int_0^\infty e^{sT^{-1}}P_-T^{-1}K\psi(s)ds, \qquad t \geq 0,$$

which implies that $P_+\psi(0) = P_+x_+ = x_+$. We conclude that ψ is a solution of the problem (6.14). □

A direct computation yields that the symbol of the Wiener-Hopf operator associated with (6.15) is the $n \times n$ matrix function W given by

$$W(\lambda) = I_n - iT^{-1}(\lambda + iT^{-1})^{-1}K,$$

where I_n is the $n \times n$ identity matrix. Thus the symbol W is not only a rational matrix function but it is already given in a concrete realized form, namely

$$W(\lambda) = I_n + C(\lambda - A)^{-1}B,$$

with

$$A = -iT^{-1}, \qquad B = K, \qquad C = -iT^{-1}. \tag{6.21}$$

Notice that A does not have eigenvalues on the real line. Thus in order to solve equation (6.15) we can apply Theorem 3.3. This requires us to analyze the spectral properties of the matrix

$$A^{\times} = A - BC = -iT^{-1}(I - K). \tag{6.22}$$

In view of (6.5) it is natural to assume $I - K$ is positive definite.

Lemma 6.2. *Assume $I - K$ is positive definite. Then the matrix $A^{\times}$ in* (6.22) *has no real eigenvalues and*

$$\mathbb{C}^n = M \dot{+} M^{\times}, \tag{6.23}$$

where M is the spectral subspace of the matrix A in (6.21) *corresponding to the eigenvalues in the upper half plane, and $M^{\times}$ is the spectral subspace of $A^{\times}$ in* (6.22) *corresponding to the eigenvalues in the lower half plane.*

Proof. Let $\langle \cdot\,, \cdot \rangle$ be the standard inner product in $\mathbb{C}^m$ and put $S = (I - K)^{-1}T$. Since $I - K$ is positive definite, S is well-defined and the sesquilinear form

$$[x, y] = \langle (I - K)x, y \rangle \tag{6.24}$$

is an inner product on $\mathbb{C}^n$. From $[Sx, y] = \langle (I - K)Sx, y \rangle = \langle Tx, y \rangle$ and the fact that T is selfadjoint, it follows that S is selfadjoint with respect to the inner product $[\cdot\,, \cdot]$. But then the same holds true for $iA^{\times} = S^{-1}$. Thus $A^{\times}$ is invertible and its eigenvalues are on the imaginary axis. In particular, $A^{\times}$ has no real eigenvalues.

Recall that P_+ is the spectral projection of T corresponding to the positive eigenvalues of T. Let $P_+^{\times}$ be the analogous projection for S. Since T and S are invertible, $T|_{\operatorname{Ker} P_+}$ is negative definite and $S|_{\operatorname{Im} P_+^{\times}}$ is positive definite. Thus

$$0 \neq x \in \operatorname{Ker} P_+ \quad \Longrightarrow \quad \langle Tx, x \rangle < 0,$$

$$0 \neq x \in \operatorname{Im} P_+^{\times} \quad \Longrightarrow \quad [Sx, x] > 0.$$

But $[Sx, x] = \langle Tx, x \rangle$ for each $x \in \mathbb{C}^n$. It follows that $\operatorname{Ker} P_+ \cap \operatorname{Im} P_+^{\times} = \{0\}$. In particular, $\operatorname{rank} P_+ \geq \operatorname{rank} P_+^{\times}$. By repeating the argument with $\operatorname{Ker} P_+$ replaced by $\operatorname{Ker} P_+^{\times}$ and $\operatorname{Im} P_+^{\times}$ by $\operatorname{Im} P_+$, we see that $\operatorname{rank} P_+^{\times} \geq \operatorname{rank} P_+$. But then we may conclude that $\mathbb{C}^n = \operatorname{Ker} P_+ \dot{+} \operatorname{Im} P_+^{\times}$. Finally, from $iA = T^{-1}$ we see that $M = \operatorname{Ker} P_+$, and from $iA^{\times} = S^{-1}$ we conclude that $M^{\times} = \operatorname{Im} P_+^{\times}$. □

We can now apply Theorem 3.3 to solve equation (6.15). Note however that the right-hand side of (6.15) is of a special form. In fact, in terms of the matrices appearing in (6.21) this right-hand side can be written as

$$g(t) = iCe^{-itA}x_+,$$

where $x_+ \in \operatorname{Im} P_+$. Thus instead of Theorem 3.3 we can also directly apply Corollary 3.4. This yields the following result.

Theorem 6.3. *Assume $I - K$ is positive definite and T is invertible. Then the matrix $(I-K)^{-1}T$ is selfadjoint with respect to the inner product (6.24) and the half range problem (6.14) has a unique solution ψ in $L_2^n(0,\infty)$, namely*

$$\psi(t) = e^{-tT^{-1}(I-K)}\Pi x_+, \qquad t \geq 0, \tag{6.25}$$

where Π is the projection of $\mathbb{C}^n$ along $\operatorname{Ker} P_+$ onto the spectral subspace $\operatorname{Im} P_+^\times$ of $(I-K)^{-1}T$ corresponding to its positive eigenvalues.

6.3 Wiener-Hopf equations with operator-valued kernel functions

It is well-known that the Wiener-Hopf integral equation

$$\psi(t) - \int_0^\infty k(t-s)\psi(s)\,dy = F(t), \qquad t \geq 0 \tag{6.26}$$

can be solved by constructing appropriate factorizations of its symbol (cf., Sections 1.1, 3.2, the papers [49], [71], or the survey article [59]). In this section we shall describe this method for the case when k is an L_1-kernel function the values of which are compact operators on a separable Hilbert space $\mathcal{H}$. So we assume that $k(t)$ is a compact operator for each real t, that $\langle k(\cdot)f, g\rangle$ is measurable on the real line for each f and g in $\mathcal{H}$, and that

$$\int_{-\infty}^\infty \|k(t)\|\,dt < 0,$$

where $\langle \cdot, \cdot \rangle$ is the inner product on $\mathcal{H}$, and $\|\cdot\|$ is the operator norm for operators on $\mathcal{H}$. Note that the kernel function h considered in the previous section falls into this category.

Recall that the *symbol* of equation (6.26) is the operator-valued function $I - K(\lambda)$, where $K(\lambda)$ is the Fourier transform of the kernel function k, i.e.,

$$K(\lambda) = \int_{-\infty}^\infty e^{i\lambda t}k(t)\,dt, \qquad -\infty < \lambda < \infty.$$

By the Riemann-Lebesgue lemma, we have $\lim_{\lambda\in\mathbb{R},\ \lambda\to\pm\infty} K(\lambda) = 0$. Here we also need the concept of canonical factorization, this time in the present infinite dimensional context. The symbol is said to admit a (*right*) *canonical factorization* with respect to the real line if

$$I - K(\lambda) = G_-(\lambda)G_+(\lambda), \qquad -\infty < \lambda < \infty, \tag{6.27}$$

where the factors G_- and G_+ meet the following requirements:

(i) the operator function G_- is analytic on the (open) lower half plane $\Im\lambda < 0$ and continuous on the closure of the left half plane in the Riemann sphere (infinity included); also for each λ in this closure (infinity included), the operator $G_-(\lambda)$ is invertible;

(ii) the operator function G_+ is analytic on the (open) upper half plane $\Im\lambda > 0$ and continuous on the closure of the right half plane in the Riemann sphere (infinity included); also for each λ in this closure (infinity included), the operator $G_+(\lambda)$ is invertible.

Note that the definition is analogous to that given earlier in the matrix-valued case (see Sections 1.1 and 3.1). According to [49], because of the fact that k is an L_1-kernel function the values of which are compact operators on $\mathcal{H}$, the inverses of the factors in the right-hand side of (6.27) can be written as

$$G_+^{-1}(\lambda) = I + \int_0^\infty e^{i\lambda t}\gamma_+(t), \qquad G_-^{-1}(\lambda) = I + \int_{-\infty}^0 e^{i\lambda t}\gamma_-(t)\,dt, \tag{6.28}$$

where, γ_+ and γ_- are L_1-functions on $[0,\infty)$ and $(-\infty,0]$, respectively, whose values are compact operators on $\mathcal{H}$.

Let $L_2(\mathbb{R}_+,\mathcal{H})$ denote the space of all L_2-integrable functions on $[0,\infty)$ with values in $\mathcal{H}$. The identities (6.28) are important, because they allow for explicit formulas for the solutions of (6.26). Indeed, by [18] equation (6.26) has a unique solution ψ in $L_2(\mathbb{R}_+,\mathcal{H})$ for each $F \in L_2(\mathbb{R}_+,\mathcal{H})$ if and only if a canonical factorization (6.27) exists, and in that case (just as in Section 1.1 for matrix-valued kernel functions) the Fourier transform $\widehat{\psi}$ of the solution ψ is given by

$$\widehat{\psi}(\lambda) = G_+^{-1}(\lambda)\mathcal{P}\big(G_-^{-1}(\lambda)\widehat{F}(\lambda)\big), \tag{6.29}$$

where $\widehat{F}$ is the Fourier transform of the right-hand side of equation (6.26), and $\mathcal{P}$ is the projection defined by

$$\mathcal{P}\left(\int_{-\infty}^\infty f(t)e^{it\lambda}\,dt\right) = \int_0^\infty f(t)e^{it\lambda}\,dt.$$

Taking inverse Fourier transforms in (6.29) one finds

$$\psi(t) = F(t) + \int_0^\infty \gamma(t,s)F(s)\,ds,$$

where $\gamma(t,s)$ is given by (1.10), i.e.,

$$\gamma(t,s) = \begin{cases} \gamma_+(t-s) + \displaystyle\int_0^s \gamma_+(t-r)\gamma_-(r-s)\,dr, & 0 \le s < t, \\[2ex] \gamma_-(t-s) + \displaystyle\int_0^t \gamma_+(t-r)\gamma_-(r-s)\,dr, & 0 \le t < s. \end{cases}$$

As we observed already, in (6.10) the kernel function $h(\cdot)$ is an L_1-function on the real line whose values are compact (in fact finite rank) operators on $L_2[-1,1]$. In the next section we shall prove that the corresponding symbol admits a canonical factorization, and we shall describe the factors explicitly.

6.4 Construction of a canonical factorization

We now return to equation (6.10). Note that its symbol is given by $I - H(\lambda)$, where

$$H(\lambda) = \int_{-\infty}^{\infty} e^{i\lambda t} h(t) K\, dt = (I - i\lambda T)^{-1} K, \qquad -\infty < \lambda < \infty.$$

Here h is given by (6.8), and the operators T and K are as in formula (6.7). The operator function H is analytic on the strip $|\Im \lambda| < 1$. Note that $\sigma(T)$ is the closed interval $[-1,1]$, so that $(I - i\lambda T)^{-1}$ is defined for all λ in the complement of the union of the subsets $i[1,\infty)$ and $i(-\infty,-1]$ of the imaginary axis. In this section we show that $I - H(\lambda)$ admits a canonical factorization with respect to the real line:

$$I - H(\lambda) = G_-(\lambda) G_+(\lambda), \qquad -\infty < \lambda < \infty,$$

where the factors and their inverses can be written as

$$\begin{aligned}
G_-(\lambda) &= I - (I - i\lambda T)^{-1}(I - P)K(I - PK)^{-1}, \\
G_+(\lambda) &= I - (I - Q^*K)^{-1}(I - Q^*)(I - i\lambda T)^{-1}K, \\
G_-^{-1}(\lambda) &= I + (I - Q^*K)^{-1}Q^*\big(I - i\lambda (T^\times)^*\big)^{-1}K, \\
G_+^{-1}(\lambda) &= I + (I - i\lambda T^\times)^{-1}PK(I - PK)^{-1}.
\end{aligned}$$

Here $T^\times = (I - K)^{-1}T$, and P and Q are projections of which the definition will be given below. With regard to the domains of the factors and their inverses, the situation is similar to what we encountered earlier for Theorems 3.2 and 5.14.

In order to make the factorization we transform the symbol of equation (6.10) into another function W which is defined and continuous on the imaginary axis. This will be done as follows.

Recall that T and K are both selfadjoint and that $I - K$, being a strictly positive operator because of (6.5), is invertible. Hence, for non-zero purely imaginary values of λ,

$$\begin{aligned}
I - H(i/\lambda)^* &= I - \big((I + \lambda^{-1}T)^{-1}K\big)^* \\
&= I - K(I - \lambda^{-1}T)^{-1} \\
&= I - \lambda K(\lambda - T)^{-1} \\
&= (I - K)\big(I - K(I - K)^{-1}T(\lambda - T)^{-1}\big).
\end{aligned}$$

We now introduce W by writing

$$W(\lambda) = I - (I-K)^{-1}KT(\lambda - T)^{-1}. \tag{6.30}$$

Note that this expression is a unital realization for W. The state space is the separable Hilbert space $\mathcal{H} = L_2[-1,1]$. The operator T is the main operator, and $(I-K)^{-1}T$ is the associate main operator, denoted above by $T^\times$ (in line with the notation adopted in Section 2.1).

Via (6.30) the function W is defined and analytic on the resolvent set of T, so on the complement of the interval $[-1,1]$. In particular, W is defined and continuous on the imaginary axis punctured at the origin. We shall now prove that by setting $W(0) = (I-K)^{-1}$ the restriction of W (now defined on the complement of the set $[-1,0) \cup (0,1]$) to the imaginary axis is a continuous function. For this we need to show that

$$\lim_{\alpha \to 0,\, \alpha \in \mathbb{R}} W(i\alpha) = (I-K)^{-1}. \tag{6.31}$$

It is convenient to establish the following lemma which will also play a role later on in this section.

Lemma 6.4. *Let S be a bounded selfadjoint operator on a given Hilbert space. Then*

$$\|S(i\alpha - S)^{-1}\| \le 1, \qquad 0 \ne \alpha \in \mathbb{R}, \tag{6.32}$$

while, furthermore,

$$\lim_{\alpha \to 0,\, \alpha \in \mathbb{R}} S(i\alpha - S)^{-1}f = -f, \qquad f \perp \operatorname{Ker} S. \tag{6.33}$$

Under the additional assumption that S is a nonnegative operator, the limit result (6.33) can be sharpened to

$$\lim_{\lambda \to 0,\, \Re\lambda \le 0} S(\lambda - S)^{-1}f = -f, \qquad f \perp \operatorname{Ker} S. \tag{6.34}$$

Proof. Let $E_S(t)$ be the spectral resolution of the identity for S, and let f be an element of the underlying Hilbert space. Then

$$\begin{aligned} \|S(i\alpha - S)^{-1}f\|^2 &\le \int_{-\infty}^{\infty} \frac{t^2}{\alpha^2 + t^2}\, d\,\|E_S(t)f\|^2 \\ &\le \int_{-\infty}^{\infty} d\,\|E_S(t)f\|^2 \\ &= \|f\|^2. \end{aligned}$$

This proves (6.32). Next, observe that

$$\|f + S(i\alpha - S)^{-1}f\|^2 \le \int_{-\infty}^{\infty} \frac{\alpha^2}{\alpha^2 + t^2}\, d\,\|E_S(t)f\|^2.$$

So by Lebesgue's dominated convergence theorem we get

$$\begin{aligned}\lim_{\alpha\to 0,\,\alpha\in\mathbb{R}} \|f + S(i\alpha - S)^{-1}f\| &\leq \|E_S(0+)f\|^2 - \|E_S(0-)f\|^2 \\ &= \|(E_S(0+) - E_S(0-))f\|^2,\end{aligned}$$

which is zero if $f \perp \operatorname{Ker} S$. Hence (6.33) is proved. The argument for (6.34), taking nonnegativity of the operator S for granted, is analogous. □

The proof of (6.31) is now as follows. As $\operatorname{Ker} T = \{0\}$, we see from Lemma 6.4 that $\lim_{\alpha\to 0,\,\alpha\in\mathbb{R}} (i\alpha + T)^{-1}Tf = f$, $f \in \mathcal{H}$. Since K is compact (actually even of finite rank), it follows that $(i\alpha + T)^{-1}TK$ tends to K in the operator norm if $\alpha \in \mathbb{R}$, $\alpha \to 0$. Taking adjoints, we obtain that the same holds true for $-KT(i\alpha - T)^{-1}$. But then we have (6.31), where the convergence is with respect to the operator norm. So with $W(0) = (I - K)^{-1}$, indeed W becomes a continuous function on the imaginary axis.

It is this operator function for which we want a (right) canonical Wiener-Hopf factorization. This time not with respect to the real line (see the definition in Section 6.3) but for the analogous situation where the curve in the Riemann sphere is the imaginary axis with infinity included. The theory concerning canonical factorization developed earlier suggests that we have to find an invariant subspace M for T such that the spectrum of T restricted to M lies in the closed right half plane, and an invariant subspace $M^\times$ for $T^\times$ such that the spectrum of $T^\times$ restricted to $M^\times$ lies in the closed left half plane. Since T is selfadjoint the choice of M is clear: $M = \mathcal{H}_+$, where $\mathcal{H}_+$ is the subspace of $\mathcal{H} = L_2[-1,1]$ consisting of all functions that are zero almost everywhere on $[-1,0]$. As we shall see below, after replacing the standard inner product on $L_2[-1,1]$ by a suitable equivalent one, the operator $T^\times$ is selfadjoint too. So for $M^\times$ we can take the spectral subspace of $T^\times$ corresponding to the part of the spectrum of $T^\times$ on $(-\infty, 0]$. The first difficulty is to prove the matching of the subspaces M and $M^\times$, i.e., to show that $\mathcal{H} = M \dot{+} M^\times$. Taking for granted that this has been established a second difficulty appears, because in the present case the imaginary axis does not split the spectra of T and $T^\times$. So we cannot apply directly the theory developed so far, but we have to prove, using the specifics of the situation, that the factors obtained have the desired boundary behavior. The purpose of this section is to show that this approach works indeed.

We begin by considering the operator $T^\times = (I - K)^{-1}T$. As $I - K$ is strictly positive, $[f, g] = \langle (I-K)f, g\rangle$ defines an inner product on $\mathcal{H} = L_2[-1,1]$ equivalent with the standard one. Writing $A^{[*]}$ for the adjoint of an operator A relative to the inner product $[\cdot\,,\cdot]$, we have

$$A^{[*]} = (I - K)^{-1}A^*(I - K). \tag{6.35}$$

In particular, we see that the operator $T^\times$ is selfadjoint with respect to the inner product $[\cdot,\cdot]$. Let $E^\times(\cdot)$ be the corresponding spectral resolution and introduce

$$\mathcal{H}_m = \operatorname{Im} E^\times(0), \qquad \mathcal{H}_p = \operatorname{Ker} E^\times(0).$$

Then $\mathcal{H}_m$ and $\mathcal{H}_p$ are both invariant under $T^\times$ and

$$\sigma(T^\times|_{\mathcal{H}_m}) \subset (-\infty, 0] \cap \sigma(T^\times), \qquad \sigma(T^\times|_{\mathcal{H}_p}) \subset [0, \infty) \cap \sigma(T^\times). \tag{6.36}$$

For T the situation is more straightforward. Indeed, T is selfadjoint with respect to the original (standard) inner product on $\mathcal{H}$ and leaves invariant the spaces $\mathcal{H}_-$ and $\mathcal{H}_+$ featured in Section 6.1. Further

$$\sigma(T|_{\mathcal{H}_-}) = [-1, 0], \qquad \sigma(T|_{\mathcal{H}_+}) = [0, 1]. \tag{6.37}$$

The subspaces M and $M^\times$ mentioned above are $\mathcal{H}_+$ and $\mathcal{H}_m$, respectively. So proving that these subspaces match amounts to showing that $\mathcal{H} = \mathcal{H}_+ \dot{+} \mathcal{H}_m$. In fact, in the next section we shall show the following stronger result:

$$\mathcal{H} = \mathcal{H}_- \dot{+} \mathcal{H}_p, \qquad \mathcal{H} = \mathcal{H}_+ \dot{+} \mathcal{H}_m. \tag{6.38}$$

Let P be the projection of $\mathcal{H}$ along $\mathcal{H}_-$ onto $\mathcal{H}_p$, and let Q be the projection of $\mathcal{H}$ along $\mathcal{H}_+$ onto $\mathcal{H}_m$. Since the subspaces $\mathcal{H}_-, \mathcal{H}_+$ are invariant under T and $\mathcal{H}_m, \mathcal{H}_p$ are invariant under $T^\times$, both P and Q are supporting projections for the realization (6.30). Associated with these projections are two factorizations:

$$W(\lambda) = \widetilde{W}_+(\lambda)\widetilde{W}_-(\lambda), \qquad W(\lambda) = \widehat{W}_-(\lambda)\widehat{W}_+(\lambda). \tag{6.39}$$

With the appropriate choice for the value of the factors at the origin, both these factorizations are canonical factorizations of W with respect to the imaginary axis; the first a left and the second a right factorization. In the sequel we only need the second factorization in (6.39).

First we give the expressions for the factors $\widehat{W}_-(\lambda)$ and $\widehat{W}_+(\lambda)$:

$$\widehat{W}_-(\lambda) \;=\; I - \big((I-K)^{-1}KT\big)|_{\mathcal{H}_+}\big(\lambda - T|_{\mathcal{H}_+}\big)^{-1}(I-Q), \tag{6.40}$$

$$\widehat{W}_+(\lambda) \;=\; I - \big((I-K)^{-1}KT\big)|_{\mathcal{H}_m}\big(\lambda - QT|_{\mathcal{H}_m}\big)^{-1}Q. \tag{6.41}$$

Note that there is slight abuse of notation here. Indeed, the operator $I - Q$ in the formula for $\widehat{W}_-(\lambda)$ should be interpreted as a mapping from $\mathcal{H}$ onto $\mathcal{H}_+$, and Q in the expression for $\widehat{W}_+(\lambda)$ must be seen as a mapping from from $\mathcal{H}$ onto $\mathcal{H}_m$. In particular $QT|_{\mathcal{H}_m}$ should be read as the compression of T to $\mathcal{H}_m$ (relative to the decomposition $\mathcal{H} = \mathcal{H}_+ \dot{+} \mathcal{H}_m$). The function $\widehat{W}_-$ is defined and analytic on the resolvent set of $T|_{\mathcal{H}_+}$ so, by the second part of (6.37) on the complement of the interval $[0, 1]$. Similarly, the function $\widehat{W}_+$ is defined and analytic on the resolvent set of the compression operator $QT|_{\mathcal{H}_m}$. Now $\mathcal{H} = \mathcal{H}_+ \dot{+} \mathcal{H}_- = \mathcal{H}_+ \dot{+} \mathcal{H}_m$, and Lemma 3.1 guarantees that $QT|_{\mathcal{H}_m}$ is similar to $T|_{\mathcal{H}_-}$. In particular the resolvent sets of $QT|_{\mathcal{H}_m}$ and $T|_{\mathcal{H}_-}$ coincide. It follows from the first part of (6.37) that function $\widehat{W}_+$ is defined and analytic on the complement of the interval $[-1, 0]$. The argument also indicates that the second factorization in (6.39) holds for all

λ outside the interval $[-1,1]$. Indeed this interval is precisely the union of the spectra of $T|_{\mathcal{H}_+}$ and $QT|_{\mathcal{H}_m}$ (cf., Theorem 2.6).

Next we deal with the invertibility of the factors $\widehat{W}_-(\lambda)$ and $\widehat{W}_+(\lambda)$. The above realization of W_- has

$$(I-Q)T^\times|_{\mathcal{H}_+} : \mathcal{H}_+ \to \mathcal{H}_+$$

as its associate main operator. From Section 2.4 we now know that $\widehat{W}_-(\lambda)$ is invertible for λ in the intersection of the resolvent sets of $T|_{\mathcal{H}_+}$ and $(I-Q)T^\times|_{\mathcal{H}_+}$. The resolvent set of $T|_{\mathcal{H}_+}$ is the complement of the interval $[0,1]$. As $\mathcal{H} = \mathcal{H}_+ \dot{+} \mathcal{H}_m = \mathcal{H}_p \dot{+} \mathcal{H}_m$, the compression operator $(I-Q)T^\times|_{\mathcal{H}_+}$ is similar to $T^\times|_{\mathcal{H}_p}$. Hence, by the second part of (6.36), the resolvent set of $(I-Q)T^\times|_{\mathcal{H}_+}$ is the complement of the set $[0,\infty) \cap \sigma(T^\times)$. It follows that $\widehat{W}_-(\lambda)$ is invertible for all non-zero λ with $\Re\lambda \le 0$, its inverse (see Theorem 2.4) being given by

$$\begin{aligned} \widehat{W}_-(\lambda)^{-1} &= I + ((I-K)^{-1}KT)|_{\mathcal{H}_+}\big(\lambda - (I-Q)T^\times|_{\mathcal{H}_+}\big)^{-1}(I-Q) \\ &= I + (KT^\times)|_{\mathcal{H}_+}\big(\lambda - (I-Q)T^\times|_{\mathcal{H}_+}\big)^{-1}(I-Q). \end{aligned}$$

In an analogous manner one proves that $\widehat{W}_+(\lambda)$ is invertible for all non-zero λ with $\Re\lambda \ge 0$, its inverse having the representation

$$\begin{aligned} \widehat{W}_+(\lambda)^{-1} &= I + ((I-K)^{-1}KT)|_{\mathcal{H}_m}\big(\lambda - T|^\times{}_{\mathcal{H}_m}\big)^{-1}Q \\ &= I + (KT^\times)|_{\mathcal{H}_m}\big(\lambda - T|^\times{}_{\mathcal{H}_m}\big)^{-1}Q. \end{aligned}$$

The above formulas contain the precise description of the factors W_-, W_+ and their inverses W_-^{-1}, W_+^{-1}. Giving up some precision but gaining in conciseness, we can also write

$$\begin{aligned} \widehat{W}_-(\lambda) &= I - (I-K)^{-1}KT(\lambda - T)^{-1}(I-Q), \\ \widehat{W}_+(\lambda) &= I - (I-K)^{-1}KTQ(\lambda - T)^{-1}, \\ \widehat{W}_-^{-1}(\lambda) &= I + KT^\times(I-Q)(\lambda - T^\times)^{-1}, \\ \widehat{W}_+^{-1}(\lambda) &= I + KT^\times(\lambda - T^\times)^{-1}Q; \end{aligned}$$

see Section 2.4 and [20], Section 2.5 for details.

We have come close to proving that the second factorization in (6.39) is a (right) canonical factorization of W with respect to the imaginary axis. To make the proof complete we need to check the behavior of the functions at infinity and at the origin. As far as the behavior at infinity is concerned the situation is simple. Indeed the functions $\widehat{W}_-$, $\widehat{W}_+$, $\widehat{W}_-^{-1}$ and $\widehat{W}_+^{-1}$ are analytic there with value the identity operator on $\mathcal{H}$. For the origin the situation is more complicated.

Earlier we completed the definition of the function W, initially introduced via the unital realization (6.30), by stipulating that $W(0) = (I - K)^{-1}$. Now we make a similar move with respect to $\widehat{W}_-$ and $\widehat{W}_+$, in the first instance given by (6.40) and (6.41), respectively. Indeed, we stipulate that

$$\widehat{W}_-(0) = (I - K)^{-1}(I - KQ), \qquad \widehat{W}_+(0) = (I - K)^{-1}(I - KP^*).$$

In this manner the closed left half plane $\Re\lambda \le 0$ is contained in the domain of $\widehat{W}_-$, and the closed right half plane $\Re\lambda \ge 0$ is contained in the domain of $\widehat{W}_+$. Our task is now threefold: to verify the invertibility of $\widehat{W}_-(0)$ and $\widehat{W}_+(0)$, to demonstrate the continuity of $\widehat{W}_-$ and $\widehat{W}_+$ on the appropriate half planes, i.e., to show that

$$\lim_{\lambda\to 0,\, \Re\lambda \le 0} \widehat{W}_-(\lambda) = (I - K)^{-1}(I - KQ), \tag{6.42}$$

$$\lim_{\lambda\to 0,\, \Re\lambda \ge 0} \widehat{W}_+(\lambda) = (I - K)^{-1}(I - KP^*), \tag{6.43}$$

and to verify that the factorization $W = \widehat{W}_-\widehat{W}_+$ holds at the origin. As a first step we present the following lemma (which will also be used in Section 6.6 below).

Lemma 6.5. *Let P, Q and K be as above. Then*

$$Q^*(I - K)P = 0, \qquad (I - Q^*)(I - P) = 0. \tag{6.44}$$

Proof. Note that $\operatorname{Im} P = \mathcal{H}_p$ is orthogonal to $\operatorname{Im} Q = \mathcal{H}_m$ with respect to the inner product $[f, g] = \langle (I - K)f, g\rangle$. Thus

$$\langle (I - K)Pf, Qg\rangle = [Pf, Qg] = 0, \qquad f, g \in L_2[-1, 1].$$

This yields the first identity in (6.44). Next observe that, relative to the usual inner product on $\mathcal{H} = L_2[-1, 1]$, the space $\operatorname{Im}(I - Q) = \mathcal{H}_+$ is orthogonal to $\operatorname{Im}(I - P) = \mathcal{H}_-$. It follows that

$$\langle (I - P)f, (I - Q)g\rangle = 0, \qquad f, g \in L_2[-1, 1],$$

which proves the second identity in (6.44). □

Corollary 6.6. *The operators $I - KQ$ and $(I - K)^{-1}(I - KP^*)$ are invertible and each other's inverse.*

Proof. As K is compact (actually even of finite rank), the operator $I - KQ$ is Fredholm of index zero. In particular $I - KQ$ is invertible if and only if $I - KQ$ is left invertible. Thus it suffices to show that the operator $(I - K)^{-1}(I - KP^*)$ is a left inverse of $I - KQ$. Now the identities in Lemma 6.5 can be rewritten as

$Q^*KP = Q^*P$ and $Q^* + P - Q^*P = I$. Combining these, one gets

$$\begin{aligned} I - K &= I - (Q^* + P - Q^*P)K \\ &= I - Q^*K - PK + Q^*PK \\ &= I - Q^*K - PK + Q^*KPK \\ &= (I - Q^*K)(I - PK). \end{aligned}$$

Taking adjoints yields $I - K = (I - KP^*)(I - KQ)$, and this identity can be rewritten as $(I - K)^{-1}(I - KP^*)(I - KQ) = I$. □

The corollary can be rephrased by saying that $\widehat{W}_+(0) = (I - K)^{-1}(I - KP^*)$ is invertible with inverse $I - KQ$. Likewise $\widehat{W}_-(0) = (I - K)^{-1}(I - KQ)$ is invertible with inverse $(I - K)^{-1}(I - KP^*)(I - K)$. It remains to verify (6.42) and (6.43). We begin with (6.42).

For $\Re\lambda \leq 0$, $\lambda \neq 0$, we have

$$\begin{aligned} \widehat{W}_-(\lambda) &= I - \big(K(I - K)^{-1}T\big)|_{\mathcal{H}_+}\big(\lambda - T|_{\mathcal{H}_+}\big)^{-1}(I - Q), \\ &= I - K_+\big(T|_{\mathcal{H}_+}\big)\big(\lambda - T|_{\mathcal{H}_+}\big)^{-1}(I - Q). \end{aligned}$$

Here K_+ is the restriction of $K(I - K)^{-1}$ to $\mathcal{H}_+$ considered as an operator from $\mathcal{H}_+$ into $\mathcal{H}$ and, as before, $I - Q$ should be read as a mapping from $\mathcal{H}$ onto $\mathcal{H}_+$. The restriction operator $T|_{\mathcal{H}_+} : \mathcal{H}_+ \to \mathcal{H}_+$ is selfadjoint and nonnegative. It also has a trivial null space. So we can apply Lemma 6.4 to show that

$$\lim_{\lambda \to 0,\, \Re\lambda \leq 0} T|_{\mathcal{H}_+}\big(\lambda - T|_{\mathcal{H}_+}\big)^{-1} f_+ = -f_+, \qquad f_+ \in \mathcal{H}_+.$$

Along with K_+, the operator $K_+^* : \mathcal{H} \to \mathcal{H}_+$ is compact (actually even of finite rank), and it follows that

$$\lim_{\lambda \to 0,\, \Re\lambda \leq 0} T|_{\mathcal{H}_+}\big(\lambda - T|_{\mathcal{H}_+}\big)^{-1} K_+^* = -K_+^*,$$

with convergence in norm. Taking adjoints we get

$$\lim_{\lambda \to 0,\, \Re\lambda \leq 0} K_+(T|_{\mathcal{H}_+})\big(\lambda - T|_{\mathcal{H}_+}\big)^{-1} = -K_+,$$

and hence

$$\lim_{\lambda \to 0,\, \Re\lambda \leq 0} K_+(T|_{\mathcal{H}_+})\big(\lambda - T|_{\mathcal{H}_+}\big)^{-1}(I - Q) = -K_+(I - Q).$$

A simple computation gives $I + K_+(I - Q) = (I - K)^{-1}(I - KQ)$, and (6.42) is immediate.

Next we turn to (6.43). By Corollary 6.6 and the continuity of the operation of taking the inverse, it suffices to show that

$$\lim_{\lambda \to 0,\, \Re\lambda \geq 0} \widehat{W}_+(\lambda)^{-1} = I - KQ.$$

For $\Re\lambda \geq 0$, $\lambda \neq 0$, we have

$$\begin{aligned} \widehat{W}_+(\lambda)^{-1} &= I + (KT^\times)|\mathcal{H}_m (\lambda - T|^\times \mathcal{H}_m)^{-1} Q, \\ &= I + K_m (T^\times|_{\mathcal{H}_m}) (\lambda - T^\times|_{\mathcal{H}_m})^{-1} Q. \end{aligned}$$

Here K_m is the restriction of K to $\mathcal{H}_m$ considered as an operator from $\mathcal{H}_m$ into $\mathcal{H}$ and, as before, Q should be read as a mapping from $\mathcal{H}$ onto $\mathcal{H}_m$. Because $T^\times = (I - K)^{-1}T$, the operator $T^\times|_{\mathcal{H}_m}$ has a trivial null space. Further it is nonpositive with respect to the alternative inner product $[\cdot\,,\cdot]$, and K_m is compact. Using Lemma 6.4 in an analogous way as in the previous paragraph, we see that

$$\lim_{\lambda \to 0,\, \Re\lambda \geq 0} K_m (T^\times|_{\mathcal{H}_m}) (\lambda - T^\times|_{\mathcal{H}_m})^{-1} = -K_m,$$

and we get $\lim_{\lambda \to 0,\, \Re\lambda \geq 0} \widehat{W}_+(\lambda)^{-1} = I - KQ$, as desired.

From what we have obtained so far and a continuity argument it is already clear that the second factorization in (6.39) holds at the origin too. The calculation

$$\begin{aligned} \widehat{W}_-(0)\widehat{W}_+(0) &= (I-K)^{-1}(I-KQ)(I-K)^{-1}(I-KP^*) \\ &= (I-K)^{-1}\big((I-KP^*)^{-1}I-K)\big)(I-K)^{-1}(I-KP^*) \\ &= (I-K)^{-1} = W(0), \end{aligned}$$

based on Corollary 6.6, corroborates this fact.

Our ultimate goal in this section is to produce a right canonical factorization with respect to the real line of the symbol $I - H(\lambda)$ of equation (6.10). For non-zero real λ we have $I - H(\lambda) = W(i/\lambda)^*(I - K)$, and with the right interpretation this identity even holds on the extended real line. Indeed, as W is given by a unital realization, the value of W at ∞ is I, and this corresponds with the fact that $H(0) = K$. Also by the Riemann-Lebesgue lemma, H vanishes at ∞, and this is in accord with $W(0) = (I - K)^{-1}$. The right canonical Wiener-Hopf factorization $W = \widehat{W}_-\widehat{W}_+$ with respect to the imaginary axis that we obtained for W now induces a right canonical Wiener-Hopf factorization with respect to the real line for the symbol. The details are given in the next two paragraphs.

We begin by defining a function G_- on the complement in $\mathbb{C}_\infty$ of the interval $i[1, \infty)$ which is situated on the imaginary axis. The determining expressions are

$$\begin{aligned} G_-(\lambda) &= \widehat{W}_+(i/\overline{\lambda})^*(I - Q^*K), \\ G_-(0) &= I - Q^*K, \\ G_-(\infty) &= I. \end{aligned}$$

Note that G_- is analytic on the complement of $i[1,\infty)$ in the finite complex plane $\mathbb{C}$. Also G_- is continuous on the closed lower half plane $\Im\lambda \le 0$, this time infinity included. Indeed,

$$\begin{aligned}\lim_{\lambda\to\infty,\,\Im\lambda\le 0} G_-(\lambda) &= \lim_{\mu\to 0,\,\Re\mu\ge 0} W_+(\mu)^*(I-Q^*K)\\ &= (I-PK)(I-K)^{-1}(I-Q^*K) = I = G_-(0).\end{aligned}$$

Here we used (6.43). Further define G_+ on the complement in $\mathbb{C}_\infty$ of the interval $i(-\infty,-1]$, again located on the imaginary axis, by

$$\begin{aligned}G_+(\lambda) &= (I-Q^*K)^{-1}\widehat{W}_-(i/\overline{\lambda})^*(I-K),\\ G_+(0) &= I-PK,\\ G_+(\infty) &= I.\end{aligned}$$

Then G_+ is analytic on the complement of $i(-\infty,-1])$ in $\mathbb{C}$. Also G_+ is continuous on the closed upper plane $\Im\lambda \ge 0$, infinity included. Indeed, using (6.42) one gets

$$\lim_{\lambda\to\infty,\,\Im\lambda\ge 0} G_+(\lambda) = \lim_{\mu\to 0,\,\Re\mu\le 0}(I-Q^*K)^{-1}W_-(\mu)^*(I-K) = I = G_+(0).$$

Observe that $I-H(\lambda) = G_-(\lambda)G_+(\lambda)$, $\lambda\in\mathbb{R}$. For non-zero λ this is clear from the corresponding factorization for W; for $\lambda = 0$ we have $G_-(0)G_+(0) = (I-Q^*K)(I-PK) = I-K = I-H(0)$. From what we saw in the preceding paragraph it is now clear that we have arrived at a right canonical factorization with respect to the real line, of the symbol $I-H(\lambda)$. Explicit formulas for the factors G_-, G_+ and their inverses G_-^{-1}, G_+^{-1} can be obtained from the descriptions of $\widehat{W}_+$, $\widehat{W}_-$, $\widehat{W}_+^{-1}$ and $\widehat{W}_-^{-1}$ given earlier in this section. In fact the formulas in question coincide with the ones already presented in the first paragraph of this section. For the verification of this we need the following intertwining result.

Lemma 6.7. *Let P and Q be as above. Then $(I-Q^*)T = TP$.*

Proof. It is sufficient to establish the identities $(I-Q^*)T(I-P) = 0$ and $Q^*TP = 0$. For the first of these we argue as follows. Clearly

$$\langle (I-Q^*)T(I-P)f, g\rangle = \langle T(I-P)f, (I-Q)g\rangle.$$

Now $(I-P)f\in\mathcal{H}_-$ and $(I-Q)g\in\mathcal{H}_+$. As $\mathcal{H}_-$ is T-invariant we also have $T(I-P)f\in\mathcal{H}_-$. But $\mathcal{H}_-\perp\mathcal{H}_+$. So $\langle (I-Q^*)T(I-P)f, g\rangle = 0$ for all f and g in $\mathcal{H}$. It follows that $(I-Q^*)T(I-P) = 0$, as desired. Next observe that

$$\langle Q^*TPf, g\rangle = \langle TPf, Qg\rangle = [(I-K)^{-1}TPf, Qg] = [T^\times Pf, Qg].$$

As $Pf\in\mathcal{H}_p$ and $\mathcal{H}_p$ is invariant under $T^\times$, we have $T^\times Pf\in\mathcal{H}_p$. Furthermore $Qg\in\mathcal{H}_m$. But $\mathcal{H}_m\perp\mathcal{H}_p$ is $\mathcal{H}$ endowed with the inner product $[\cdot,\cdot]$. It follows that $(Q^*TPf, g) = 0$ for all f and g. Hence $Q^*TP = 0$, which is the second identity we wanted to establish. □

We proceed by deriving the state space formulas for G_-, G_+ and their inverses G_-^{-1}, G_+^{-1}. Recall that $\widehat{W}_+(\lambda) = I - (I-K)^{-1}KTQ(\lambda - T)^{-1}$. Hence, for $\lambda \neq 0$,

$$\begin{aligned} G_-(\lambda) &= \widehat{W}_+(i/\overline{\lambda})^*(I - Q^*K) \\ &= \big(I - (I-K)^{-1}KTQ(i/\overline{\lambda} - T)^{-1}\big)^*(I - Q^*K) \\ &= \big(I - i\lambda(I - i\lambda T)^{-1}Q^*TK(I-K)^{-1}\big)(I - Q^*K). \end{aligned}$$

On account of Lemma 6.7, we have $Q^*T = T(I-P)$. Also $(I-K)^{-1}(I-Q^*K) = (I-PK)^{-1}$, and we get

$$\begin{aligned} G_-(\lambda) &= \big(I - i\lambda\,(I - i\lambda T)^{-1}\,T(I-P)K(I-K)^{-1}\big)(I - Q^*K) \\ &= \big(I - K - i\lambda\,(I - i\lambda T)^{-1}\,T(I-P)K\big)(I-PK)^{-1}. \end{aligned}$$

But then, proceeding in a straightforward manner,

$$\begin{aligned} G_-(\lambda) &= \big(I - K - i\lambda T(I - i\lambda T)^{-1}(I-P)K\big)(I-PK)^{-1} \\ &= \big(I - K + (I-P)K - (I - i\lambda T)^{-1}(I-P)K\big)(I-PK)^{-1} \\ &= \big(I - PK - (I - i\lambda T)^{-1}(I-P)K\big)(I-PK)^{-1} \\ &= I - (I - i\lambda T)^{-1}(I-P)K(I-PK)^{-1}. \end{aligned}$$

In this computation λ was of course taken to be non-zero. For $\lambda = 0$, the last expression in the above series of identities reduces to $I - (I-P)K(I-PK)^{-1}$ and this is easily seen to be equal to $(I-K)(I-PK)^{-1}$. The latter can be rewritten as $I - Q^*K$ which was earlier identified as the value $G_-(0)$ of G_- in the origin. So in the final analysis the zero value of λ is admissible too.

Next we turn to G_+ which was defined using W_-. For the latter we have the expression $\widehat{W}_-(\lambda) = I - (I-K)^{-1}KT(\lambda - T)^{-1}(I-Q)$ and we can carry out a similar computation as the one presented above:

$$\begin{aligned} G_+(\lambda) &= (I - Q^*K)^{-1}\widehat{W}_-(i/\overline{\lambda})^*(I-K) \\ &= (I - Q^*K)^{-1}\big(I - (I-K)^{-1}KT(i/\overline{\lambda} - T)^{-1}(I-Q)\big)^*(I-K) \\ &= (I - Q^*K)^{-1}\big(I - i\lambda(I-Q^*)(I - i\lambda T)^{-1}TK(I-K)^{-1}\big)(I-K) \\ &= (I - Q^*K)^{-1}\big(I - K - i\lambda(I-Q^*)T\,(I - i\lambda T)^{-1}\,K\big) \\ &= (I - Q^*K)^{-1}\big(I - K + (I-Q^*)(I - i\lambda T - I)(I - i\lambda T)^{-1}K\big) \\ &= (I - Q^*K)^{-1}\big(I - K + (I-Q^*)K - (I-Q^*)(I - i\lambda T)^{-1}K\big) \end{aligned}$$

$$
\begin{aligned}
&= (I - Q^*K)^{-1}\big(I - Q^*K - (I - Q^*)(I - i\lambda T)^{-1}K\big) \\
&= I - (I - Q^*K)^{-1}(I - Q^*)(I - i\lambda T)^{-1}K.
\end{aligned}
$$

For $\lambda = 0$, the last expression comes down to $I - (I - Q^*K)^{-1}(I - Q^*)K$ and this is easily seen to be equal to $(I - QK^*)^{-1}(I - K)$, so to $I - PK$. The latter was earlier identified as the value $G_+(0)$ of G_+ in the origin. So here the zero value of λ is admissible too.

Let us now deal with G_-^{-1} and G_+^{-1}. The first of these functions is tied to $\widehat{W}_+^{-1}$ for which we have the expression $\widehat{W}_+(\lambda)^{-1} = I + KT^\times(\lambda - T^\times)^{-1}Q$. From this we get

$$
\begin{aligned}
G_-^{-1}(\lambda) &= (I - Q^*K)^{-1}\big(\widehat{W}_+(i/\overline{\lambda})^*\big)^{-1} \\
&= (I - Q^*K)^{-1}\big(I + KT^\times(i/\overline{\lambda} - T^\times)^{-1}Q\big)^* \\
&= (I - Q^*K)^{-1}\big(I + Q^*\big((i/\overline{\lambda} - T^\times)^{-1}\big)^*(T^\times)^*K\big) \\
&= (I - Q^*K)^{-1}\big(I + i\lambda Q^*\big(I - i\lambda(T^\times)^*\big)^{-1}(T^\times)^*K\big) \\
&= (I - Q^*K)^{-1}\big(I - Q^*K + Q^*\big(I - i\lambda(T^\times)^*\big)^{-1}K\big) \\
&= I + (I - Q^*K)^{-1}Q^*\big(I - i\lambda(T^\times)^*\big)^{-1}K.
\end{aligned}
$$

Putting $\lambda = 0$ in the last expression gives $I + (I - Q^*K)^{-1}Q^*K$ which is obviously equal to $(I - Q^*K)^{-1}$, the value of G_-^{-1} at the origin.

Finally we consider G_+^{-1}. For the appropriate values of λ, we have

$$
\begin{aligned}
G_+^{-1}(\lambda) &= (I - K)^{-1}\big(\widehat{W}_-(i/\overline{\lambda})^*\big)^{-1}(I - Q^*K) \\
&= \big((I - K)^{-1}\big(\widehat{W}_-(i/\overline{\lambda})\big)^*(I - K)\big)^{-1}(I - K)^{-1}(I - Q^*K) \\
&= \Big(\big(\widehat{W}_-(i/\overline{\lambda})\big)^{[*]}\Big)^{-1}(I - PK)^{-1} \\
&= \Big(\big(\widehat{W}_-(i/\overline{\lambda})\big)^{-1}\Big)^{[*]}(I - PK)^{-1}.
\end{aligned}
$$

Here we have used (6.35) and the fact, already noted above, that $I - PK$ and $(I - K)^{-1}(I - Q^*K)$ are each other's inverse. Recall now that

$$
\widehat{W}_-(\lambda)^{-1} = I + KT^\times(I - Q)(\lambda - T^\times)^{-1}.
$$

Thus, as $T^\times$ and K are $[\cdot\,,\cdot]$-selfadjoint,

$$
\begin{aligned}
G_+^{-1}(\lambda) &= \big(I + KT^\times(I - Q)(i/\overline{\lambda} - T^\times)^{-1}\big)^{[*]}(I - PK)^{-1} \\
&= \big(I + (1/i\lambda - T^\times)^{-1}(I - Q)^{[*]}T^\times K\big)(I - PK)^{-1}
\end{aligned}
$$

$$= \big(I + i\lambda(I - i\lambda T^\times)^{-1}(I - Q^{[*]})T^\times K\big)(I - PK)^{-1}.$$

As an intermediate step, we note that the identity in Lemma 6.7 can be rewritten as $\big(I - Q^{[*]}\big)T^\times = T^\times P$. Indeed,

$$\begin{aligned} T^\times P &= (I-K)^{-1}TP \\ &= (I-K)^{-1}(I-Q^*)T \\ &= (I-K)^{-1}(I-Q)^*(I-K)T^\times \\ &= (I-Q)^{[*]}T^\times \\ &= \big(I-Q^{[*]}\big)T^\times. \end{aligned}$$

This makes it possible to proceed as follows:

$$\begin{aligned} G_+^{-1}(\lambda) &= \big(I + i\lambda(I - i\lambda T^\times)^{-1}T^\times PK\big)(I-PK)^{-1} \\ &= \big(I - PK + (I - i\lambda T^\times)^{-1}PK\big)(I-PK)^{-1} \\ &= I + (I - i\lambda T^\times)^{-1}PK(I-PK)^{-1}. \end{aligned}$$

The check for $\lambda = 0$ yields the desired result, namely $I + PK(I-PK)^{-1} = (I-PK)^{-1}$ which is the value of G_+^{-1} at the origin.

6.5 The matching of the subspaces

In the canonical factorization carried out in the previous section, we used that

$$\mathcal{H} = \mathcal{H}_- \dot{+} \mathcal{H}_p, \qquad \mathcal{H} = \mathcal{H}_+ \dot{+} \mathcal{H}_m. \tag{6.45}$$

In this section we shall prove that, indeed, the space $\mathcal{H}$ may be decomposed in these two ways.

Let P_- and P_+ be the orthogonal projections of $\mathcal{H}$ onto $\mathcal{H}_-$ and $\mathcal{H}_+$, respectively. Also, put $P_m = E^\times(0)$ and $P_p = I - E^\times(0)$, where $E^\times(t)$ is the spectral resolution of the identity for the operator $T^\times = (I-K)^{-1}T$ with respect to the inner product $[f,g] = \langle (I-K)f, g\rangle$. By definition

$$\mathcal{H}_- = \operatorname{Im} P_-, \qquad \mathcal{H}_+ = \operatorname{Im} P_+, \qquad \mathcal{H}_m = \operatorname{Im} P_m, \qquad \mathcal{H}_p = \operatorname{Im} P_p.$$

We claim that

$$\mathcal{H} = \mathcal{H}_- \dot{+} \mathcal{H}_p \iff P_+|_{\mathcal{H}_p} : \mathcal{H}_p \to \mathcal{H}_+ \text{ is bijective,} \tag{6.46}$$

$$\mathcal{H} = \mathcal{H}_+ \dot{+} \mathcal{H}_m \iff P_-|_{\mathcal{H}_m} : \mathcal{H}_m \to \mathcal{H}_- \text{ is bijective.} \tag{6.47}$$

The argument for this is simple and in a different context (involving a different notation too) spelled out in the beginning of *Part 2* of the proof of Theorem 4.4.

For the convenience of the reader we give it here too. Note that $\operatorname{Ker}(P_+|_{\mathcal{H}_p}) = \mathcal{H}_- \cap \mathcal{H}_p$, and thus $P_+|_{\mathcal{H}_p}$ is injective if and only if $\mathcal{H}_- \cap \mathcal{H}_p = \{0\}$. Next, observe that for each $y \in \mathcal{H}_p$ we have $y = (I - P_+)y + P_+|_{\mathcal{H}_p} y \in \mathcal{H}_- + \operatorname{Im}(P_+|_{\mathcal{H}_p})$. Thus $\mathcal{H}_- + \mathcal{H}_p \subset \mathcal{H}_- + \operatorname{Im}(P_+|_{\mathcal{H}_p})$. The reverse inclusion also holds. Indeed, for $z \in \mathcal{H}_p$ we have $P_+ z = (P_+ z - z) + z \in \operatorname{Ker} P_+ + \mathcal{H}_p = \mathcal{H}_- + \mathcal{H}_p$. It follows that $\mathcal{H}_- + \operatorname{Im}(P_+|_{\mathcal{H}_p}) = \mathcal{H}_- + \mathcal{H}_p$, and hence $P_+|_{\mathcal{H}_p}$ is surjective if and only if $\mathcal{H} = \mathcal{H}_- + \mathcal{H}_p$. This proves (6.46). The proof of (6.47) is similar. Now $\mathcal{H} = \mathcal{H}_m \dot{+} \mathcal{H}_p$ and $\mathcal{H} = \mathcal{H}_- \dot{+} \mathcal{H}_+$. Combining this with (6.46) and (6.47), we see that (6.45) holds if and only if the operator $V = P_- P_m + P_+ P_p$ is bijective.

It is not difficult to prove that V is injective. Indeed, take $f \in \mathcal{H}$ and assume $Vf = 0$. Put $f_m = P_m f$ and $f_p = P_p f$. Then $P_- f_m + P_+ f_p = Vf = 0$, and hence $P_+ f_p = 0$ and $P_- f_m = 0$. The latter gives $f_m = P_+ f_m$, and we get

$$0 \geq [T^\times f_m, f_m] = \langle T f_m, f_m \rangle = \langle T P_+ f_m, P_+ f_m \rangle \geq 0.$$

It follows that $P_+ f_m \in \operatorname{Ker} T$. But T is injective. So $P_+ f_m = 0$. As $P_- f_m = 0$ too, we have $f_m = 0$. In the same way one proves that $f_p = 0$. Hence $f = 0$, as desired.

To prove that V is surjective too, we use that $I - V$ is compact. Indeed, as soon as we know that this is the case, the Fredholm alternative implies that $V = I - (I - V)$ is surjective if and only if V is injective.

Lemma 6.8. *The operator $I - V$ is compact.*

Proof. The compact operators form an ideal and

$$\begin{aligned} I - V &= P_- + P_+ - P_- P_m - P_+ P_p \\ &= P_- + P_+ P_m + P_+ P_p - P_- P_m - P_+ P_p \\ &= P_- + P_+ P_m - P_- P_m \\ &= (P_+ - P_-)(P_m - P_-). \end{aligned}$$

Hence it suffices to prove that $P_m - P_-$ is compact. Now $P_m = E^\times(0)$, where $E^\times(t)$ is the spectral resolution of the identity for $T^\times$ with respect to the inner product $[\cdot,\cdot]$. Similarly, $P_- = E(0)$, where $E(t)$ is the spectral resolution of the identity for T. As T and $T^\times$ are injective, in both cases the spectral resolutions are continuous at zero. So, using a standard formula for the spectral resolution (see [99], Problem VI.5.7) we may write, for each $f \in \mathcal{H}$,

$$(P_m - P_-)f = \lim_{h \downarrow 0} \frac{1}{2\pi i} \int_{\Gamma_h} \left((\lambda - T)^{-1} - (\lambda - T^\times)^{-1}\right) f \, d\lambda. \tag{6.48}$$

Here h is a (sufficiently small) positive number and Γ_h is the union of two non-closed oriented curves as in the following picture:

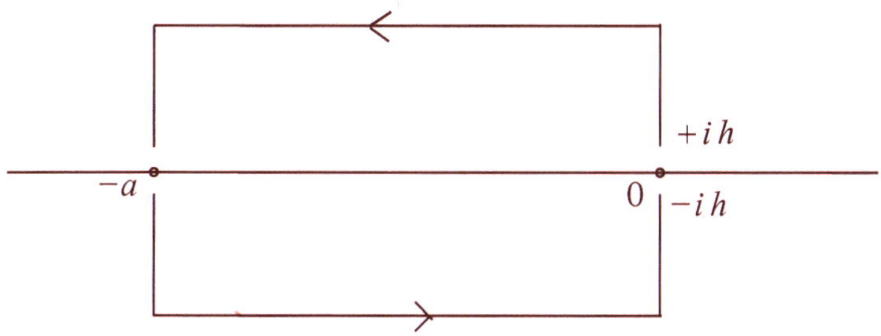

The positive number a is chosen in such a way that the spectra of T and $T^\times$ both are in the open half-line $(-a, \infty)$. For the difference of the resolvents of T and $T^\times$ appearing in (6.48) we have

$$\begin{aligned}(\lambda - T)^{-1} - (\lambda - T^\times)^{-1} &= (\lambda - T)^{-1}(I - (\lambda - T)(\lambda - T^\times)^{-1}) \\ &= (\lambda - T)^{-1}(T - T^\times)(\lambda - T^\times)^{-1} \\ &= -(\lambda - T)^{-1}KT^\times(\lambda - T^\times)^{-1},\end{aligned}$$

and from the latter expression we see that it is a finite rank (hence compact) operator.

Let Δ be the closed contour obtained from Γ_h by letting the positive number h go to zero. As $T^\times$ is selfadjoint in $\mathcal{H}$ endowed with the inner product $[\cdot\,,\cdot]$, we know from (6.32) in Lemma 6.4 and the choice of a that $T^\times(\lambda - T^\times)^{-1}$ is bounded in norm on $\Delta \setminus \{0\}$.

Next, let us investigate $(\lambda - T)^{-1}K$. First we shall prove that

$$\|(ic - T)^{-1}K\| \le \frac{q_0}{\sqrt{|c|}}, \qquad 0 \neq c \in \mathbb{R}, \tag{6.49}$$

where q is some positive constant. To prove this, recall that K is the finite rank operator given by the right-hand side of (6.7), and hence

$$\|(ic - T)^{-1}K\| \le \sum_{j=0}^{n} |a_j|\,\|p_j\|\,\|(ic - T)^{-1}p_j\|, \qquad 0 \neq c \in \mathbb{R}.$$

For each j the function p_j is a normalized Legendre polynomial in t (and so the norm of p_j appearing in the above expression is actually equal to 1). Also T is the multiplication operator given by the left-hand side of (6.7). So to find an upper bound for $\|(ic - T)^{-1}p_j\|$, we need to estimate

$$\sqrt{\int_{-1}^{1} \frac{t^{2k}}{c^2 + t^2}\,dt}\,. \tag{6.50}$$

As $t^{2k+2} \leq t^{2k}$ for $|t| \leq 1$, it suffices to find an upper bound for (6.50) for the case $k = 0$. But

$$\sqrt{\int_{-1}^{1} \frac{dt}{c^2+t^2}} = \sqrt{\frac{2}{|c|}\arctan\frac{1}{|c|}}, \qquad 0 \neq c \in \mathbb{R}.$$

This proves (6.49) for an appropriate choice of q_0.

Note that the function $(\lambda - T)^{-1}KT^{\times}(\lambda - T^{\times})^{-1}$ is continuous on $\Delta \setminus \{0\}$. Also, for some positive constant q,

$$\|(ic - T)^{-1}KT^{\times}(ic - T^{\times})^{-1}\| \leq \frac{q}{\sqrt{|c|}}, \qquad 0 \neq c \in \mathbb{R}.$$

A straightforward Cauchy argument now gives that

$$\lim_{h\downarrow 0} \int_{\Gamma_h} (\lambda - T)^{-1}KT^{\times}(\lambda - T^{\times})^{-1}\, d\lambda$$

exists in norm. But then the same is true for

$$\lim_{h\downarrow 0} \int_{\Gamma_h} \left((\lambda - T)^{-1} - (\lambda - T^{\times})^{-1}\right) d\lambda.$$

As the integrand in this expression is a compact operator-valued function, we can use (6.48) to show that $P_m - P_-$ is compact too. □

Close inspection of the above proof shows that $I - V$ is in fact a trace class operator (cf., Lemma 6.3 in [11]).

6.6 Formulas for solutions

Let $I - H(\lambda)$ be the symbol of the Wiener-Hopf integral equation (6.10). From the results of the previous sections we know that $I - H(\lambda)$ admits a right canonical factorization with respect to the real line:

$$I - H(\lambda) = G_-(\lambda)G_+(\lambda), \qquad -\infty < \lambda < \infty. \tag{6.51}$$

As we have seen in Section 6.3, this implies that equation (6.10) is uniquely solvable in $L^1([0,\infty),\mathcal{H})$, where $\mathcal{H} = L_2[-1,1]$. This fact and the equivalence (explained in the first section of this chapter) of equations (6.1) and (6.10), allows us to prove the following result.

Theorem 6.9. *Consider equation* (6.1) *with the kernel function k being given by* (6.4). *Let T and K be the operators on $L_2[-1,1]$ defined by* (6.7), *and assume that $I - K$ is strictly positive. Then equation* (6.1) *has a unique solution ψ satisfying the initial condition* (6.2) *and*

$$\int_0^{\infty} \int_{-1}^{1} |\psi(t,\mu)|^2 \, d\mu \, dt < \infty. \tag{6.52}$$

This solution is given by

$$\psi(t,\cdot) = e^{-t(T_p^\times)^{-1}} Pf_+, \qquad t \geq 0. \tag{6.53}$$

Here f_+ is the given function appearing in the initial condition (6.2)*, the operator P is the projection of $L_2[-1,1]$ defined directly after* (6.38)*, and $T_p^\times$ is the restriction of $T^\times = (I-K)^{-1}T$ to $\mathcal{H}_p = \operatorname{Im} P$.*

Note that (6.53) is the natural analogue of (6.25) in Theorem 6.3. Formula (6.53) features the inverse of the injective operator

$$T_p^\times = T^\times|_{\mathcal{H}_p} : \mathcal{H}_p \to \mathcal{H}_p.$$

This operator has dense range and is nonnegative with respect to the inner product $[\cdot,\cdot]$. Hence its inverse $(T_p^\times)^{-1}(\mathcal{H}_p \to \mathcal{H}_p)$ is an unbounded operator which has $\operatorname{Im} T_p^\times$ as its (dense) domain and is nonnegative with regard to the inner product $[\cdot,\cdot]$. Thus the expression

$$e^{-t(T_p^\times)^{-1}} \tag{6.54}$$

is well-defined via the operational calculus for selfadjoint operators based on the notion of the resolution of the identity. One can view (6.54) also as the operator semigroup generated by the unbounded infinitesimal generator $-(T_p^\times)^{-1}$.

Proof. Recall that $I-H(\lambda)$ is the symbol of equation (6.10). Since $I-H(\lambda)$ admits the canonical Wiener-Hopf factorization (6.51) we can use the general theory of Wiener-Hopf equations (see the one but last paragraph in Section 6.3) to show that equation (6.10) has a unique solution ψ in $L^1([0,\infty),\mathcal{H})$, where $\mathcal{H} = L_2[-1,1]$. Moreover, the Fourier transform $\widehat{\psi}$ of ψ is given by

$$\widehat{\psi}(\lambda) = G_+^{-1}(\lambda)\mathcal{P}\big(G_-^{-1}(\lambda)\widehat{F}(\lambda)\big), \tag{6.55}$$

where $\widehat{F}$ is the Fourier transform of the right-hand side of equation (6.10), and $\mathcal{P}$ is the projection defined by

$$\mathcal{P}\left(\int_{-\infty}^{\infty} e^{it\lambda} f(t)\,dt\right) = \int_0^{\infty} e^{it\lambda} f(t)\,dt.$$

Since $\psi \in L^1([0,\infty),\mathcal{H})$, condition (6.52) is fulfilled. To derive formula (6.53), we first compute $\widehat{\psi}$ using equation (6.55).

Recall that F is given by (6.9). It follows that

$$\widehat{F}(\lambda) = (I - i\lambda T)^{-1}Tf_+, \qquad \Im\lambda \geq 0. \tag{6.56}$$

As we know from Section 6.4 the inverses of the factors $G_-(\lambda)$ and $G_+(\lambda)$ in (6.51) are given by

$$G_-^{-1}(\lambda) = I + (I - Q^*K)^{-1}Q^*\big(I - i\lambda(T^\times)^*\big)^{-1}K,$$

$$G_+^{-1}(\lambda) = I + (I - i\lambda T^\times)^{-1}PK(I-PK)^{-1}.$$

Here $T^\times = (I-K)^{-1}T$. Let us use these formulas to compute $\widehat{\psi}(\lambda)$ from (6.55). As a first step we have $G_-^{-1}(\lambda)\widehat{F}(\lambda) = \widehat{F}(\lambda) + X(\lambda)\widehat{F}(\lambda)$, where

$$\begin{aligned} X(\lambda) &= (I-Q^*K)^{-1}Q^*\big(I - i\lambda(T^\times)^*\big)^{-1}K \\ &= (I-Q^*K)^{-1}Q^*\big(I - i\lambda T(I-K)^{-1}\big)^{-1}K \\ &= (I-Q^*K)^{-1}Q^*(I-K)\big(I-K-i\lambda T\big)^{-1}K. \end{aligned}$$

Thus $X(\lambda)\widehat{F}(\lambda) = (I-Q^*K)^{-1}Q^*(I-K)R(\lambda)Tf_+$, where

$$\begin{aligned} R(\lambda) &= (I-K-i\lambda T)^{-1}K(I-i\lambda T)^{-1} \\ &= (I-K-i\lambda T)^{-1}\big((I-i\lambda T)-(I-K-i\lambda T)\big)(I-i\lambda T)^{-1} \\ &= (I-K-i\lambda T)^{-1} - (I-i\lambda T)^{-1}. \end{aligned}$$

Hence

$$\begin{aligned} X(\lambda)\widehat{F}(\lambda) = {}& (I-Q^*K)^{-1}Q^*(I-K)(I-K-i\lambda T)^{-1}Tf_+ \\ & -(I-Q^*K)^{-1}Q^*(I-K)(I-i\lambda T)^{-1}Tf_+. \end{aligned}$$

We conclude that

$$\begin{aligned} G_-^{-1}(\lambda)\widehat{F}(\lambda) = {}& \widehat{F}(\lambda) - (I-Q^*K)^{-1}Q^*(I-K)\widehat{F}(\lambda) \\ & + (I-Q^*K)^{-1}Q^*\big(I-i\lambda(T^\times)^*\big)^{-1}Tf_+. \end{aligned}$$

Now apply the projection $\mathcal{P}$. Since $f_+ \in \mathcal{H}_+$ and $T|_{\mathcal{H}_+}$ is nonnegative, we have $\mathcal{P}(\widehat{F}) = \widehat{F}$. Furthermore, using the spectral properties of $T^\times$ and the definition of Q, we see that the function $Q^*(I-i\lambda(T^\times)^*)^{-1}$ is annihilated by $\mathcal{P}$. Therefore

$$\begin{aligned} \mathcal{P}\big(G_-^{-1}(\lambda)\widehat{F}(\lambda)\big) &= \widehat{F}(\lambda) - (I-Q^*K)^{-1}Q^*(I-K)\widehat{F}(\lambda) \\ &= (I-Q^*K)^{-1}\big(I-Q^*K-Q^*(I-K)\big)\widehat{F}(\lambda) \\ &= (I-Q^*K)^{-1}(I-Q^*)\widehat{F}(\lambda). \end{aligned}$$

Put $Z(\lambda) = (I-Q^*K)^{-1}(I-Q^*)\widehat{F}(\lambda)$. Recall from the previous section that $I-PK$ is invertible with inverse $(I-K)^{-1}(I-Q^*K)$. Hence

$$(I-PK)^{-1}(I-Q^*K)^{-1} = (I-K)^{-1},$$

and it follows that $G_+^{-1}(\lambda)\mathcal{P}\big(G_-^{-1}(\lambda)\widehat{F}(\lambda)\big) = Z(\lambda) + H(\lambda)$, where

$$\begin{aligned} H(\lambda) &= (I-i\lambda T^\times)^{-1}PK(I-K)^{-1}(I-Q^*)\widehat{F}(\lambda) \\ &= (I-i\lambda T^\times)^{-1}P\big(I-(I-K)\big)(I-K)^{-1}(I-Q^*)\widehat{F}(\lambda) \\ &= A(\lambda) - B(\lambda), \end{aligned}$$

with

$$A(\lambda) = (I - i\lambda T^\times)^{-1} P(I-K)^{-1}(I-Q^*)T(I-i\lambda T)^{-1} f_+,$$

$$B(\lambda) = (I - i\lambda T^\times)^{-1} P(I-Q^*)T(I-i\lambda T)^{-1} f_+.$$

Using $(I-Q^*)T = TP$ and $T^\times = (I-K)^{-1}T$ we get

$$\begin{aligned} A(\lambda) &= (I - i\lambda T^\times)^{-1} P(I-K)^{-1} TP(I-i\lambda T)^{-1} f_+ \\ &= (I - i\lambda T^\times)^{-1} PT^\times P(I-i\lambda T)^{-1} f_+ \\ &= (I - i\lambda T^\times)^{-1} T^\times P(I-i\lambda T)^{-1} f_+, \end{aligned}$$

and

$$\begin{aligned} B(\lambda) &= (I - i\lambda T^\times)^{-1} PTP(I-i\lambda T)^{-1} f_+ \\ &= (I - i\lambda T^\times)^{-1} PT(I-i\lambda T)^{-1} f_+. \end{aligned}$$

Thus (with $\lambda \neq 0$ in the intermediate steps)

$$\begin{aligned} H(\lambda) &= (I - i\lambda T^\times)^{-1}(T^\times P - PT)(I-i\lambda T)^{-1} f_+ \\ &= \frac{1}{i\lambda}\big((I - i\lambda T^\times)^{-1}\big(P(I-i\lambda T) - (I-i\lambda T^\times)P\big)(I-i\lambda T)^{-1} f_+\big) \\ &= \frac{1}{i\lambda}(I - i\lambda T^\times)^{-1} Pf_+ - \frac{1}{i\lambda} P(I-i\lambda T)^{-1} f_+ \\ &= \frac{1}{i\lambda} Pf_+ + T^\times(I - i\lambda T^\times)^{-1} Pf_+ - \frac{1}{i\lambda} Pf_+ - PT(I-i\lambda T)^{-1} f_+ \\ &= T^\times(I - i\lambda T^\times)^{-1} Pf_+ - PT(I-i\lambda T)^{-1} f_+. \end{aligned}$$

Therefore

$$\begin{aligned} G_+^{-1}(\lambda)\mathcal{P}\big(G_-^{-1}(\lambda)\widehat{F}(\lambda)\big) &= T^\times(I - i\lambda T^\times)^{-1} Pf_+ \\ &\quad + \big((I-Q^*K)^{-1}(I-Q^*) - P\big)T(I-i\lambda T)^{-1} f_+. \end{aligned}$$

From Lemma 6.5 we get

$$\begin{aligned} (I-Q^*K)^{-1}(I-Q^*) - P &= (I-Q^*K)^{-1}(I - Q^* - P + Q^*KP) \\ &= (I-Q^*K)^{-1}(I - Q^* - P + Q^*P) \\ &= (I-Q^*K)^{-1}(I-Q^*)(I-P) = 0, \end{aligned}$$

and we conclude that

$$\widehat{\psi}(\lambda) = G_+^{-1}(\lambda)\mathcal{P}\big(G_-^{-1}(\lambda)\widehat{F}(\lambda)\big) = T^\times(I - i\lambda T^\times)^{-1} Pf_+. \tag{6.57}$$

Now $T^\times$ maps $\mathcal{H}_p$ into $\mathcal{H}_p$, and so the operator $T_p^\times = T^\times|_{\mathcal{H}_p} : \mathcal{H}_p \to \mathcal{H}_p$ is well-defined. Since $T_p^\times$ is injective, the expression (6.57) can be rewritten as

$$\widehat{\psi}(\lambda) = -\big(i\lambda - (T_p^\times)^{-1}\big)^{-1} P f_+. \tag{6.58}$$

As was already observed, $(T_p^\times)^{-1}\big(\mathcal{H}_p \to \mathcal{H}_p\big)$ is an unbounded operator which has $\operatorname{Im} T_p^\times$ as its (dense) domain and is nonnegative with regard to the inner product $[\cdot\,,\cdot]$. Hence we can take the inverse Fourier transform in (6.58), to get the desired formula (6.53).

From (6.53) we see that $\psi(0) = \psi(0, \cdot\,) = Pf_+$. Now let P_+ be the orthogonal projection of $L_2[-1,1]$ onto $\mathcal{H}_+$. Since $\operatorname{Ker} P = \mathcal{H}_-$, we have $P_+(I-P) = 0$, and thus

$$P_+\psi(0) = P_+(Pf_+) = P_+(Pf_+ + (I-P)f_+) = P_+f_+ = f_+.$$

Therefore ψ satisfies the initial condition (6.2). Finally, the uniqueness statement follows from the general theory of Wiener-Hopf equations. □

Notes

The theory of the linear transport equation has a long history. For this see the books [28] and [96] which also contain extensive lists of references. The material in Section 6.2 is taken from Section XIII.9 of [51] where the reader can also find an illustrative example. The other sections in this chapter follow basically Chapter 6 in [11] which was inspired by the dissertation [80] and the papers [81], [82]. In [108] one can also find an analytic description of the subspaces concerned. Later results based on [110] and [124] are also included here. Further developments using the method described in this chapter can be found in [110], where the case of non-degenerate kernel functions $k(\mu, \mu')$ is treated. See also the book [78], and the paper [124]. For an alternative proof of Theorem 6.9, not using Wiener-Hopf factorization, we refer to Section XIX.7 in [51].

The results presented in Sections 6.4 – 6.6 can also be understood from the point of view described in Chapter 5. Note, however, that in Sections 6.4–6.6 the symbol is an operator-valued function (and not a matrix-valued function as in Chapter 5). On the other hand, the operator $(T^\times)^{-1}$ appearing in Theorem 6.9 is exponentially dichotomous. This has been proved in Section 5.2 of the recent monograph [111]. The latter book also contains many new additions related to the analysis of equation (6.6). See also the notes to Chapter 5.

Chapter 7

Wiener-Hopf factorization and factorization indices

This chapter concerns canonical as well as non-canonical Wiener-Hopf factorization of an operator-valued function which is analytic on a Cauchy contour. Such an operator function is given by a realization with a possibly infinite dimensional Banach space as state space, and with a bounded state operator and with bounded input-output operators. The first main result is a generalization to operator-valued functions of the canonical factorization theorem for rational matrix functions presented earlier in Section 3.1. In terms of the given realization, necessary and sufficient conditions are also presented in order that the operator function involved admits a (possibly non-canonical) Wiener-Hopf factorization. The corresponding factorization indices are described in terms of certain spectral invariants which are defined in terms of the realization but do only depend on the operator function and not on the particular choice of the realization. The analysis of these spectral invariants is one of the main themes of this chapter.

The chapter consists of three sections. Section 7.1 describes the main result for canonical factorization and introduces the spectral invariants involved. The proof that the spectral invariants do not depend on the particular realization is given in Section 7.2. The final section of the chapter, Section 7.3, deals with non-canonical Wiener-Hopf factorization and the corresponding factorization indices.

7.1 Canonical factorization of operator functions

Throughout this chapter, W is an operator function, analytic on an open neighborhood of a given Cauchy contour Γ, and with values that are operators on a possibly infinite dimensional Banach space Y. Anticipating the results to be presented below, we note that in this situation W admits a realization on Γ involving a possibly infinite dimensional state space X and having I_Y as external operator:

$$W(\lambda) = I_Y + C(\lambda I_X - A)^{-1}B, \tag{7.1}$$

where Γ splits the spectrum of A, that is $\Gamma \subset \rho(A)$. This is immediate from Theorem 2.2.

As before, we denote by F_+ the interior domain of Γ, and by F_- the complement of $\overline{F}_+$ in the Riemann sphere $\mathbb{C}_\infty$. By a *right canonical factorization* of W with respect to Γ we mean a factorization

$$W(\lambda) = W_-(\lambda)W_+(\lambda), \qquad \lambda \in \Gamma, \tag{7.2}$$

where W_- and W_+ are functions with values in $\mathcal{L}(Y)$ satisfying

(i) W_- is analytic on F_- and continuous on $\overline{F}_-$,

(ii) W_+ is analytic on F_+ and continuous on $\overline{F}_+$,

(iii) W_- and W_+ take invertible values on $\overline{F}_-$ and $\overline{F}_+$, respectively.

If in (7.2) the factors W_- and W_+ are interchanged, we speak of a *left canonical factorization*. A necessary condition for a right or left canonical factorization with respect to Γ to exist is that W takes invertible values on Γ. In terms of the realization (7.1) this means that Γ also splits the spectrum of the associate main operator $A^\times = A - BC$ (see Theorem 2.4).

We now extend Theorem 3.2 to a possibly infinite dimensional context.

Theorem 7.1. *Let W be an operator function, analytic on an open neighborhood of a Cauchy contour Γ, and with values that are operators on a Banach space Y. Let (7.1) be a realization of W, i.e.,*

$$W(\lambda) = I_Y + C(\lambda I_X - A)^{-1}B,$$

and suppose Γ splits the spectrum of A. Then W admits a right canonical factorization with respect to Γ if and only if the following two conditions are satisfied:

(a) *Γ splits the spectrum of $A^\times = A - BC$,*

(b) $X = \operatorname{Im} P(A;\Gamma) \dot{+} \operatorname{Ker} P(A^\times;\Gamma)$.

In that case, a right canonical factorization of W is given by

$$W(\lambda) = W_-(\lambda)W_+(\lambda), \qquad \lambda \in \Gamma,$$

where the factors and their inverses can be written as

$$\begin{aligned} W_-(\lambda) &= I_m + C(\lambda I_X - A)^{-1}(I_X - \Pi)B, \\ W_+(\lambda) &= I_m + C\Pi(\lambda I_X - A)^{-1}B, \\ W_-^{-1}(\lambda) &= I_m - C(I_X - \Pi)(\lambda I_X - A^\times)^{-1}B, \\ W_+^{-1}(\lambda) &= I_m - C(\lambda I_X - A^\times)^{-1}\Pi B. \end{aligned}$$

Here Π is the projection of $\mathbb{C}^n$ along $\operatorname{Im} P(A;\Gamma)$ *onto* $\operatorname{Ker} P(A^\times;\Gamma)$.

For left canonical factorizations an analogous theorem holds. In the result in question, (b) is replaced by $X = \operatorname{Ker} P(A;\Gamma) \dotplus \operatorname{Im} P(A^\times;\Gamma)$. The theorem also has an analogue for appropriate closed contours in the Riemann sphere $\mathbb{C}_\infty$ like the extended real line or the extended imaginary axis.

Proof. To establish the theorem, we can rely for a large part on the proof of Theorem 3.2. In fact, we only have to add an argument for the following assertion: if W admits a right canonical factorization with respect to Γ, then the decomposition in (b) holds. The first step consists in showing that if W admits a right canonical factorization with respect to Γ, then there is a way of representing W in the form $W(\lambda) = I_Y + \widetilde{C}(\lambda I_{\widetilde{X}} - \widetilde{A})^{-1}\widetilde{B}$ such that Γ splits the spectra of $\widetilde{A}$ and $\widetilde{A}^\times$ while, in addition, $\widetilde{X} = \operatorname{Im} P(\widetilde{A};\Gamma) \dotplus \operatorname{Ker} P(\widetilde{A}^\times;\Gamma)$.

Let $W(\lambda) = W_-(\lambda)W_+(\lambda)$, $\lambda \in \Gamma$, be a right canonical factorization of W. Recall that ∞ belongs to F_-. Since $W_-(\infty)$ is invertible we may assume without loss of generality that $W_-(\infty) = I_Y$. From the identity $W_-(\lambda) = W(\lambda)W_+(\lambda)^{-1}$ and the fact that W is analytic on a neighborhood of Γ, it follows that W_- has an analytic extension, again denoted by W_-, to some open neighborhood Ω_- of the closed set $F_- \cup \Gamma$. Taking Ω_- sufficiently small, we have that W_- assumes only invertible values on Ω_-. But then Theorems 2.3 and 2.4 can be applied to show that W_- admits a realization of the form

$$W_-(\lambda) = I_Y + C_-(\lambda I_{X_-} - A_-)^{-1}B_-, \qquad \lambda \in \Omega_-, \tag{7.3}$$

where $\sigma(A_-) \subset F_+$ and $\sigma(A_-^\times) \subset F_+$. Here $A_-^\times = A_- - B_-C_-$.

A similar reasoning holds for W_+. This function has an analytic extension, again denoted by W_+, to some open neighborhood Ω_+ of the closed set $F_+ \cup \Gamma$. Taking Ω_+ sufficiently small, we have that $W_+(\lambda)$ is invertible for all $\lambda \in \Omega_+$. But then Theorems 2.2 and 2.4 yield that W_+ admits a realization

$$W_+(\lambda) = I_Y + C_+(\lambda I_{X_+} - A_+)^{-1}B_+, \qquad \lambda \in \Omega_+, \tag{7.4}$$

such that $\sigma(A_+) \subset F_-$ and $\sigma(A_+^\times) \subset F_-$. Here $A_+^\times = A_+ - B_+C_+$.

On Γ we have the factorization $W(\lambda) = W_-(\lambda)W_+(\lambda)$, and so we can apply the product rule of Section 2.5 to show that $W(\lambda) = I_Y + \widetilde{C}(\lambda I_{\widetilde{X}} - \widetilde{A})^{-1}\widetilde{B}$, $\lambda \in \Gamma$, where $\widetilde{X} = X_- \dotplus X_+$ and $A : \widetilde{X} \to \widetilde{X}$, $B : Y \to \widetilde{X}$ and $C : \widetilde{X} \to Y$ are given by the operator matrices

$$\widetilde{A} = \begin{bmatrix} A_- & B_-C_+ \\ 0 & A_+ \end{bmatrix}, \qquad \widetilde{B} = \begin{bmatrix} B_- \\ B_+ \end{bmatrix}, \qquad \widetilde{C} = \begin{bmatrix} C_- & C_+ \end{bmatrix}.$$

The realization $I_Y + \widetilde{C}(\lambda I_{\widetilde{X}} - \widetilde{A})^{-1}\widetilde{B}$ has the desired properties. This can be seen as follows.

From the operator matrix representation for $\tilde{A}$, and the corresponding one for $A^\times = A - BC : \widetilde{X} \to \widetilde{X}$, namely

$$\widetilde{A}^\times = \begin{bmatrix} A_-^\times & 0 \\ -B_+C_- & A_+^\times \end{bmatrix},$$

it is immediate that Γ splits the spectra of both $\widetilde{A}$ and $\widetilde{A}^\times$. Furthermore, the spectral projections $P(\widetilde{A};\Gamma)$ and $P(\widetilde{A}^\times,\Gamma)$ are of the form

$$P(\widetilde{A};\Gamma) = \begin{bmatrix} I_{X_-} & \star \\ 0 & 0 \end{bmatrix}, \qquad P(\widetilde{A}^\times,\Gamma) = \begin{bmatrix} I_{X_-} & 0 \\ \star & 0 \end{bmatrix}.$$

Hence $\operatorname{Im} P(\widetilde{A};\Gamma) = X_- \dotplus \{0\}$ and $\operatorname{Ker} P(\widetilde{A}^\times;\Gamma) = \{0\} \dotplus X_+$, and from this $\widetilde{X} = \operatorname{Im} P(\widetilde{A};\Gamma) \dotplus \operatorname{Ker} P(\widetilde{A}^\times;\Gamma)$ is immediate.

The proof can now be finished by verifying the following two identities:

$$\begin{aligned} \dim\big(\operatorname{Im} P(\widetilde{A};\Gamma) \cap \operatorname{Ker} P(\widetilde{A}^\times;\Gamma)\big) &= \dim\big(\operatorname{Im} P(A;\Gamma) \cap \operatorname{Ker} P(A^\times;\Gamma)\big), \\ \dim\left(\frac{\widetilde{X}}{\operatorname{Im} P(\widetilde{A};\Gamma) + \operatorname{Ker} P(\widetilde{A}^\times;\Gamma)}\right) &= \dim\left(\frac{X}{\operatorname{Im} P(A;\Gamma) + \operatorname{Ker} P(A^\times;\Gamma)}\right). \end{aligned}$$

In other words, we are ready once it has been shown that the right-hand side of these identities depend only on W and Γ and are independent of the realization (7.1) of W. This is indeed the case as is seen from Theorem 7.2 below which even exhibits several other spectral invariants. □

Theorem 7.2. *Let W be an operator function, analytic on an open neighborhood of a Cauchy contour Γ, and with values that are operators on a Banach space Y. Let (7.1) be a realization of W, i.e., $W(\lambda) = I_Y + C(\lambda I_X - A)^{-1}B$, and suppose Γ splits the spectrum of A. In addition, assume that Γ also splits the spectrum of $A^\times = A - BC$. Introduce*

$$P = P(A;\Gamma), \qquad M = \operatorname{Im} P, \qquad P^\times = P(A^\times;\Gamma), \qquad M^\times = \operatorname{Ker} P^\times.$$

Then the quantities

$$\dim(M \cap M^\times), \qquad \dim\left(\frac{X}{M + M^\times}\right),$$

$$\dim\left(\frac{M \cap M^\times \cap \operatorname{Ker} C \cap \operatorname{Ker} CA \cap \cdots \cap \operatorname{Ker} CA^{k-1}}{M \cap M^\times \cap \operatorname{Ker} C \cap \operatorname{Ker} CA \cap \cdots \cap \operatorname{Ker} CA^{k}}\right), \qquad k = 0,1,2\ldots,$$

$$\dim\left(\frac{M + M^\times + \operatorname{Im} B + \operatorname{Im} AB + \cdots + \operatorname{Im} AB^{k}}{M + M^\times + \operatorname{Im} B + \operatorname{Im} AB + \cdots + \operatorname{Im} AB^{k-1}}\right), \qquad k = 0,1,2\ldots,$$

depend on W only and do not depend on the realization (7.1) of W.

The theorem has an analogue for appropriate closed contours in the Riemann sphere $\mathbb{C}_\infty$ like the extended real line or the extended imaginary axis.

To put Theorem 7.2 in context, consider a proper rational matrix function W having the value I_m at infinity. With a realization $W(\lambda) = I_m + C(\lambda I_n - A)^{-1}B$ of W, one can associate the numbers

$$\dim\bigl(\operatorname{Ker} C \cap \operatorname{Ker} CA \cap \cdots \cap \operatorname{Ker} CA^{k-1}\bigr), \qquad k = 0, 1, 2, \ldots, \tag{7.5}$$

$$\operatorname{codim}\bigl(\operatorname{Im} B + \operatorname{Im} AB + \cdots + \operatorname{Im} AB^{k-1}\bigr), \qquad k = 0, 1, 2, \ldots. \tag{7.6}$$

Here the codimension is taken with respect to $\mathbb{C}^n$. Now realizations of rational matrix functions are not unique and the above numbers, as well as their differences, generally vary with different choices of A, B and C in the realization for W. The above theorem shows that this dependence on the specific form of (7.1) disappears when one combines the spaces appearing in (7.5) and (7.6) with certain spectral subspaces of A and $A^\times$. We will meet the subspaces featuring in (7.5) and (7.6) again in Section 8.1.

The proof of Theorem 7.2 is rather complicated and we will devote a separate section to it.

7.2 Proof of Theorem 7.2

Let W and Γ be as in Theorem 7.2, and suppose we have the realizations

$$W(\lambda) = I_Y + \widetilde{C}(\lambda I_{\widetilde{X}} - \widetilde{A})^{-1}\widetilde{B}, \tag{7.7}$$

$$W(\lambda) = I_Y + \widehat{C}(\lambda I_{\widehat{X}} - \widehat{A})^{-1}\widehat{B}, \tag{7.8}$$

where Γ splits the spectra of $\widetilde{A}$ and $\widetilde{A}^\times$ as well as those of $\widehat{A}$ and $\widehat{A}^\times$. In other words $\Gamma \subset \rho(\widetilde{A}) \cap \rho(\widetilde{A}^\times) \cap \rho(\widehat{A}) \cap \rho(\widehat{A}^\times)$. Writing

$$\widetilde{P} = P(\widetilde{A};\Gamma), \qquad \widetilde{M} = \operatorname{Im} \widetilde{P}, \qquad \widetilde{P}^\times = P(\widetilde{A}^\times;\Gamma), \qquad \widetilde{M}^\times = \operatorname{Ker} \widetilde{P}^\times,$$

$$\widehat{P} = P(\widehat{A};\Gamma), \qquad \widehat{M} = \operatorname{Im} \widehat{P}, \qquad \widehat{P}^\times = P(\widehat{A}^\times;\Gamma), \qquad \widehat{M}^\times = \operatorname{Ker} \widehat{P}^\times,$$

we need to show that

$$\begin{aligned}
&\dim(\widetilde{M} \cap \widetilde{M}^\times) = \dim(\widehat{M} \cap \widehat{M}^\times),\\
&\dim\left(\frac{\widetilde{X}}{\widetilde{M} + \widetilde{M}^\times}\right) = \dim\left(\frac{\widehat{X}}{\widehat{M} + \widehat{M}^\times}\right),\\
&\dim\left(\frac{\widetilde{M} \cap \widetilde{M}^\times \cap \operatorname{Ker} \widetilde{C} \cap \operatorname{Ker} \widetilde{C}\widetilde{A} \cap \cdots \cap \operatorname{Ker} \widetilde{C}\widetilde{A}^{k-1}}{\widetilde{M} \cap \widetilde{M}^\times \cap \operatorname{Ker} \widetilde{C} \cap \operatorname{Ker} \widetilde{C}\widetilde{A} \cap \cdots \cap \operatorname{Ker} \widetilde{C}\widetilde{A}^{k-1} \cap \operatorname{Ker} \widetilde{C}\widetilde{A}^{k}}\right)\\
&= \dim\left(\frac{\widehat{M} \cap \widehat{M}^\times \cap \operatorname{Ker} \widehat{C} \cap \operatorname{Ker} \widehat{C}\widehat{A} \cap \cdots \cap \operatorname{Ker} \widehat{C}\widehat{A}^{k-1}}{\widehat{M} \cap \widehat{M}^\times \cap \operatorname{Ker} \widehat{C} \cap \operatorname{Ker} \widehat{C}\widehat{A} \cap \cdots \cap \operatorname{Ker} \widehat{C}\widehat{A}^{k-1} \cap \operatorname{Ker} \widehat{C}\widehat{A}^{k}}\right),
\end{aligned}$$

$$\dim\left(\frac{\widetilde{M}+\widetilde{M}^{\times}+\operatorname{Im}\widetilde{B}+\operatorname{Im}\widetilde{A}\widetilde{B}+\cdots+\operatorname{Im}\widetilde{A}\widetilde{B}^{k-1}+\operatorname{Im}\widetilde{A}\widetilde{B}^{k}}{\widetilde{M}+\widetilde{M}^{\times}+\operatorname{Im}\widetilde{B}+\operatorname{Im}\widetilde{A}\widetilde{B}+\cdots+\operatorname{Im}\widetilde{A}\widetilde{B}^{k-1}}\right)$$
$$=\dim\left(\frac{\widehat{M}+\widehat{M}^{\times}+\operatorname{Im}\widehat{B}+\operatorname{Im}\widehat{A}\widehat{B}+\cdots+\operatorname{Im}\widehat{A}\widehat{B}^{k-1}+\operatorname{Im}\widehat{A}\widehat{B}^{k}}{\widehat{M}+\widehat{M}^{\times}+\operatorname{Im}\widehat{B}+\operatorname{Im}\widehat{A}\widehat{B}+\cdots+\operatorname{Im}\widehat{A}\widehat{B}^{k-1}}\right).$$

Here $k = 0, 1, 2, \ldots$.

It is convenient to first present a series of auxiliary results. These concern the operators $\widetilde{\Psi}$ and $\widehat{\Psi}$ given by the integrals

$$\widetilde{\Psi} = \frac{1}{2\pi i}\int_{\Gamma}(\lambda-\widehat{A}^{\times})^{-1}\widehat{B}\widetilde{C}(\lambda-\widetilde{A})^{-1}\,d\lambda, \tag{7.9}$$

$$\widehat{\Psi} = \frac{1}{2\pi i}\int_{\Gamma}(\lambda-\widetilde{A}^{\times})^{-1}\widetilde{B}\widehat{C}(\lambda-\widehat{A})^{-1}\,d\lambda. \tag{7.10}$$

Note that $\widetilde{\Psi} : \widetilde{X} \to \widehat{X}$ and $\widehat{\Psi} : \widehat{X} \to \widetilde{X}$.

Lemma 7.3. *The operators $\widetilde{\Psi}$ and $\widehat{\Psi}$ also admit the representation:*

$$\widetilde{\Psi} = \frac{1}{2\pi i}\int_{\Gamma}(\lambda-\widehat{A})^{-1}\widehat{B}\widetilde{C}(\lambda-\widetilde{A}^{\times})^{-1}\,d\lambda, \tag{7.11}$$

$$\widehat{\Psi} = \frac{1}{2\pi i}\int_{\Gamma}(\lambda-\widetilde{A})^{-1}\widetilde{B}\widehat{C}(\lambda-\widehat{A}^{\times})^{-1}\,d\lambda, \tag{7.12}$$

Proof. From Theorem 2.4 we know that

$$W(\lambda)\widetilde{C}(\lambda-\widetilde{A}^{\times})^{-1} = \widetilde{C}(\lambda-\widetilde{A})^{-1}, \qquad W(\lambda)\widehat{C}(\lambda-\widehat{A}^{\times})^{-1} = \widehat{C}(\lambda-\widehat{A})^{-1},$$
$$(\lambda-\widetilde{A}^{\times})^{-1}\widetilde{B}W(\lambda) = (\lambda-\widetilde{A})^{-1}\widetilde{B}, \qquad (\lambda-\widehat{A}^{\times})^{-1}\widehat{B}W(\lambda) = (\lambda-\widehat{A})^{-1}\widehat{B}.$$

Now make the appropriate substitutions. □

Lemma 7.4. *For the products of $\widetilde{\Psi}$ and $\widehat{\Psi}$ the following identities hold:*

$$\widehat{\Psi}\widetilde{\Psi} = (\widetilde{P}^{\times}-\widetilde{P})^2, \qquad \widetilde{\Psi}\widehat{\Psi} = (\widehat{P}^{\times}-\widehat{P})^2.$$

Proof. It is assumed that $\Gamma \subset \rho(\widetilde{A}) \cap \rho(\widehat{A})$. For $\lambda \in \Gamma$, we have

$$\widetilde{C}(\lambda-\widetilde{A})^{-1}(\mu-\widetilde{A})^{-1}\widetilde{B} = \widehat{C}(\lambda-\widehat{A})^{-1}(\mu-\widehat{A})^{-1}\widehat{B}. \tag{7.13}$$

Indeed, taking advantage of the resolvent identity, we get for $\lambda \in \Gamma$,

$$\begin{aligned}(\mu-\lambda)&\widetilde{C}(\lambda-\widetilde{A})^{-1}(\mu-\widetilde{A})^{-1}\widetilde{B}\\ &= \widetilde{C}\big((\lambda-\widetilde{A})^{-1} - (\mu-\widetilde{A})^{-1}\big)\widetilde{B}\\ &= \widetilde{C}(\lambda-\widetilde{A})^{-1}\widetilde{B} - \widetilde{C}(\mu-\widetilde{A})^{-1}\widetilde{B}\end{aligned}$$

$$
\begin{aligned}
&= \big(W(\lambda) - I\big) - \big(W(\mu) - I\big) \\
&= \widehat{C}(\lambda - \widehat{A})^{-1}\widehat{B} - \widehat{C}(\mu - \widehat{A})^{-1}\widehat{B} \\
&= \widehat{C}\big((\lambda - \widehat{A})^{-1} - (\mu - \widehat{A})^{-1}\big)\widehat{B} \\
&= (\mu - \lambda)\widehat{C}(\lambda - \widehat{A})^{-1}(\mu - \widehat{A})^{-1}\widehat{B}.
\end{aligned}
$$

Now, when $\lambda \neq \mu$, divide by $\mu - \lambda$; for $\lambda = \mu$, employ a continuity argument.

To compute $\widehat{\Psi}\widetilde{\Psi}$, we use the expression (7.10) for $\widehat{\Psi}$, formula (7.11) for $\widetilde{\Psi}$, and the identity (7.13):

$$
\begin{aligned}
\widehat{\Psi}\widetilde{\Psi} &= \left(\frac{1}{2\pi i}\right)^2 \int_\Gamma \int_\Gamma (\lambda - \widetilde{A}^\times)^{-1}\widetilde{B}\widehat{C}(\lambda - \widehat{A})^{-1} \\
&\qquad\qquad \cdot(\mu - \widehat{A})^{-1}\widehat{B}\widetilde{C}(\mu - \widetilde{A}^\times)^{-1}\, d\lambda\, d\mu \\
&= \left(\frac{1}{2\pi i}\right)^2 \int_\Gamma \int_\Gamma (\lambda - \widetilde{A}^\times)^{-1}\widetilde{B}\widetilde{C}(\lambda - \widetilde{A})^{-1} \\
&\qquad\qquad \cdot(\mu - \widetilde{A})^{-1}\widetilde{B}\widetilde{C}(\mu - \widetilde{A}^\times)^{-1}\, d\lambda\, d\mu \\
&= \left(\frac{1}{2\pi i}\right)^2 \int_\Gamma \int_\Gamma (\lambda - \widetilde{A}^\times)^{-1}(\widetilde{A} - \widetilde{A}^\times)(\lambda - \widetilde{A})^{-1} \\
&\qquad\qquad \cdot(\mu - \widetilde{A})^{-1}(\widetilde{A} - \widetilde{A}^\times)(\mu - \widetilde{A}^\times)^{-1}\, d\lambda\, d\mu \\
&= \left(\frac{1}{2\pi i}\right)^2 \int_\Gamma \int_\Gamma \Big((\lambda - \widetilde{A}^\times)^{-1} - (\lambda - \widetilde{A})^{-1})\Big) \\
&\qquad\qquad \cdot\Big((\mu - \widetilde{A}^\times)^{-1} - (\mu - \widetilde{A})^{-1}\Big)\, d\lambda\, d\mu \\
&= \left(\frac{1}{2\pi i}\int_\Gamma (\lambda - \widetilde{A}^\times)^{-1} - (\lambda - \widetilde{A})^{-1}\, d\lambda\right)^2 \\
&= (\widetilde{P}^\times - \widetilde{P})^2.
\end{aligned}
$$

For $\widetilde{\Psi}\widehat{\Psi} = (\widehat{P}^\times - \widehat{P})^2$, interchange the roles of the realizations (7.7) and (7.8). □

Lemma 7.5. *The operators* $\widetilde{\Psi}$ *and* $\widehat{\Psi}$ *satisfy the following intertwining relations:*

$$
\widetilde{\Psi}\widetilde{P} = (I - \widehat{P}^\times)\widetilde{\Psi}, \qquad \widetilde{\Psi}\widetilde{P}^\times = (I - \widehat{P})\widetilde{\Psi}, \tag{7.14}
$$

$$
\widehat{\Psi}\widehat{P} = (I - \widetilde{P}^\times)\widehat{\Psi}, \qquad \widehat{\Psi}\widehat{P}^\times = (I - \widetilde{P})\widehat{\Psi}. \tag{7.15}
$$

Proof. Focussing on the first identity in (7.14), note that the function

$$
\widehat{P}^\times(\lambda - \widehat{A}^\times)^{-1}\widehat{B}\widetilde{C}(\lambda - \widetilde{A})^{-1}\widetilde{P}
$$

is analytic on an open neighborhood of $F_- \cup \Gamma$. Here F_- is the exterior domain of Γ (including ∞). Furthermore, the expansion of this function at infinity is of the form $\lambda^{-2}\widehat{P}^\times \widehat{B}\widetilde{C}\widetilde{P}$ plus lower order terms. Hence

$$\frac{1}{2\pi i}\int_\Gamma \widehat{P}^\times(\lambda - \widehat{A}^\times)^{-1}\widehat{B}\widetilde{C}(\lambda - \widetilde{A})^{-1}\widetilde{P}d\lambda = 0.$$

On the other hand $(I - \widehat{P}^\times)(\lambda - \widehat{A}^\times)^{-1}\widehat{B}\widetilde{C}(\lambda - \widetilde{A})^{-1}(I - \widetilde{P})$ is analytic on an open neighborhood of $F_+ \cup \Gamma$, where F_+ is the interior domain of Γ, and so

$$\frac{1}{2\pi i}\int_\Gamma (I - \widehat{P}^\times)(\lambda - \widehat{A}^\times)^{-1}\widehat{B}\widetilde{C}(\lambda - \widetilde{A})^{-1}(I - \widetilde{P})d\lambda = 0.$$

It follows that

$$\begin{aligned}
\widetilde{\Psi}\widetilde{P} &= \frac{1}{2\pi i}\int_\Gamma (\lambda - \widehat{A}^\times)^{-1}\widehat{B}\widetilde{C}(\lambda - \widetilde{A})^{-1}\widetilde{P}d\lambda \\
&= \frac{1}{2\pi i}\int_\Gamma (I - \widehat{P}^\times)(\lambda - \widehat{A}^\times)^{-1}\widehat{B}\widetilde{C}(\lambda - \widetilde{A})^{-1}\widetilde{P}d\lambda \\
&= \frac{1}{2\pi i}\int_\Gamma (I - \widehat{P}^\times)(\lambda - \widehat{A}^\times)^{-1}\widehat{B}\widetilde{C}(\lambda - \widetilde{A})^{-1}d\lambda \\
&= (I - \widehat{P}^\times)\widetilde{\Psi},
\end{aligned}$$

as desired.

This proves the first identity in (7.14). The second identity in (7.14) is proved in a similar way using the formula for $\widetilde{\Psi}$ given by (7.11). The identities in (7.15) follow from those in (7.14) by interchanging the roles of the realizations (7.7) and (7.8). □

Lemma 7.6. *The operators $\widetilde{\Psi}$ and $\widehat{\Psi}$ satisfy the following Lyapunov equations:*

$$\widetilde{\Psi}\widetilde{A} - \widehat{A}^\times\widetilde{\Psi} = \widehat{B}\widetilde{C}\widetilde{P} - \widehat{P}^\times\widehat{B}\widetilde{C}, \tag{7.16}$$

$$\widetilde{\Psi}\widetilde{A}^\times - \widehat{A}\widetilde{\Psi} = \widehat{B}\widetilde{C}\widetilde{P}^\times - \widehat{P}\widehat{B}\widetilde{C}, \tag{7.17}$$

$$\widehat{\Psi}\widehat{A} - \widetilde{A}^\times\widehat{\Psi} = \widetilde{B}\widehat{C}\widehat{P} - \widetilde{P}^\times\widetilde{B}\widehat{C}, \tag{7.18}$$

$$\widehat{\Psi}\widehat{A}^\times - \widetilde{A}\widehat{\Psi} = \widetilde{B}\widehat{C}\widehat{P}^\times - \widetilde{P}\widetilde{B}\widehat{C}. \tag{7.19}$$

Proof. Using the definition of $\widetilde{\Psi}$ via (7.9), we have

$$\widetilde{\Psi}\widetilde{A} = \frac{1}{2\pi i}\int_\Gamma (\lambda - \widehat{A}^\times)^{-1}\widehat{B}\widetilde{C}(\lambda - \widetilde{A})^{-1}\widetilde{A}\,d\lambda$$

$$
\begin{aligned}
&= \frac{1}{2\pi i}\int_\Gamma (\lambda - \widehat{A}^\times)^{-1}\widehat{B}\widetilde{C}(\lambda - \widetilde{A})^{-1}(\widetilde{A} - \lambda I + \lambda I)\, d\lambda \\
&= \frac{1}{2\pi i}\int_\Gamma \lambda(\lambda - \widehat{A}^\times)^{-1}\widehat{B}\widetilde{C}(\lambda - \widetilde{A})^{-1}\, d\lambda - \frac{1}{2\pi i}\int_\Gamma (\lambda - \widehat{A}^\times)^{-1}\widehat{B}\widetilde{C}\, d\lambda \\
&= \frac{1}{2\pi i}\int_\Gamma (\lambda I - \widehat{A}^\times + \widehat{A}^\times)(\lambda - \widehat{A}^\times)^{-1}\widehat{B}\widetilde{C}(\lambda - \widetilde{A})^{-1}\, d\lambda - \widehat{P}^\times\widehat{B}\widetilde{C} \\
&= \widehat{B}\widetilde{C}\widetilde{P} + \widehat{A}^\times\widetilde{\Psi} - \widehat{P}^\times\widehat{B}\widetilde{C}.
\end{aligned}
$$

This gives (7.16). The identity (7.17) can be proved similarly by using the alternative expression for $\widetilde{\Psi}$ of Lemma 7.3. For (7.18) and (7.19), use (7.16) and (7.17) and interchange the roles of of the realizations (7.7) and (7.8). Direct computations as the one above of course also work. □

Lemma 7.7. *The operators $\widetilde{\Psi}$, $\widehat{\Psi}$, $\widetilde{B}$, $\widehat{B}$, $\widetilde{C}$ and $\widehat{C}$ are related as follows:*

$$
\widetilde{\Psi}\widetilde{B} = (\widehat{P} - \widehat{P}^\times)\widehat{B}, \qquad \widehat{C}\widetilde{\Psi} = \widetilde{C}(\widetilde{P} - \widetilde{P}^\times), \tag{7.20}
$$

$$
\widehat{\Psi}\widehat{B} = (\widetilde{P} - \widetilde{P}^\times)\widetilde{B}, \qquad \widetilde{C}\widehat{\Psi} = \widehat{C}(\widehat{P} - \widehat{P}^\times). \tag{7.21}
$$

Proof. Using the expression (7.9) for $\widetilde{\Psi}$, we have

$$
\begin{aligned}
\widetilde{\Psi}\widetilde{B} &= \frac{1}{2\pi i}\int_\Gamma (\lambda - \widehat{A}^\times)^{-1}\widehat{B}\widetilde{C}(\lambda - \widetilde{A})^{-1}\widetilde{B}\, d\lambda \\
&= \frac{1}{2\pi i}\int_\Gamma (\lambda - \widehat{A}^\times)^{-1}\widehat{B}(W_1(\lambda) - I)\, d\lambda \\
&= \frac{1}{2\pi i}\int_\Gamma (\lambda - \widehat{A}^\times)^{-1}\widehat{B}(W_2(\lambda) - I)\, d\lambda
\end{aligned}
$$

$$
= \frac{1}{2\pi i}\int_\Gamma (\lambda - \widehat{A}^\times)^{-1}\widehat{B}W_2(\lambda)\, d\lambda \; - \; \frac{1}{2\pi i}\int_\Gamma (\lambda - \widehat{A}^\times)^{-1}\widehat{B}\, d\lambda.
$$

By Theorem 2.4, we may replace $(\lambda - \widehat{A}^\times)^{-1}\widehat{B}W(\lambda)$ by $(\lambda - \widehat{A})^{-1}\widehat{B}$. Hence

$$
\widetilde{\Psi}\widetilde{B} = \frac{1}{2\pi i}\int_\Gamma (\lambda - \widehat{A})^{-1}\widehat{B}\, d\lambda - \frac{1}{2\pi i}\int_\Gamma (\lambda - \widehat{A}^\times)^{-1}\widehat{B}\, d\lambda,
$$

and this can be rewritten as the first part of (7.20). The second part can be proved via a similar computation. The identities in (7.21) follow by interchanging the roles of the realizations (7.7) and (7.8). □

Proof of Theorem 7.2. The proof will be divided into three parts. The first contains some preliminary observations about the spaces $\widetilde{M}$, $\widetilde{M}^\times$, $\widehat{M}$ and $\widehat{M}^\times$, ending up in an argument establishing the identities

$$\dim(\widetilde{M} \cap \widetilde{M}^\times) = \dim(\widehat{M} \cap \widehat{M}^\times), \qquad \dim\left(\frac{\widetilde{X}}{\widetilde{M} + \widetilde{M}^\times}\right) = \dim\left(\frac{\widehat{X}}{\widehat{M} + \widehat{M}^\times}\right).$$

Part 1. We begin by noting that

$$\widetilde{\Psi}[\widetilde{M} \cap \widetilde{M}^\times] \subset \widehat{M} \cap \widehat{M}^\times, \qquad \widetilde{\Psi}[\widetilde{M} + \widetilde{M}^\times] \subset \widehat{M} + \widehat{M}^\times,$$

$$\widehat{\Psi}[\widehat{M} \cap \widehat{M}^\times] \subset \widetilde{M} \cap \widetilde{M}^\times, \qquad \widehat{\Psi}[\widehat{M} + \widehat{M}^\times] \subset \widetilde{M} + \widetilde{M}^\times.$$

To prove this, it suffices to show that

$$\widetilde{\Psi}\widetilde{M} \subset \widehat{M}^\times, \qquad \widetilde{\Psi}\widetilde{M}^\times \subset \widehat{M}, \qquad \widehat{\Psi}\widehat{M} \subset \widetilde{M}^\times, \qquad \widehat{\Psi}\widehat{M}^\times \subset \widetilde{M}.$$

These inclusions, however, are obvious from (7.14) and (7.15).

Next observe that

$$\widetilde{M} \cap \widetilde{M}^\times \subset \operatorname{Ker}(I - \widehat{\Psi}\widetilde{\Psi}), \qquad \widehat{M} \cap \widehat{M}^\times \subset \operatorname{Ker}(I - \widetilde{\Psi}\widehat{\Psi}), \tag{7.22}$$

$$\widetilde{M} + \widetilde{M}^\times \supset \operatorname{Im}(I - \widehat{\Psi}\widetilde{\Psi}), \qquad \widehat{M} + \widehat{M}^\times \supset \operatorname{Im}(I - \widetilde{\Psi}\widehat{\Psi}). \tag{7.23}$$

The formulas concerning the product $\widehat{\Psi}\widetilde{\Psi}$, follow from

$$I - \widehat{\Psi}\widetilde{\Psi} = \widetilde{P}\widetilde{P}^\times + (I - \widetilde{P}^\times)(I - \widetilde{P}),$$

which, in turn, is immediate from Lemma 7.4. The two expressions involving the product $\widetilde{\Psi}\widehat{\Psi}$ are obtained by interchanging the roles of the realizations (7.7) and (7.8).

Consider the restriction operators

$$\widetilde{\Psi}|_{\widetilde{M}\cap\widetilde{M}^\times} : \widetilde{M} \cap \widetilde{M}^\times \to \widehat{M} \cap \widehat{M}^\times,$$
$$\widehat{\Psi}|_{\widehat{M}\cap\widehat{M}^\times} : \widehat{M} \cap \widehat{M}^\times \to \widetilde{M} \cap \widetilde{M}^\times.$$

From (7.22) it is clear that these operators are each others inverse. Hence $\widetilde{M}\cap\widetilde{M}^\times$ and $\widehat{M}\cap\widehat{M}^\times$ are linearly isomorphic and so they have the same (possibly infinite) dimension. Next we turn to the operators

$$\widetilde{\Phi} : \frac{\widetilde{X}}{\widetilde{M} + \widetilde{M}^\times} \to \frac{\widehat{X}}{\widehat{M} + \widehat{M}^\times}, \qquad \widehat{\Phi} : \frac{\widehat{X}}{\widehat{M} + \widehat{M}^\times} \to \frac{\widetilde{X}}{\widetilde{M} + \widetilde{M}^\times},$$

induced by $\widetilde{\Psi}$ and $\widehat{\Psi}$, respectively. These are well-defined because of the inclusions $\widetilde{\Psi}[\widetilde{M}+\widetilde{M}^\times] \subset \widehat{M}+\widehat{M}^\times$ and $\widehat{\Psi}[\widehat{M}+\widehat{M}^\times] \subset \widetilde{M}+\widetilde{M}^\times$. Also it follows from (7.23) that $\widetilde{\Phi}$ and $\widehat{\Phi}$ are each other's inverse. Thus the quotient spaces $\widetilde{X}/(\widetilde{M}+\widetilde{M}^\times)$ and $\widehat{X}/(\widehat{M}+\widehat{M}^\times)$ are linearly isomorphic. In particular they have the same (possibly infinite) dimension.

Part 2. In this part of the proof we shall verify that for all nonnegative integers k the following identities hold:

$$\dim\left(\frac{\widetilde{M}\cap\widetilde{M}^\times\cap\operatorname{Ker}\widetilde{C}\cap\operatorname{Ker}\widetilde{C}\widetilde{A}\cap\cdots\cap\operatorname{Ker}\widetilde{C}\widetilde{A}^{k-1}}{\widetilde{M}\cap\widetilde{M}^\times\cap\operatorname{Ker}\widetilde{C}\cap\operatorname{Ker}\widetilde{C}\widetilde{A}\cap\cdots\cap\operatorname{Ker}\widetilde{C}\widetilde{A}^{k-1}\cap\operatorname{Ker}\widetilde{C}\widetilde{A}^{k}}\right)$$
$$=\dim\left(\frac{\widehat{M}\cap\widehat{M}^\times\cap\operatorname{Ker}\widehat{C}\cap\operatorname{Ker}\widehat{C}\widehat{A}\cap\cdots\cap\operatorname{Ker}\widehat{C}\widehat{A}^{k-1}}{\widehat{M}\cap\widehat{M}^\times\cap\operatorname{Ker}\widehat{C}\cap\operatorname{Ker}\widehat{C}\widehat{A}\cap\cdots\cap\operatorname{Ker}\widehat{C}\widehat{A}^{k-1}\cap\operatorname{Ker}\widetilde{C}\widetilde{A}^{k}}\right).$$

This will be done by showing that the quotient spaces appearing in these identities are linearly isomorphic. To facilitate the discussion, we adopt the notation

$$\operatorname{Ker}_k(\widetilde{C}|\widetilde{A}) = \operatorname{Ker}\widetilde{C}\cap\operatorname{Ker}\widetilde{C}\widetilde{A}\cap\cdots\cap\operatorname{Ker}\widetilde{C}\widetilde{A}^{k-1},$$

where, following standard convention, $\operatorname{Ker}_0(\widetilde{C}|\widetilde{A})$ is read as $\widetilde{X}$. Of course the notation $\operatorname{Ker}_k(\widehat{C}|\widehat{A})$ is defined similarly. First we shall prove that the operator $\widetilde{\Psi}$ maps $\widetilde{M}\cap\widetilde{M}^\times\cap\operatorname{Ker}_k(\widetilde{C}|\widetilde{A})$ into $\widehat{M}\cap\widehat{M}^\times\cap\operatorname{Ker}_k(\widehat{C}|\widehat{A})$.

This has already been established for $k=0$ (Part 1). For $k=1$ it must be proved that

$$\widetilde{\Psi}\big[\widetilde{M}\cap\widetilde{M}^\times\cap\operatorname{Ker}\widetilde{C}\big]\subset\widehat{M}\cap\widehat{M}^\times\cap\operatorname{Ker}\widehat{C}.$$

We know already that $\widetilde{\Psi}\big[\widetilde{M}\cap\widetilde{M}^\times\big]\subset\widehat{M}\cap\widehat{M}^\times$, and so it is enough to derive the inclusion $\widetilde{\Psi}\big[\widetilde{M}\cap\widetilde{M}^\times\cap\operatorname{Ker}\widetilde{C}\big]\subset\operatorname{Ker}\widehat{C}$ or, what comes down to the same, $\widetilde{M}\cap\widetilde{M}^\times\cap\operatorname{Ker}\widetilde{C}\subset\operatorname{Ker}\widehat{C}\widetilde{\Psi}$. The latter, however, is immediate from the identity $\widehat{C}\widetilde{\Psi}=-\widetilde{C}(I-\widetilde{P})-\widetilde{C}\widetilde{P}^\times+\widetilde{C}$ for which we refer to Lemma 7.7.

We proceed by induction. Let k be a nonnegative integer and suppose that the operator $\widetilde{\Psi}$ maps $\widetilde{M}\cap\widetilde{M}^\times\cap\operatorname{Ker}_k(\widetilde{C}|\widetilde{A})$ into $\widehat{M}\cap\widehat{M}^\times\cap\operatorname{Ker}_k(\widehat{C}|\widehat{A})$. We shall show that the same is true with k replaced by $k+1$. Clearly

$$\widetilde{M}\cap\widetilde{M}^\times\cap\operatorname{Ker}_{k+1}(\widetilde{C}|\widetilde{A}) = \widetilde{M}\cap\widetilde{M}^\times\cap\operatorname{Ker}_k(\widetilde{C}|\widetilde{A})\cap\operatorname{Ker}\widetilde{C}\widetilde{A}^k,$$

and similarly with $\widetilde{M}$, $\widetilde{M}^\times$, $\widetilde{A}$ and $\widetilde{C}$ replaced by $\widehat{M}$, $\widehat{M}^\times$, $\widehat{A}$ and $\widehat{C}$, respectively. Hence, in view of the induction hypothesis, it is sufficient to verify that $\widetilde{\Psi}$ maps the space $\widetilde{M}\cap\widetilde{M}^\times\cap\operatorname{Ker}_{k+1}(\widetilde{C}|\widetilde{A})$ into $\operatorname{Ker}\widehat{C}\widehat{A}^k$. In other words, what we need is the inclusion

$$\operatorname{Ker}\widehat{C}\widehat{A}^k\widetilde{\Psi}\subset\widetilde{M}\cap\widetilde{M}^\times\cap\operatorname{Ker}_{k+1}(\widetilde{C}|\widetilde{A}). \tag{7.24}$$

With the help of (the second identity in) Lemma 7.6, the operator $\widehat{C}\widehat{A}^k\widetilde{\Psi}$ can be written as

$$\begin{aligned}\widehat{C}\widehat{A}^k\widetilde{\Psi} &= \widehat{C}\widehat{A}^{k-1}(-\widehat{B}\widetilde{C}\widetilde{P}^\times+\widehat{P}\widehat{B}\widetilde{C}+\widetilde{\Psi}\widetilde{A}^\times)\\ &= \widehat{C}\widehat{A}^{k-1}\big(-\widehat{B}\widetilde{C}\widetilde{P}^\times+(\widehat{P}\widehat{B}-\widetilde{\Psi}\widetilde{B})\widetilde{C}+\widetilde{\Psi}\widetilde{A}\big)\end{aligned}$$

and we may conclude that

$$\operatorname{Ker} \widehat{C}\widehat{A}^k\widetilde{\Psi} \supset \widetilde{M}^\times \cap \operatorname{Ker} \widetilde{C} \cap \operatorname{Ker} \widehat{C}\widehat{A}^{k-1}\widetilde{\Psi}\widetilde{A}. \tag{7.25}$$

Now $\operatorname{Ker} \widehat{C}\widehat{A}^{k-1} \supset \widehat{M} \cap \widehat{M}^\times \cap \operatorname{Ker}_k(\widehat{C}|\widehat{A})$. Employing the induction hypothesis once again gives $\operatorname{Ker} \widehat{C}\widehat{A}^{k-1} \supset \widetilde{\Psi}\big[\widetilde{M} \cap \widetilde{M}^\times \cap \operatorname{Ker}_k(\widetilde{C}|\widetilde{A})\big]$, i.e.,

$$\operatorname{Ker} \widehat{C}\widehat{A}^{k-1}\widetilde{\Psi} \supset \widetilde{M} \cap \widetilde{M}^\times \cap \operatorname{Ker}_k(\widetilde{C}|\widetilde{A}).$$

But then

$$\begin{aligned}
\operatorname{Ker} \widehat{C}\widehat{A}^{k-1}\widetilde{\Psi}\widetilde{A} &= \widetilde{A}^{-1}\big[\operatorname{Ker} \widehat{C}\widehat{A}^{k-1}\widetilde{\Psi}\big] \\
&\supset \widetilde{A}^{-1}\big[\widetilde{M} \cap \widetilde{M}^\times \cap \operatorname{Ker}_k(\widetilde{C}|\widetilde{A})\big] \\
&= \widetilde{A}^{-1}\big[\widetilde{M}\big] \cap \widetilde{A}^{-1}\big[\widetilde{M}^\times\big] \cap \widetilde{A}^{-1}\big[\operatorname{Ker}_k(\widetilde{C}|\widetilde{A})\big] \\
&\supset \widetilde{M} \cap \big(\widetilde{A}^\times + \widetilde{B}\widetilde{C}\big)^{-1}\big[\widetilde{M}^\times\big] \cap \widetilde{A}^{-1}\big[\operatorname{Ker}_k(\widetilde{C}|\widetilde{A})\big] \\
&\supset \widetilde{M} \cap \widetilde{A}^{\times -1}\big[\widehat{M}^\times\big] \cap \operatorname{Ker} \widetilde{C} \cap \widetilde{A}^{-1}\big[\operatorname{Ker}_k(\widetilde{C}|\widetilde{A})\big] \\
&\supset \widetilde{M} \cap \widetilde{M}^\times \cap \operatorname{Ker} \widetilde{C} \cap \widetilde{A}^{-1}\big[\operatorname{Ker}_k(\widetilde{C}|\widetilde{A})\big],
\end{aligned}$$

and hence, taking into account (7.25),

$$\operatorname{Ker} \widehat{C}\widehat{A}^k\widetilde{\Psi} \supset \widetilde{M} \cap \widetilde{M}^\times \cap \operatorname{Ker} \widetilde{C} \cap \widetilde{A}^{-1}\big[\operatorname{Ker}_k(\widetilde{C}|\widetilde{A})\big].$$

As $\operatorname{Ker} \widetilde{C} \cap \widetilde{A}^{-1}\big[\operatorname{Ker}_k(\widetilde{C}|\widetilde{A})\big] = \operatorname{Ker}_{k+1}(\widetilde{C}|\widetilde{A})$, the inclusion (7.24) follows.

Fix the nonnegative integer k. As we have seen, the linear operator $\widetilde{\Psi}$ maps $\widetilde{M} \cap \widetilde{M}^\times \cap \operatorname{Ker}_k(\widetilde{C}|\widetilde{A})$ into $\widehat{M} \cap \widehat{M}^\times \cap \operatorname{Ker}_k(\widehat{C}|\widehat{A})$. Likewise $\widehat{\Psi}$ maps the space $\widehat{M} \cap \widehat{M}^\times \cap \operatorname{Ker}_k(\widehat{C}|\widetilde{A})$ into $\widetilde{M} \cap \widetilde{M}^\times \cap \operatorname{Ker}_k(\widetilde{C}|\widetilde{A})$. The same is true with k replaced by $k+1$. But then the linear operators

$$\widetilde{\Theta}_k \; : \; \frac{\widetilde{M} \cap \widetilde{M}^\times \cap \operatorname{Ker}_k(\widetilde{C}|\widetilde{A})}{\widetilde{M} \cap \widetilde{M}^\times \cap \operatorname{Ker}_{k+1}(\widetilde{C}|\widetilde{A})} \to \frac{\widehat{M} \cap \widehat{M}^\times \cap \operatorname{Ker}_k(\widehat{C}|\widehat{A})}{\widehat{M} \cap \widehat{M}^\times \cap \operatorname{Ker}_{k+1}(\widehat{C}|\widehat{A})}, \tag{7.26}$$

$$\widehat{\Theta}_k \; : \; \frac{\widehat{M} \cap \widehat{M}^\times \cap \operatorname{Ker}_k(\widehat{C}|\widehat{A})}{\widehat{M} \cap \widehat{M}^\times \cap \operatorname{Ker}_{k+1}(\widehat{C}|\widehat{A})} \to \frac{\widetilde{M} \cap \widetilde{M}^\times \cap \operatorname{Ker}_k(\widetilde{C}|\widetilde{A})}{\widetilde{M} \cap \widetilde{M}^\times \cap \operatorname{Ker}_{k+1}(\widetilde{C}|\widetilde{A})}, \tag{7.27}$$

induced by $\widetilde{\Psi}$ and $\widehat{\Psi}$, respectively, are well-defined. They are also each others inverse. This can be deduced easily from

$$\begin{aligned}
\widetilde{M} \cap \widetilde{M}^\times \cap \operatorname{Ker}_k(\widetilde{C}|\widetilde{A}) &\subset \operatorname{Ker}(I - \widehat{\Psi}\widetilde{\Psi}), \\
\widehat{M} \cap \widehat{M}^\times \cap \operatorname{Ker}_k(\widehat{C}|\widehat{A}) &\subset \operatorname{Ker}(I - \widetilde{\Psi}\widehat{\Psi}),
\end{aligned}$$

two inclusions which are immediate from (7.22). Thus the quotient spaces appearing in (7.26) and (7.27) are linearly isomorphic. In particular they have the same (possibly infinite) dimension.

Part 3. Finally we shall prove that the identities

$$\dim\left(\frac{\widetilde{M}+\widetilde{M}^{\times}+\operatorname{Im}\widetilde{B}+\operatorname{Im}\widetilde{A}\widetilde{B}+\cdots+\operatorname{Im}\widetilde{A}\widetilde{B}^{k-1}+\operatorname{Im}\widetilde{A}\widetilde{B}^{k}}{\widetilde{M}+\widetilde{M}^{\times}+\operatorname{Im}\widetilde{B}+\operatorname{Im}\widetilde{A}\widetilde{B}+\cdots+\operatorname{Im}\widetilde{A}\widetilde{B}^{k-1}}\right)$$
$$=\dim\left(\frac{\widehat{M}+\widehat{M}^{\times}+\operatorname{Im}\widehat{B}+\operatorname{Im}\widehat{A}\widehat{B}+\cdots+\operatorname{Im}\widehat{A}\widehat{B}^{k-1}+\operatorname{Im}\widehat{A}\widehat{B}^{k}}{\widehat{M}+\widehat{M}^{\times}+\operatorname{Im}\widehat{B}+\operatorname{Im}\widehat{A}\widehat{B}+\cdots+\operatorname{Im}\widehat{A}\widehat{B}^{k-1}}\right)$$

are valid for all nonnegative integers k. This will be done by showing that the quotient spaces appearing in these identities are linearly isomorphic. To facilitate the discussion, we adopt the notation

$$\operatorname{Im}_k(\widetilde{A}|\widetilde{B}) = \operatorname{Im}\widetilde{B} + \operatorname{Im}\widetilde{A}\widetilde{B} + \cdots + \operatorname{Im}\widetilde{A}\widetilde{B}^{k-1},$$

where, following standard convention, $\operatorname{Im}_0(\widetilde{A}|\widetilde{B})$ is read as $\{0\}$. Of course the notation $\operatorname{Im}_k(\widehat{A}|\widehat{B})$ is defined similarly. First we shall verify that the operator $\widetilde{\Psi}$ maps $\widetilde{M}+\widetilde{M}^{\times}+\operatorname{Im}_k(\widetilde{A}|\widetilde{B})$ into $\widehat{M}+\widehat{M}^{\times}+\operatorname{Im}_k(\widehat{A}|\widehat{B})$.

This has already been established for $k=0$ (Part 1). For $k=1$ it must be proved that

$$\widetilde{\Psi}\big[\widetilde{M}+\widetilde{M}^{\times}+\operatorname{Im}\widetilde{B}\big] \subset \widehat{M}+\widehat{M}^{\times}+\operatorname{Im}\widehat{B}.$$

We know already that $\widetilde{\Psi}\big[\widetilde{M}+\widetilde{M}^{\times}\big] \subset \widehat{M}+\widehat{M}^{\times}$, and so it is enough to derive the inclusion $\widetilde{\Psi}\big[\operatorname{Im}\widetilde{B}\big] \subset \widehat{M}+\widehat{M}^{\times}+\operatorname{Im}\widehat{B}$ or, what comes down to the same, $\operatorname{Im}\widetilde{\Psi}\widetilde{B} \subset \widehat{M}+\widehat{M}^{\times}+\operatorname{Im}\widehat{B}$. The latter, however, is immediate from the identity $\widetilde{\Psi}\widetilde{B} = \widehat{P}\widehat{B}+(I-\widehat{P}^{\times})\widehat{B}-\widehat{B}$ for which we refer to Lemma 7.7.

We proceed by induction. Let k be a positive integer and suppose that the operator $\widetilde{\Psi}$ maps the space $\widetilde{M}+\widetilde{M}^{\times}+\operatorname{Im}_k(\widetilde{A}|\widetilde{B})$ into $\widehat{M}+\widehat{M}^{\times}+\operatorname{Im}_k(\widehat{A}|\widehat{B})$. We shall show that the same is true with k replaced by $k+1$. Clearly

$$\widetilde{M}+\widetilde{M}^{\times}+\operatorname{Im}_{k+1}(\widetilde{A}|\widetilde{B}) = \widetilde{M}+\widetilde{M}^{\times}+\operatorname{Im}_k(\widetilde{A}|\widetilde{B})+\operatorname{Im}\widetilde{A}^k\widetilde{B},$$

and similarly with $\widetilde{M}$, $\widetilde{M}^{\times}$, $\widetilde{A}$ and $\widetilde{B}$ replaced by $\widehat{M}$, $\widehat{M}^{\times}$, $\widehat{A}$ and $\widehat{B}$, respectively. Hence, in view of the induction hypothesis, it suffices to verify that $\widetilde{\Psi}$ maps $\operatorname{Im}\widetilde{A}^k\widetilde{B}$ into $\widehat{M}+\widehat{M}^{\times}+\operatorname{Im}_{k+1}(\widehat{A}|\widehat{B})$. In other words, what we need is the inclusion

$$\operatorname{Im}\widetilde{\Psi}\widetilde{A}^k\widetilde{B} \subset \widehat{M}+\widehat{M}^{\times}+\operatorname{Im}_k(\widehat{A}|\widehat{B}). \tag{7.28}$$

With the help of (the first identity in) Lemma 7.6, the operator $\widetilde{\Psi}\widetilde{A}^k\widetilde{B}$ can be written as

$$\begin{aligned}\widetilde{\Psi}\widetilde{A}^k\widetilde{B} &= (\widehat{B}\widetilde{C}\widetilde{P}-\widehat{P}^{\times}\widehat{B}\widetilde{C}+\widehat{A}^{\times}\widetilde{\Psi})\widetilde{A}^{k-1}\widetilde{B}\\ &= ((I-\widehat{P}^{\times})\widehat{B}\widetilde{C}+\widehat{B}(\widetilde{C}\widetilde{P}-\widetilde{C}-\widehat{C}\widetilde{\Psi})+\widehat{A}\widetilde{\Psi})\widetilde{A}^{k-1}\widetilde{B},\end{aligned}$$

and we may conclude that

$$\operatorname{Im}\widetilde{\Psi}\widetilde{A}^k\widetilde{B} \subset \widehat{M}^{\times} + \operatorname{Im}\widehat{B} + \operatorname{Im}\widehat{A}\widetilde{\Psi}\widetilde{A}^{k-1}\widetilde{B}. \tag{7.29}$$

Now $\operatorname{Im}\widetilde{A}^{k-1}\widetilde{B} \subset \widetilde{M} + \widetilde{M}^{\times} + \operatorname{Im}_k(\widetilde{A}|\widetilde{B})$. Employing the induction hypothesis once again gives $\widetilde{\Psi}\big[\operatorname{Im}\widetilde{A}^{k-1}\widetilde{B}\big] \subset \widehat{M} + \widehat{M}^{\times} + \operatorname{Im}_k(\widehat{A}|\widehat{B})$, i.e.,

$$\operatorname{Im}\widetilde{\Psi}\widetilde{A}^{k-1}\widetilde{B} \subset \widehat{M} + \widehat{M}^{\times} + \operatorname{Im}_k(\widehat{A}|\widehat{B}).$$

But then

$$\begin{aligned}
\operatorname{Im}\widehat{A}\widetilde{\Psi}\widetilde{A}^{k-1}\widetilde{B} &= \widehat{A}\big[\operatorname{Im}\widetilde{\Psi}\widetilde{A}^{k-1}\widetilde{B}\big]\\
&\subset \widehat{A}\big[\widehat{M} + \widehat{M}^{\times} + \operatorname{Im}_k(\widehat{A}|\widehat{B})\big]\\
&= \widehat{A}\big[\widehat{M}\big] + \widehat{A}\big[\widehat{M}^{\times}\big] + \widehat{A}\big[\operatorname{Im}_k(\widehat{A}|\widehat{B})\big]\\
&\subset \widehat{M} + (\widehat{A}^{\times} + \widehat{B}\widehat{C})\big[\widehat{M}^{\times}\big] + \widehat{A}\big[\operatorname{Im}_k(\widehat{A}|\widehat{B})\big]\\
&\subset \widehat{M} + \widehat{A}^{\times}\big[\widehat{M}^{\times}\big] + \operatorname{Im}\widehat{B} + \widehat{A}\big[\operatorname{Im}_k(\widehat{A}|\widehat{B})\big]\\
&\subset \widehat{M} + \widehat{M}^{\times} + \operatorname{Im}\widehat{B} + \widehat{A}\big[\operatorname{Im}_k(\widehat{A}|\widehat{B})\big],
\end{aligned}$$

and hence, taking into account (7.29),

$$\operatorname{Im}\widetilde{\Psi}\widetilde{A}^k\widetilde{B} \subset \widehat{M} + \widehat{M}^{\times} + \operatorname{Im}\widehat{B} + \widehat{A}\big[\operatorname{Im}_k(\widehat{A}|\widehat{B})\big].$$

As $\operatorname{Im}\widehat{B} + \widehat{A}\big[\operatorname{Im}_k(\widehat{A}|\widehat{B})\big] = \operatorname{Im}_{k+1}(\widehat{A}|\widehat{B})$, the inclusion (7.28) follows.

Fix the nonnegative integer k. As we have seen, the linear operator $\widetilde{\Psi}$ maps $\widetilde{M} + \widetilde{M}^{\times} + \operatorname{Im}_k(\widetilde{A}|\widetilde{B})$ into $\widehat{M} + \widehat{M}^{\times} + \operatorname{Im}_k(\widehat{A}|\widehat{B})$. Likewise $\widehat{\Psi}$ maps the space $\widehat{M} + \widehat{M}^{\times} + \operatorname{Im}_k(\widehat{A}|\widehat{B})$ into $\widetilde{M} + \widetilde{M}^{\times} + \operatorname{Im}_k(\widetilde{A}|\widetilde{B})$. The same is true with k replaced by $k+1$. But then the linear operators

$$\widetilde{\Phi}_k \ : \ \frac{\widetilde{M} + \widetilde{M}^{\times} + \operatorname{Im}_k(\widetilde{A}|\widetilde{B})}{\widetilde{M} + \widetilde{M}^{\times} \cap \operatorname{Im}_{k+1}(\widetilde{A}|\widetilde{B})} \ \to \ \frac{\widehat{M} + \widehat{M}^{\times} + \operatorname{Im}_k(\widehat{A}|\widehat{B})}{\widehat{M} + \widehat{M}^{\times} + \operatorname{Im}_{k+1}(\widehat{A}|\widehat{B})}, \tag{7.30}$$

$$\widehat{\Phi}_k \ : \ \frac{\widehat{M} + \widehat{M}^{\times} + \operatorname{Im}_k(\widehat{A}|\widehat{B})}{\widehat{M} + \widehat{M}^{\times} \cap \operatorname{Im}_{k+1}(\widehat{A}|\widehat{B})} \ \to \ \frac{\widetilde{M} + \widetilde{M}^{\times} + \operatorname{Im}_k(\widetilde{A}|\widetilde{B})}{\widetilde{M} + \widetilde{M}^{\times} + \operatorname{Im}_{k+1}(\widetilde{A}|\widetilde{B})} \tag{7.31}$$

induced by $\widetilde{\Psi}$ and $\widehat{\Psi}$, respectively, are well-defined. They are also each other's inverse. This can be deduced easily from

$$\begin{aligned}
\widetilde{M} + \widetilde{M}^{\times} + \operatorname{Im}_k(\widetilde{C}|\widetilde{A}) \ &\supset \ \operatorname{Im}(I - \widehat{\Psi}\widetilde{\Psi}),\\
\widehat{M} + \widehat{M}^{\times} + \operatorname{Im}_k(\widehat{C}|\widehat{A}), \ &\supset \ \operatorname{Ker}(I - \widetilde{\Psi}\widehat{\Psi}),
\end{aligned}$$

two inclusion relations which are immediate from (7.23). Thus the quotient spaces appearing in (7.30) and (7.31) are linearly isomorphic. In particular they have the same (possibly infinite) dimension. □

The symmetry in the arguments employed in the above proof (Parts 2 and 3 especially) suggests the possible use of a duality reasoning. Working in a finite dimensional context this line of approach is indeed possible. In the infinite dimensional situation, however, it does not work, an obstacle being that (sums of) operator ranges need not be closed.

7.3 Wiener-Hopf factorization and spectral invariants

Let Y, W, Γ, F_+ and F_- be as in the preceding two sections, and let $\varepsilon_+, \varepsilon_- \in \mathbb{C}$ be points in F_+ and F_-, respectively. By a *right Wiener-Hopf factorization* of W with respect to Γ (and the points ε_+ and ε_-) we mean a factorization

$$W(\lambda) = W_-(\lambda)D(\lambda)W_+(\lambda), \qquad \lambda \in \Gamma, \tag{7.32}$$

where the factors W_- and W_+ are operator-valued functions, the values being operators on Y, such that

(i) W_- is analytic on F_- and continuous on $\overline{F}_-$,

(ii) W_+ is analytic on F_+ and continuous on $\overline{F}_+$,

(iii) W_- and W_+ take invertible values on $\overline{F}_-$ and $\overline{F}_+$, respectively,

(iv) the middle term D in (7.32) has the form

$$D(\lambda) = \Pi_0 + \sum_{j=1}^{r} \left(\frac{\lambda - \varepsilon_+}{\lambda - \varepsilon_-} \right)^{\kappa_j} \Pi_j, \qquad \lambda \in \Gamma, \tag{7.33}$$

where $\kappa_1, \ldots, \kappa_r$ are non-zero integers, $\kappa_1 \leq \kappa_2 \leq \cdots \leq \kappa_r$, the operators $\Pi_1, \ldots, \Pi_r$ are mutually disjoint rank 1 projections on Y, and

$$\Pi_0 = I_Y - (\Pi_1 + \cdots + \Pi_r)$$

so Π_0 is a projection disjoint from $\Pi_1, \ldots, \Pi_r$.

A necessary condition for such a factorization to exist is that W takes invertible values on Γ. In terms of a realization of W on Γ this means that Γ splits the spectrum of the associate main operator (see again Theorem 2.4). If in (7.32) the factors W_- and W_+ are interchanged, we speak of a *left Wiener-Hopf factorization*. We will focus on the right version; for the left variant analogous results hold.

A few remarks are in order. Suppose W admits a right Wiener-Hopf factorization with respect to Γ and the points $\varepsilon_+ \in F_+$ and $\varepsilon_- \in F_-$. Then W also admits a right Wiener-Hopf factorization with respect to Γ and any other two points $\gamma_+ \in F_+$ and $\gamma_- \in F_-$. For γ_- in the finite complex plane this is clear from the simple identity

$$\left(\frac{\lambda-\varepsilon_+}{\lambda-\varepsilon_-}\right) = \left(\frac{\lambda-\varepsilon_+}{\lambda-\gamma_+}\right)\left(\frac{\lambda-\gamma_+}{\lambda-\gamma_-}\right)\left(\frac{\lambda-\gamma_-}{\lambda-\varepsilon_-}\right).$$

For $\gamma_- = \infty$, use

$$\left(\frac{\lambda-\varepsilon_+}{\lambda-\varepsilon_-}\right) = \left(\frac{\lambda-\varepsilon_+}{\lambda-\gamma_+}\right)(\lambda-\gamma_+)\left(\frac{1}{\lambda-\varepsilon_-}\right).$$

This brings the middle term $D(\lambda)$ into the form

$$D(\lambda) = \Pi_0 + \sum_{j=1}^{r} (\lambda-\gamma_+)^{\kappa_j}\Pi_j, \qquad \lambda \in \Gamma. \tag{7.34}$$

Note that the scalar functions $(\lambda-\gamma_+)^{\kappa_j}$ featured in the latter expression have their zeros and poles in γ_+ and ∞. When the origin belongs to F_+, one can take $\gamma_+ = 0$ and (7.34) becomes

$$D(\lambda) = \Pi_0 + \sum_{j=1}^{r} \lambda^{\kappa_j}\Pi_j, \qquad \lambda \in \Gamma.$$

This type of middle term plays a role in the study of Toeplitz equations where Γ is taken to be the unit circle (see [52], Chapter XXIV).

Although a right Wiener-Hopf factorization is (generally) not unique, the non-zero integers $\kappa_1, \ldots, \kappa_r$ are. They are called the *right (Wiener-Hopf) factorization indices of W with respect to* Γ. Left factorization indices are defined similarly. Sometimes the term *partial indices* is used instead of factorization indices. Finally, we mention that right (left) canonical factorization corresponds to the case when the right (left) factorization indices are all zero.

For the convenience of the reader, we recall (from the previous section) that $\operatorname{Ker}_k(C|A)$ and $\operatorname{Im}_k(A|B)$ are defined as

$$\begin{aligned}\operatorname{Ker}_k(C|A) &= \operatorname{Ker} C \cap \operatorname{Ker} CA \cap \cdots \cap \operatorname{Ker} CA^{k-1},\\ \operatorname{Im}_k(A|B) &= \operatorname{Im} B + \operatorname{Im} AB + \cdots + \operatorname{Im} A^{k-1}B.\end{aligned}$$

Theorem 7.8. *Let the function W be given by the realization* (7.1), *i.e,*

$$W(\lambda) = I_Y + C(\lambda I_X - A)^{-1}B,$$

where Γ splits the spectrum of A. Then W admits a right Wiener-Hopf factorization with respect to Γ if and only if the following two conditions are satisfied:

(a) *Γ splits the spectrum of $A^\times = A - BC$,*

(b) $\dim\left(M \cap M^\times\right) < \infty$ *and* $\dim\left(\dfrac{X}{M + M^\times}\right) < \infty$,

where $M = \operatorname{Im} P(A;\Gamma)$ *and* $M^\times = \operatorname{Ker} P(A^\times;\Gamma)$. *In that case, the right factorization indices of* W *can be described in terms of the operators appearing in* (7.1) *as follows:*

(c) *the number* s *of negative right factorization indices and the negative right factorization indices* $-\alpha_1, \ldots, -\alpha_s$ *(in the ordinary order:* $-\alpha_1 \leq \cdots \leq -\alpha_s$*) themselves are given by*

$$s = \dim\left(\frac{M \cap M^\times}{M \cap M^\times \cap \operatorname{Ker} C}\right),$$

$$\alpha_j = \sharp\left\{k = 1, 2, \ldots \mid \dim\left(\frac{M \cap M^\times \cap \operatorname{Ker}_{k-1}(C|A)}{M \cap M^\times \cap \operatorname{Ker}_k(C|A)}\right) \geq j\right\}, \qquad j = 1, \ldots, s,$$

(d) *the number* t *of positive right factorization indices and the positive right factorization indices* $\omega_1, \ldots, \omega_t$ *(in reversed order:* $\omega_t \leq \cdots \leq \omega_1$*) themselves are given by*

$$t = \dim\left(\frac{M + M^\times + \operatorname{Im} B}{M + M^\times}\right),$$

$$\omega_j = \sharp\left\{k = 1, 2, \ldots \mid \dim\left(\frac{M + M^\times + \operatorname{Im}_k(A|B)}{M + M^\times + \operatorname{Im}_{k-1}(A|B)}\right) \geq j\right\}, \qquad j = 1, \ldots, t.$$

As was already indicated above, for left Wiener-Hopf factorizations an analogous theorem holds. The theorem also has an analogue for appropriate closed contours in the Riemann sphere $\mathbb{C}_\infty$ like the extended real line or the extended imaginary axis.

Proof. For the (long and complicated) proof of the "if part" of Theorem 7.8 we refer to [17]. Here we shall concentrate on the "only if part" and the description of the right factorization indices. So we shall assume that W admits a Wiener-Hopf factorization (7.32) with respect to the contour Γ and, say, the points $\varepsilon_+ \in F_+$ and $\varepsilon_- \in F_-$. According to Theorem 7.2 it suffices to prove that there exists a special realization for W, for convenience also written as (7.1), such that Γ splits the spectra of A and $A^\times$ and for which (b)–(d) hold. The argument consists of several steps.

Step 1. Write the negative right factorization indices of W in the ordinary order (so from small to large) as $-\alpha_1, \ldots, -\alpha_s$, and the positive right factorization indices in the reversed order (so from large to small) as $\omega_1, \ldots, \omega_t$:

$$-\alpha_1 \le \cdots \le -\alpha_s < 0 < \omega_t \le \cdots \le \omega_1. \tag{7.35}$$

Then D can be written in the form

$$D(\lambda) = P_0 + \sum_{j=1}^{s} \left(\frac{\lambda - \varepsilon_-}{\lambda - \varepsilon_+}\right)^{\alpha_j} P_{-j} + \sum_{j=t}^{1} \left(\frac{\lambda - \varepsilon_+}{\lambda - \varepsilon_-}\right)^{\omega_j} P_j, \tag{7.36}$$

where $P_{-1}, \ldots, P_{-s}, P_t, \ldots, P_1$ are mutually disjoint rank 1 projections on Y, and $P_0 = I_Y - (P_{-1} + \cdots + P_{-s} + P_t + \cdots + P_1)$, so P_0 is a projection disjoint from $P_{-1}, \ldots, P_{-s}, P_t, \ldots, P_1$. For definiteness, we shall assume that s and t are both positive.

Step 2. Fix j among the integers $1, \ldots, s$, and let $D_j^-(\lambda)$ be the scalar function given by

$$D_j^-(\lambda) = \left(\frac{\lambda - \varepsilon_-}{\lambda - \varepsilon_+}\right)^{\alpha_j}, \qquad \lambda \neq \varepsilon_+.$$

Write J_j^- for the lower triangular Jordan block with eigenvalue ε_+ and order α_j, so that $\sigma(J_j^-) = \{\varepsilon_+\}$. Further introduce

$$B_j^- = \begin{bmatrix} (\varepsilon_+ - \varepsilon_-)^{\alpha_j} \binom{\alpha_j}{\alpha_j} \\ \vdots \\ (\varepsilon_+ - \varepsilon_-)^2 \binom{\alpha_j}{2} \\ (\varepsilon_+ - \varepsilon_-) \binom{\alpha_j}{1} \end{bmatrix},$$

$$C_j^- = \begin{bmatrix} 0 & \ldots & 0 & 1 \end{bmatrix}.$$

Then $D_j^-(\lambda) = 1 + C_j^-(\lambda - J_j^-)^{-1}B_j^-$ is a (minimal) realization of D_j^-. Now $J_j^{-\times} - \varepsilon_- I_{\alpha_j}$ is similar with the lower triangular nilpotent Jordan block of order α_j and having eigenvalue ε_+, a similarity being given by the upper triangular matrix

$$\left[(\varepsilon_+ - \varepsilon_-)^{\nu-\mu} \binom{\nu - 1}{\nu - \mu} \right]_{\mu,\nu=1}^{\alpha_j},$$

where $\binom{\nu-1}{\mu-1}$ is read as zero for $\mu > \nu$. Thus $\sigma(J_j^{-\times}) = \{\varepsilon_-\}$.

Clearly $P(J_j^-;\Gamma) = I$ and $P(J_j^{-\times};\Gamma) = 0$. Hence

$$\operatorname{Im} P(J_j^-;\Gamma) = \operatorname{Ker} P(J_j^{-\times};\Gamma) = \mathbb{C}^{\alpha_j},$$

and so, trivially,

$$\operatorname{Im}_k\big(J_j^- B_j^-\big) = \mathbb{C}^{\alpha_j}, \qquad k = 0, 1, \dots . \tag{7.37}$$

Furthermore, as is easily verified,

$$\operatorname{Ker}_k\big(C_j^- | J_j^-\big) = \mathbb{C}^{\alpha_j - k} \dotplus \{0\}^k, \qquad k = 0, 1, \dots, \tag{7.38}$$

where the right-hand side of the equality is read as $\{0\}^{\alpha_j}$ for $k \geq \alpha_j$.

Step 3. Take j among the integers $1, \dots, t$, and let $D_j^+(\lambda)$ be the scalar function given by

$$D_j^+(\lambda) = \left(\frac{\lambda - \varepsilon_+}{\lambda - \varepsilon_-}\right)^{\omega_j}, \qquad \lambda \neq \varepsilon_-.$$

Write J_j^+ for the lower triangular Jordan block with eigenvalue ε_- and order ω_j, so that $\sigma(J_j^-) = \{\varepsilon_-\}$. Further introduce

$$B_j^+ = \begin{bmatrix} 1 \\ 0 \\ \vdots \\ 0 \end{bmatrix},$$

$$C_j^+ = \begin{bmatrix} (\varepsilon_- - \varepsilon_+)\binom{\omega_j}{1} & (\varepsilon_- - \varepsilon_+)^2\binom{\omega_j}{2} & \cdots & (\varepsilon_- - \varepsilon_+)^{\omega_j}\binom{\omega_j}{\omega_j} \end{bmatrix}.$$

Then $D_j^+(\lambda) = 1 + C_j^+(\lambda - J_j^+)^{-1}B_j^+$ is a (minimal) realization of D_j^+. Analogously to what we saw in the previous step for the matrix $J_j^{-\times} - \varepsilon_- I_{\alpha_j}$, the matrix $J_j^{+\times} - \varepsilon_+ I_{\omega_j}$ is similar with the lower triangular nilpotent Jordan block of order ω_j and having ε_+ as eigenvalue. Thus $\sigma\big(J_j^{+\times}\big) = \{\varepsilon_+\}$.

Clearly $P\big(J_j^+;\Gamma\big) = 0$ and $P\big(J_j^{+\times};\Gamma\big) = I$. Hence

$$\operatorname{Im} P\big(J_j^-;\Gamma\big) = \operatorname{Ker} P\big(J_j^{-\times};\Gamma\big) = \{0\}^{\omega_j},$$

and so, trivially,

$$\operatorname{Ker}_k\big(C_j^+ | J_j^+\big) = \{0\}^{\omega_j}, \qquad k = 0, 1, \dots . \tag{7.39}$$

Furthermore, as is easily verified,

$$\operatorname{Im}_k\big(J_j^+ | B_j^+\big) = \mathbb{C}^k \dotplus \{0\}^{\omega_j - k}, \qquad k = 0, 1 \dots, \tag{7.40}$$

where the right-hand side of the equality is read as $\mathbb{C}^{\omega_j}$ for $k \geq \omega_j$.

Step 4. Let $D_0(\lambda)$ be the diagonal matrix given by

$$D_0(\lambda) = \begin{bmatrix} D_1^-(\lambda) & & & & & \\ & \ddots & & & & \\ & & D_s^-(\lambda) & & & \\ & & & D_t^+(\lambda) & & \\ & & & & \ddots & \\ & & & & & D_1^+(\lambda) \end{bmatrix}, \tag{7.41}$$

i.e., $D_0(\lambda)$ is the direct sum of the matrices $D_1^-(\lambda), \ldots, D_s^-(\lambda), D_t^+(\lambda), \ldots, D_1^+(\lambda)$. Then D_0 is a rational $m \times m$ matrix function, where $m = s + t$. To obtain a realization for D_0, we introduce $n = \alpha_1 + \cdots + \alpha_s + \omega_t + \cdots + \omega_1$, and introduce an $n \times n$ matrix A_0, an $n \times m$ matrix B_0 and an $m \times n$ matrix C_0 as follows: A_0 is the direct sum of the matrices $J_1^-, \ldots, J_s^-, J_t^+, \ldots, J_1^+$, B_0 is the direct sum of the matrices $B_1^-, \ldots, B_s^-, B_t^+, \ldots, B_1^+$, and C_0 is the direct sum of the matrices $C_1^-, \ldots, C_s^-, C_t^+, \ldots, C_1^+$. Then, indeed, $D_0(\lambda) = I_m + C_0(\lambda I_n - A_0)^{-1}B_0$ is a (minimal) realization.

Obviously, Γ splits the spectra of A_0 and $A_0^\times = A_0 - B_0C_0$. In fact, these spectra coincide with $\{\varepsilon_+, \varepsilon_-\}$. (Without the assumption introduced in *Step 1* that s and t are both positive, we would have that the spectra of A_0 and $A_0^\times$ are subsets of $\{\varepsilon_+, \varepsilon_-\}$, and these inclusions are both proper if and only if one of the integers s or t equals zero.) Put $M_0 = \operatorname{Im} P(A_0; \Gamma)$ and $M_0^\times = \operatorname{Ker} P(A_0^\times; \Gamma)$. Then

$$M_0 = M_0^\times = \mathbb{C}^{\alpha_1} \dotplus \cdots \dotplus \mathbb{C}^{\alpha_s} \dotplus \{0\}^{\omega_t} \dotplus \cdots \dotplus \{0\}^{\omega_1}. \tag{7.42}$$

Further we have, for $k = 1, 2, \ldots,$

$$\begin{aligned} \operatorname{Ker}_k(C_0|A_0) &= \operatorname{Ker}_k(C_1^-|J_1^-) \dotplus \cdots \dotplus \operatorname{Ker}_k(C_s^-|J_s^-) \dotplus \{0\}^{\omega_t} \dotplus \cdots \dotplus \{0\}^{\omega_1}, \\ \operatorname{Im}_k(A_0|B_0) &= \mathbb{C}^{\alpha_1} \dotplus \cdots \dotplus \mathbb{C}^{\alpha_s} \dotplus \operatorname{Im}_k(J_t^+|B_t^+)) \dotplus \cdots \dotplus \operatorname{Im}_k(J_1^+|B_1^+), \end{aligned}$$

and

$$M_0 \cap M_0^\times \cap \operatorname{Ker}_k(C_0|A_0) = \operatorname{Ker}_k(C_0|A_0), \tag{7.43}$$

$$M_0 + M_0^\times + \operatorname{Im}_k(A_0|B_0) = \operatorname{Im}_k(A_0|B_0). \tag{7.44}$$

Here we used (7.39) and (7.37).

It is clear from (7.42) that

$$\dim\left(M_0 \cap M_0^\times\right) = \alpha_1 + \cdots + \alpha_s.$$

Combining (7.43) and (7.38), we get

$$\dim\left(M_0 \cap M_0^\times \cap \operatorname{Ker}_k(C_0|A_0)\right) = \max\{0, \alpha_1 - k\} + \cdots + \max\{0, \alpha_s - k\}.$$

In particular

$$\dim\big(M_0 \cap M_0^\times \cap \operatorname{Ker} C_0)\big) = (\alpha_1 - 1) + \cdots + (\alpha_s - 1),$$

and it follows that

$$\dim\left(\frac{M_0 \cap M_0^\times}{M_0 \cap M_0^\times \cap \operatorname{Ker} C_0}\right) = s.$$

Thus, with M, $M^\times$, C replaced by M_0, $M_0^\times$, C_0, respectively, the first identity in Theorem 7.8, item (c) is satisfied .

We also have

$$\dim\left(\frac{M_0 \cap M_0^\times \cap \operatorname{Ker}_{k-1}(C_0|A_0)}{M_0 \cap M_0^\times \cap \operatorname{Ker}_k(C_0|A_0)}\right) = \sum_{l\in\{1,\ldots,s\}} \big(\max\{\alpha_l - k + 1\} - \max\{\alpha_l - k\}\big)$$

$$= \sum_{l\in\{1,\ldots,s\},\ \alpha_l\geq k} 1 = \sharp\{l = 1,\ldots,s \,|\, \alpha_l \geq k\}.$$

Now, fix $j \in \{1,\ldots,s\}$. Then

$$\sharp\{l = 1,\ldots,s \,|\, \alpha_l \geq k\} \geq j \quad \Leftrightarrow \quad k \in \{1,\ldots,\alpha_j\},$$

and hence

$$\sharp\Big\{k = 1,2,\ldots \,|\, \sharp\{l = 1,\ldots,s \,|\, \alpha_l \geq k\} \geq j\Big\} = \alpha_j.$$

Combining these elements we see that, with $M, M^\times, A, B, C$ replaced by $M_0, M_0^\times, A_0, B_0, C_0$, respectively, the second identity in Theorem 7.8, item (c) holds too.

For the two identities in Theorem 7.8, item (d), the analogous observation is true. The arguments are basically the same as the ones presented for item (b).

Step 5. Next we deal with the middle term D in the factorization (7.32), written in the form (7.36) with $-\alpha_1,\ldots,-\alpha_s,\omega_t,\ldots,\omega_1$ satisfying (7.35), $P_{-1},\ldots,P_{-s}$, $P_t,\ldots,P_1$ mutually disjoint rank 1 projections on Y and

$$P_0 = I_Y - (P_{-1} + \cdots + P_{-s} + P_t + \cdots + P_1)\,.$$

Clearly P_0 and $P_{-1} + \cdots + P_{-s} + P_t + \cdots + P_1$ are complementary projections. Put $Y_0 = \operatorname{Ker} P_0$. Then $Y_0 = \operatorname{Im} P_{-1} \dotplus \cdots \dotplus P_{-s} \dotplus \operatorname{Im} P_t \dotplus \cdots \dotplus P_1$ and so Y_0 can be identified with $\mathbb{C}^m$ where, as before, $m = s + t$. Thus $Y = \mathbb{C}^m \dotplus \operatorname{Im} P_0$ and with respect to this decomposition $D(\lambda)$ can be written as an operator matrix

$$D(\lambda) = \begin{bmatrix} D_0(\lambda) & 0 \\ 0 & I \end{bmatrix}.$$

Here D_0 is given by (7.41) and I is the identity operator on $\operatorname{Im} P_0$. Now let

$$A_D = A_0, \qquad B_D = \begin{bmatrix} B_0 & 0 \end{bmatrix}, \qquad C_D = \begin{bmatrix} C_0 \\ 0 \end{bmatrix},$$

where A_0, B_0 and C_0 are as in *Step 4*. Then we have the realization

$$D(\lambda) = I_Y + C_D(\lambda I_n - A_D)^{-1}B_D,$$

$n = \alpha_1 + \cdots + \alpha_s + \omega_t + \cdots + \omega_1$, with Γ splitting the spectra of $A_D = A_0$ and $A_D^\times = A_0 - B_0C_0 = A_0^\times$. Write $M_D = \operatorname{Im} P(A_D;\Gamma)$ and $M_D^\times = \operatorname{Ker} P(A_D^\times;\Gamma)$. In other words, $M_D = M_0$ and $M_D^\times = M_0^\times$ where, again, we use the notation of the previous step. For $k = 1, 2, \ldots$, clearly, $\operatorname{Ker}_k(C_D|A_D) = \operatorname{Ker}_k(C_0|A_0)$ and $\operatorname{Im}_k(A_D|B_D) = \operatorname{Im}_k(A_0|B_0$. It follows that, with M, $M^\times, A$, B, C replaced by M_D, $M_D^\times, A_D$, B_D, C_D, respectively, (b)–(d) in Theorem 7.8 are satisfied.

Step 6. We begin this sixth and final step by representing the factors W_- and W_+ in the Wiener-Hopf factorization (7.32) in the form

$$\begin{aligned} W_-(\lambda) &= I_Y + C_-(\lambda I_{X_-} - A_-)^{-1}B_-, \qquad \lambda \in \Omega_-, \\ W_+(\lambda) &= I_Y + C_+(\lambda_I X_+ - A_+)^{-1}B_+, \qquad \lambda \in \Omega_+, \end{aligned}$$

with

$$\sigma(A_-) \subset F_+, \qquad \sigma(A_-^\times) \subset F_+, \qquad \sigma(A_+) \subset F_-, \qquad \sigma(A_+^\times) \subset F_-.$$

Why this can be done is explained in the proof of Theorem 7.1. On Γ we have the factorization (7.32), and so we can apply the product rule of Section 2.5 to show that $W(\lambda) = I_Y + C(\lambda I_X - A)^{-1}B,\ \lambda \in \Gamma$, where

$$X = X_- \dotplus \mathbb{C}^n \dotplus X_+, \tag{7.45}$$

$n = \alpha_1 + \cdots + \alpha_s + \omega_t + \cdots + \omega_1$, and $A : X \to X$, $B : Y \to X$ and $C : X \to Y$ are given by

$$A = \begin{bmatrix} A_- & B_-C_D & B_-C_+ \\ 0 & A_D & B_DC_+ \\ 0 & 0 & A_+ \end{bmatrix}, \qquad B = \begin{bmatrix} B_- \\ B_D \\ B_+ \end{bmatrix}, \qquad C = \begin{bmatrix} C_- & C_D & C_+ \end{bmatrix}.$$

Here the operator matrices are taken with respect to the decomposition (7.45). Now the realization obtained for W this way has the desired properties. This can be seen as follows.

Obviously Γ splits the spectrum of A and the same is true for $A^\times = A - BC$ which has the matrix representation

$$A^\times = \begin{bmatrix} A_-^\times & 0 & 0 \\ -B_DC_- & A_D^\times & 0 \\ -B_+C_- & -B_+C_D & A_+^\times \end{bmatrix}.$$

Let $M = \operatorname{Im} P(A;\Gamma)$ and $M^{\times} = \operatorname{Ker} P(A^{\times};\Gamma)$. Assume for the moment that we have established the identities

$$M \cap M^{\times} \cap \operatorname{Ker}_k(C|A) = \{0_-\} \dotplus \big(M_D \cap M_D^{\times} \cap \operatorname{Ker}_k(C_D|A_D)\big) \dotplus \{0_+\}, \quad (7.46)$$

$$M + M^{\times} + \operatorname{Im}_k(A|B) = X_- \dotplus \big(M_D + M_D^{\times} + \operatorname{Im}_k(A_D|B_D)\big) \dotplus X_+, \quad (7.47)$$

where 0_- is the zero element in X_-, 0_+ is the zero element in X_+ and k is allowed to take the values $0, 1, 2, \ldots$. Then it would be clear from the conclusions obtained in the previous step that (b)–(d) in Theorem 7.8 are met and we would be ready. So we need to concentrate on (7.46) and (7.47).

Clearly $P = P(A;\Gamma)$ has the form

$$P = \begin{bmatrix} I_{X_-} & P_1 & P_2 \\ 0 & P(A_D;\Gamma) & P_3 \\ 0 & 0 & 0 \end{bmatrix}.$$

From the fact that $P(A;\Gamma)$ is a projection one gets the relations

$$P_1 P(A_D;\Gamma) = 0, \qquad P_1 P_3 = 0, \qquad P(A_D;\Gamma) P_3 = P_3,$$

(where the two outer ones imply the middle). In turn these give

$$P = \begin{bmatrix} I_{X_-} & -P_1 & -P_2 \\ 0 & I_n & -P_3 \\ 0 & 0 & I_{X_+} \end{bmatrix} \begin{bmatrix} I_{X_-} & 0 & 0 \\ 0 & P(A_D;\Gamma) & 0 \\ 0 & 0 & 0 \end{bmatrix} \begin{bmatrix} I_{X_-} & P_1 & P_2 \\ 0 & I_n & P_3 \\ 0 & 0 & I_{X_+} \end{bmatrix},$$

with the first and last factor in the right-hand side invertible and being each other's inverse. Hence

$$M = \begin{bmatrix} I_{X_-} & -P_1 & -P_2 \\ 0 & I_n & -P_3 \\ 0 & 0 & I_{X_+} \end{bmatrix} \big[X_- \dotplus M_D \dotplus \{0_+\}\big] = X_- \dotplus M_D \dotplus \{0_+\}.$$

In the same way one gets $M^{\times} = \{0_-\} \dotplus M_D^{\times} \dotplus X_+$, and it follows that

$$M \cap M^{\times} = \{0_-\} \dotplus (M_D \cap M_D^{\times}) \dotplus \{0_+\}, \quad (7.48)$$

$$M + M^{\times} = X_- \dotplus (M_D + M_D^{\times}) \dotplus X_+. \quad (7.49)$$

Thus (7.46) and (7.47) are valid for $k = 0$.

To prove (7.46) for arbitrary k we argue as follows. A simple induction argument shows that CA^l is of the form

$$CA^l = \begin{bmatrix} * & \left(\sum_{\nu=0}^{l-1} Q_{\nu,l} C_D A_D^\nu\right) + C_D A_D^l & * \end{bmatrix}, \qquad l = 0, 1, \ldots, \tag{7.50}$$

where $Q_{0,l}, \ldots Q_{l-1,l}$ and the stars denote appropriate but here not explicitly specified operators. Together with (7.48) this gives that the right-hand side of (7.46) is contained in the left-hand side. The reverse inclusion can be proved by an induction argument in which (7.50) is employed once more.

Finally let us turn to (7.47). For $A^l B$ there is an expression analogous to (7.50), namely

$$A^l B = \begin{bmatrix} * \\ \left(\sum_{\nu=0}^{l-1} A_D^\nu B_D R_{\nu,l}\right) + A_D^l B_D \\ * \end{bmatrix}, \qquad l = 0, 1, \ldots, \tag{7.51}$$

where $R_{0,l}, \ldots R_{l-1,l}$ and the stars stand for certain operators. Together with (7.49) this yields that the left-hand side of (7.47) is contained in the right-hand side. The reverse inclusion can be proved by an induction argument in which (7.51) is used once again. □

We close this section with a couple of observations on the dimension numbers featuring in Theorems 7.2 and 7.8. For shortness sake, introduce

$$\begin{aligned} \widehat{\alpha}_k &= \dim\left(\frac{M \cap M^\times \cap \operatorname{Ker}_{k-1}(C|A)}{M \cap M^\times \cap \operatorname{Ker}_k(C|A)}\right), \\ \widehat{\omega}_k &= \dim\left(\frac{M + M^\times + \operatorname{Im}_k(A|B)}{M + M^\times + \operatorname{Im}_{k-1}(A|B)}\right). \end{aligned}$$

Here k may run through the positive integers $1, 2, \ldots$. Recall that $\operatorname{Ker}_0(C|A)$ is read as X and $\operatorname{Im}_0(A|B)$ as $\{0\}$, so

$$\begin{aligned} \widehat{\alpha}_1 &= \dim\left(\frac{M \cap M^\times}{M \cap M^\times \cap \operatorname{Ker} C}\right), \\ \widehat{\omega}_1 &= \dim\left(\frac{M + M^\times + \operatorname{Im} B}{M + M^\times}\right). \end{aligned}$$

Using standard linear algebra arguments it can be shown that the sequences $\widehat{\alpha}_1, \widehat{\alpha}_2, \ldots$ and $\widehat{\omega}_1, \widehat{\omega}_2, \ldots$ are decreasing, i.e.,

$$\widehat{\alpha}_k \geq \widehat{\alpha}_{k+1}, \qquad \widehat{\omega}_k \geq \widehat{\omega}_{k+1}, \qquad k = 1, 2, \ldots.$$

In addition it can be proved that $\widehat{\alpha}_k$ and $\widehat{\omega}_k$ vanish for k sufficiently large, provided that $M \cap M^\times$ and $M + M^\times$ have finite dimension and codimension, respectively. In fact we then even have,

$$\begin{aligned} M \cap M^\times \cap \operatorname{Ker}_k(C|A) &= \{0\}, \\ M + M^\times + \operatorname{Im}_k(A|B) &= X, \end{aligned}$$

again holding for k sufficiently large. The considerations in *Step 4* in the above proof corroborate these facts.

Here are some details for the integers $\widehat{\alpha}_1, \widehat{\alpha}_2, \ldots$; for $\widehat{\omega}_1, \widehat{\omega}_2, \ldots$ the situation is analogous. The mapping

$$\frac{M \cap M^\times \cap \operatorname{Ker}_k(C|A)}{M \cap M^\times \cap \operatorname{Ker}_{k+1}(C|A)} \mapsto \frac{M \cap M^\times \cap \operatorname{Ker}_{k-1}(C|A)}{M \cap M^\times \cap \operatorname{Ker}_k(C|A)}$$

induced by A is easily seen to be injective. Hence $\widehat{\alpha}_{k+1} \leq \widehat{\alpha}_k$. Assume now that $M \cap M^\times$ has finite dimension. Then there exists a positive integer r such that

$$M \cap M^\times \cap \operatorname{Ker}_k(C|A) = M \cap M^\times \cap \operatorname{Ker}_r(C|A), \qquad k = r, r+1, \ldots .$$

Evidently $M \cap M^\times \cap \operatorname{Ker}_r(C|A)$ is invariant under both A and $A^\times$. Also A and $A^\times$ coincide on $M \cap M^\times \cap \operatorname{Ker}_r(C|A)$. As the restriction of A to M and that of $A^\times$ to $M^\times$ have no eigenvalue in common, it follows that $M \cap M^\times \cap \operatorname{Ker}_r(C|A) = \{0\}$.

Notes

This chapter is based on the papers [17] and [18]. The material of these papers relevant for this book has been reorganized and several of the arguments have been improved. The details are as follows. The "if part" of Theorem 7.1 is a special case of Theorem 3.1 in [17]; it also has the first part of Theorem 1.5 in [11] as a less general predecessor. Theorem 7.2 combines Theorems 5.1 and 6.1 of [18] in a more appropriate formulation. The proof of Theorem 7.2 given in Section 7.2 is a significant improvement over the argument given in [18]. The results from [17] and [18] to be mentioned in connection with Theorem 7.8 are Theorem 3.1 and Corollary 3.2 in [17] and Theorem 1.2 in [18].

The spectral invariants appearing in Theorem 7.2 are closely related to the block similarity invariants of operator blocks of the first or third kind; see [58], Section XI.5 in particular. For a review of the theory of possibly non-canonical Wiener-Hopf factorization of matrix-valued functions taking invertible values, we refer to the book [29] and the more recent survey article [59]. Wiener-Hopf factorization of operator-valued functions goes back to [71] and [72]; see also the recent book [73] . The fact that the Wiener-Hopf factorization indices depend on the given function only (and not on the particular Wiener-Hopf factorization) is well-known for continuous matrix-valued functions (see [60]) and for certain classes of continuous operator-valued functions (see [49]). The latter do not cover the class of operator-valued functions considered in this chapter.

Part IV
Factorization of selfadjoint rational matrix functions

This part deals with factorization problems for rational matrix functions that have Hermitian values on the real line, the imaginary axis, or the unit circle. Included are problems of spectral factorization and pseudo-spectral factorization. The emphasis is on positive definite and nonnegative functions. In general, the factorizations considered are canonical or pseudo-canonical, and they are symmetric in the sense that they consist of two factors, where the first factor is the adjoint of the second (relative to the given curve). This part consists of four chapters.

Minimal realizations play an important role in the analysis of rational matrix functions that have Hermitian values on a curve. These are realization of which the order of the state matrix is equal to the MacMillan degree of the function. In the first chapter (Chapter 8) we review the theory of such realizations. Included are the state space similarity theorem and the minimal factorization theorem. In this first chapter we also introduce the notion of pseudo-canonical factorization and describe such factorizations in state space terms. In Chapter 9 we study in a state space setting spectral factorizations, that is, symmetric canonical factorizations for rational matrix functions that are positive definite on the unit circle, the real line or the imaginary axis. Chapter 10 carries out a similar program for nonnegative functions. In this case one has to consider symmetric pseudo-canonical factorization. In the final chapter (Chapter 11) we present (without proofs) some background material on matrices in finite dimensional indefinite inner product spaces, and review the main results from this area that are used in this part and the other remaining parts.

Chapter 8

Preliminaries concerning minimal factorization

In this chapter we gather together several results concerning minimal realizations and minimal factorizations that will play an important role in the sequel. Most of these results can also be found in Part II of the book [20]. For the reader's convenience we have chosen to summarize them here (without proofs). Special attention is given to the notion of pseudo-canonical factorization, which is a generalization of canonical factorization by allowing singularities on the curve.

This chapter consists of three sections. Sections 8.1 and 8.2 deal with minimal realizations and minimal factorizations, respectively. Section 8.3 is devoted to pseudo-canonical factorization.

8.1 Minimal realizations

Let W be a proper rational $m \times m$ matrix function, and let

$$W(\lambda) = D + C(\lambda I_n - A)^{-1}B \tag{8.1}$$

be a realization of W. The realization is said to be *minimal* if the dimension n of the state space has the smallest possible value. This smallest possible value is equal to the *McMillan degree* of W (see Section 8.5 in [20] for details). The McMillan degree of W will be denoted by $\delta(W)$.

For a characterization of minimality in terms of the matrices A, B and C, we need some more terminology. Let A be an $n \times n$ matrix, let B be an $n \times m$ matrix, and let C be an $m \times n$ matrix. The pair (A, B) is called *controllable* if

$$\operatorname{Im}(A|B) = \operatorname{Im} B + \operatorname{Im} AB + \cdots + \operatorname{Im} AB^{n-1} = \mathbb{C}^n.$$

So (A, B) is controllable if and only if $\mathbb{C}^n$ is the unique A-invariant subspace containing $\operatorname{Im} B$. The pair (C, A) is said to be *observable* if

$$\operatorname{Ker}(C|A) = \operatorname{Ker} C \cap \operatorname{Ker} CA \cap \cdots \cap \operatorname{Ker} CA^{n-1} = \{0\}.$$

Thus (C, A) is observable if and only if $\{0\}$ is the unique A-invariant subspace contained in $\operatorname{Ker} C$.

In line with these definitions, the realization (8.1) is called *controllable*, respectively *observable*, if the pair (A, B) is controllable, respectively the pair (C, A) is observable. From Sections 7.1 and 7.3 in [20] we now recall the main results on minimal realizations in the following two theorems.

Theorem 8.1. *A realization of a proper rational matrix function is minimal if and only if it is controllable and observable.*

Theorem 8.2. *Let W be a proper rational matrix function and suppose*

$$W(\lambda) = D_1 + C_1(\lambda I_n - A_1)^{-1}B_1, \tag{8.2}$$

$$W(\lambda) = D_2 + C_2(\lambda I_n - A_2)^{-1}B_2, \tag{8.3}$$

are minimal realizations of W. Then $D_1 = D_2$ and there exists a unique invertible $n \times n$ matrix S such that

$$S^{-1}A_1S = A_2, \qquad S^{-1}B_1 = B_2, \qquad C_1S = C_2. \tag{8.4}$$

This second theorem is known as the *state space similarity theorem*; the operator S is called a (*state space*) *similarity between the realizations* (8.2) and (8.3).

In the situation where (8.1) is a minimal realization, there is a close connection between the poles of W and the eigenvalues of A. Obviously, whether or not the realization is minimal, the poles of W form a subset of $\sigma(A)$. However, when the realization is minimal, the spectrum of A coincides with the set of poles of W. In addition, when W is a square matrix-valued function, and D is invertible so that $A^\times = A - BD^{-1}C$ is well-defined, $\sigma(A^\times)$ is precisely equal to the set of zeros of W. Here a *zero* of W is a pole of the inverse W^{-1} of W . For further details, including a more intrinsic definition of the notion of a zero of a rational matrix function, taking into account multiplicities and pole orders too, see Chapter 8 in [20]. From Chapter 7 in [20] we also recall that (8.1) is minimal when $\sigma(A) \cap \sigma(A^\times) = \emptyset$.

Next we consider the concept of local minimality. Let λ_0 be a point in the complex plane. The realization (8.1) is called *locally minimal at* λ_0 if

$$\operatorname{Im} PB + \operatorname{Im} PAB + \cdots + \operatorname{Im} PAB^{n-1} = \operatorname{Im} P, \tag{8.5}$$

$$\operatorname{Ker} CP \cap \operatorname{Ker} CAP \cap \cdots \cap \operatorname{Ker} CA^{n-1}P = \operatorname{Ker} P, \tag{8.6}$$

where P is the Riesz projection of A at λ_0. There is a local version of the observation given at the end of the previous paragraph: if λ_0 is not a common eigenvalue of A and $A^\times$, then (8.1) is minimal at λ_0. For details see Section 8.4 in [20]) where it is also shown that the realization (8.1) is minimal if and only if it is minimal at each point in the complex plane.

We finish this section by reviewing some results on Jordan chains and co-pole functions. Let W be a rational square matrix-valued function, and let φ be a $\mathbb{C}^m$-valued function which is analytic at λ_0 with $\varphi(\lambda_0) = 0$. We call φ a *co-pole function* of W at λ_0 if $W(\lambda)\varphi(\lambda)$ is analytic at λ_0 and $\lim_{\lambda\to\lambda_0} W(\lambda)\varphi(\lambda)$ is non-zero. For this to happen, it is necessary that $\det W(\lambda)$ does not vanish identically. As before, let W^{-1} denote the pointwise inverse of W, i.e., the function determined by the expression $W^{-1}(\lambda) = W(\lambda)^{-1}$. Now, if φ is a co-pole function of W at λ_0, then the function $\psi(\lambda) = W(\lambda)\varphi(\lambda)$ is a so-called root function of W^{-1} at λ_0, that is, ψ is analytic at λ_0 with $\psi(\lambda_0) \neq 0$ and $\lim_{\lambda\to\lambda_0} W(\lambda)^{-1}\psi(\lambda) = 0$. The converse is also true. A root function of W^{-1} at λ_0 is also referred to as a pole function of W at λ_0 (see [7], page 67).

The next two results have been taken from [20], Section 8.4 (Proposition 8.21 and Corollary 8.22).

Proposition 8.3. *Let the rational square matrix-valued function W be given by the realization* (8.1)*, and let λ_0 be an eigenvalue of A. Assume the realization is minimal at λ_0. Let $k \geq 1$, and let*

$$\varphi(\lambda) = (\lambda - \lambda_0)^k \varphi_k + (\lambda - \lambda_0)^{k+1} \varphi_{k+1} + \cdots$$

be a co-pole function of W at λ_0. Put

$$x_j = \sum_{\nu=k}^{\infty} P(A - \lambda_0)^{\nu-j-1} B\varphi_\nu, \qquad j = 0, \ldots, k-1, \tag{8.7}$$

where P is the Riesz projection of A corresponding to λ_0. Then $x_0, \ldots, x_{k-1}$ is a Jordan chain of A at λ_0, that is, $x_0 \neq 0$ and

$$(A - \lambda_0)x_0 = 0, \qquad (A - \lambda_0)^r x_{k-1} = x_{k-1-r}, \quad r = 0, \ldots, k-1. \tag{8.8}$$

Moreover, each Jordan chain of A at λ_0 is obtained in this way. Finally, if the chain $x_0, \ldots, x_{k-1}$ given by (8.7) *is maximal, that is, $x_{k-1} \notin \operatorname{Im}(A - \lambda_0)$, then $\varphi_k \neq 0$.*

With respect to (8.7) there is no convergence issue; actually only a finite number of terms in the sum are non-zero.

Proposition 8.4. *Let the rational square matrix-valued function W be given by the realization* (8.1)*, and suppose $\det W(\lambda) \not\equiv 0$. Let λ_0 be an eigenvalue of A, and assume that* (8.1) *is minimal at λ_0. If $x_0, \ldots, x_{k-1}$ is a Jordan chain of A at λ_0, then $Cx_0, \ldots, Cx_{k-1}$ is a Jordan chain of W^{-1} at λ_0, and each Jordan chain of W^{-1} at λ_0 is obtained in this way.*

For later use (see Section 10.1) we introduce the following terminology suggested by Proposition 8.3. Let W be given by the realization (8.1). If $x_0, \ldots, x_{k-1}$ is a Jordan chain of A at λ_0, any co-pole function $\varphi(\lambda) = \sum_{j=k}^{\infty}(\lambda - \lambda_0)^j \varphi_j$ satisfying (8.7) will be called a *co-pole function corresponding to the Jordan chain* $x_0, \ldots, x_{k-1}$. In this case Cx_j is precisely the coefficient of $(\lambda - \lambda_0)^r$ in the Taylor expansion of $W(\lambda)\varphi(\lambda)$ at λ_0. To see this, use (8.7) and the fact that the coefficients in the principal part of the Laurent expansion of W at λ_0 are given by the expression $CP(A - \lambda_0)^{j-1}B$, where P is the Riesz projection of A corresponding to the eigenvalue λ_0. These observations lie also behind Proposition 8.4 above.

8.2 Minimal factorization

The McMillan degree features a sublogarithmic property. Indeed, if W_1 and W_2 are rational matrix functions and $W = W_1 W_2$, that is

$$W(\lambda) = W_1(\lambda)W_2(\lambda),$$

then the McMillan degree of W is less than or equal to the sum of the McMillan degrees of W_1 and W_2:

$$\delta(W_1 W_2) \leq \delta(W_1) + \delta(W_2). \tag{8.9}$$

This is clear from Theorem 2.5 and the definition of the McMillan degree given in the beginning of the previous section. A factorization $W = W_1 W_2$ is called a *minimal factorization (involving two factors)* if equality occurs, that is, when $\delta(W) = \delta(W_1) + \delta(W_2)$. Intuitively, this means that there is no pole-zero cancellation in the product $W_1 W_2$; this is made precise in Theorem 9.1 in [20].

Let $W(\lambda) = D + C(\lambda I_n - A)^{-1}B$ be a realization of an $m \times m$ rational matrix function, assume that D is invertible, and let $D = D_1 D_2$ with D_1, D_2 $m \times m$ matrices (automatically invertible). Put $A^\times = A - BD^{-1}C$. Suppose $M, M^\times$ is a pair of subspaces of $\mathbb{C}^n$ satisfying

$$AM \subset M, \qquad A^\times M^\times \subset M^\times, \qquad M \dot{+} M^\times = \mathbb{C}^n. \tag{8.10}$$

In that case we know (see Section 2.6) that W admits a factorization $W = W_1 W_2$ where the factors can be described using the projection Π onto $M^\times$ along M as follows:

$$W_1(\lambda) = D_1 + C(\lambda I_n - A)^{-1}(I - \Pi)BD_2^{-1}, \tag{8.11}$$

$$W_2(\lambda) = D_2 + D_1^{-1}C\Pi(\lambda I_n - A)^{-1}B. \tag{8.12}$$

The next theorem, which is a reformulation of the main result in [20], Section 9.1, shows that the above factorization principle yields all minimal factorizations of W whenever the given realization is minimal.

Theorem 8.5. *Let $W(\lambda) = D + C(\lambda I_n - A)^{-1}B$ be a minimal realization of the $m \times m$ rational matrix-valued function W, and assume D is invertible.*

(i) *Let $D = D_1 D_2$ with D_1, D_2 (invertible) $m \times m$ matrices. If a pair of subspaces M and $M^\times$ of $\mathbb{C}^n$ satisfies* (8.10)*, then the factorization $W = W_1 W_2$, with the factors W_1 and W_2 given by* (8.11) *and* (8.12)*, is a minimal factorization.*

(ii) *If $W = W_1 W_2$ is a minimal factorization of W involving proper rational $m \times m$ matrix functions W_1 and W_2, then there is a unique pair of subspaces M and $M^\times$ satisfying* (8.10) *such that the factors W_1 and W_2 are given by* (8.11) *and* (8.12) *where D_1 and D_2 are the* (*invertible*) *values of W_1 and W_2 at ∞, respectively.*

The notion of minimal factorization can be extended to products involving an arbitrary number of factors. Indeed, a factorization $W = W_1 \cdots W_k$ is called a *minimal factorization* if

$$\delta(W) = \delta(W_1) + \cdots + \delta(W_k). \tag{8.13}$$

In general all we can say is that the left-hand side of (8.13) does not exceed the right-hand side.

The special case of complete factorization is of particular interest. Let W be a rational $m \times m$ matrix-valued function which is *biproper*, that is, W is analytic at infinity and has an invertible value there. A minimal factorization of W into biproper rational $m \times m$ matrix functions, each having McMillan degree 1, is called a *complete factorization* of W. The number of factors in such a complete factorization is necessarily equal to the McMillan degree of W. If $W(\lambda) = D + C(\lambda I_n - A)^{-1}B$ is a minimal realization of W, then W admits a complete factorization if and only if the matrices A and $A^\times$ can be brought into complementary triangular form, i.e., there is a basis such that, with respect to this basis, A has upper triangular form and $A^\times$ has lower triangular form. For further details, see Chapter 10 in [20]. We shall meet complete factorization later in Section 17.3.

We conclude this section with some remarks on a local version of minimal factorization. First we introduce the local (McMillan) degree. Let W be a proper rational matrix function, let

$$W(\lambda) = D + C(\lambda I_n - A)^{-1}B \tag{8.14}$$

be a minimal realization of W, and let $\mu \in \mathbb{C}$. The algebraic multiplicity af μ as an eigenvalue of A is called the *local* (*McMillan*) *degree of W at μ*, written $\delta(W; \mu)$. By the state space similarity theorem, this definition does not depend on the choice of the minimal realization (8.14). For an alternative definition of the local degree, we refer to Section 8.4 in [20] where the square case is considered. In that situation, when $\det W(\lambda)$ does not vanish identically, the local degree of W

at μ coincides with the pole-multiplicity of W at μ in the sense of [20], Section 8.2.

It is obvious, again from Theorem 2.5, that the global sublogarithmic property (8.9) has the following local counterpart:

$$\delta(W_1W_2;\mu) \leq \delta(W_1\ \mu) + \delta(W_2;\mu). \tag{8.15}$$

A factorization $W = W_1W_2$ is said to be *locally minimal at* μ if equality occurs in (8.15), that is, when $\delta(W_1W_2;\mu) = \delta(W_1\ \mu) + \delta(W_2;\mu)$. Intuitively, this means that in the product W_1W_2 no pole-zero cancellation occurs at the point μ (see again Theorem 9.1 in [20]). For the case of proper rational matrix functions (as considered here), the minimality of a factorization comes down to local minimality at each point in the complex plane. Thus $W = W_1W_2$ is a minimal factorization if and only if

$$\delta(W_1W_2;\lambda) = \delta(W_1\ \lambda) + \delta(W_2;\lambda), \qquad \lambda \in \mathbb{C};$$

see Section 9.1 in [20].

8.3 Pseudo-canonical factorization

Let Γ be a Cauchy contour in $\mathbb{C}$. As before, the interior domain of Γ is denoted by F_+, and the exterior domain by F_-. By definition (see Chapter 0), $\infty \in F_-$. Let W be an $m \times m$ rational matrix function, possibly having poles and zeros on Γ. By a *right pseudo-canonical factorization* of W with respect to Γ we mean a factorization

$$W(\lambda) = W_-(\lambda)W_+(\lambda), \qquad \lambda \in \Gamma,\ \lambda \text{ not a pole of } W, \tag{8.16}$$

where W_- and W_+ are rational $m \times m$ matrix functions such that W_- is analytic and takes invertible values on F_- (i.e., W_- has neither poles nor zeros there), W_+ is analytic and takes invertible values on F_+ (i.e., W_- has neither poles nor zeros there), and the factorization (8.16) is locally minimal at each point of Γ. If in (8.16) the factors W_- and W_+ are interchanged, we speak of a *left pseudo-canonical factorization.*

In passing we mention that the definition of pseudo-canonical factorization given in the second paragraph of [20], Section 9.2 is not quite correct. The point is that the function W is allowed to have poles and zeros on Γ. This is explicitly stated in the third paragraph of the section in question, but the formal definition referred to above in the second paragraph erroneously suggests otherwise.

As for canonical factorization, the notion of pseudo-canonical factorization extends to factorization with respect to the real line and the imaginary axis. To be more specific, if Γ is the closure of the real line on the Riemann sphere, then F_+ is the open upper half plane, and F_- is the open lower half plane. Replacing $\mathbb{R}$

by $i\mathbb{R}$ means only replacing the open upper half plane by the open left half plane, and the open lower half plane by the open right half plane.

A pseudo-canonical factorization is not only minimal at each point of Γ but also at all other points of $\mathbb{C}$ and at infinity. This follows from the conditions on the poles and zeros of the factors W_- and W_+ in (8.16). Thus a pseudo-canonical factorization is a minimal factorization. In combination with Theorem 8.5 this fact makes it possible to describe all right pseudo-canonical factorizations of a biproper rational matrix function W in terms of a minimal realization of W. The resulting theorem (which is taken from Section 9.2 in [20]) is given below. In contrast to the main theorem on canonical factorization (Theorem 3.2) we are forced here to work with minimal realizations.

Theorem 8.6. *Let $W(\lambda) = D + C(\lambda I_n - A)^{-1}B$ be a minimal realization of a biproper rational matrix-valued function W, and put $A^\times = A - BD^{-1}C$. Let Γ be a Cauchy contour. Let $D = D_1D_2$, with D_1 and D_2 invertible square matrices. Then there is a one-to-one correspondence between the right pseudo-canonical factorizations $W = W_-W_+$ of W with respect to Γ with $W_-(\infty) = D_1$ and $W_+(\infty) = D_2$, and the pairs of subspaces $M, M^\times$ of $\mathbb{C}^n$ with the following properties:*

(i) *M is an A-invariant subspace such that the restriction $A|_M$ of A to M has no eigenvalues in F_-, and M contains the span of all eigenvectors and generalized eigenvectors of A corresponding to eigenvalues in F_+,*

(ii) *$M^\times$ is an $A^\times$-invariant subspace such that the restriction $A^\times|_{M^\times}$ of $A^\times$ to $M^\times$ has no eigenvalues in F_+, and $M^\times$ contains the span of all eigenvectors and generalized eigenvectors of $A^\times$ corresponding to eigenvalues in F_-,*

(iii) $\mathbb{C}^n = M \dot{+} M^\times$.

The correspondence is as follows: given a pair of subspaces $M, M^\times$ of $\mathbb{C}^n$ with the properties (i), (ii) *and* (iii), *a right pseudo-canonical factorization of W with respect to Γ is given by $W(\lambda) = W_-(\lambda)W_+(\lambda)$, where*

$$W_-(\lambda) = D_1 + C(\lambda I_n - A)^{-1}(I - \Pi)BD_2^{-1}, \tag{8.17}$$

$$W_+(\lambda) = D_2 + D_1^{-1}C\Pi(\lambda I_n - A)^{-1}B, \tag{8.18}$$

where Π is the projection along M onto $M^\times$. Conversely, given a right pseudo-canonical factorization of W with respect to Γ and with $W_-(\infty) = D_1$, $W_+(\infty) = D_2$, there exists a unique pair of subspaces $M, M^\times$ with the properties (i), (ii) *and* (iii) *above, such that the factors W_- and W_+ are given by* (8.17) *and* (8.18), *respectively.*

The span of all eigenvectors and generalized eigenvectors of A corresponding to eigenvalues in F_+ mentioned in (i) is just the spectral subspace of A associated with the part of the spectrum of A lying in F_+. Similarly, the span of all eigenvectors and generalized eigenvectors of $A^\times$ featuring in (ii) corresponding to eigenvalues in F_- is the spectral subspace of $A^\times$ associated with the part of $\sigma(A^\times)$ lying in F_-.

A pair of subspaces $M, M^\times$ for which (i), (ii) and (iii) hold need not be unique. In line with this, pseudo-canonical factorizations are generally not unique either. An example illustrating this is given in [133]; see also Section 9.2 in [20].

Note that for an $m \times m$ rational matrix function W, a canonical factorization of W with respect to the curve Γ is a pseudo-canonical factorization with the additional property that the factors have no poles or zeros on the curve. In that case, W has no poles or zeros on Γ also. Conversely, if W has no poles or zeros on Γ, then any pseudo-canonical factorization $W = W_1 W_2$ of W is automatically a canonical factorization. Indeed, if W has no poles or zeros on Γ, then the fact that the factorization $W = W_1 W_2$ is locally minimal at each point of Γ, implies that W_1 and W_2 have no poles or zeros on Γ, and thus the pseudo-canonical factorization $W = W_1 W_2$ is a canonical one. As a result we have the following special case of Theorem 8.6.

Theorem 8.7. *Let $W(\lambda) = D + C(\lambda I_n - A)^{-1}B$ be a minimal realization of a biproper rational matrix-valued function W, and put $A^\times = A - BD^{-1}C$. Let Γ be a Cauchy contour. Assume that A has no eigenvalues on Γ. Then W admits a right canonical factorization with respect to Γ if and only if the following two conditions are satisfied:*

(i) *$A^\times$ has no eigenvalues on Γ,*

(ii) $\mathbf{C}^n = \operatorname{Im} P(A;\Gamma) \dot{+} \operatorname{Ker} P(A^\times;\Gamma)$.

In that case, the right canonical factorizations with respect to Γ are of the form $W = W_- W_+$, with W_- and W_+ given by (8.17) and (8.18), where Π is the projection along $\operatorname{Im} P(A;\Gamma)$ onto $\operatorname{Ker} P(A^\times;\Gamma)$, and where $D = D_1 D_2$, with D_1 and D_2 invertible square matrices. This correspondence is a one-to-one correspondence between the right canonical factorizations of W and the factorizations of D into square factors.

Observe that the above theorem is a modest refinement of Theorem 3.2 in the sense that we allow the value of W at infinity to be an arbitrary invertible matrix here. The result of the theorem also holds for non-minimal realizations. The argument for this consists of a straightforward modification of the proof of Theorem 3.2. Theorem 8.7 allows for analogues in which the Cauchy contour Γ is replaced by the extended real or imaginary axis.

Notes

The material in the first section is standard and can be found in many textbooks; see, e.g., [94], or the more recent [33], [85]. The idea of minimal factorization originates from mathematical systems theory and has been developed systematically in Chapter 4 of [11] (see also [21]), and with further details in Part II of [20]. An extensive analysis of factorization into square degree 1 factors can be found in Part III of [20]. The analysis involves a connection with a problem of job scheduling from operations research. Minimal factorization into possibly non-square factors of McMillan degree 1 is always possible. This has been established in [143]. The notion of a pseudo-canonical factorization is introduced and developed in [132], [133].

Chapter 9

Factorization of positive definite rational matrix functions

The central theme of this chapter is the state space analysis of rational matrix functions with Hermitian values either on the real line, on the imaginary axis, or on the unit circle. The main focus will be on rational matrix functions that take positive definite values on one of these contours. It will be shown that if W is such a function, then W admits a spectral factorization, i.e., a canonical factorization $W = W_-W_+$ with an additional symmetry between the corresponding factors, depending on the contour.

This chapter consists of three sections. In Section 9.1 we analyze selfadjointness of a rational matrix function relative to the real line, the imaginary axis or the unit circle. The analysis is done in terms of (minimal) realizations of the functions involved. Elements of the theory of matrices that are selfadjoint with respect to an indefinite inner product enter into the analysis in a natural way. Section 9.2 deals with rational matrix functions that are positive definite on the real line or on the imaginary axis. The results of Section 9.1 are used to show that such a function admits a spectral factorization and in terms of a given realization an explicit formula for the corresponding spectral factor is given. Section 9.3 presents an analogous result for rational matrix functions that are positive definite on the unit circle.

9.1 Preliminaries on selfadjoint rational matrix functions

Let Γ be one of the following two contours in the complex plane: the real line $\mathbb{R}$, or the imaginary axis $i\mathbb{R}$. A rational $m \times m$ matrix function W is called *selfadjoint on* Γ or *Hermitian on* Γ if for each $\lambda \in \Gamma$, λ not a pole of W, the matrix $W(\lambda)$

is selfadjoint or, which is the same, Hermitian. By the uniqueness theorem for analytic functions, a rational matrix function W is selfadjoint on $\mathbb{R}$ if and only if $W(\lambda) = W(\bar{\lambda})^*$ for all $\lambda \in \mathbb{C}$, λ not a pole of W. Similarly, W is selfadjoint on $i\mathbb{R}$ if and only if $W(\lambda) = W(-\bar{\lambda})^*$, λ not a pole of W. From these characterizations it follows that if W is selfadjoint on Γ and $\det W(\lambda)$ does not vanish identically, then W^{-1} is also selfadjoint on Γ.

This section is concerned with the problem how selfadjointness of a rational matrix function is reflected in properties of the matrices in a minimal realization of the function. For proper rational matrix functions this is described in the following theorem.

Theorem 9.1. *Let $W(\lambda) = D + C(\lambda I_n - A)^{-1}B$ be a minimal realization of an $m \times m$ rational matrix function. Then the following statements hold:*

(i) *W is Hermitian on the real line if and only if $D = D^*$ and there exists an $n \times n$ matrix H such that*

$$HA = A^*H, \qquad HB = C^*, \qquad H = H^*; \tag{9.1}$$

(ii) *W is Hermitian on the imaginary axis if and only if $D = D^*$ and there exists an $n \times n$ matrix H such that*

$$HA = -A^*H, \qquad HB = C^*, \qquad H = -H^*. \tag{9.2}$$

In both cases (because of the minimality of the realization), the matrix H is uniquely determined by the matrices in the given realization of W and invertible.

A matrix H such that $H = -H^*$ is called *skew-Hermitian*. For such a matrix iH is Hermitian.

Proof. We first prove (i). Assume the matrix function W is Hermitian on $\mathbb{R}$, so the rational matrix functions $W(\lambda)$ and $W(\bar{\lambda})^*$ coincide. Hence

$$W(\lambda) = D^* + B^*(\lambda - A^*)^{-1}C^*$$

is also a minimal realization for W. By the state space similarity theorem (Theorem 8.2) we obtain the existence of a unique (invertible) $n \times n$ matrix H such that

$$HA = A^*H, \qquad HB = C^*, \qquad B^*H = C.$$

Taking adjoints one gets

$$H^*A^* = AH, \qquad C = B^*H^*, \qquad H^*B = C^*.$$

Comparing these two sets of equations and employing the uniqueness of H, we see that $H = H^*$. Clearly $D = D^*$ as $D = W(\infty)$ must be selfadjoint.

For the converse, suppose $D = D^*$ and there exist an $n\times n$ matrix H for which (9.1) holds. From the first equality in (9.1) we see that $H(\lambda-A)^{-1} = (\lambda-A^*)^{-1}H$. Then, using the second equality in (9.1), one computes

$$\begin{aligned} W(\bar{\lambda})^* &= D^* + B^*(\lambda - A^*)^{-1}C^* = D + B^*(\lambda - A^*)^{-1}HB \\ &= D + B^*H(\lambda - A)^{-1}B = D + C(\lambda - A)^{-1}B = W(\lambda). \end{aligned}$$

So W is selfadjoint on $\mathbb{R}$.

Next we show that (because of minimality) the identities in (9.1) imply that H is invertible. Indeed, assume $Hx = 0$ for some $x \in \mathbb{C}^n$. Then the first equality in (9.1) yields $HAx = 0$. Repeating the argument, using induction, we obtain $HA^kx = 0$ for $k = 0, 1, 2, \ldots$. Using the two other equalities in (9.1) we see that $CA^kx = B^*HA^kx = 0$ for $k = 0, 1, 2, \ldots$. Since the given realization is minimal, the pair (C, A) is observable, and hence $x = 0$. Thus H is invertible.

The proof of (ii) can be given using the same type of reasoning as for (i). On the other hand (ii) also follows directly from (i) by using the transformation $\lambda \to -i\lambda$. Indeed, put $\widetilde{W}(\lambda) = W(-i\lambda)$. Since W is assumed to be selfadjoint on $i\mathbb{R}$, the function $\widetilde{W}$ is selfadjoint on $\mathbb{R}$. Moreover, $\widetilde{W}$ admits the minimal realization

$$\widetilde{W}(\lambda) = D + \widetilde{C}(\lambda - \widetilde{A})^{-1}B,$$

where $\widetilde{C} = iC$ and $\widetilde{A} = iA$. By (i), there exists an (invertible) selfadjoint matrix $\tilde{H}$ such that $\widetilde{H}\widetilde{A} = \widetilde{A}^*\widetilde{H}$ and $\widetilde{H}B = \widetilde{C}^*$. Setting $H = -i\widetilde{H}$ we derive the desired equalities in (9.2). □

In the proof of the "if parts" of (i) and (ii), minimality does not play a role. Thus, if (9.1) holds and $D = D^*$, then $W(\lambda) = D + C(\lambda - A)^{-1}B$ is selfadjoint on $\mathbb{R}$. Similarly, if $D = D^*$ and (9.2) holds, then W is selfadjoint on $i\mathbb{R}$.

In the next proposition we consider the case when the rational matrix function in Theorem 9.1 is biproper, and we describe the effect of the matrices H on the associate main operator $A^\times = A - BD^{-1}C$.

Proposition 9.2. *Let $W(\lambda) = D + C(\lambda I_n - A)^{-1}B$ be a realization of an $m \times m$ rational matrix function, and let H be an $n \times n$ matrix. Assume D is invertible, and put $A^\times = A - BD^{-1}C$. Then the following statements hold:*

(i) *If $D = D^*$ and (9.1) is satisfied, then $HA^\times = (A^\times)^*H$;*

(ii) *If $D = D^*$ and (9.2) is satisfied, then $HA^\times = -(A^\times)^*H$.*

Proof. Assume $D = D^*$ and the identities (9.1). Then

$$HA^\times = HA - HBD^{-1}C = A^*H - C^*D^{-*}B^*H = (A^\times)^*H,$$

so (i) holds. Statement (ii) is proved analogously. □

Next we analyze how the matrix H appearing in Theorem 9.1 behaves under a state space similarity transformation on the realization of W.

Theorem 9.3. *For $i = 1, 2$, let $W(\lambda) = D + C_i(\lambda_n - A_i)^{-1}B_i$ be a minimal realization of the rational matrix function W, and let S be the (unique invertible) $n \times n$ matrix such that*

$$SA_1 = A_2S, \qquad C_1 = C_2S, \qquad B_2 = SB_1.$$

Then the following statements hold:

(i) *Let W be selfadjoint on the real line. For $i = 1, 2$, write H_i for the (unique invertible) Hermitian $n \times n$ matrix such that $H_iA_i = A_i^*H_i$ and $H_iB_i = C_i^*$. Then $H_1 = S^*H_2S$;*

(ii) *Let W be selfadjoint on $i\mathbb{R}$. For $i = 1, 2$, write H_i for the (unique invertible) skew-Hermitian $n \times n$ matrix such that $H_iA_i = -A_i^*H_i$ and $H_iB_i = C_i^*$. Then $H_1 = S^*H_2S$.*

Proof. We shall only prove (i); statement (ii) can be verified analogously. One easily checks that S^*H_2S satisfies (9.1):

$$S^*H_2SA_1 = S^*H_2A_2S = S^*A_2^*H_2S = A_1^*S^*H_2S,$$

$$S^*H_2SB_1 = S^*H_2B_2 = S^*C_2^* = C_1^*.$$

By the uniqueness of H_1 the assertion (i) follows. □

We conclude this section with a comment on the theory of matrices acting in an indefinite inner product space. Elements of this theory play an important role in the study of selfadjoint rational matrix functions. To see the connection, let H be an invertible Hermitian $n \times n$ matrix, and consider on $\mathbb{C}^n$ the sesquilinear form

$$[x, y] = \langle Hx, y \rangle.$$

If $HA = A^*H$, then $[Ax, y] = [x, Ay]$, and hence A is selfadjoint in the indefinite inner product $[\cdot\,, \cdot]$ on $\mathbb{C}^n$ induced by H. Thus the first part and third identity in (9.1) imply that A is selfadjoint in an indefinite inner product space. In the sequel we call A *H-selfadjoint* if $H = H^*$ and $HA = A^*H$. Notice that the third identity in (9.2) implies that iH is Hermitian, and hence the first identity in (9.2) can be summarized by saying that iA is iH-selfadjoint.

In Section 11.2 we review the results from the theory of matrices acting in an indefinite inner product space insofar as they are useful to us in this and the next chapters.

9.2 Spectral factorization

The first factorization result to be presented in this section concerns an important class of rational matrix functions, namely those which are positive definite on the contour Γ under consideration (again, either $\mathbb{R}$ or $i\mathbb{R}$). A rational $m \times m$ matrix function W is called *positive definite on* Γ if for each $\lambda \in \Gamma$, λ not a pole of W, the matrix $W(\lambda)$ is positive definite.

Suppose W is a rational $m \times m$ matrix function. A factorization

$$W(\lambda) = L(\bar{\lambda})^* L(\lambda) \tag{9.3}$$

is called a *right spectral factorization with respect to the real line* if L and L^{-1} are rational $m \times m$ matrix functions which are analytic on the closed upper half plane (infinity included). In that case the function $L(\bar{\lambda})^*$ and its inverse are analytic on the closed lower half plane (including infinity). Thus a right spectral factorization with respect to $\mathbb{R}$ is a right canonical factorization with respect to the real line featuring an additional symmetry property between the factors. A factorization (9.3) is called a *left spectral factorization with respect to the real line* if L and L^{-1} are rational $m \times m$ matrix functions which are analytic on the closed lower half plane (infinity included), in which case the function $L(\bar{\lambda})^*$ and its inverse are analytic on the closed upper half plane including infinity). Such a factorization is a left canonical factorization with respect to $\mathbb{R}$.

A factorization

$$W(\lambda) = L(-\bar{\lambda})^* L(\lambda) \tag{9.4}$$

is called a *right spectral factorization with respect to the imaginary axis* if L and L^{-1} are rational $m \times m$ matrix functions which are analytic on the closed left half plane (infinity included). Such a factorization is, in particular, a right canonical factorization with respect to $i\mathbb{R}$. Analogously, (9.4) is called a *left spectral factorization with respect to the imaginary axis* if L and L^{-1} are rational $m \times m$ matrix functions which are analytic on the closed right half plane (infinity included).

The factors in a spectral factorization are uniquely determined up to multiplication with a constant unitary matrix. More precisely, if $W(\lambda) = L(\bar{\lambda})^* L(\lambda)$ is a spectral factorization with respect to the real line, and E is an $m \times m$ unitary matrix, then $W(\lambda) = \tilde{L}(\bar{\lambda})^* \tilde{L}(\lambda)$ with $\tilde{L}(\lambda) = EL(\lambda)$ is again a spectral factorization of W, and this is all the freedom one has. To see the latter, assume that $W(\lambda) = L(\bar{\lambda})^* L(\lambda)$ and $W(\lambda) = \tilde{L}(\bar{\lambda})^* \tilde{L}(\lambda)$ are right spectral factorizations with respect to $\mathbb{R}$, then

$$\tilde{L}(\lambda) L(\lambda)^{-1} = \tilde{L}(\bar{\lambda})^{-*} L(\bar{\lambda})^*.$$

The left-hand side of this identity is an $m \times m$ rational matrix function which is analytic on the closed upper half plane and the right-hand side is analytic on the closed lower half plane (in both cases the point infinity included). By Liouville's theorem neither side depends on λ, that is, there exists an $m \times m$ matrix E such that $E = L(\lambda)\tilde{L}(\lambda)^{-1}$ and $L(\bar{\lambda})^{-*}\tilde{L}(\bar{\lambda})^*$. But this implies that E is invertible and $E^* = E^{-1}$. Hence E is unitary and $E = \tilde{L}(\lambda) = EL(\lambda)$, as desired.

If (9.4) is a right (respectively, left) spectral factorization of W with respect to the real line, we refer to L as the *right* (respectively, *left*) *spectral factor*. Without further explanation a similar terminology will be used in comparable circumstances.

Note that existence of a spectral factorization implies that W has no poles or zeros on the given contour and on the contour it is positive definite. The converse also holds: for positive definite rational matrix functions, both left and right spectral factorizations exist. This will now be proved for the case when W is a proper rational $m \times m$ matrix function. Moreover, explicit formulas for the factors will be given in terms of a realization of W. First we consider the situation where W is positive definite on the real line.

Theorem 9.4. *Let $W(\lambda) = D + C(\lambda I_n - A)^{-1}B$ be a realization of the rational $m \times m$ matrix function W. Suppose A has no real eigenvalues, W is positive definite on the real line, and $W(\infty) = D$ is positive definite too. Further assume there exists an invertible Hermitian $n \times n$ matrix H for which $HA = A^*H$ and $HB = C^*$. Then, with respect to the real line, W admits right and left spectral factorization. Such factorizations can be obtained in the following way. Let M_+ and M_- be the spectral subspaces of A associated with the parts of $\sigma(A)$ lying in the lower and upper half plane, respectively, and let $M_+^\times$ and $M_-^\times$ be the spectral subspaces of $A^\times$ associated with the parts of $\sigma(A^\times)$ lying in the lower and upper half plane, respectively. Then*

$$\mathbb{C}^n = M_- \dotplus M_+^\times, \qquad \mathbb{C}^n = M_+ \dotplus M_-^\times. \tag{9.5}$$

Write Π_+ for the projection of $\mathbb{C}^n$ along M_- onto $M_+^\times$, Π_- for the projection of $\mathbb{C}^n$ along M_+ onto $M_-^\times$, and introduce

$$L_+(\lambda) = D^{1/2} + D^{-1/2}C\Pi_+(\lambda I_n - A)^{-1}B, \tag{9.6}$$
$$L_-(\lambda) = D^{1/2} + D^{-1/2}C\Pi_-(\lambda I_n - A)^{-1}B. \tag{9.7}$$

Then

$$W(\lambda) = L_+(\bar{\lambda})^*L_+(\lambda), \qquad W(\lambda) = L_-(\bar{\lambda})^*L_-(\lambda),$$

are right and left spectral factorizations with respect to the real line, respectively. These spectral factorizations are uniquely determined by the fact that they have the value $D^{1/2}$ at infinity.

The conditions of the theorem are satisfied in case W has no poles on the real line, $W(\lambda)$ is positive definite for all real λ, and the given (biproper) realization of W is a minimal one.

Proof. The invertibility of $W(\lambda)$ for real λ combined with the fact that A has no real eigenvalues implies that $A^\times$ does not have real eigenvalues either (see Theorem 2.4). So the subspaces M_+, M_-, $M_+^\times$ and $M_-^\times$ are well-defined. Let P and $P^\times$ be the Riesz projections of A, and $A^\times$, respectively, with respect to the upper half plane. From $HA = A^*H$ and $HA^\times = (A^\times)^*H$ one easily computes that

$$HP = (I - P^*)H, \qquad HP^\times = (I - P^\times)^*H.$$

It follows that the spaces M_+, M_-, $M_+^\times$ and $M_-^\times$ satisfy

$$HM_+ = M_+^\perp, \qquad HM_- = M_-^\perp, \qquad HM_+^\times = M_+^{\times\perp}, \qquad HM_-^\times = M_-^{\times\perp}. \quad (9.8)$$

First it will be shown that $M_+ \cap M_-^\times = \{0\}$. Suppose $x \in M_+ \cap M_-^\times$. As M_+ is invariant under A, we have $Ax \in M_+$. But then the first identity in (9.8) shows that $\langle HAx, x\rangle = 0$. The space $M_-^\times$ is invariant under $A^\times$. Thus $A^\times x$ belongs to $M_-^\times$, and the last identity in (9.8) yields $\langle HA^\times x, x\rangle = 0$. Hence

$$0 = \langle H(A - A^\times)x, x\rangle = \langle HBD^{-1}Cx, x\rangle = \langle D^{-1}Cx, Cx\rangle = \|D^{-1/2}Cx\|^2.$$

As $D > 0$, it follows that $Cx = 0$. Thus $A^\times x = (A - BD^{-1}C)x = Ax$. We conclude that $M_+ \cap M_-^\times$ is invariant under both A and $A^\times$, and we have $A|_{M_+\cap M_-^\times} = A^\times|_{M_+\cap M_-^\times}$. However,

$$\sigma(A|_{M_+\cap M_-^\times}) \subset \sigma(A|_{M_+}) \subset \{\lambda \mid \Im\lambda > 0\},$$

$$\sigma(A^\times|_{M_+\cap M_-^\times}) \subset \sigma(A^\times|_{M_-^\times}) \subset \{\lambda \mid \Im\lambda < 0\}.$$

Thus $A|_{M_+\cap M_-^\times} = A^\times|_{M_+\cap M_-^\times}$ implies that $M_+ \cap M_-^\times = \{0\}$.

Proving (9.5) is now done via a dimension argument. Since H is invertible, the first identity in (9.8) shows that M_+ and $M_+^\perp$ have the same dimension. In particular, $\dim M_+ = n/2$. Similarly, the last identity in (9.8) yields $\dim M_-^\times = n/2$. Hence the first identity in (9.5) holds. Let Π_- be the projection along M_+ onto $M_-^\times$. The second identity in (9.5) is established in a similar way.

Let Π_- be the projection along M_+ onto $M_-^\times$. Then Π_- is a supporting projection, and by Theorem 3.2 the corresponding factorization is a left canonical factorization given by

$$W(\lambda) = K_-(\lambda)L_-(\lambda),$$

where L_- is given by (9.7), and

$$K_-(\lambda) = D^{1/2} + C(\lambda - A)^{-1}(I - \Pi_-)BD^{-1/2}.$$

It remains to prove that $K_-(\lambda) = L_-(\bar\lambda)^*$. Using (9.7) and (9.1) we have

$$\begin{aligned} L_-(\bar\lambda)^* &= D^{1/2} + B^*(\lambda - A^*)^{-1}\Pi_-^* C^* D^{-1/2} \\ &= D^{1/2} + C(\lambda - A)^{-1}H^{-1}\Pi_-^* HBD^{-1/2}. \end{aligned}$$

Thus in order to get $K_-(\lambda) = L_-(\bar\lambda)^*$, it suffices to show that $H(I - \Pi_-) = \Pi_-^* H$.

Using the definition of Π_-, together with the first and the last identity in (9.8), we see that $\langle H(I-\Pi_-)x,(I-\Pi_-)y\rangle = 0$ and $\langle H\Pi_- x,\Pi_- y\rangle = 0$ for all x and y in $\mathbb{C}^n$. Hence for all x, y,

$$\langle H(I-\Pi_-)x, y\rangle = \langle H(I-\Pi_-)x, \Pi_- y\rangle = \langle Hx, \Pi_- y\rangle,$$

which yields the desired identity $H(I-\Pi_-) = \Pi_-^* H$.

As for the last statement in the theorem, recall that the factors in a spectral factorization are uniquely determined up to multiplication with a constant unitary matrix. This settles the theorem as far as left spectral factorization is concerned. For right spectral factorizations the reasoning is similar. □

With minor modifications one proves the following theorem concerning left and right spectral factorizations with respect to the imaginary axis.

Theorem 9.5. *Let $W(\lambda) = D + C(\lambda I_n - A)^{-1}B$ be a realization of the rational $m \times m$ matrix function W. Suppose A has no pure imaginary eigenvalues, W is positive definite on the imaginary axis, and $W(\infty) = D$ is positive definite too. Further assume there exists an invertible skew-Hermitian $n \times n$ matrix H for which $HA = -A^*H$ and $HB = C^*$. Then, with respect to the imaginary axis, W admits right and left spectral factorization. Such factorizations can be obtained in the following way. Let M_+ and M_- be the spectral subspaces of A associated with the parts of $\sigma(A)$ lying in the right and left half plane, respectively, and let $M_+^\times$ and $M_-^\times$ be the spectral subspaces of $A^\times$ associated with the parts of $\sigma(A^\times)$ lying in the right and left half plane, respectively. Then*

$$\mathbb{C}^n = M_- \dot{+} M_+^\times, \qquad \mathbb{C}^n = M_+ \dot{+} M_-^\times.$$

Write Π_+ for the projection of $\mathbb{C}^n$ along M_- onto $M_+^\times$, Π_- for the projection of $\mathbb{C}^n$ along M_+ onto $M_-^\times$, and introduce

$$\begin{aligned} L_+(\lambda) &= D^{1/2} + D^{-1/2}C\Pi_+(\lambda I_n - A)^{-1}B, \\ L_-(\lambda) &= D^{1/2} + D^{-1/2}C\Pi_-(\lambda I_n - A)^{-1}B. \end{aligned}$$

Then

$$W(\lambda) = L_+(-\bar{\lambda})^* L_+(\lambda), \qquad W(\lambda) = L_-(-\bar{\lambda})^* L_-(\lambda),$$

are right and left spectral factorizations with respect to the imaginary axis, respectively. These spectral factorizations are uniquely determined by the fact that they have the value $D^{1/2}$ at infinity.

The conditions of the theorem are satisfied in case W has no poles on the imaginary axis, $W(\lambda)$ is positive definite for $\lambda \in i\mathbb{R}$, and the given (biproper) realization of W is a minimal one. In terms of the theory of spaces with an indefinite metric (see the appendix at the end of this chapter), the identities in (9.8) say that the spectral subspaces M_+, M_-, $M_+^\times$ and $M_-^\times$ are Lagrangian subspaces in the indefinite inner product induced by H.

9.3 Positive definite functions on the unit circle

In this section we shall discuss rational matrix functions that take positive definite values on the unit circle $\mathbb{T}$ and their spectral factorizations. This class of functions is more complicated than the ones discussed in the previous sections, the main reason being that infinity is not on the contour, and so the value at infinity is not necessarily a selfadjoint matrix.

A rational $m \times m$ matrix function W is called *selfadjoint on the unit circle* or *Hermitian on the unit circle* if for each $\lambda \in \mathbb{T}$, λ not a pole of W, the matrix $W(\lambda)$ is selfadjoint or, which is the same, Hermitian. By the uniqueness theorem for analytic functions, a rational matrix function W is selfadjoint on $\mathbb{T}$ if and only if $W(\lambda) = W(\bar{\lambda}^{-1})^*$, for all $\lambda \in \mathbb{C}$, λ not a pole of W. It follows that if W is selfadjoint on $\mathbb{T}$ and $\det W(\lambda)$ does not vanish identically, then W^{-1} is also selfadjoint on $\mathbb{T}$.

We first discuss how selfadjointness of W is reflected in properties of the matrices in a minimal realization of the function. For proper rational matrix functions this is described in the following theorem, a counterpart of Theorem 9.1 for the unit circle.

Theorem 9.6. *Let $W(\lambda) = D + C(\lambda I_n - A)^{-1}B$ be a minimal realization of an $m \times m$ rational matrix function. Then W is Hermitian on $\mathbb{T}$ if and only if A is invertible, $D^* = D - CA^{-1}B$, and there exists an $n \times n$ matrix H such that*

$$HA = A^{-*}H, \qquad HB = A^{-*}C^*, \qquad H = -H^*. \tag{9.9}$$

The matrix H is uniquely determined by the matrices in the given realization of W and invertible.

Recall that A^{-*} stands for $(A^*)^{-1}$ or, which amounts to the same, $(A^{-1})^*$. The first part of (9.9) means that A is iH-unitary, that is, A is unitary with respect to the indefinite inner product induced by the selfadjoint matrix iH (cf., Chapter 11 and Section 17.1). The first part of (9.9) can be rewritten as $A^*HA = H$. Note that, given the invertibility of H, the identity $A^*HA = H$ implies the invertibility of A.

Proof. First observe that if W is Hermitian on $\mathbb{T}$, then W has no pole at 0, as $W(\infty) = D$ and $W(0) = W(\infty)^*$. Because of minimality, this shows that A is invertible and $D^* = D - CA^{-1}B$. But then

$$\begin{aligned} W(\bar{\lambda}^{-1})^* &= D^* + B^*(\lambda^{-1} - A^*)^{-1}C^* \\ &= D^* - B^*A^{-*}(\lambda - A^{-*})^{-1}\lambda C^* \\ &= D^* - B^*A^{-*}C^* - B^*A^{-*}(\lambda - A^{-*})^{-1}A^{-*}C^*. \end{aligned}$$

Now the rational matrix functions $W(\lambda)$ and $W(\bar{\lambda}^{-1})^*$ coincide. Thus, again by the state space similarity theorem (Theorem 8.2), there exists a unique invertible matrix H such that

$$HA = A^{-*}H, \qquad HB = A^{-*}C^*, \qquad -B^*A^{-*}H = C.$$

Taking adjoints and employing the uniqueness of H, one finds $H = -H^*$.

This settles the "only if part" of the theorem; the "if part" is obtained via a straightforward computation (not using minimality). Because of minimality, the identities in (9.9) imply that H is invertible. The argument is similar to that given in the third paragraph of the proof of Theorem 9.1. □

Next, we consider the associate main operator.

Proposition 9.7. *Let $W(\lambda) = D + C(\lambda I_n - A)^{-1}B$ be a realization of an $m \times m$ rational matrix function and assume D is invertible. Suppose A is invertible too, $D^* = D - CA^{-1}B$, and there exists an $n \times n$ matrix H such that (9.9) holds. Then $A^\times = A - BD^{-1}C$ is invertible and $HA^\times = (A^\times)^{-*}H$.*

Proof. From the invertibility of A and D, and the assumption $D^* = D - CA^{-1}B$, we get

$$\begin{bmatrix} A & B \\ C & D \end{bmatrix} = \begin{bmatrix} I & 0 \\ CA^{-1} & I \end{bmatrix} \begin{bmatrix} A & 0 \\ 0 & D^* \end{bmatrix} \begin{bmatrix} I & A^{-1}B \\ 0 & I \end{bmatrix}$$

$$= \begin{bmatrix} I & BD^{-1} \\ 0 & I \end{bmatrix} \begin{bmatrix} A^\times & 0 \\ 0 & D \end{bmatrix} \begin{bmatrix} I & 0 \\ D^{-1}C & I \end{bmatrix}.$$

As both A and D^* are invertible, $A^\times$ must be invertible too. Furthermore, by (9.9), we have

$$\begin{aligned} (A^\times)^*HA^\times &= (A^* - C^*D^{-*}B^*)H(A - BD^{-1}C) \\ &= H - C^*D^{-*}B^*HA - A^*HBD^{-1}C + C^*D^{-*}B^*HBD^{-1}C \\ &= H + C^*D^{-*}(D - D^* + B^*HB)D^{-1}C \\ &= H + C^*D^{-*}(CA^{-1}B + B^*A^{-*}C^*)D^{-1}C. \end{aligned}$$

However, as $D - D^* = CA^{-1}B$, we have $B^*A^{-*}C^* = -CA^{-1}B$. Therefore, $(A^\times)^*HA^\times = H$. □

Next we analyze how the matrix H appearing in Theorem 9.6 behaves under a state space similarity transformation on the realization of W. The proof of the next theorem is analogous to the proof of Theorem 9.3.

Theorem 9.8. *For $i = 1, 2$, let $W(\lambda) = D + C_i(\lambda I_n - A_i)^{-1}B_i$ be minimal realizations of the rational $m \times m$ matrix function W, and let S be the (unique invertible) $n \times n$ matrix such that*

$$SA_1 = A_2S, \qquad C_1 = C_2S, \qquad B_2 = SB_1.$$

Suppose W is Hermitian on the unit circle. For $i = 1, 2$, write H_i for the (unique invertible) skew-Hermitian $n \times n$ matrix such that $A_i^* H_i A_i = H_i$ and $H_i B_i = A_i^{-*} C_i^*$. Then $H_1 = S^* H_2 S$.

The above results can also be obtained by reduction to the real line results of Section 9.1. To illustrate this, let $W(\lambda) = D + C(\lambda I_n - A)^{-1}B$ be a minimal realization of an $m \times m$ rational matrix function, and let $\alpha \in \mathbb{T}$ be a regular point for A, that is, α is not an eigenvalue of A. Consider the Möbius transformation

$$\phi(\lambda) = \alpha(\lambda - i)(\lambda + i)^{-1}. \tag{9.10}$$

Note that ϕ maps the upper half plane in a one-to-one way onto the open unit disc $\mathbb{D}$, and the extended real line is mapped in a one-to-one way onto the unit circle $\mathbb{T}$, with $\phi(\infty) = \alpha$. Put $\widetilde{W}(\lambda) = W(\phi(\lambda))$. Then $\widetilde{W}$ is again an $m \times m$ rational matrix function and (see Section 3.6 in [20]) the function $\widetilde{W}$ admits the realization $\widetilde{W}(\lambda) = \widetilde{D} + \widetilde{C}(\lambda I_n - \widetilde{A})^{-1}\widetilde{B}$, where

$$\widetilde{A} = (i\alpha + iA)(\alpha - A)^{-1} = \phi^{-1}(A), \qquad \widetilde{B} = (\alpha - A)^{-1}B,$$
$$\widetilde{C} = 2i\alpha C(\alpha - A)^{-1}, \qquad \widetilde{D} = W(\alpha) = D + C(\alpha - A)^{-1}B.$$

Moreover this realization is again minimal. Now assume that W is selfadjoint on $\mathbb{T}$, then $\widetilde{W}$ is selfadjoint on $\mathbb{R}$ and by Theorem 9.1 there exists an invertible $n \times n$ matrix $\widetilde{H}$ such that

$$\widetilde{H}\widetilde{A} = \widetilde{A}^*\widetilde{H}, \qquad \widetilde{H}\widetilde{B} = \widetilde{C}^*, \qquad \widetilde{H} = \widetilde{H}^*.$$

Observe that

$$\begin{aligned}
\widetilde{H}\widetilde{A} = \widetilde{A}^*\widetilde{H} \quad &\Longleftrightarrow \quad \widetilde{H}(i\alpha + iA)(\alpha - A)^{-1} = (\bar{\alpha} - A^*)^{-1}(-i\bar{\alpha} - iA^*)\widetilde{H} \\
&\Longleftrightarrow \quad (\bar{\alpha} - A^*)\widetilde{H}(i\alpha + iA) = (-i\bar{\alpha} - iA^*)\widetilde{H}(\alpha - A) \\
&\Longleftrightarrow \quad i\widetilde{H} + i\bar{\alpha}\widetilde{H}A - i\alpha A^*\widetilde{H} - iA^*\widetilde{H}A \\
&\qquad\qquad = -i\widetilde{H} + i\bar{\alpha}\widetilde{H}A - i\alpha A^*\widetilde{H} + iA^*\widetilde{H}A \\
&\Longleftrightarrow \quad 2i\widetilde{H} = 2iA^*\widetilde{H}A.
\end{aligned}$$

We already know (see the first paragraph of the proof of Theorem 9.6) that the operator A is invertible, and thus we may conclude that $\widetilde{H}A = A^{-*}\widetilde{H}$. Using this and the invertibility of H, one gets

$$\begin{aligned}
\widetilde{H}\widetilde{B} &= \widetilde{H}(\alpha - A)^{-1}B = (\alpha\widetilde{H}^{-1} - A\widetilde{H}^{-1})^{-1}B \\
&= (\alpha\widetilde{H}^{-1} - \widetilde{H}^{-1}A^{-*})^{-1}B = (\alpha - A^{-*})^{-1}\widetilde{H}B \\
&= (\alpha A^* - I_n)^{-1}A^*\widetilde{H}B = -\bar{\alpha}(\bar{\alpha} - A^*)^{-1}A^*\widetilde{H}B.
\end{aligned}$$

From the definition of $\widetilde{C}$ we know that $\widetilde{C}^* = -2i\bar{\alpha}(\bar{\alpha} - A^*)^{-1}C^*$, and hence

$$\widetilde{H}\widetilde{B} = \widetilde{C}^* \iff A^*\widetilde{H}B = 2iC^*.$$

Now define H by $2iH = \widetilde{H}$. Then H has the properties listed in (9.9).

In a similar way it can be shown that Proposition 9.7 and Theorem 9.8 follow from the analogous results in Section 9.1.

We now turn to spectral factorization. Suppose W is a rational $m \times m$ matrix function. A factorization

$$W(\lambda) = L(\bar{\lambda}^{-1})^* L(\lambda) \tag{9.11}$$

is called a *right spectral factorization with respect to the unit circle* if L and L^{-1} are rational matrix functions which are analytic on the closure of the (open) unit disc $\mathbb{D}$. In that case the function $L(\bar{\lambda}^{-1})^*$ and its inverse are analytic on the closure of $\mathbb{D}_{\rm ext}$, the exterior domain of the unit circle in $\mathbb{C}$ (infinity included). Thus, in particular, a right spectral factorization with respect to the unit circle is a right canonical factorization with respect to $\mathbb{T}$. Analogously, (9.11) is called a *left spectral factorization with respect to the unit circle* if L and L^{-1} are analytic on the closure of $\mathbb{D}_{\rm ext}$ (infinity included), in which case the function $L(\bar{\lambda}^{-1})^*$ and its inverse are analytic on the closed unit disc. Such a factorization is a left canonical factorization with respect to $\mathbb{T}$. Observe that the existence of a spectral factorization implies that W has positive definite values on the unit circle. As we will see in the next theorem, the converse is also true.

A rational $m \times m$ matrix function W is called *positive definite on the unit circle* if for each $\lambda \in \mathbb{T}$, λ not a pole of W, the matrix $W(\lambda)$ is positive definite. Left and right spectral factorization of functions which are positive definite on the unit circle is slightly more complicated than spectral factorization of functions which are positive definite on either the real line or the imaginary axis. This is mainly caused by the fact that the value at infinity generally is no longer positive definite.

Theorem 9.9. *Let $W(\lambda) = D + C(\lambda I_n - A)^{-1}B$ be a realization of a rational $m \times m$ matrix function such that $W(\lambda)$ is positive definite for $|\lambda| = 1$. Suppose D is invertible, A is invertible, and A has no eigenvalues on the unit circle. Furthermore, assume there exists an invertible skew-Hermitian $n \times n$ matrix H such that $HA = A^{-*}H$ and $HB = A^{-*}C^*$. Then, with respect to the unit circle, W admits right and left spectral factorization. Such factorizations can be obtained in the following way. Let M_+ and M_- be the spectral subspaces of A associated with the parts of $\sigma(A)$ lying in $\mathbb{D}_{\rm ext}$ and $\mathbb{D}$, respectively, and let $M_+^\times$ and $M_-^\times$ be the spectral subspaces of $A^\times$ associated with the parts of $\sigma(A^\times)$ lying in $\mathbb{D}_{\rm ext}$ and $\mathbb{D}$, respectively. Then*

$$\mathbb{C}^n = M_- \dot{+} M_+^\times, \qquad \mathbb{C}^n = M_+ \dot{+} M_-^\times. \tag{9.12}$$

Write Π_+ for the projection of $\mathbb{C}^n$ along M_- onto $M_+^\times$, Π_- for the projection of $\mathbb{C}^n$ along M_+ onto $M_-^\times$. Then $D_+ = D - CA^{-1}(I - \Pi_+)B$ and $D_- = D - CA^{-1}(I - \Pi_-)B$ are selfadjoint. Further there are unique rational matrix functions L_+ and L_- such that

$$W(\lambda) = L_+(\bar{\lambda}^{-1})L_+(\lambda), \qquad W(\lambda) = L_-(\bar{\lambda}^{-1})L_-(\lambda)$$

are right and left spectral factorizations with respect to the unit circle, respectively, and such that $L_+(\infty) = D_+^{1/2}$, $L_-(\infty) = D_-^{1/2}$. These functions are given by

$$L_+(\lambda) = D_+^{1/2} + D_+^{1/2}D^{-1}C\Pi_+(\lambda I_n - A)^{-1}B, \tag{9.13}$$

$$L_-(\lambda) = D_-^{1/2} + D_-^{1/2}D^{-1}C\Pi_-(\lambda I_n - A)^{-1}B. \tag{9.14}$$

The conditions of the theorem are satisfied in case W has no poles on the unit circle, takes positive definite values there, and the given (biproper) realization of W is a minimal one.

Proof. Our hypotheses imply that A and $A^\times$ do not have eigenvalues on the unit circle. Let P and $P^\times$ be the Riesz projections of A and $A^\times$, respectively, corresponding to the eigenvalues in $\mathbb{D}_{\text{ext}}$. As in the proof of Theorem 9.4, one first shows that $HP = (I - P^*)H$ and $HP^\times = (I - P^\times)^*H$, using $A^*HA = H$ and $(A^\times)^*HA^\times = H$. Hence for the subspaces M_+, M_-, $M_+^\times$ and $M_-^\times$ we again have the identities

$$HM_+ = M_+^\perp, \qquad HM_- = M_-^\perp, \qquad HM_+^\times = M_+^{\times\perp}, \qquad HM_-^\times = M_-^{\times\perp}.$$

Now introduce $\varphi(\lambda) = -i(\lambda - i)(\lambda + i)^{-1}$ (i.e., (9.10) with $\alpha = -i$). Observe that $\varphi^{-1} = \varphi$, and φ maps the circle to the real line, $\mathbb{D}$ to the open upper half plane and $\mathbb{D}_{\text{ext}}$ to the open lower half plane. Consider $V(\lambda) = W(\varphi(\lambda))$. Then $V(\lambda)$ is positive definite on the real line and

$$V(\lambda) = W(-i) + \widetilde{C}(\lambda - \widetilde{A})^{-1}\widetilde{B}$$

with $\widetilde{A} = (I + iA)(-A - iI)^{-1}$. Since $W(-i)$ is invertible, we can use Proposition 3.4 in [20] to show that the associate main matrix in the above realization of V is given by $\widetilde{A}^\times = (I + iA^\times)(-A^\times - iI)^{-1}$. Using $A^*HA = H$ and $(A^\times)^*HA^\times = H$ one computes that $\widetilde{A}$ and $\widetilde{A}^\times$ are H-selfadjoint. The spectral subspaces of $\widetilde{A}$ with respect to the upper and lower half planes are M_- and M_+, respectively, while the spectral subspaces of $\widetilde{A}^\times$ with respect to the upper and lower half planes are $M_-^\times$ and $M_+^\times$, respectively. From the proof of Theorem 9.4, it now follows that (9.12) holds. So the projections Π_+ and Π_- are well-defined, and they are supporting projections giving rise to right and left canonical factorizations, respectively. Moreover $H(I - \Pi_+) = \Pi_+^*H$, and $H(I - \Pi_-) = \Pi_-^*H$.

A canonical factorization corresponding to Π_+ is given by $W = W_- W_+$ where

$$\begin{aligned} W_-(\lambda) &= D + C(\lambda - A)^{-1}(I - \Pi_+)B, \\ W_+(\lambda) &= I + D^{-1}C\Pi_+(\lambda - A)^{-1}B. \end{aligned}$$

For later use, recall that the factors W_- and W_+ in a canonical factorization are uniquely determined by their values at infinity. It remains to show that from $L_+(\lambda) = D_+^{1/2}W_+(\lambda)$ it follows that $W_-(\lambda)D_+^{-1/2} = L_+(\bar{\lambda}^{-1})^*$. We shall in fact prove that $W(\lambda) = W_+(\bar{\lambda}^{-1})^* D_+ W_+(\lambda)$.

Observe that $D_+ = W_-(0)$. To see that D_+ is selfadjoint, just carry out the calculation

$$\begin{aligned} D_+^* &= D^* - B^*(I - \Pi_+^*)A^{-*}C^* = D^* - B^*A^{-*}C^* + B^*\Pi_+^*A^{-*}C^* \\ &= D - CH^{-1}A^*\Pi_+^*HB = D - CA^{-1}H^{-1}\Pi_+^*HB \\ &= D - CA^{-1}(I - \Pi_+)B \\ &= D_+. \end{aligned}$$

Then write $W(\lambda) = K(\lambda)D_+W_+(\lambda)$ with

$$K(\lambda) = DD_+^{-1} + C(\lambda - A)^{-1}(I - \Pi_+)BD_+^{-1}.$$

Now compute $W_+(\bar{\lambda}^{-1})^*$:

$$\begin{aligned} W_+(\bar{\lambda}^{-1})^* &= I + B^*(\lambda^{-1} - A^*)^{-1}\Pi_+^*C^*D^{-*} \\ &= I - B^*A^{-*}\Pi_+^*C^*D^{-*} \\ &\qquad - B^*A^{-*}(\lambda - A^{-*})^{-1}A^{-*}\Pi_+^*C^*D^{-*} \\ &= (D^* - B^*A^{-*}\Pi_+^*C^*)D^{-*} \\ &\qquad + C(\lambda - A)^{-1}AH^{-1}\Pi_+^*HA^{-1}BD^{-*}. \end{aligned}$$

We claim that

$$(D^* - B^*A^{-*}\Pi_+^*C^*)D^{-*} = DD_+^{-1}, \tag{9.15}$$

$$AH^{-1}\Pi_+^*HA^{-1}BD^{-*} = (I - \Pi_+)BD_+^{-1}. \tag{9.16}$$

Indeed, for (9.15), observe that D_+ is invertible because $W(0) = D^*$ is invertible, and

$$D^* = D_+W_+(0) = D_+(I - D^{-1}C\Pi_+A^{-1}B).$$

So $D_+^{-1}D^* = D^{-1}(D - C\Pi_+A^{-1}B)$. Taking adjoints yields (9.15).

To verify (9.16), compute $(I - \Pi_+)BD_+^{-1}D^*$, using what we just have proved:

$$\begin{aligned}(I - \Pi_+)BD_+^{-1}D^* &= (I - \Pi_+)B(I - D^{-1}C\Pi_+A^{-1}B)\\ &= (I - \Pi_+)(A - BD^{-1}C\Pi_+)A^{-1}B\\ &= (I - \Pi_+)(A - (A - A^\times)\Pi_+)A^{-1}B\\ &= (I - \Pi_+)\{A(I - \Pi_+) + A^\times\Pi_+\}A^{-1}B.\end{aligned}$$

Now $\operatorname{Im}\Pi_+$ is $A^\times$-invariant, so $(I - \Pi_+)A^\times\Pi_+ = 0$. Hence

$$\begin{aligned}(I - \Pi_+)BD_+D^* &= (I - \Pi_+)A(I - \Pi_+)A^{-1}B\\ &= A(I - \Pi_+)A^{-1}B\\ &= AH^{-1}\Pi_+^*HA^{-1}B,\end{aligned}$$

as $\operatorname{Ker}\Pi$ is A-invariant. Thus (9.16) holds.

Now using (9.15) and (9.16) we see

$$W_+(\bar{\lambda}^{-1})^* = DD_+^{-1} + C(\lambda - A)^{-1}(I - \Pi_+)BD_+^{-1} = K(\lambda).$$

As $W(\lambda) = W_+(\bar{\lambda}^{-1})^*D_+W_+(\lambda)$ we see that D_+ must be positive definite. Since the factors W_+ and W_- in a canonical factorization are uniquely determined by their values at infinity, it follows that the factor L_+ in a right spectral factorization is also uniquely determined by its value at infinity. Thus the part of the theorem concerned with right spectral factorization follows. For the other part dealing with left spectral factorization the reasoning is similar. □

Notes

The results of Section 9.1 can be found in several sources, e.g., [26] and [45]. The factorization results of Sections 9.2 and 9.3 are based on [119] (see also Chapter 1 in [120]). Spectral factorizations play an important role in mathematical systems theory, see e.g., [4]. In [4], [41] and [147] spectral factorizations of a selfadjoint rational matrix function W are studied in state space form, starting from different representations of W.

Part IV of [20] is devoted to stability of minimal factorizations of rational matrix functions. The issue of stability of factorizations within the class of spectral factorizations has also been studied. This requires the analysis of perturbations of H-selfadjoint matrices and stability of their invariant Lagrangian subspaces. For instance, from Theorem 14.12 in [20] it follows straightforwardly that canonical factorizations are Lipschitz stable under small perturbations of the matrices in the realization. Restricting attention to spectral factorizations of positive definite rational matrix functions, and to perturbations of the matrices in the realizations that make the perturbed rational matrix function also positive definite, it still holds that spectral factorization is Lipschitz stable in this sense. For these and related results we refer to [123], see also [127].

Chapter 10

Pseudo-spectral factorizations of selfadjoint rational matrix functions

In this chapter we consider rational matrix functions on a contour having values that are selfadjoint matrices, but not necessarily positive definite ones. Whereas in the previous chapter we studied spectral factorization, in the present chapter the focus will be on functions that have poles or zeros on the contour, and so we will consider pseudo-spectral factorization here.

This chapter consists of two sections. Section 10.1 develops the notion of pseudo-spectral factorization for nonnegative rational matrix functions. The contours considered are the real line, the imaginary axis and the unit circle. In Section 10.2 the main result of the first section is generalized to the case of arbitrary selfadjoint rational matrix functions with positive definite value at infinity.

10.1 Nonnegative rational matrix functions

In this section we consider rational matrix functions W having nonnegative values on either the real line, the imaginary axis or the unit circle. The section may be viewed as a continuation of the discussion in Chapter 9. However, in contrast to the situation there, in this section we consider cases where W may have poles or zeros on the contour.

A rational $m \times m$ matrix function W is called *nonnegative on the real line* if for each $\lambda \in \mathbb{R}$, λ not a pole of W, the matrix $W(\lambda)$ is nonnegative. Without further explanation, the analogous terminology will be used for rational matrix functions having nonnegative values on the imaginary axis or on the unit circle, respectively.

As in Section 9.2 we shall start by considering the case of nonnegative rational matrix functions W on the real line, and continue with the situation where W is nonnegative on the imaginary axis. However, it is the latter case that we shall use frequently in the subsequent chapters. Therefore only for this case shall we provide a detailed proof. The real line situation can then be dealt with by using the Möbius transformation $\lambda \mapsto -i\lambda$. The section is concluded by presenting the results for the case of the unit circle. Again, the proof may be obtained by using a Möbius transformation (cf., the proofs of Theorems 9.4 and 9.9).

A factorization

$$W(\lambda) = L(\bar{\lambda})^* L(\lambda) \tag{10.1}$$

is called a *right pseudo-spectral factorization with respect to the real line* if L has no poles or zeros in the open upper half plane and the factorization is locally minimal at each point of the real line. Analogously, (10.1) is called a *left pseudo-spectral factorization with respect to the real line* if L has no poles or zeros in the open lower half plane and the factorization is locally minimal at each point of the real line. Such right or left pseudo-spectral factorizations are pseudo-canonical factorizations with respect to $i\mathbb{R}$ in the sense of Section 8.3.

Although a nonnegative rational matrix function generally does not allow for a left or right spectral factorization, it does admit left and right pseudo-spectral factorization.

Theorem 10.1. *Let $W(\lambda) = D + C(\lambda I_n - A)^{-1}B$ be a minimal realization of a rational $m \times m$ matrix function which is nonnegative on the real line, and assume D is positive definite. Then, with respect to the real line, W admits left and right pseudo-spectral factorization. Such factorizations can be obtained in the following way. Let H be the (unique invertible) Hermitian $n \times n$ matrix with $HA = A^*H$ and $HB = C^*$. Then there are unique A-invariant subspaces M_+ and M_-, and unique $A^\times$-invariant subspaces $M_+^\times$ and $M_-^\times$, such that*

(i) *M_+ contains the spectral subspace of A associated with the part of $\sigma(A)$ lying in the open lower half plane, and $\sigma(A|_{M_+}) \subset \{\lambda \mid \Im\lambda \leq 0\}$,*

(ii) *M_- contains the spectral subspace of A associated with the part of $\sigma(A)$ lying in the open upper half plane, and $\sigma(A|_{M_-}) \subset \{\lambda \mid \Im\lambda \geq 0\}$,*

(iii) *$M_+^\times$ contains the spectral subspace of $A^\times$ associated with the part of $\sigma(A^\times)$ lying in the open lower half plane, and $\sigma(A^\times|_{M_+^\times}) \subset \{\lambda \mid \Im\lambda \leq 0\}$,*

(iv) *$M_-^\times$ contains the spectral subspace of $A^\times$ associated with the part of $\sigma(A^\times)$ lying in the open upper half plane, and $\sigma(A^\times|_{M_-^\times}) \subset \{\lambda \mid \Im\lambda \geq 0\}$,*

(v) $H[M_+] = M_+^\perp, \quad H[M_-] = M_-^\perp, \quad H[M_+^\times] = M_+^{\times\perp}, \quad H[M_-^\times] = M_-^{\times\perp}.$

The subspaces in question also satisfy the matching conditions

$$\mathbb{C}^n = M_- \dot{+} M_+^\times, \qquad \mathbb{C}^n = M_+ \dot{+} M_-^\times. \tag{10.2}$$

Let Π_+ be the projection along M_- onto $M_+^\times$, let Π_- be the projection along M_+ onto $M_-^\times$, and introduce

$$L_+(\lambda) = D^{1/2} + D^{-1/2}C\Pi_+(\lambda I_n - A)^{-1}B, \tag{10.3}$$

$$L_-(\lambda) = D^{1/2} + D^{-1/2}C\Pi_-(\lambda I_n - A)^{-1}B. \tag{10.4}$$

Then

$$W(\lambda) = L_+(\bar{\lambda})^* L_+(\lambda), \qquad W(\lambda) = L_-(\bar{\lambda})^* L_-(\lambda),$$

are right and left pseudo-spectral factorizations with respect to the real line, respectively. These pseudo-spectral factorizations are uniquely determined by the fact that they have the value $D^{1/2}$ at infinity.

All possible right pseudo-spectral factors can be obtained from L_+ as given in (10.3) by multiplying on the left with a unitary matrix, and likewise, all possible left pseudo-spectral factors are obtained from L_- as given in (10.4) by multiplication on the left with a unitary matrix. Indeed, suppose $W(\lambda) = \widetilde{L}_-(\bar{\lambda})^*\widetilde{L}_-(\lambda)$ is another left pseudo-spectral factorization of W. Put $E(\lambda) = \widetilde{L}_-(\bar{\lambda})^{-*}L_-(\bar{\lambda})^* = \widetilde{L}_-(\lambda)L_-(\lambda)^{-1}$. Then $E(\lambda)$ is analytic outside the real line, and on the real line it is unitary, except for possible poles. So for all values of λ concerned, the norm of $E(\lambda)$ is 1. But then E cannot have poles. Indeed, in the vicinity of a pole the norm of $E(\lambda)$ cannot be bounded (cf., [134], Chapter 10, page 211). It follows that E is analytic on the whole complex plane. But then it must be a constant function by Liouville's theorem. As it is unitary for real λ, we conclude that the sole value of E is a unitary matrix.

Let W be a rational $m \times m$, and suppose W is nonnegative on the real line. A factorization

$$W(\lambda) = L(-\bar{\lambda})^* L(\lambda)$$

is called a *right pseudo-spectral factorization with respect to the imaginary axis* if L has no poles or zeros in the open left half plane and the factorization is locally minimal at each point of the imaginary axis. *Left pseudo-spectral factorizations with respect to the imaginary axis* are defined by replacing the upper half plane by the lower half plane.

Theorem 10.2. *Let $W(\lambda) = D + C(\lambda I_n - A)^{-1}B$ be a minimal realization of an $m \times m$ rational matrix function which is nonnegative on the imaginary axis, and assume D is positive definite. Put $A^\times = A - BD^{-1}C$. Then, with respect to the imaginary axis, W admits left and right pseudo-spectral factorization. Such factorizations can be obtained in the following way. Let H be the (unique invertible) skew-Hermitian $n \times n$ matrix with $HA = -A^*H$ and $HB = C^*$. Then there are unique A-invariant subspaces M_+ and M_-, and unique $A^\times$-invariant subspaces $M_+^\times$ and $M_-^\times$, such that*

(i) *M_+ contains the spectral subspace of A associated with the part of $\sigma(A)$ lying in the open right half plane, and $\sigma(A|_{M_+}) \subset \{\lambda \mid \Re\lambda \geq 0\}$,*

(ii) M_- *contains the spectral subspace of* A *associated with the part of* $\sigma(A)$ *lying in the open left half plane, and* $\sigma(A|_{M_-}) \subset \{\lambda \mid \Re\lambda \leq 0\}$,

(iii) $M_+^\times$ *contains the spectral subspace of* $A^\times$ *associated with the part of* $\sigma(A^\times)$ *lying in the open right half plane, and* $\sigma(A^\times|_{M_+^\times}) \subset \{\lambda \mid \Re\lambda \geq 0\}$,,

(iv) $M_-^\times$ *contains the spectral subspace of* $A^\times$ *associated with the part of* $\sigma(A^\times)$ *lying in the open left half plane, and* $\sigma(A^\times|_{M_-^\times}) \subset \{\lambda \mid \Re\lambda \leq 0\}$,

(v) $H[M_+] = M_+^\perp, \quad H[M_-] = M_-^\perp, \quad H[M_+^\times] = M_+^{\times\perp}, \quad H[M_-^\times] = M_-^{\times\perp}.$

The subspaces in question also satisfy the matching conditions

$$\mathbb{C}^n = M_- \dotplus M_+^\times, \qquad \mathbb{C}^n = M_+ \dotplus M_-^\times.$$

Let Π_+ *be the projection of* $\mathbb{C}^n$ *along* M_- *onto* $M_+^\times$, *let* Π_- *be the projection of* $\mathbb{C}^n$ *along* M_+ *onto* $M_-^\times$, *and define* L_+ *and* L_- *by* (10.3) *and* (10.4), *that is*

$$\begin{aligned} L_+(\lambda) &= D^{1/2} + D^{-1/2}C\Pi_+(\lambda I_n - A)^{-1}B, \\ L_-(\lambda) &= D^{1/2} + D^{-1/2}C\Pi_-(\lambda I_n - A)^{-1}B. \end{aligned}$$

Then

$$W(\lambda) = L_+(-\bar{\lambda})^* L_+(\lambda), \qquad W(\lambda) = L_-(-\bar{\lambda})^* L_-(\lambda), \tag{10.5}$$

are right and left pseudo-spectral factorizations with respect to the imaginary axis, respectively. These pseudo-spectral factorizations are the unique ones for which $L_+(\infty) = D^{1/2}$ *and* $L_-(\infty) = D^{1/2}$.

As was noted before, Theorem 10.1 can be derived from Theorem 10.2 via the transformation $\lambda \mapsto -i\lambda$. Conversely, Theorem 10.2 obtained from Theorem 10.1 by the transformation $\lambda \mapsto i\lambda$.

Before we prove Theorem 10.2 we need some preparations concerning the spectral properties of nonnegative rational matrix functions. First we discuss the partial pole-multiplicities and partial zero-multiplicities of W. These notions have been defined in Sections 8.2 and 8.1 of [20], respectively. We start with a minimal realization

$$W(\lambda) = D + C(\lambda - A)^{-1}B. \tag{10.6}$$

Assume that W is biproper, i.e., D is invertible. Then the eigenvalues of A coincide with the poles of W and the eigenvalues of $A^\times$ coincide with the zeros of W. More precisely, the partial multiplicities of λ as an eigenvalue of A coincide with the partial pole-multiplicities of λ as a pole of W, and the multiplicities of λ as an eigenvalue of $A^\times$ coincide with the partial zero-multiplicities of λ as a zero of W (cf., [20], Section 8.4, in particular Proposition 8.23).

We also need the connection between the Jordan chains of A at an eigenvalue λ_0 and the co-pole functions of W at λ_0 described in Proposition 8.3. For a nonnegative rational matrix function, we have the following addition to that proposition.

Proposition 10.3. *Let $W(\lambda) = D + C(\lambda I_n - A)^{-1}B$ be a minimal realization for a rational $m \times m$ matrix function which is selfadjoint on the imaginary axis, and let H be the (unique) invertible skew-Hermitian $n \times n$ matrix such that*

$$HA = -A^*H, \qquad HB = C^*.$$

Let $\lambda_0 \in i\mathbb{R}$ be an eigenvalue of A, let $x_0, \ldots, x_{k-1}$ be a Jordan chain for A at λ_0, and let φ be a co-pole function of W at λ_0 corresponding to the Jordan chain $x_0, \ldots, x_{k-1}$. Then the function $\langle W(\lambda)\varphi(\lambda), \varphi(-\bar{\lambda})\rangle$ has a zero of order at least k at λ_0 and its Taylor expansion at λ_0 has the following form:

$$\langle W(\lambda)\varphi(\lambda), \varphi(-\bar{\lambda})\rangle = (-1)^k \langle x_0, Hx_{k-1}\rangle(\lambda - \lambda_0)^k + \cdots$$
$$\cdots + (-1)^k \langle x_{k-1}, Hx_{k-1}\rangle(\lambda - \lambda_0)^{2k-1} + h.o.t.,$$

where h.o.t. stands for higher order terms.

Proof. The fact that φ is a co-pole function of W at λ_0 implies that $W(\lambda)\varphi(\lambda)$ is analytic at λ_0. This together with the fact that φ has a zero of order at least k at λ_0 shows that the function $\langle W(\lambda)\varphi(\lambda), \varphi(-\bar{\lambda})\rangle$ has a zero of order at least k at λ_0 too. The property that φ is a co-pole function of W at λ_0 corresponding to the Jordan chain $x_0, \ldots, x_{k-1}$ means that

$$x_j = \sum_{\nu=k}^{\infty} P_0(A - \lambda_0)^{\nu-j-1}B\varphi_\nu, \qquad j = 0, \ldots, k-1 \tag{10.7}$$

(where the sum in the right-hand side of the identity is actually finite so that there is no convergence issue). Here P_0 is the Riesz projection of A corresponding to the eigenvalue λ_0, and φ_ν is the coefficient of $(\lambda - \lambda_0)^\nu$ in the Taylor expansion of φ at λ_0. We use this connection to compute $\langle Hx_i, x_{k-1}\rangle$. The fact that λ_0 is in $i\mathbb{R}$ yields $HP_0 = P_0^*H$. Indeed, since $HAH^{-1} = -A^*$, we have that HPH^{-1} is the Riesz projection of $-A^*$ for the eigenvalue λ_0. Thus, using Proposition I.2.5 in [51], we get $HP_0H^{-1} = P(-A^*; \{\lambda_0\}) = P(A^*; \{-\lambda_0\}) = P(A^*; \{\overline{\lambda_0}\}) = P(A; \{\lambda_0\})^* = P_0^*$. Also, note that the vectors $x_0, \ldots, x_{k-1}$ belong to $\operatorname{Im} P_0$. In particular, $P_0x_{k-1} = x_{k-1}$. Now use (10.7) and the identities $HA = -A^*H$ and $HB = C^*$. This gives, for $i = 0, \ldots, k-1$,

$$\begin{aligned}
\langle Hx_i, x_{k-1}\rangle &= \sum_{\nu=k}^{\infty} \langle HP_0(A - \lambda_0)^{\nu-i-1}B\varphi_\nu, x_{k-1}\rangle \\
&= \sum_{\nu=k}^{\infty} \langle H(A - \lambda_0)^{\nu-i-1}B\varphi_\nu, P_0x_{k-1}\rangle \\
&= \sum_{\nu=k}^{\infty} (-1)^{\nu-i-1}\langle \varphi_\nu, C(A - \lambda_0)^{\nu-i-1}x_{k-1}\rangle \\
&= \sum_{\nu=k}^{k+i} (-1)^{\nu-i-1}\langle \varphi_\nu, Cx_{k-\nu+i}\rangle.
\end{aligned}$$

From the final paragraph of Section 8.1 we know that the vector $Cx_{k-\nu+1}$ is given by $Cx_{k-\nu+1} = (W\varphi)_{k-\nu+i}$, where $(W\varphi)_j$ is the coefficient of $(\lambda-\lambda_0)^j$ in the Taylor expansion of $W(\lambda)\varphi(\lambda)$ at λ_0. So

$$\langle Hx_i, x_{k-1}\rangle = \sum_{\nu=k}^{k+i}(-1)^{\nu-i-1}\langle \varphi_\nu, (W\varphi)_{k-\nu+i}\rangle, \qquad i=0,\dots,k-1. \tag{10.8}$$

On the other hand we have

$$\langle W(\lambda)\varphi(\lambda), \varphi(-\bar\lambda)\rangle = \sum_{\ell=k}^{\infty}\Big(\sum_{\nu=k}^{\ell}(-1)^\nu\langle (W\varphi)_{\ell-\nu}, \varphi_\nu\rangle\Big)(\lambda-\lambda_0)^\ell. \tag{10.9}$$

Comparing formulas (10.8) and (10.9), we see that for $i=0,\dots,k-1$ the coefficient of $(\lambda-\lambda_0)^{k+i}$ in the Taylor expansion of $W(\lambda)\varphi(\lambda)$ at λ_0 is given by $(-1)^{i+1}\langle x_{k-1}, Hx_i\rangle$. Now note that

$$\begin{aligned}\langle Hx_i, x_{k-1}\rangle &= \langle H(A-\lambda_0)^{k-i-1}x_{k-1}, x_{k-1}\rangle\\ &= (-1)^{k-1-i}\langle Hx_{k-1}, (A-\lambda_0)^{k-1-i}x_{k-1}\rangle\\ &= (-1)^{k-1-i}\langle Hx_{k-1}, x_i\rangle = (-1)^{k-1-i}\langle x_{k-1}, Hx_i\rangle.\end{aligned}$$

We conclude that $(-1)^{i+1}\langle x_{k-1}, Hx_i\rangle = (-1)^k\langle Hx_i, x_{k-1}\rangle$, which completes the proof. □

Specializing to the case when W is nonnegative on $i\mathbb{R}$ we obtain the following result.

Proposition 10.4. *Let $W(\lambda) = D + C(\lambda I_n - A)^{-1}B$ be a minimal realization for a rational $m\times m$ matrix function which is nonnegative on $i\mathbb{R}$. Assume D is positive definite, and let H be the (unique invertible) skew-Hermitian $n\times n$ matrix such that $HA = -A^*H$ and $HB = C^*$. Then the partial multiplicities corresponding to pure imaginary eigenvalues of A and $A^\times$ are all even, the sign characteristic of (iA, iH) consists of the integers $+1$ only, and the sign characteristic of the pair $(iA^\times, iH)$ consists of the integers -1 only.*

For the definition of the notion of sign characteristic the reader is referred to Section 11.2 below.

Proof. Let us first prove the proposition for the matrix A. Let $\lambda_0 = i\mu_0$ be a pure imaginary eigenvalue of A, and let $x_0,\dots,x_{k-1}$ be a maximal Jordan chain for A at λ_0. Then x_0, $-ix_1$, $(-i)^2x_2$, …, $(-i)^{k-1}x_{k-1}$ is a Jordan chain of iA for its eigenvalue $-\mu_0$. In fact, all Jordan chains of iA for $-\mu_0$ can be obtained in this way. Choose a Jordan basis for A such that relative to it the pair (iA, iH) is in canonical form (see Section 11.2). This means, in particular, that if $x_0,\dots,x_{k-1}$ is a maximal Jordan chain of A for λ_0, which is part of this basis, then $\langle iHx_0, (-i)^{k-1}x_{k-1}\rangle =$

$i^k\langle Hx_0, x_{k-1}\rangle$ is either $+1$ or -1. The sequence of $+1$'s and -1's, obtained in this manner, is the sign characteristic of the pair (iA, iH).

Let $x_0, \ldots, x_{k-1}$ be as in the previous paragraph, and let

$$\varphi(\lambda) = (\lambda - \lambda_0)^k \varphi_k + (\lambda - \lambda_0)^{k+1}\varphi_{k+1} + \cdots$$

be a corresponding co-pole function for W at λ_0. From Proposition 10.3 we know that on a neighborhood of λ_0

$$\langle W(\lambda)\varphi(\lambda), \varphi(-\bar{\lambda})\rangle = (\lambda - \lambda_0)^k h(\lambda),$$

where the scalar function h is analytic at λ_0 and $h(\lambda_0) = (-1)^k\langle Hx_0, x_{k-1}\rangle$. Consider the pure imaginary $\lambda = i\mu$ in this neighborhood. Rewriting the expression above in terms of $\mu - \mu_0$, and using the fact that W is nonnegative, one sees that k is even and $(-i)^k\langle Hx_0, x_{k-1}\rangle > 0$. This proves that the partial multiplicities corresponding to pure imaginary eigenvalues of A are even, and that the sign characteristic of the pair (iA, iH) consists of $+1$'s only.

To prove the part of the proposition concerning $A^\times$, note that the function $W(\lambda)^{-1} = D^{-1} - D^{-1}C(\lambda - A^\times)^{-1}BD^{-1}$ is nonnegative on $i\mathbb{R}$ too. Moreover, for this realization we have $(-H)A^\times = -(A^\times)^*(-H)$ and $(-H)BD^{-1} = (-D^{-1}C)^*$. So, the corresponding indefinite inner product is given by $-H$ rather than H. The desired result now follows by basically repeating the argument given above. $\square$

We now have all the equipment necessary for the proof of Theorem 10.2.

Proof of Theorem 10.2. Based on Proposition 10.4 the existence and uniqueness of A-invariant subspaces M_+, M_- and $A^\times$-invariant subspaces $M_+^\times$, $M_-^\times$ such that (i), (ii) and (iii) hold follow from Theorem 11.5 in Section 11.2 below.

To prove the first equality in (10.2) one establishes $M_+ \cap M_-^\times \subset \operatorname{Ker} C$ as in the proof of Theorem 9.4: use (9.2) instead of (9.1). Hence $M_+ \cap M_-^\times$ is invariant for both A and $A^\times$. However, as the realization is minimal, an A-invariant subspace contained in $\operatorname{Ker} C$ must be the zero space. Thus $M_+ \cap M_-^\times = \{0\}$. To show $\mathbb{C}^n = M_+ \dot{+} M_-^\times$ it remains to note that $\dim M_+ = \dim M_-^\times = n/2$. In a similar manner one gets $\mathbb{C}^n = M_- \dot{+} M_+^\times$.

Denote by Π_+ the projection along M_- onto $M_+^\times$, then Π_+ is a supporting projection, and by Theorem 8.5 the factorization

$$W(\lambda) = K(\lambda)L_+(\lambda),$$

with L_+ given by (10.3) and

$$K(\lambda) = D^{1/2} + C(\lambda - A)^{-1}(I - \Pi_+)BD^{-1/2},$$

is minimal. Moreover, L_+ has no poles in the open left half plane because $\Pi_+ A = \Pi_+ A \Pi_+$. So

$$L_+(\lambda) = D^{1/2} + D^{-1/2}C\Pi_+(\lambda - \Pi_+ A\Pi_+)^{-1}\Pi_+ B.$$

Also

$$L_+^{-1}(\lambda) = D^{-1/2} - C\Pi_+(\lambda - \Pi_+ A^\times \Pi_+)^{-1}\Pi_+ B D^{-1/2},$$

thus L_+ has no zeros in the open left half plane. Finally, $K(\lambda) = L_+(-\bar{\lambda})^*$. Indeed,

$$\begin{aligned} L_+(-\bar{\lambda})^* &= D^{1/2} - B^*(\lambda + A^*)^{-1}\Pi_+^* C^* D^{-1/2} \\ &= D^{1/2} + C(\lambda - A)^{-1} H^{-1}\Pi_+^* H B D^{-1/2}. \end{aligned}$$

As $H[\operatorname{Ker}\Pi_+] = (\operatorname{Ker}\Pi_+)^\perp$ and $H[\operatorname{Im}\Pi_+] = (\operatorname{Im}\Pi_+)^\perp$, we have $H^{-1}(\Pi_+)^* H = I - \Pi_+$. But then the factorization corresponding to Π_+ is a right pseudo-spectral factorization. One proves in a similar way that Π_- gives rise to a left pseudo-spectral factorization. □

Next, we introduce the notion of left and right pseudo-spectral factorizations with respect to the unit circle, Let W be a rational matrix function having nonnegative values on $\mathbb{T}$. A factorization

$$W(\lambda) = L(\bar{\lambda}^{-1})^* L(\lambda)$$

is called a *right pseudo-spectral factorization with respect to the unit circle* if L has no poles or zeros in the open unit disc and the factorization is locally minimal at each point of the unit circle. *Left pseudo-spectral factorizations with respect to the unit circle* are defined by replacing the open unit disc $\mathbb{D}$ by $\mathbb{D}_{\text{ext}}$.

In dealing with pseudo-spectral factorizations with respect to the unit circle, we discuss only a restricted class of rational matrix functions that are nonnegative on the unit circle, namely those which are biproper. Because of symmetry, this forces the function to have an invertible value at zero too. The restriction is induced by our methods, rather than by the problem itself.

The following theorem can be obtained from using an appropriate Möbius transformation (cf., the proof of Theorem 9.9).

Theorem 10.5. *Let $W(\lambda) = D + C(\lambda I_n - A)^{-1}B$ be a minimal realization of a rational $m \times m$ matrix function which is nonnegative on the unit circle, and assume D and A are invertible. Then, with respect to the unit circle, W admits left and right pseudo-spectral factorization. Such factorizations can be obtained in the following way. Let H be the (unique invertible) skew-Hermitian $n \times n$ matrix satisfying $A^*HA = H$ and $A^*HB = C^*$. Then there are unique A-invariant subspaces M_+, M_- and unique $A^\times$-invariant subspaces $M_+^\times$, $M_-^\times$, such that*

(i) *M_+ contains the spectral subspace of A associated with the part of $\sigma(A)$ lying in the open exterior of the unit disc, and $\sigma(A|_{M_+}) \subset \{\lambda \mid |\lambda| \geq 1\}$,*

(ii) *M_- contains the spectral subspace of A associated with the part of $\sigma(A)$ lying in the open unit disc, and $\sigma(A|_{M_-}) \subset \{\lambda \mid |\lambda| \leq 1\}$,*

(iii) *$M_+^\times$ contains the spectral subspace of $A^\times$ associated with the part of $\sigma(A^\times)$ lying in the open exterior of the unit disc, and $\sigma(A^\times|_{M_+^\times}) \subset \{\lambda \mid |\lambda| \geq 1\}$,*

(iv) $M_-^\times$ *contains the spectral subspace of* $A^\times$ *associated with the part of* $\sigma(A^\times)$ *lying in the open unit disc, and* $\sigma(A^\times|_{M_-^\times}) \subset \{\lambda \mid |\lambda| \leq 1\}$,

(v) $H[M_+] = M_+^\perp, \quad H[M_-] = M_-^\perp, \quad H[M_+^\times] = M_+^{\times\perp}, \quad H[M_-^\times] = M_-^{\times\perp}.$

The subspaces in question also satisfy (10.2), *i.e.*,

$$\mathbb{C}^n = M_+ \dot{+} M_-^\times, \qquad \mathbb{C}^n = M_- \dot{+} M_+^\times.$$

Let Π_+ *be the projection of* $\mathbb{C}^n$ *along* M_- *onto* $M_+^\times$, *and let* Π_- *be the projection of* $\mathbb{C}^n$ *along* M_+ *onto* $M_-^\times$, *and define* L_+ *and* L_- *by* (9.13) *and* (9.14), *so*

$$\begin{aligned} L_+(\lambda) &= D_+^{1/2} + D_+^{1/2} D^{-1} C \Pi_+ (\lambda I_n - A)^{-1} B, \\ L_-(\lambda) &= D_-^{1/2} + D_-^{1/2} D^{-1} C \Pi_- (\lambda I_n - A)^{-1} B, \end{aligned}$$

where $D_+ = D - CA^{-1}(I - \Pi_+)B$ *and* $D_- = D - CA^{-1}(I - \Pi_-)B$. *Then*

$$W(\lambda) = L_+(\bar{\lambda}^{-1})^* L_+(\lambda), \qquad W(\lambda) = L_-(\bar{\lambda}^{-1})^* L_-(\lambda),$$

are right and left pseudo-spectral factorizations with respect to the unit circle, respectively. The functions L_+ *and* L_- *are the unique right and left pseudo-spectral factors, respectively, such that* $L_+(\infty) = D_+^{1/2}$ *and* $L_-(\infty) = D_-^{1/2}$.

10.2 Selfadjoint rational matrix functions and further generalizations

The main result of Section 10.1 will be generalized here to the case of an arbitrary selfadjoint rational matrix function with positive definite value at infinity. We start with the case of selfadjoint functions on the real line.

Theorem 10.6. *Let* $W(\lambda) = D + C(\lambda I_n - A)^{-1}B$ *be a minimal realization of an* $m \times m$ *rational matrix function which is selfadjoint on the real line, and assume* D *is positive definite. Then, with respect to the real line,* W *admits right and left pseudo-canonical factorization. Such factorizations can be obtained in the following way. Let* H *be the (unique invertible) Hermitian* $n \times n$ *matrix such that* $HA = A^*H$ *and* $HB = C^*$. *Then there exist* A*-invariant subspaces* M_+ *and* M_-, *and* $A^\times$*-invariant subspaces* $M_+^\times$ *and* $M_-^\times$ *such that*

(i) M_+ *contains the spectral subspace of* A *associated with the part of* $\sigma(A)$ *lying in the open lower half plane, and* $\sigma(A|_{M_+}) \subset \{\lambda \mid \Im\lambda \leq 0\}$,

(ii) M_- *contains the spectral subspace of* A *associated with the part of* $\sigma(A)$ *lying in the open upper half plane, and* $\sigma(A|_{M_-}) \subset \{\lambda \mid \Im\lambda \geq 0\}$,

(iii) $M_+^\times$ *contains the spectral subspace of* $A^\times$ *associated with the part of* $\sigma(A^\times)$ *lying in the open lower half plane, and* $\sigma(A^\times|_{M_+^\times}) \subset \{\lambda \mid \Im\lambda \leq 0\}$,

(iv) $M_-^\times$ *contains the spectral subspace of* $A^\times$ *associated with the part of* $\sigma(A^\times)$ *lying in the open upper half plane, and* $\sigma(A^\times|_{M_-^\times}) \subset \{\lambda \mid \Im\lambda \geq 0\}$,

(v) M_+ *and* M_- *are maximal* H*-nonnegative, and* $M_+^\times$ *and* $M_-^\times$ *are maximal* H*-nonpositive.*

The subspaces in question also satisfy

$$\mathbb{C}^n = M_+ \dot{+} M_-^\times, \qquad \mathbb{C}^n = M_- \dot{+} M_+^\times.$$

Let Π_+ *be the projection of* $|BC^n$ *onto* $M_+^\times$ *along* M_-, *and let* Π_- *be the projection of* $\mathbb{C}^n$ *onto* $M_-^\times$ *along* M_+, *and introduce*

$$L_-(\lambda) = D^{1/2} + C(\lambda I_n - A)^{-1}(I - \Pi_+)BD^{-1/2}, \tag{10.10}$$

$$L_+(\lambda) = D^{1/2} + D^{-1/2}C\Pi_+(\lambda I_n - A)^{-1}B, \tag{10.11}$$

$$K_+(\lambda) = D^{1/2} + C(\lambda I_n - A)^{-1}(I - \Pi_-)BD^{-1/2}, \tag{10.12}$$

$$K_-(\lambda) = D^{1/2} + D^{-1/2}C\Pi_-(\lambda I_n - A)^{-1}B. \tag{10.13}$$

Then

$$W(\lambda) = L_-(\lambda)L_+(\lambda), \qquad W(\lambda) = K_+(\lambda)K_-(\lambda), \tag{10.14}$$

are right and left pseudo-canonical factorizations with respect to the real line, respectively.

The subspaces $M_+, M_-, M_+^\times$ and $M_-^\times$ are not unique. In line with this, the uniqueness of the factorizations that we had at earlier occasions is lacking here. Also, not all pseudo-canonical factorizations for selfadjoint rational matrix functions are obtained in the way described in Theorem 10.6 .

The theorem will be obtained from the more general result stated below.

Theorem 10.7. *Let* $W(\lambda) = D + C(\lambda I_n - A)^{-1}B$ *be a minimal realization of an* $m \times m$ *rational matrix function which is selfadjoint on the real line, and assume* D *is positive definite. Suppose* $D = D_+D_-$ *with* D_+ *and* D_- $m \times m$ *matrices (automatically invertible). Let* H *be the (unique invertible) Hermitian* $n \times n$ *matrix for which* $HA = A^*H$ *and* $HB = C^*$. *Let* M_+ *be an* A*-invariant maximal* H*-nonnegative subspace, and let* M_- *be an* $A^\times$*-invariant maximal* H*-nonpositive subspace. Then*

$$\mathbb{C}^n = M_+ \dot{+} M_-. \tag{10.15}$$

In that case, the projection Π *of* $\mathbb{C}^n$ *along* M_+ *onto* M_- *is a supporting projection, and (hence)* W *admits a minimal factorization* $W(\lambda) = W_+(\lambda)W_-(\lambda)$ *with* W_+ *and* W_- *given by*

$$W_+(\lambda) = D_+ + C(\lambda I_n - A)^{-1}(I - \Pi)BD_-^{-1},$$

$$W_-(\lambda) = D_- + D_+^{-1}C\Pi(\lambda I_n - A)^{-1}B.$$

For the existence of A-invariant maximal H-nonnegative and maximal H-nonpositive subspaces, see Section 11.2 below.

Proof. First we show that $M_+ \cap M_- = \{0\}$. Choose $x \in M_+ \cap M_-$. As M_+ is nonnegative and M_- is nonpositive, we have $\langle Hx, x\rangle = 0$. On M_+ the Schwartz inequality holds for the H-inner product. Since $x \in M_+$ and $Ax \in M_+$, we get

$$|\langle HAx, x\rangle|^2 \le \langle HAx, Ax\rangle \cdot \langle Hx, x\rangle = 0.$$

So for all $x \in M_+ \cap M_-$ we have $\langle HAx, x\rangle = 0$. In the same way one shows that for all $x \in M_+ \cap M_-$ we have $\langle HA^\times x, x\rangle = 0$. It follows that

$$0 = \langle H(A - A^\times)x, x\rangle = \langle HBD^{-1}Cx, x\rangle = \langle C^*D^{-1}Cx, x\rangle = \|D^{-1/2}Cx\|^2,$$

and hence $M_+ \cap M_- \subset \operatorname{Ker} C$. But then $A^\times x = Ax - BCx = Ax$ for all x belonging to $M_+ \cap M_-$, and so $M_+ \cap M_-$ is A-invariant. Hence $CA^n x = 0$ for all $x \in M_+ \cap M_-$ and $n = 0, 1, 2, \ldots$. So

$$M_+ \cap M_- \subset \bigcap_{j=0}^{\infty} \operatorname{Ker} CA^j = \{0\}.$$

Now (see Section 11.2) every maximal nonnegative subspace has the same dimension as M_+. Also, for a maximal H-nonpositive subspace M_-, the subspace $H^{-1}[M_-^\perp]$ is maximal H-nonnegative. Hence

$$\dim M_+ = \dim H^{-1}[M_-^\perp] = \dim M_-^\perp = n - \dim M_-,$$

and from this we get (10.15), i.e, the first part of the theorem. To obtain the second part, apply Theorem 8.5. □

Proof of Theorem 10.6. For the existence of A-invariant subspaces M_+, M_- and $A^\times$-invariant subspaces $M_+^\times, M_-^\times$ such that (i), (ii) and (iii) hold we refer to Section 11.2. The matching of the appropriate subspaces is an immediate consequence of Theorem 10.7. The factorizations (10.14), where the factors are given by (10.10)–(10.13) are minimal by Theorem 8.5. As in the proof of Theorem 10.2 one shows that L_+ and K_+ have no zeros or poles in the open upper half plane. In the same vein, L_- and K_- have no zeros or poles in the open lower half plane. Hence the factorizations in (10.14) are right and left pseudo-canonical factorizations, respectively. □

Analogues of Theorems 10.6 and 10.7 concerning rational matrix functions which are selfadjoint on the unit circle or imaginary axis can be derived too. An analogue of Theorem 10.7 also holds true if one takes M_+ to be A-invariant maximal H-nonpositive (instead of maximal H-nonnegative) and M_- to be $A^\times$-invariant maximal H-nonnegative (instead of maximal H-nonpositive). A similar remark can be made concerning Theorem 10.6.

We finish this section with a theorem concerning symmetric factorization of rational matrix functions which are nonnegative. Here we shall present only the case involving the imaginary axis.

Theorem 10.8. *Let $W(\lambda) = D + C(\lambda I_n - A)^{-1}B$ be a minimal realization of an $m \times m$ rational matrix function which is nonnegative on $i\mathbb{R}$. Assume D is positive definite, and let H be the (unique invertible) skew-Hermitian $n \times n$ matrix such that $HA = -A^*H$ and $HB = C^*$. Suppose M and $M^\times$ are subspaces of $\mathbb{C}^N$ for which*

$$A[M] \subset M, \quad A^\times[M^\times] \subset M^\times, \quad H[M] = M^\perp, \quad H[M^\times] = M^{\times\perp}. \tag{10.16}$$

Then $\mathbb{C}^n = M \dot{+} M^\times$. Let Π be the projection of $\mathbb{C}^n$ along M onto $M^\times$, and introduce

$$L(\lambda) = D^{1/2} + D^{-1/2}C\Pi(\lambda I_n - A)^{-1}B. \tag{10.17}$$

Then

$$W(\lambda) = L(-\bar{\lambda})^* L(\lambda) \tag{10.18}$$

is a minimal factorization. Conversely, given a minimal factorization (10.18), *with $L(\infty) = D^{1/2}$, the factor L is as in* (10.17) *for a supporting projection Π such that $M = \operatorname{Ker}\Pi$ and $M^\times = \operatorname{Im}\Pi$ satisfy* (10.16).

Proof. Let M and $M^\times$ be as in the theorem. We shall show that $\mathbb{C}^n = M \dot{+} M^\times$. The argument follows a (by now) familiar pattern. One first shows that the intersection $M \cap M^\times$ is contained in $\operatorname{Ker} C$ (see, e.g., the proof of Theorem 9.4, or the proof of Theorem 10.7). Then $M \cap M^\times$ is both A-invariant and $A^\times$-invariant and contained in $\operatorname{Ker} C$. By minimality (in fact observability) it follows that $M \cap M^\times = \{0\}$. Since $\dim M = \dim M^\times = n/2$, we have the desired matching.

Denote by Π the projection along M onto $M^\times$. Then Π is a supporting projection. Write the factorization of W corresponding to Π and the factorization $D = D^{1/2}D^{1/2}$ as $W(\lambda) = K(\lambda)L(\lambda)$, where

$$\begin{aligned} K(\lambda) &= D^{1/2} + C(\lambda - A)^{-1}(I - \Pi)BD^{-1/2}, \\ L(\lambda) &= D^{1/2} + D^{-1/2}C\Pi(\lambda - A)^{-1}B. \end{aligned}$$

Arguing as in the proof of Theorem 9.4 we have $\Pi^*H = H(I - \Pi)$. Using also (9.2) it then follows easily that $L(-\bar{\lambda})^* = K(\lambda)$.

Conversely, suppose $W(\lambda) = L(-\bar{\lambda})^*L(\lambda)$ is a minimal factorization with $L(\infty) = D^{1/2}$. Let Π be the corresponding supporting projection (which exists by Theorem 8.5). From the fact that the left-hand factor $K(\lambda)$ is $L(-\bar{\lambda})^*$, where $L(\lambda)$ is the right-hand factor, and using (9.2), we have $\Pi^*H = H(I - \Pi)$. Thus both $M = \operatorname{Ker}\Pi$ and $M^\times = \operatorname{Im}\Pi$ satisfy (i) and (ii). □

Notes

This chapter originates from [119] which deals with rational matrix functions that are selfadjoint on the real line. The term pseudo-canonical is from a later date, and is taken from [132]. The results presented here for nonnegative rational matrix functions on the unit circle are based on Section 3 of [104]. In this case, the restriction to W being invertible at infinity and at zero may be lifted by considering a different type of realization, namely, realizations of the type discussed in [79].

In mathematical systems theory also the following problem is of interest: given is a nonnegative rational matrix function W as in Theorem 10.8, without poles on the imaginary axis. One is looking for all possible factorizations $W(\lambda) = L(-\bar{\lambda})^* L(\lambda)$, where L has all its poles in the open left half plane, but there is no condition on the zeros of L. This problem too sometimes goes by the name of "spectral factorization problem" and such factors L are sometimes also called "spectral factors". The problem of parametrizing such factors is considered in many papers and books, see, e.g., [116] and [46] and the references given there. The papers [30], [31], provide a discussion involving computational aspects.

For matrix polynomials a similar problem is considered in the literature, see e.g., [88] and [66]. For later developments on factorization of selfadjoint matrix polynomials, see [103], [125].

In [20] stability of factorizations of rational matrix functions under small perturbations of the matrices in a realization is studied. For the particular case where the function is positive semidefinite on the real line, and the factorizations are of the type (10.1), stability under small perturbations is treated in [123]. This involves stability of invariant Lagrangian subspaces for matrices that are selfadjoint in a space with an indefinite inner product. It turns out that the left and right pseudo-spectral factorizations are stable (see Theorem 2.5 in [123]).

Chapter 11

Review of the theory of matrices in indefinite inner product spaces

In this chapter we present some background material on matrices in indefinite inner product spaces, and review the main results from this area that are used in this book. No proofs will be provided; we refer to the literature for more information. Good sources are [68] and [70]. The material is not only useful for understanding of the results of the preceding two chapters, but is also intended for use in subsequent chapters.

This chapter consists of three sections. Section 11.1 considers subspaces that are negative, positive or neutral relative to an indefinite inner product and various generalizations of such subspaces. Section 11.2 deals with matrices that are selfadjoint relative to an indefinite inner product, and Section 11.3 with matrices that are dissipative relative to an indefinite inner product.

11.1 Subspaces of indefinite inner product spaces

Let H be an invertible Hermitian $n \times n$ matrix. On $\mathbb{C}^n$ we denote the usual inner product with $\langle \cdot, \cdot \rangle$. The *indefinite inner product* given by H is defined as follows:

$$[x, y] = \langle Hx, y \rangle.$$

A vector $x \in \mathbb{C}^n$ is called *H-positive*, *H-negative*, or *H-neutral*, respectively, if $[x, x] > 0$, $[x, x] < 0$, or $[x, x] = 0$, respectively. A subspace M of $\mathbb{C}^n$ is called *H-nonnegative*, *H-nonpositve*, or *H-neutral*, respectively, if $[x, x] \geq 0$, $[x, x] \leq 0$, or $[x, x] = 0$, respectively, for all $x \in M$. Observe that an H-neutral subspace is at the same time H-nonnegative and H-nonpositive.

Although the Cauchy-Schwarz inequality does not hold for just any two vectors x, y in an indefinite inner product space, it does hold for vectors x, y which are both in an H-nonnegative subspace, or both in an H-nonpositive subspace. Note that it follows from this that M is H-neutral if and only if $H[M] \subset M^{\perp}$.

A subspace M of $\mathbb{C}^n$ will be called *maximal H-nonnegative* whenever it is H-nonnegative and not properly contained in a larger H-nonnegative subspace. Similarly, M will be called a *maximal H-nonpositive subspace* if it is H-nonpositive and not properly contained in a larger H-nonpositive subspace. The first part of the following proposition can be found in Theorem 2.3.2 in [70], the second part is Lemma 6.3 in [25].

Proposition 11.1. *The dimension of any maximal H-nonnegative subspace coincides with the number of positive eigenvalues of H, while the dimension of any maximal H-nonpositive subspace coincides with the number of negative eigenvalues of H. Also, if M is maximal H-nonpositive then $H^{-1}[M^{\perp}]$ is maximal H-nonnegative.*

A subspace M of $\mathbb{C}^n$ is said to be *H-Lagrangian* if $H[M] = M^{\perp}$. Such a subspace is both maximal H-nonnegative and maximal H-nonpositive, and hence such a subspace can exist only if H has as many positive eigenvalues as it has negative ones. As an example, suppose n is even, $n = 2k$ say, and let

$$H = i \begin{bmatrix} 0 & I_k \\ -I_k & 0 \end{bmatrix}.$$

Then any subspace of the form $M = \operatorname{Im} [\, P \quad I \,]^*$ with P Hermitian will be a Lagrangian subspace.

The concepts involving ordinary orthogonality have straightforward analogues for *H-orthogonality*. For instance, vectors x and y in $\mathbb{C}^n$ are *H-orthogonal* if $[x, y] = 0$.

A subspace M is called *H-nondegenerate* in case there is no non-zero vector $x \in M$ that is H-orthogonal to all vectors in M. An equivalent requirement is that $M \cap H[M]^{\perp} = \{0\}$. It follows that for H-nondegenerate subspaces M, one has

$$\mathbb{C}^n = M \dot{+} H[M]^{\perp}.$$

Conversely, each subspace M of $\mathbb{C}^n$ with this property is H-nondegenerate.

11.2 H-selfadjoint matrices

Let the indefinite inner product on $\mathbb{C}^n$ be given by the invertible Hermitian matrix H. An $n \times n$ matrix A has an *H-adjoint* $A^{[*]}$ defined by

$$[Ax, y] = [x, A^{[*]}y].$$

Thus $A^{[*]} = H^{-1}A^*H$. The matrix A is called *H-selfadjoint* if $A = A^{[*]}$ or which amounts to the same, $HA = A^*H$.

As an example, let $A = J_n(\lambda)$ be the $n \times n$ upper triangular Jordan block with a real eigenvalue λ, and let $H = \varepsilon P_n$, where ε is $+1$ or -1, and P_n is the standard $n \times n$ involutary matrix (also called the $n \times n$ reversed identity matrix). Thus P_n is the $n \times n$ matrix with 1s on the diagonal running from the lower left corner to the upper right corner, and 0s elsewhere. Clearly H is invertible and selfadjoint while, moreover, $HA = A^*H$. Hence A is H-selfadjoint.

As a second example, suppose n is even, $n = 2k$ say, let λ be non-real, and let $A = \operatorname{diag}\big(J_k(\lambda),\, J_k(\overline{\lambda})\big)$ be the block diagonal sum of two Jordan blocks of size k with eigenvalues λ and $\overline{\lambda}$, respectively. Further, let $H = P_{2k}$. Then again $HA = A^*H$, so A is H-selfadjoint.

It turns out that these two examples can serve as the building blocks for any pair (A, H), where A is H-selfadjoint. To state this more precisely, first observe that if A is H-selfadjoint, and if S is an invertible matrix, then $S^{-1}AS$ is S^*HS-selfadjoint. The map $(A, H) \mapsto (S^{-1}AS, S^*HS)$ defines an equivalence relation on the set of pairs (A, H) with A being H-selfadjoint. The following result, which can be found in [70], Theorem 5.1.1, describes a canonical form for pairs of matrices of this type.

Theorem 11.2. *Let A be an H-selfadjoint matrix. Then there exists an invertible matrix S such that $S^{-1}AS$ is equal to the block-diagonal matrix*

$$\operatorname{diag}\big(J_{k_1}(\lambda_1), \ldots, J_{k_m}(\lambda_m), J_{k_{m+1}}(\lambda_{m+1}), J_{k_{m+1}}(\overline{\lambda_{m+1}}), \ldots, J_{k_l}(\lambda_l), J_{k_l}(\overline{\lambda_l})\big),$$

while

$$S^*HS = \operatorname{diag}\big(\varepsilon_1 P_{k_1}, \ldots, \varepsilon_m P_{k_m}, P_{2k_{m+1}}, \ldots, P_{2k_l}\big).$$

Here $\lambda_1, \ldots, \lambda_m$ are the real eigenvalues of A, geometric multiplicities counted, $\lambda_{m+1}, \overline{\lambda}_{m+1}, \ldots, \lambda_l, \overline{\lambda}_l$ are the non-real eigenvalues of A, geometric multiplicities counted too, and the numbers $\varepsilon_1, \ldots, \varepsilon_m$ take the values $+1$ and -1.

Behind the theorem is the fact that if A is H-selfadjoint, then the spectrum of A is closed under complex conjugation, taking (partial) multiplicities into account. By slight abuse of terminology, the ordered m-tuple $(\varepsilon_1, \ldots, \varepsilon_m)$ is called the *sign characteristic* of the pair (A, H). It is uniquely determined by the pair (A, H) up to permutations of signs corresponding to equal Jordan blocks.

Next, we consider invariant maximal H-nonnegative and invariant maximal H-nonpositive subspaces. We start again with examples. Let A be a single Jordan block of size $n \times n$ with a real eigenvalue, and take $H = \varepsilon P_n$. Denote the standard basis of $\mathbb{C}^n$ by $e_1, \dots, e_n$. Introduce

$$M^+ = \begin{cases} \text{span}\,\{e_1, \dots, e_{n/2}\} \text{ in case } n \text{ is even,} \\ \text{span}\,\{e_1, \dots, e_{(n+1)/2}\} \text{ in case } n \text{ is odd and } \varepsilon = +1, \\ \text{span}\,\{e_1, \dots, e_{(n-1)/2}\} \text{ in case } n \text{ is odd and } \varepsilon = -1, \end{cases}$$

$$M^- = \begin{cases} \text{span}\,\{e_1, \dots, e_{n/2}\} \text{ in case } n \text{ is even,} \\ \text{span}\,\{e_1, \dots, e_{(n+1)/2}\} \text{ in case } n \text{ is odd and } \varepsilon = -1, \\ \text{span}\,\{e_1, \dots, e_{(n-1)/2}\} \text{ in case } n \text{ is odd and } \varepsilon = +1. \end{cases}$$

Then M^+ is A-invariant and maximal H-nonnegative, while M^- is A-invariant and maximal H-nonpositive.

As a second example, suppose n is even, $n = 2k$ say, let $A = J_k(\lambda) \oplus J_k(\overline{\lambda})$ with λ non-real, let $H = P_{2k}$, and write $e_1, \dots, e_{2k}$ for the standard basis of $\mathbb{C}^{2k}$. Then, for $l = 0, \dots, k$, we have that $M = \text{span}\,\{e_1, \dots, e_l, e_{k+1}, \dots, e_{2k-l}\}$ is an A-invariant H-Lagrangian subspace.

If A is H-selfadjoint, and λ is a real eigenvalue of A, then the spectral invariant subspace of A corresponding to λ is H-orthogonal to the spectral invariant subspace of A corresponding to all other eigenvalues. A similar statement holds for a pair of complex conjugate non-real eigenvalues $\lambda, \overline{\lambda}$. This allows one to build up A-invariant maximal H-nonnegative subspaces by taking direct sums of subspaces constructed "locally" as in the previous two examples. In particular the following holds, see Theorem 5.12.1 in [70].

Theorem 11.3. *Let A be H-selfadjoint. The following statements hold:*

(i) *There exists an A-invariant maximal H-nonnegative subspace M_u^+ such that $\sigma(A|_{M_u^+})$ is in the closed upper half plane. Furthermore, any such M_u^+ contains the spectral invariant subspace of A corresponding to the open upper half plane.*

(ii) *There exists an A-invariant maximal H-nonpositive subspace M_u^- such that $\sigma(A|_{M_u^-})$ is in the closed upper half plane. Furthermore, any such M_u^- contains the spectral invariant subspace of A corresponding to the open upper half plane.*

(iii) *There exists an A-invariant maximal H-nonnegative subspace M_l^+ such that $\sigma(A|_{M_l^+})$ is in the closed lower half plane. Furthermore, any such M_l^+ contains the spectral invariant subspace of A corresponding to the open lower half plane.*

(iv) *There exists an A-invariant maximal H-nonpositive subspace M_l^- such that $\sigma(A|_{M_l^-})$ is in the closed lower half plane. Furthermore, any such M_l^- contains the spectral invariant subspace of A corresponding to the open lower half plane.*

Our next concern is the existence of A-invariant H-Lagrangian subspaces. These do not always exist. The next theorem gives a necessary and sufficient condition.

Theorem 11.4. *Let A be H-selfadjoint. There exists an A-invariant H-Lagrangian subspace if and only if for each real eigenvalue μ of A the following two conditions hold:*

(i) *the number of odd partial multiplicities associated with μ is even,*

(ii) *exactly half of those odd partial multiplicities associated with μ have sign $+1$ corresponding to them in the sign characteristic of (A, H), the other half have sign -1 corresponding to them.*

In particular, if all the partial multiplicities associated with the real eigenvalues of A are even, there does exist an A-invariant H-Lagrangian subspace.

To elucidate what is said in Theorem 11.4, let us return to Theorem 11.2. With the notation employed there, write $s(1), \ldots, s(t)$ for the positive integers such that $\lambda_{s(j)} = \mu$, $j = 1, \ldots, t$. Then the numbers $k_{s(1)}, \ldots, k_{s(t)}$ are the partial multiplicities associated with μ, and the corresponding signs in the sign characteristic of (A, H) are $\varepsilon_{s(1)}, \ldots, \varepsilon_{s(t)}$. Item (i) of the above theorem declares that the number of j for which $k_{s(j)}$ is odd is even, $2p$ say. Suppose $k_{s(r_1)}, \ldots, k_{s(r_{2p})}$ are odd. Then item (ii) of the theorem says that among the signs $\varepsilon_{s(r_1)}, \ldots, \varepsilon_{s(r_{2p})}$ there are p having the value $+1$ and p with the value -1.

We now state a result on the uniqueness of A-invariant H-Lagrangian subspaces. In one direction, this result can be found in Theorem 5.12.4 in [70], the other direction is proved in [122].

Theorem 11.5. *Assume that A is H-selfadjoint. The following two statements are equivalent:*

(i) *There exist unique A-invariant H-Lagrangian subspaces M_u and M_l such that $\sigma(A|_{M_u})$ is in the closed upper half plane and $\sigma(A|_{M_l})$ is in the closed lower half plane;*

(ii) *The real eigenvalues of A have even partial multiplicities, and for each real eigenvalue μ of A the signs in the sign characteristic of the pair (A, H) corresponding to the partial multiplicities associated with μ are all the same.*

In particular, the existence of subspaces M_u and M_l with the properties mentioned in (i) *is guaranteed when A has no real eigenvalues. In this case M_u and M_l are the spectral subspaces of A associated with the part of $\sigma(A)$ lying in the open upper and open lower half plane, respectively.*

11.3 H-dissipative matrices

Next, we turn to another class of matrices. An $n \times n$ matrix is *H-dissipative* if $\frac{1}{2i}(HA - A^*H)$ is nonnegative. It can be shown that the spectral subspace of an

H-dissipative matrix A associated with the part of $\sigma(A)$ lying in the open upper half plane is H-nonnegative, while the spectral subspace corresponding to the part of $\sigma(A)$ lying in the open lower half plan is H-nonpositive.

Theorem 11.6. *Let A be H-dissipative. Then the following statements hold:*

(i) *There exists an A-invariant maximal H-nonnegative subspace M_+ such that $\sigma(A|_{M_+})$ is in the closed upper half plane. Furthermore, any such M_+ contains the spectral subspace of A associated with the part of $\sigma(A)$ lying in the open upper half plane.*

(ii) *There exists an A-invariant maximal H-nonpositive subspace M_- such that $\sigma(A|_{M_-})$ is in the closed lower half plane. Furthermore, any such M_- contains the spectral subspace of A associated with the part of $\sigma(A)$ lying in the open lower half plane.*

The usual proof of this result is quite involved, uses a fixed point argument, and holds in an infinite dimensional setting as well, see [6], [87]. A constructive argument for the finite dimensional case can be found in [129], [137].

The matrix A is said to be *strictly* H-dissipative if $\frac{1}{2i}(HA - A^*H)$ is positive definite. In that case A cannot have real eigenvalues. Hence, for a strictly H-dissipative matrix A, the spectral subspace of A associated with the part of $\sigma(A)$ lying in the open upper half plane is maximal H-positive, and, similarly, the spectral subspace of A corresponding to the part of $\sigma(A)$ contained in the open lower half plane is maximal H-negative.

Notes

The material in this chapter is taken from the books [68] and [70]. For other books in this area, with an emphasis on infinite dimensional spaces, see [87], [25], and [6].

Part V
Riccati equations and factorization

In this part the canonical factorization theorem is presented in a different way using the notion of an angular subspace and Riccati equations. In this case one has to look for angular subspaces that are also spectral subspaces, and the solutions of the Riccati equation must have additional spectral properties. Spectral factorization as well as pseudo-spectral factorization are described in terms of Hermitian solutions of such a Riccati equation. The study of rational matrix functions that take Hermitian values on certain curves, started in the previous part, is continued with an analysis of rational matrix functions that have Hermitian values for which the inertia is independent of the point on the curve. Such functions may still admit a symmetric canonical factorization, provided one allows for a constant Hermitian invertible matrix as a middle factor. A factorization of this type is commonly known as a J-spectral factorization.

This part consists of three chapters. The first chapter (Chapter 12), which has a preliminary character, introduces the (non-symmetric) algebraic Riccati equation and presents the state space canonical factorization theorem in terms of solutions of such an equation. Pseudo-canonical factorization is treated in an analogous way. In the second chapter (Chapter 13) the symmetric algebraic Riccati equation is introduced, and spectral factorization as well as pseudo-spectral factorization are described using such Riccati equations. In the third chapter (Chapter 14) the notion of a J-spectral factorization of a rational matrix function is introduced. Necessary and sufficient conditions for the existence of a such factorization are given, first in terms of invariant subspaces and then in terms of solutions of a corresponding symmetric algebraic Riccati equation. The connection between left and right J-spectral factorization is also studied.

Chapter 12

Canonical factorization and Riccati equations

In this chapter the canonical factorization theorem from Section 7.1 is presented in a different way using the notion of an angular subspace and Riccati equations. In this case one has to look for solutions of the Riccati equation that have additional spectral properties. Section 12.1, which has a preliminary character, deals with angular subspaces, and in particular those that are also spectral subspaces. Section 12.2 deals with the connection between factorization and Riccati equations in general, while Section 12.3 contains the main result. It specifies further the main theorem of the second section for the case of canonical factorization. In Section 12.4, as an application, we solve in state space form the problem of obtaining a right canonical factorization when a left one is given (or reversely).

12.1 Preliminaries on spectral angular subspaces

Let X be a complex Banach space, let X_1 and X_2 be closed subspaces of X, and suppose

$$X = X_1 \dot{+} X_2. \tag{12.1}$$

A closed subspace N of X is said to be *angular* relative to the decomposition (12.1) if $X = X_1 \dot{+} N$. In that case there is a unique operator $R : X_2 \to X_1$, called the *angular operator* for N, such that

$$N = \{Rx + x \mid x \in X_2\} = \operatorname{Im} \begin{bmatrix} R \\ I \end{bmatrix},$$

where I, as always in this section, stands for the identity operator on the appropriate space which can be easily identified from the context (in this case X_2).

Let N be an angular subspace of X relative to (12.1), and let

$$T = \begin{bmatrix} T_{11} & T_{12} \\ T_{21} & T_{22} \end{bmatrix} : X_1 \dot{+} X_2 \to X_1 \dot{+} X_2 \tag{12.2}$$

be an operator on X. We consider the question when N is invariant under T. For this purpose, set

$$E = \begin{bmatrix} I & R \\ 0 & I \end{bmatrix} : X_1 \dot{+} X_2 \to X_1 \dot{+} X_2.$$

This operator is invertible, and maps X_2 in a one-to-one way onto N. It follows that T leaves N invariant if and only if $E^{-1}TE$ leaves X_2 invariant. A direct computation yields

$$E^{-1}TE = \begin{bmatrix} T_{11} - RT_{21} & -RT_{21}R - RT_{22} + T_{11}R + T_{12} \\ T_{21} & T_{22} + T_{21}R \end{bmatrix}. \tag{12.3}$$

This formula shows that $E^{-1}TE$ leaves X_2 invariant if and only if the angular operator R for N satisfies the *algebraic Riccati equation*

$$RT_{21}R + RT_{22} - T_{11}R - T_{12} = 0. \tag{12.4}$$

More precisely, this equation is usually referred to as a nonsymmetric algebraic Riccati equation. In the next chapter we shall encounter symmetric algebraic Riccati equations. The 2×2 operator matrix (12.2) is often referred to as the *Hamiltonian* corresponding to the algebraic Riccati equation (12.4).

Next, let E_2 be the restriction of E to X_2 considered as an operator from X_2 into N. Then E_2 is invertible. In fact, E_2^{-1} is the restriction of E^{-1} to N viewed as an operator from N into X_2. Using (12.3) we see that $E_2^{-1}(T|_N)E_2 = T_{22} + T_{21}R$, and hence $T|_N$ and $T_{22} + T_{21}R$ are similar.

In this section we want additionally that N is a spectral subspace of T. The next proposition shows in terms of the angular operator when this happens.

Proposition 12.1. *Let N be an angular subspace of X relative to the decomposition* (12.1), *and let T be the operator on X given by* (12.2). *Then N is a spectral subspace for T if and only if the angular operator R for N satisfies the algebraic Riccati equation* (12.4) *and*

$$\sigma(T_{11} - RT_{21}) \cap \sigma(T_{22} + T_{21}R) = \emptyset.$$

More precisely the following holds. If $N = \operatorname{Im} P(T;\Gamma)$, where Γ is a Cauchy contour that splits $\sigma(T)$, then $\sigma(T_{22} + T_{21}R)$ is inside Γ and $\sigma(T_{11} - RT_{21})$ is outside Γ. Conversely, if Γ is a Cauchy contour such that $\sigma(T_{22} + T_{21}R)$ is inside Γ and $\sigma(T_{11} - RT_{21})$ is outside Γ, then the spectrum of T does not intersect with Γ and $N = \operatorname{Im} P(T;\Gamma)$.

Proof. We use the operator E introduced before. The operator E is invertible and maps X_2 in a one-to-one way onto N. Since a spectral subspace of T is invariant under T, we may assume without loss of generality that the angular operator R for N satisfies the Riccati equation (12.4). Then formula (12.3) shows that

$$E^{-1}TE = \begin{bmatrix} T_{11} - RT_{21} & 0 \\ T_{21} & T_{22} + T_{21}R \end{bmatrix}. \tag{12.5}$$

Since E maps X_2 in a one-to-one way onto N, the space N is a spectral subspace for T if and only if X_2 is a spectral subspace for $E^{-1}TE$, and we can apply Lemma 3.1 to get the desired result. □

12.2 Angular operators and factorization

In this section we use the concepts introduced in the previous section to bring the factorization theorem (see Section 2.6) for realizations in a different form. The main point is that throughout we work with a fixed decomposition $X = X_1 \dot{+} X_2$ of the state space X of the realization that has to be factorized and the factors are described with respect to this decomposition. In the finite dimensional case this corresponds to working with a fixed coordinate system.

Theorem 12.2. *Let $W(\lambda) = D + C(\lambda I_X - A)^{-1}B$ be a biproper realization with state space X and input-output space Y. Let X_1 and X_2 be closed subspaces of X such that* (12.1) *holds, i.e., $X = X_1 \dot{+} X_2$, let N be a closed subspace of X which is angular relative to this decomposition, so $X = X_1 \dot{+} N$, and denote the corresponding angular operator by R. Assume*

$$A[X_1] \subset X_1, \qquad A^{\times}[N] \subset N, \tag{12.6}$$

and let $D = D_1 D_2$ with D_1 and D_2 invertible operators on Y. Write

$$\begin{aligned} A &= \begin{bmatrix} A_{11} & A_{12} \\ 0 & A_{22} \end{bmatrix} : X_1 \dot{+} X_2 \to X_1 \dot{+} X_2, \\ B &= \begin{bmatrix} B_1 \\ B_2 \end{bmatrix} : Y \to X_1 \dot{+} X_2, \\ C &= \begin{bmatrix} C_1 & C_2 \end{bmatrix} : X_1 \dot{+} X_2 \to Y. \end{aligned}$$

Then R satisfies the algebraic Riccati equation

$$\begin{aligned} RB_2D^{-1}C_1R - R(A_{22} - B_2D^{-1}C_2) + (A_{11} - B_1D^{-1}C_1)R \\ + (A_{12} - B_1D^{-1}C_2) = 0. \end{aligned} \tag{12.7}$$

Introduce the functions W_1 and W_2 via the biproper realizations

$$\begin{aligned} W_1(\lambda) &= D_1 + C_1(\lambda I_{X_1} - A_{11})^{-1}B_1D_2^{-1}, \\ W_2(\lambda) &= D_2 + D_1^{-1}C_2(\lambda I_{X_2} - A_{22})^{-1}B_2. \end{aligned}$$

Then W admits the factorization

$$W(\lambda) = W_1(\lambda)W_2(\lambda), \qquad \lambda \in \rho(A_{11}) \cap \rho(A_{22}) \subset \rho(A).$$

Also put

$$A_{11}^\times = A_{11} - (B_1 - RB_2)D^{-1}C_1, \qquad A_{22}^\times = A_{22} - B_2D^{-1}(C_1R + C_2). \quad (12.8)$$

Then, for $\lambda \in \rho(A_{11}^\times) \cap \rho(A_{22}^\times) \cap \rho(A_{11}) \cap \rho(A_{22})$, the operators $W(\lambda), W_1(\lambda)$ and $W_2(\lambda)$ are invertible, and

$$W(\lambda)^{-1} = W_2(\lambda)^{-1}W_1(\lambda)^{-1},$$

where

$$\begin{aligned} W_1^{-1}(\lambda) &= D_1^{-1} - D_1^{-1}C_1(\lambda I_{X_1} - A_{11}^\times)^{-1}(B_1 - RB_2)D^{-1}, \\ W_2^{-1}(\lambda) &= D_2^{-1} - D^{-1}(C_1R + C_2)(\lambda I_{X_2} - A_{22}^\times)^{-1}B_2D_2^{-1}. \end{aligned}$$

Proof. The first part of the theorem is a direct consequence of the observations presented before Proposition 12.1, applied to $A^\times$. Indeed, let E be the invertible operator

$$E = \begin{bmatrix} I & R \\ 0 & I \end{bmatrix},$$

and write $\widehat{A} = E^{-1}AE$, $\widehat{B} = E^{-1}B$, $\widehat{C} = CE$. Then

$$\begin{aligned} \widehat{A} &= \begin{bmatrix} A_{11} & A_{12} - RA_{22} + A_{11}R \\ 0 & A_{22} \end{bmatrix}, \\ \widehat{B} &= \begin{bmatrix} B_1 - RB_2 \\ B_2 \end{bmatrix}, \\ \widehat{C} &= \begin{bmatrix} C_1 & C_1R + C_2 \end{bmatrix} \end{aligned}$$

and it follows that

$$\widehat{A}^\times = E^{-1}A^\times E = \begin{bmatrix} A_{11}^\times & H \\ -B_2D^{-1}C_1 & A_{22}^\times \end{bmatrix},$$

where $A_{11}^{\times}$ and $A_{22}^{\times}$ are defined by (12.8), and where H is equal to the left-hand side of (12.7). Now E maps X_1 onto X_1 and X_2 onto N. Thus (12.6) implies that

$$\widehat{A}[X_1] \subset X_1, \qquad \widehat{A}^{\times}[X_2] \subset X_2.$$

Hence (12.7) is satisfied.

It remains to prove the factorization $W = W_1W_2$ and to establish the formulas for W_1, W_2 and their inverses. We have $W(\lambda) = D + \widehat{C}(\lambda I - \widehat{A})^{-1}\widehat{B}$. On the other hand, by the product rule for realizations,

$$W_1(\lambda)W_2(\lambda) = D + \widehat{C}(\lambda I - \widetilde{A})^{-1}\widehat{B},$$

where

$$\widetilde{A} = \begin{bmatrix} A_{11} & (B_1 - RB_2)D^{-1}(C_1R + C_2) \\ 0 & A_{22} \end{bmatrix}.$$

It remains to observe that by (12.7)

$$(B_1 - RB_2)D^{-1}(C_1R + C_2) = A_{12} - RA_{22} + A_{11}R.$$

So $W = W_1W_2$. The formulas for the inverses are immediate. □

The next theorem is a symmetric version of Theorem 12.2.

Theorem 12.3. *Let $W(\lambda) = D + C(\lambda I_X - A)^{-1}B$ be a biproper realization with state space X and input-output space Y. Let X_1 and X_2 be closed subspaces of X with $X = X_1 \dot{+} X_2$. Further, let N_1 and N_2 be closed subspaces of X for which*

$$X = X_1 \dot{+} N_2, \qquad X = N_1 \dot{+} X_2,$$

that is, N_2 is angular relative to the decomposition $X = X_1 \dot{+} X_2$ while N_1 is angular relative to $X = X_2 \dot{+} X_1$. Let $R_{12} : X_2 \to X_1$ and $R_{21} : X_1 \to X_2$ be the corresponding angular operators. Assume

$$X = N_1 \dot{+} N_2, \qquad A[N_1] \subset N_1, \qquad A^{\times}[N_2] \subset N_2, \tag{12.9}$$

and let $D = D_1D_2$ with D_1 and D_2 invertible operators on Y. Write

$$A = \begin{bmatrix} A_{11} & A_{12} \\ A_{21} & A_{22} \end{bmatrix} : X_1 \dot{+} X_2 \to X_1 \dot{+} X_2,$$

$$B = \begin{bmatrix} B_1 \\ B_2 \end{bmatrix} : Y \to X_1 \dot{+} X_2,$$

$$C = \begin{bmatrix} C_1 & C_2 \end{bmatrix} : X_1 \dot{+} X_2 \to Y,$$

and put $R_1 = I_{X_1} - R_{12}R_{21}$ *and* $R_2 = I_{X_2} - R_{21}R_{12}$. *Then* $R_1 : X_1 \to X_1$ *and* $R_2 : X_2 \to X_2$ *are invertible. Introduce the functions* W_1 *and* W_2 *via the biproper realizations*

$$W_1(\lambda) = D_1 + (C_1 + C_2R_{21})\big(\lambda I_{X_1} - (A_{11} + A_{12}R_{21})\big)^{-1}R_1^{-1}(B_1 - R_{12}B_2)D_2^{-1},$$

$$W_2(\lambda) = D_2 + D_1^{-1}(C_1R_{12} + C_2)R_2^{-1}\big(\lambda I_{X_2} - (A_{22} - R_{21}A_{12})\big)^{-1}(B_2 - R_{21}B_1).$$

Then W *admits the factorization*

$$W(\lambda) = W_1(\lambda)W_2(\lambda), \qquad \lambda \in \rho(A_{11} + A_{12}R_{21}) \cap \rho(A_{22} - R_{21}A_{12}) \subset \rho(A).$$

Also put

$$A_{11}^{\times} = A_{11} - B_1D^{-1}C_1 - R_{12}A_{21} + R_{12}B_2D^{-1}C_1,$$
$$A_{22}^{\times} = A_{22} - B_2D^{-1}C_2 + A_{21}R_{12} - B_2D^{-1}C_1R_{12}.$$

Then, for $\lambda \in \rho(A_{11}+A_{12}R_{21}) \cap \rho(A_{22}-R_{21}A_{12}) \cap \rho(A_{11}^{\times}) \cap \rho(A_{22}^{\times})$, *the operators* $W(\lambda), W_1(\lambda)$ *and* $W_2(\lambda)$ *are invertible, and*

$$W(\lambda)^{-1} = W_2(\lambda)^{-1}W_1(\lambda)^{-1},$$

where

$$\begin{aligned} W_1^{-1}(\lambda) &= D_1^{-1} - D_1^{-1}(C_1 + C_2R_{21})R_1^{-1}(\lambda I_{X_1} - A_{11}^{\times})^{-1}(B_1 - R_{12}B_2)D^{-1}, \\ W_2^{-1}(\lambda) &= D_2^{-1} - D^{-1}(C_1R_{12} + C_2)(\lambda I_{X_2} - A_{22}^{\times})^{-1}R_2^{-1}(B_2 - R_{21}B_1)D_2^{-1}. \end{aligned}$$

We prepare for the proof of the theorem with a lemma.

Lemma 12.4. *Let* X *be a Banach space, and let* X_1 *and* X_2 *be closed subspaces of* X *with* $X = X_1 \dot{+} X_2$. *Further, let* N_1 *and* N_2 *be closed subspaces of* X *for which*

$$X = X_1 \dot{+} N_2, \qquad X = N_1 \dot{+} X_2,$$

i.e., N_2 *is angular relative to the decomposition* $X = X_1 \dot{+} X_2$ *while* N_1 *is angular relative to* $X = X_2 \dot{+} X_1$. *Let* $R_{12} : X_2 \to X_1$ *and* $R_{21} : X_1 \to X_2$ *be the corresponding angular operators. Then the following statements are equivalent:*

(i) $X = N_1 \dot{+} N_2$;

(ii) $I - R_{21}R_{12}$ *is invertible;*

(iii) $I - R_{12}R_{21}$ *is invertible;*

(iv) $F = \begin{bmatrix} I & R_{12} \\ R_{21} & I \end{bmatrix} : X_1 \dot{+} X_2 \to X_1 \dot{+} X_2$ *is invertible.*

In case the equivalent conditions (i)–(iv) *hold, the projection* P_N *of* X *along* N_1 *onto* N_2 *is given by*

$$P_N = \begin{bmatrix} R_{12} \\ I \end{bmatrix} (I - R_{21}R_{12})^{-1} \begin{bmatrix} -R_{21} & I \end{bmatrix},$$

while the complementary projection $I - P_N$ *can be written as*

$$I - P_N = \begin{bmatrix} I \\ R_{21} \end{bmatrix} (I - R_{12}R_{21})^{-1} \begin{bmatrix} I & -R_{12} \end{bmatrix}.$$

Proof. The equivalence of (ii), (iii) and (iv) is straightforward. Observe that F maps X_1 and X_2 in a one-to-one manner onto N_1 and N_2, respectively. Since $X = X_1 \dot{+} X_2$, it is clear that $X = N_1 \dot{+} N_2$ if and only if F is invertible. So (i) and (iv) are equivalent.

To complete the proof it remains to prove the formula for P_N. Observe that the expression in the right-hand side of the claimed identity for P_N does define a projection. Its image and kernel are given by

$$\operatorname{Im} \begin{bmatrix} R_{12} \\ I \end{bmatrix}, \qquad \operatorname{Im} \begin{bmatrix} I \\ R_{21} \end{bmatrix},$$

respectively, so it is indeed equal to the projection P_N. □

Proof of Theorem 12.3. From Lemma 12.4 we know that the operator

$$F = \begin{bmatrix} I & R_{12} \\ R_{21} & I \end{bmatrix} : X_1 \dot{+} X_2 \to X_1 \dot{+} X_2$$

is invertible. Introduce $\widehat{A} = F^{-1}AF$, $\widehat{B} = F^{-1}B$ and $\widehat{C} = CF$. Then $W(\lambda) = D + \hat{C}(\lambda I - \hat{A})^{-1}\hat{B}$. Note that $\widehat{A}[X_1] \subset X_1$ and $\widehat{A}^\times[X_2] \subset X_2$, where, following standard convention $\widehat{A}^\times = \widehat{A} - \widehat{B}D^{-1}\widehat{C}$, and so $\widehat{A}^\times = F^{-1}A^\times F$. Write

$$\widehat{A} = \begin{bmatrix} \widehat{A}_{11} & \widehat{A}_{12} \\ 0 & \widehat{A}_{22} \end{bmatrix}, \qquad \widehat{B} = \begin{bmatrix} \widehat{B}_1 \\ \widehat{B}_2 \end{bmatrix}, \qquad \widehat{C} = \begin{bmatrix} \widehat{C}_1 & \widehat{C}_2 \end{bmatrix},$$

and put

$$\begin{aligned} \widehat{W}_1(\lambda) &= D_1 + \widehat{C}_1(\lambda - \widehat{A}_{11})^{-1}\widehat{B}_1 D_2^{-1}, \\ \widehat{W}_2(\lambda) &= D_2 + D_1^{-1}\widehat{C}_2(\lambda - \widehat{A}_{22})^{-1}\widehat{B}_2. \end{aligned}$$

Then on $\rho(\widehat{A}_{11}) \cap \rho(\widehat{A}_{22}) \subset \rho(\widehat{A}) = \rho(A)$, the function W is the product of $\widehat{W}_1$ and $\widehat{W}_2$.

The inverse of F is given by

$$F^{-1} = \begin{bmatrix} R_1^{-1} & -R_1^{-1}R_{12} \\ -R_{21}R_1^{-1} & I + R_{21}R_1^{-1}R_{12} \end{bmatrix} : X_1 \dot{+} X_2 \to X_1 \dot{+} X_2.$$

Using this and the expression for F, one easily sees that

$$\begin{aligned} \widehat{A}_{11} &= R_1^{-1}(A_{11} + A_{12}R_{21} - R_{12}A_{21} - R_{12}A_{22}R_{21}), \\ \widehat{B}_1 D_2^{-1} &= R_1^{-1}(B_1 - R_{12}B_2)D_2^{-1}, \\ \widehat{C}_1 &= C_1 + C_2R_{21}. \end{aligned}$$

Now R_{21} satisfies the algebraic Riccati equation

$$R_{21}A_{12}R_{21} + R_{21}A_{11} - A_{22}R_{21} - A_{21} = 0,$$

and it follows that $\widehat{A}_{11} = A_{11} + A_{12}R_{21}$. Thus, for the function $\widehat{W}_1$, we have

$$\begin{aligned} \widehat{W}_1(\lambda) &= D_1 + \widehat{C}_1(\lambda - \widehat{A}_{11})^{-1}\widehat{B}_1 D_2^{-1} \\ &= D_1 + (C_1 + C_2R_{21})\big(\lambda - (A_{11} + A_{12}R_{21})\big)^{-1} R_1^{-1}(B_1 - R_{12}B_2)D_2^{-1}, \end{aligned}$$

as desired.

Next we compute the function $\widehat{W}_2$. Using the alternative formula

$$F^{-1} = \begin{bmatrix} I + R_{12}R_2^{-1}R_{21} & -R_{12}R_2^{-1} \\ -R_2^{-1}R_{21} & R_2^{-1} \end{bmatrix} : X_1 \dot{+} X_2 \to X_1 \dot{+} X_2$$

for the inverse of F, we obtain

$$\begin{aligned} \widehat{A}_{22} &= R_2^{-1}(A_{22} - R_{21}A_{12})R_2^{-1}, \\ \widehat{B}_2 &= R_2^{-1}(B_2 - R_{21}B_1), \\ D_1^{-1}\widehat{C}_1 &= D_1^{-1}(C_1R_{12} + C_2). \end{aligned}$$

Hence, for the function $\widehat{W}_2$ we get

$$\begin{aligned} \widehat{W}_2(\lambda) &= D_2 + D_1^{-1}\widehat{C}_2(\lambda - \widehat{A}_{22})^{-1}\widehat{B}_2 \\ &= D_2 + D_1^{-1}(C_1R_{12} + C_2)R_2^{-1}\big(\lambda - (A_{22} - R_{21}A_{12})\big)^{-1}(B_1 - R_{12}B_2)D_2^{-1}, \end{aligned}$$

again as desired.

This proves that the factorization claimed in the theorem holds on

$$\rho(A_{11} + A_{12}R_{21}) \cap \rho(A_{22} - R_{21}A_{12})$$

which is a subset of $\rho(A)$. What remains to be done is to deduce the formulas for the inverses. But this amounts to repeating the work with W replaced by W^{-1}. In doing so, one employs the Riccati equation

$$R_{12}(A_{21} - B_2D^{-1}C_1)R_{12} + R_{12}(A_{22} - B_2D^{-1}C_2)$$
$$-(A_{11} - B_1D^{-1}C_1)R_{12} - (A_{12} - B_1D^{-1}C_2) = 0$$

for R_{12} instead of the one for R_{21} used above. The details are omitted. □

12.3 Riccati equations and canonical factorization

In this section Theorem 12.2 is specified further for the case of canonical factorization. As usual, Γ is a Cauchy contour in the complex plane, F_+ is its interior domain, and F_- its exterior domain (infinity included).

Theorem 12.5. *Let $W(\lambda) = D + C(\lambda I_X - A)^{-1}B$ be a biproper realization with state space X and input-output space Y. Assume that the spectrum of A does not intersect Γ. Put $X_1 = \operatorname{Im} P(A;\Gamma)$ and let X_2 be a closed subspace of X such that $X = X_1 \dot{+} X_2$, so*

$$X = \operatorname{Im} P(A;\Gamma) \dot{+} X_2.$$

Let $D = D_1D_2$ with D_1 and D_2 invertible operators on Y, and write

$$A = \begin{bmatrix} A_{11} & A_{12} \\ 0 & A_{22} \end{bmatrix} : X_1 \dot{+} X_2 \to X_1 \dot{+} X_2,$$

$$B = \begin{bmatrix} B_1 \\ B_2 \end{bmatrix} : Y \to X_1 \dot{+} X_2,$$

$$C = \begin{bmatrix} C_1 & C_2 \end{bmatrix} : X_1 \dot{+} X_2 \to Y.$$

Then W admits a right canonical factorization with respect to Γ if and only if the Riccati equation

$$RB_2D^{-1}C_1R - R(A_{22} - B_2D^{-1}C_2) + (A_{11} - B_1D^{-1}C_1)R + (A_{12} - B_1D^{-1}C_2) = 0 \tag{12.10}$$

has a (unique) solution R satisfying the constraints

$$\sigma\big(A_{11} - (B_1 - RB_2)D^{-1}C_1\big) \subset F_+, \tag{12.11}$$

$$\sigma\big(A_{22} - B_2D^{-1}(C_1R + C_2)\big) \subset F_-. \tag{12.12}$$

In that case a right canonical factorization $W(\lambda) = W_-(\lambda)W_+(\lambda)$ of W with respect to Γ is obtained by taking

$$\begin{aligned} W_-(\lambda) &= D_1 + C_1(\lambda - A_{11})^{-1}(B_1 - RB_2)D_2^{-1}, \\ W_+(\lambda) &= D_2 + D_1^{-1}(C_1R + C_2)(\lambda - A_{22})^{-1}B_2. \end{aligned}$$

Moreover, the inverses of W_- and W_+ are given by

$$\begin{aligned} W_-^{-1}(\lambda) &= D_1^{-1} - D_1^{-1}C_1(\lambda - A_{11}^\times)^{-1}(B_1 - RB_2)D^{-1}, \\ W_+^{-1}(\lambda) &= D_2^{-1} - D^{-1}(C_1R + C_2)(\lambda - A_{22}^\times)^{-1}B_2D_2^{-1}, \end{aligned}$$

where

$$A_{11}^\times = A_{11} - (B_1 - RB_2)D^{-1}C_1, \qquad A_{22}^\times = A_{22} - B_2D^{-1}(C_1R + C_2).$$

With the appropriate modifications, the theorem also holds for certain contours in the Riemann sphere. For instance, if for Γ one takes the (extended) imaginary axis, one has to take for F_+ the open left half plane and for F_- the open right half plane. For left canonical factorizations analogous results hold: just interchange the roles of inner and outer domains (see the comment after Theorem 3.2).

Proof. The subspace $X_1 = \operatorname{Im} P(A;\Gamma)$ is invariant under A, and hence the zero entry in the left lower corner of the operator matrix representation of A is justified. Furthermore $\sigma(A_{11}) \subset F_+$ and $\sigma(A_{22}) \subset F_-$.

Next note that relative to the decomposition $X = X_1 \dot{+} X_2$ we have

$$A^\times = A - BD^{-1}C = \begin{bmatrix} A_{11} - B_1D^{-1}C_1 & A_{12} - B_1D^{-1}C_2 \\ -B_2D^{-1}C_1 & A_{22} - B_2D^{-1}C_2 \end{bmatrix}.$$

Thus $-A^\times$ is precisely the Hamiltonian of the Riccati equation (12.10).

Assume that W admits a right canonical factorization with respect to Γ. Then, in particular, $W(\lambda)$ is invertible for each $\lambda \in \Gamma$; hence, by Theorem 2.4, the spectrum of the operator $A^\times$ does not intersect Γ. Thus we can use Theorem 7.1 to show that $N = \operatorname{Ker} P(A^\times;\Gamma)$ is an angular subspace for the decomposition $X = X_1 \dot{+} X_2$. Let R be the corresponding angular operator. Since $A^\times$ leaves N invariant, we know that R satisfies the Riccati equation

$$\begin{aligned} -RB_2D^{-1}C_1R + R(A_{22} - B_2D^{-1}C_2) - (A_{11} - B_1D^{-1}C_1)R \qquad (12.13) \\ -(A_{12} - B_1D^{-1}C_2) = 0, \end{aligned}$$

which is equivalent to (12.10). Now Proposition 12.1, applied to $A^\times$ and with the roles of the interior and exterior domain of the contour Γ being reversed, shows that (12.11) and (12.12) are fulfilled.

Conversely, let R be a solution of the Riccati equation (12.10) for which (12.11) and (12.12) are satisfied. Thus R satisfies the Riccati equation (12.13) which has $A^\times$ as its Hamiltonian. Hence the corresponding angular subspace N is invariant under $A^\times$. Next we again use Proposition 12.1 with $T = A^\times$ and with the roles of the interior and exterior domain of the contour Γ being reversed. This yields that the spectrum of $A^\times$ does not intersect Γ and that $N = \operatorname{Ker} P(A^\times;\Gamma)$. Since N is an angular subspace of X relative to $X = X_1 \dot{+} X_2$, the latter implies that $X = \operatorname{Im} P(A;\Gamma) \dot{+} \operatorname{Ker} P(A^\times;\Gamma)$. But then Theorem 3.2 implies that W admits a right canonical factorization with respect to the contour Γ.

To show uniqueness of the solution R of (12.10) for which the spectral inclusions (12.11) and (12.12) are satisfied, it suffices to note that these spectral inclusions imply that $N = \operatorname{Ker} P(A^\times;\Gamma)$. Indeed, in that case the angular operator R for N relative to $X = X_1 \dot{+} X_2$ is uniquely determined.

It remains to get the formulas for the factors. First note that Theorem 12.2 shows that $W(\lambda) = W_-(\lambda)W_+(\lambda)$ with the factors $W_-(\lambda)$, $W_+(\lambda)$ and their inverses being of the desired form. The spectral properties of A_{11} and A_{22}, together with those of $A_{11}^\times$ and $A_{22}^\times$, show that the factorization $W(\lambda) = W_-(\lambda)W_+(\lambda)$ is a right canonical factorization with respect to Γ. □

12.4 Left versus right canonical factorization

In this section we answer the following question: if a rational matrix function W admits a left canonical factorization, under what conditions does it also have a right canonical factorization? And, if so, how can the right factorization be obtained from the left one?

Our starting point is a given biproper operator function W, a Cauchy contour Γ, and a left canonical factorization

$$W(\lambda) = Y_+(\lambda)Y_-(\lambda), \quad \lambda \in \Gamma. \tag{12.14}$$

The biproper factors Y_+ and Y_- are given in terms of realizations, that is,

$$Y_+(\lambda) = D_+ + C_+(\lambda I_{X_+} - A_+)^{-1}B_+, \tag{12.15}$$

$$Y_-(\lambda) = D_- + C_-(\lambda I_{X_-} - A_-)^{-1}B_-. \tag{12.16}$$

We are looking for a right canonical factorization $W(\lambda) = W_-(\lambda)W_+(\lambda)$. The key idea for solving this problem is the following: combine the realizations of Y_+ and Y_- into a realization for W using the product rule for realizations, then apply the canonical factorization theorem (Theorem 7.1) to see if a right canonical factorization exists and, if so, produce formulas for the factors.

As before the interior of Γ will be denoted by F_+, the exterior by F_-. We (may and) shall assume that the operators in the realizations are chosen in such a way that the operators D_+ and D_- are invertible, the spectra of the operators

A_+ and $A_+^\times = A_+ - B_+D_+^{-1}C_+$ are contained in F_-, and those of A_- and $A_-^\times = A_- - B_-D_-^{-1}C_-$ in F_+. Then, in particular, the spectra of A_- and A_+ are disjoint and the Lyapunov equation

$$A_+Z - ZA_- = -B_+C_- \tag{12.17}$$

has a unique solution $Z : X_- \to X_+$(see Section I.4 in [51]). Similarly, the Lyapunov equation

$$A_-^\times Z - ZA_+^\times = B_-D_-^{-1}D_+^{-1}C_+ \tag{12.18}$$

has a unique solution $Z : X_+ \to X_-$. These facts are used in the following theorem and its proof.

Theorem 12.6. *Let $W(\lambda) = Y_+(\lambda)Y_-(\lambda)$ be a left canonical factorization of W with respect to the Cauchy contour Γ, and let the factors be given by* (12.15) *and* (12.16). *Let $Q : X_- \to X_+$ and $P : X_+ \to X_-$ be the unique solutions of the Lyapunov equations* (12.17) *and* (12.18), *respectively, that is,*

$$A_+Q - QA_- = -B_+C_-, \qquad A_-^\times P - PA_+^\times = B_-D_-^{-1}D_+^{-1}C_+. \tag{12.19}$$

Then W has a right canonical factorization $W(\lambda) = W_-(\lambda)W_+(\lambda)$ with respect to Γ if and only if $I_{X_+} - QP$ is invertible, or, which amounts to the same, $I_{X_-} - PQ$ is invertible. In that case, on the appropriate domains, the factors W_- and W_+, and their inverses W_-^{-1} and W_+^{-1}, are given by

$$\begin{aligned}
W_-(\lambda) &= D_+ + (D_+C_- + C_+Q)(\lambda I_{X_-} - A_-)^{-1} \\
&\qquad \cdot (I_{X_-} - PQ)^{-1}(B_-D_-^{-1} - PB_+), \\
W_+(\lambda) &= D_- + (D_+^{-1}C_+ + C_-P)(I_{X_+} - QP)^{-1} \\
&\qquad \cdot (\lambda I_{X_+} - A_+)^{-1}(B_+D_- - QB_-), \\
W_-^{-1}(\lambda) &= D_+^{-1} - D_+^{-1}(D_+C_- + C_+Q)(I_{X_-} - PQ)^{-1} \\
&\qquad \cdot (\lambda I_{X_-} - A_-^\times)^{-1}(B_-D_-^{-1} - PB_+)D_+^{-1}, \\
W_+^{-1}(\lambda) &= D_-^{-1} - D_-^{-1}(D_+^{-1}C_+ + C_-P)(\lambda I_{X_+} - A_+^\times)^{-1} \\
&\qquad \cdot (I_{X_+} - QP)^{-1}(B_+D_- - QB_-)D_-^{-1}.
\end{aligned}$$

Proof. First we use (12.15) and (12.16) to obtain a realization for W given in the form (12.14). So we write $X = X_- \dotplus X_+$ and define $A : X \to X$ by

$$A = \begin{bmatrix} A_- & 0 \\ B_+C_- & A_+ \end{bmatrix} : X_- \dotplus X_+ \to X_- \dotplus X_+.$$

Then, by the product rule (see Section 2.5),

$$W(\lambda) = D_+D_- + \begin{bmatrix} D_+C_- & C_+ \end{bmatrix} (\lambda I_X - A)^{-1} \begin{bmatrix} B_- \\ B_+D_- \end{bmatrix}.$$

The associate main operator of this realization is

$$A^\times = \begin{bmatrix} A_-^\times & -B_-D_-^{-1}D_+^{-1}C_+ \\ 0 & A_+^\times \end{bmatrix} : X_-\dot{+}X_+ \to X_-\dot{+}X_+.$$

The spectra of A and $A^\times$ do not intersect Γ. Put

$$M = \operatorname{Im} P(A;\Gamma), \qquad M^\times = \operatorname{Ker} P(A^\times;\Gamma).$$

In order that W admits a right canonical factorization with respect to Γ it is necessary and sufficient (see Theorem 7.1) that $X = M\dot{+}M^\times$.

From the matrix representation of A given above we see that $\operatorname{Ker} P(A;\Gamma)$ coincides with X_+. So $X = M\dot{+}X_+$, and hence for some $Z : X_- \to X_+$ we have

$$M = \operatorname{Im} \begin{bmatrix} I \\ Z \end{bmatrix}.$$

The fact that M is invariant under A now amounts to (12.17). But then the operator Z must be equal to Q. In a similar way one shows that

$$M^\times = \operatorname{Im} \begin{bmatrix} P \\ I \end{bmatrix},$$

where $P : X_+ \to X_-$ is the unique solution of (12.18). From Lemma 12.4 we know that the condition $X = M^\times\dot{+}M$ is equivalent to the invertibility of the matrix

$$\begin{bmatrix} I & P \\ Q & I \end{bmatrix},$$

which, in turn, is equivalent to the invertibility of $I - QP$ or, which amounts to the same, the invertibility of $I - PQ$. This proves the first part of the theorem.

The formulas for the factors follow by applying Theorem 12.3 with X_-, X_+, M, $M^\times$, Q and P in the role of X_1, X_2, N_1, N_2, R_{21} and R_{12}, respectively. □

With the obvious modifications, Theorem 12.6 holds true for canonical factorizations with respect to the usual contours in the Riemann sphere (real line and imaginary axis).

Notes

This chapter is a rewritten and enriched version of Chapter 5 in [11]. Theorem 12.5 in Section 12.3 seems to be new. The material in the final section can be found in [8]. The notion of an angular operator is standard in operator theory and goes back to [101]. The theory of Riccati equations is important in system theory; see, e.g., the text books [94], [33]. For more details on this subject we also refer to the monograph [106] and to Section 1.6 in [69].

Chapter 13

The symmetric algebraic Riccati equation

As we know from the previous part there is an intimate connection between canonical factorization and Riccati equations. In this chapter this connection is developed further for the case when the rational matrix functions involved have Hermitian values on the imaginary axis. In this case the corresponding Riccati equation has additional symmetry properties too.

The chapter consists of three sections. In Section 13.1 we discuss two special cases, which both lead to symmetric algebraic Riccati equations of a special type. In a somewhat more general form, this symmetric version of the algebraic Riccati equation is studied in Section 13.2, with special attention for stabilizing solutions. The study is completed in Section 13.3 where we consider Hermitian solutions of the symmetric algebraic Riccati equation and related pseudo-spectral factorizations.

13.1 Spectral factorization and Riccati equations

In this section we present two illustrative special cases of spectral factorization. In both cases the corresponding Riccati equations are symmetric.

For our first case, the starting point is a rational $m \times m$ matrix function G given in realized form $G(\lambda) = I_m + C(\lambda I_n - A)^{-1}B$, with $\sigma(A)$ in the open left half plane, and we consider the product $W(\lambda) = G(-\bar{\lambda})^* G(\lambda)$. Clearly W is a nonnegative rational $m \times m$ matrix function on the imaginary axis. We shall assume additionally that $G(\lambda)$ is invertible for each $\lambda \in i\mathbb{R}$, which in the present situation is equivalent to the requirement that $A^\times = A - BC$ has no eigenvalue on $i\mathbb{R}$. The fact that $G(\lambda)$ is invertible for each $\lambda \in i\mathbb{R}$ means that W is positive definite on $\mathbb{R}$ and, as we shall see, Theorem 9.5 can be applied to show that the function W admits a left spectral factorization with respect to $i\mathbb{R}$. We shall use

Theorem 12.5 to obtain such a factorization explicitly in terms of the matrices A, B and C appearing in the realization of G.

Theorem 13.1. *Let $G(\lambda) = I_m + C(\lambda I_n - A)^{-1}B$ be a realization of a rational $m \times m$ matrix function G such that A has all its eigenvalues in the open left half plane. Put $A^\times = A - BC$, and assume that $A^\times$ has no eigenvalue on $i\mathbb{R}$. Then the Riccati equation*

$$-PBB^*P + PA^\times + (A^\times)^*P = 0 \tag{13.1}$$

has a unique Hermitian solution P such that $A^\times - BB^*P$ has all its eigenvalues in the left half plane. Furthermore, the rational matrix function $W(\lambda) = G(-\bar{\lambda})^*G(\lambda)$ admits a left spectral factorization of W with respect to the imaginary axis. In fact, $W(\lambda) = L_-(-\bar{\lambda})^*L_-(\lambda)$ with

$$L_-(\lambda) = I_m + (C + B^*P)(\lambda I_n - A)^{-1}B,$$

is such a factorization.

By Theorem 2.4, the inverse L_-^{-1} of the spectral factor L_- in the above theorem is given by

$$L_-^{-1}(\lambda) = I_m - (C + B^*P)(\lambda I_n - A^\times + BB^*P)^{-1}B.$$

In comparable situations later on in the book, where obtaining descriptions of inverses of factors would involve only a routine application of Theorem 2.4, we will refrain from giving the expressions.

Proof. We split the proof into two parts. In the first part we show that equation (13.1) has a unique Hermitian solution P such that $A^\times - BB^*P$ has all its eigenvalues in the left half plane.

Part 1. From the given realization of G we get $G(-\bar{\lambda})^* = I_m - B^*(\lambda I_n + A^*)^{-1}C^*$. Now apply the product rule from Section 2.5). This gives

$$W(\lambda) = I + \begin{bmatrix} -B^* & C \end{bmatrix} \left(\lambda - \begin{bmatrix} -A^* & C^*C \\ 0 & A \end{bmatrix}\right)^{-1} \begin{bmatrix} C^* \\ B \end{bmatrix}. \tag{13.2}$$

It is easy to check that the hypotheses of Theorem 9.5 are satisfied with the skew-Hermitian matrix H given by

$$H = \begin{bmatrix} 0 & -I_n \\ I_n & 0 \end{bmatrix}. \tag{13.3}$$

Hence W admits both a left and a right spectral factorization with respect to $i\mathbb{R}$. In particular W admits both a left and a right canonical factorization with respect to the imaginary axis.

Put $F_- = \mathbb{C}_{\text{left}}$ and $F_+ = \mathbb{C}_{\text{right}}$, where $\mathbb{C}_{\text{left}}$ and $\mathbb{C}_{\text{right}}$ are the open left and right half planes, respectively. By hypothesis $\sigma(A) \subset \mathbb{C}_{\text{left}}$. So $\sigma(-A^*) \subset \mathbb{C}_{\text{right}}$. Thus the realization of W in (13.2) is of the form required in Theorem 12.5, and the Riccati equation (12.10) in the theorem reduces here to

$$-RBB^*R - RA^\times - (A^\times)^*R = 0, \tag{13.4}$$

where, as usual, $A^\times = A - BC$. Since W admits a left canonical factorization with respect to the imaginary axis, (the appropriate version of) Theorem 12.5 (see the remark made between the theorem and its proof) shows that (13.4) has a unique solution R satisfying

$$\sigma\big(A^\times + BB^*R\big) \subset \mathbb{C}_{\text{left}}, \qquad \sigma\big((A^\times)^* + RBB^*\big) \subset \mathbb{C}_{\text{left}}. \tag{13.5}$$

Here we used that $\sigma\big(-(A^\times)^* - RBB^*\big) \subset \mathbb{C}_{\text{right}}$ is equivalent to the second inclusion in (13.5). Taking adjoints in (13.4) and (13.5) we see that (13.4) and (13.5) remain true if R is replaced by R^*. But then the uniqueness of the solution implies $R = R^*$. Note that for $R = R^*$ the two inclusions in (13.5) are equivalent. Thus we see that (13.4) has a unique Hermitian solution R satisfying the first inclusion in (13.5). When R is replaced $-P$, equation (13.4) transforms into equation (13.1). Thus (13.1) has a unique Hermitian solution P satisfying $\sigma(A^\times - BB^*P) \subset \mathbb{C}_{\text{left}}$.

Part 2. Theorem 12.5 also yields a canonical factorization of the rational matrix function given by (13.2). In fact, such a factorization is given by $W(\lambda) = W_-(\lambda)W_+(\lambda)$ where the factors and their inverses are given by

$$\begin{aligned}
W_-(\lambda) &= I - B^*(\lambda + A^*)^{-1}(C^* + PB),\\
W_+(\lambda) &= I + (B^*P + C)(\lambda - A)^{-1}B,\\
W_-^{-1}(\lambda) &= I + B^*\big(\lambda + (A^\times)^* - PBB^*\big)^{-1}(C^* + PB),\\
W_+^{-1}(\lambda) &= I - (B^*P + C)\big(\lambda - A^\times + BB^*P\big)^{-1}B.
\end{aligned}$$

Comparing the first two expressions we see that $W_-(\lambda) = W_+(-\bar{\lambda})^*$, and hence the factorization $W(\lambda) = W_-(\lambda)W_+(\lambda)$ is a left spectral factorization with respect to $i\mathbb{R}$. Now put $L_- = W_+$ to arrive at the desired result. □

For our second special case, we assume that W is proper, Hermitian on the imaginary axis, and has no poles there. This implies that W can be written in the form

$$W(\lambda) = D + C(\lambda I_n - A)^{-1}B - B^*(\lambda I_n + A^*)^{-1}C^*, \tag{13.6}$$

where D is Hermitian and A has all its eigenvalues in the open left half plane. On the basis of this representation we shall prove the following theorem.

Theorem 13.2. *Let the rational $m \times m$ function W be given by* (13.6), *where D is positive definite and A has all its eigenvalues in the open left half plane. Assume* *additionally that W has no zeros on the imaginary axis, and put $A^\times = A - BD^{-1}C$. Then the Riccati equations*

$$PBD^{-1}B^*P - PA^\times - (A^\times)^*P + C^*D^{-1}C = 0, \tag{13.7}$$

$$QC^*D^{-1}CQ - Q(A^\times)^* - A^\times Q + BD^{-1}B^* = 0 \tag{13.8}$$

have unique Hermitian solutions P and Q that satisfy

$$\sigma\big(A^\times - BD^{-1}B^*P\big) \subset \mathbb{C}_{\text{left}}, \qquad \sigma\big((A^\times)^* - C^*D^{-1}CQ\big) \subset \mathbb{C}_{\text{left}}. \tag{13.9}$$

Furthermore, with respect to the imaginary axis, W admits left and right spectral factorizations,

$$W(\lambda) = L_-(-\bar{\lambda})^*L_-(\lambda), \qquad W(\lambda) = L_+(-\bar{\lambda})^*L_+(\lambda), \tag{13.10}$$

respectively, with the factors L_- and L_+ being given by

$$L_-(\lambda) = D^{1/2} + D^{-1/2}(C + B^*P)(\lambda I_n - A)^{-1}B, \tag{13.11}$$

$$L_+(\lambda) = D^{1/2} - D^{-1/2}(CQ + B^*)(\lambda I_n + A^*)^{-1}C^*. \tag{13.12}$$

Proof. We split the proof into four parts. In the first three parts the attention is focussed on equation (13.7) and the first parts of (13.9) and (13.10).

Part 1. From (13.6) we get

$$W(\lambda) = D + \begin{bmatrix} B^* & C \end{bmatrix}\left(\lambda - \begin{bmatrix} -A^* & 0 \\ 0 & A \end{bmatrix}\right)^{-1}\begin{bmatrix} -C^* \\ B \end{bmatrix}.$$

The main matrix of this realization has no pure imaginary eigenvalues. This follows from the assumption on the eigenvalues of A. Clearly W is selfadjoint on the imaginary axis and takes invertible values there. It follows that for $\lambda \in i\mathbb{R}$ the signature of the matrix $W(\lambda)$, that is, the difference between the number of positive and negative eigenvalues of $W(\lambda)$, does not depend on λ. As $W(\infty) = D$ is positive definite, we obtain that $W(\lambda)$ is positive definite for $\lambda \in i\mathbb{R}$. So the hypotheses of Theorem 9.5 are satisfied with the skew-Hermitian matrix H given by (13.3). Hence W admits both a left and a right spectral factorization with respect to $i\mathbb{R}$. To get the formulas for the factors we will apply (the appropriate version of) Theorem 12.5 (see the remark made between the theorem and its proof)

Part 2. For the case considered here the Riccati equation (12.10) in Theorem 12.5 has the form

$$RBD^{-1}B^*R - RA^\times - (A^\times)^*R + C^*D^{-1}C = 0.$$

This is precisely equation (13.7) with R in place of P. Since W admits a left canonical factorization with respect to the imaginary axis, Theorem 12.5 tells us that equation (13.7) has a unique solution P satisfying

$$\sigma\big(-(A^\times)^* + PBD^{-1}B^*\big) \subset \mathbb{C}_{\text{right}}, \qquad \sigma\big(A^\times - BD^{-1}B^*P\big) \subset \mathbb{C}_{\text{left}}. \tag{13.13}$$

Using the symmetry properties in (13.7) and (13.13), we see that P^* is also a solution of (13.7) satisfying (13.13). Because of the uniqueness of P, we have $P = P^*$, and hence P is a Hermitian solution of (13.7) satisfying the first inclusion in (13.9). On the other hand, if $\widetilde{P}$ is a Hermitian solution of (13.7) satisfying the first inclusion in (13.9), then $\widetilde{P}$ actually satisfies both inclusions in (13.13), and hence $P = \widetilde{P}$.

Part 3. Next, we derive the first factorization in (13.10). By Theorem 12.5 the matrix function W admits a right canonical factorization, $W(\lambda) = W_-(\lambda)W_+(\lambda)$, with respect to $i\mathbb{R}$. The factors in this factorization are given by

$$\begin{aligned} W_-(\lambda) &= D^{1/2} + B^*(\lambda + A^*)^{-1}(-C^* - PB)D^{-1/2}, \\ W_+(\lambda) &= D^{1/2} + D^{-1/2}(B^*P + C)(\lambda - A)^{-1}B. \end{aligned}$$

Put $L_-(\lambda) = W_+(\lambda)$. Then $L_-(-\bar{\lambda})^* = W_-(\lambda)$, and hence the first identity in (13.10) holds. Moreover, the function $L_-(\lambda)$ is given by (13.11). Since the factorization $W(\lambda) = W_-(\lambda)W_+(\lambda)$ is a canonical one, we also know that the factorization $W(\lambda) = L_-(-\bar{\lambda})^*L_-(\lambda)$ is a left spectral factorization of W with respect to $i\mathbb{R}$.

Part 4. Finally, to get the corresponding result for the Riccati equation (13.8) and the second factorization in (13.10), we apply the results obtained in the preceding paragraphs to $V(\lambda) = W(-\lambda)$, that is, to

$$V(\lambda) = D + B^*(\lambda - A^*)^{-1}C^* - C(\lambda + A)^{-1}B.$$

Note that A^* has all its eigenvalues in $\mathbb{C}_{\text{left}}$. Furthermore, if the function V admits a left spectral factorization with respect to the imaginary axis, $V(\lambda) = K_-(-\bar{\lambda})^*K_-(\lambda)$ say, then $W(\lambda) = K_-(\bar{\lambda})^*K_-(-\lambda)$ is a right spectral factorization of W with respect to $i\mathbb{R}$. □

We conclude this section with a few remarks about the Hermitian solutions of the Riccati equations appearing in Theorem 13.2. Let W be given by (13.6) with D positive definite.

First we show that any Hermitian solution P of (13.7) is invertible whenever the pair (C, A) is observable. Suppose $Px = 0$. Since P is Hermitian, we also have $x^*P = 0$. Then (13.7) yields $x^*C^*D^{-1}Cx = 0$. As D is positive definite, this gives $Cx = 0$. But then, again using (13.7), we get $PA^\times = 0$, and hence $PAx = PA^\times x + PBD^{-1}Cx = 0$. So $\operatorname{Ker} P$ is A-invariant and is contained in $\operatorname{Ker} C$. Hence $\operatorname{Ker} P$ is contained in $\operatorname{Ker}(C|A)$, and thus $\operatorname{Ker} P = \{0\}$ when $\operatorname{Ker}(C|A) = \{0\}$.

In a similar way one shows that controllability of the pair (A, B) implies that every Hermitian solution Q of (13.8) is invertible. Thus, if the realization $C(\lambda - A)^{-1}B$ is minimal, then the Hermitian solutions of the Riccati equations (13.7) and (13.8) are automatically invertible.

Now let P be an invertible Hermitian solution of (13.7). Multiplying (13.7) from both sides by P^{-1} shows that $Q = P^{-1}$ is an invertible Hermitian solution of

(13.8). The converse is also true, that is, if Q is an invertible Hermitian solution of (13.8), then $P = Q^{-1}$ is an invertible Hermitian solution of (13.7). Thus the map $P \mapsto Q = P^{-1}$ provides a one-to-one correspondence between the invertible Hermitian solutions P of (13.7) and the invertible Hermitian solutions Q of (13.8). Furthermore, in this case (with $Q = P^{-1}$) we have

$$\sigma\big(A^\times - BD^{-1}B^*P\big) = \sigma\big(-(A^\times)^* + C^*D^{-1}CQ\big).$$

Indeed, by (13.7) we have $PA^\times - PBD^{-1}B^*P = -(A^\times)^*P + C^*D^{-1}C$, and so

$$\begin{aligned} A^\times - BD^{-1}B^*P &= P^{-1}(PA^\times - PBD^{-1}B^*P) \\ &= P^{-1}\big(-(A^\times)^*P + C^*D^{-1}C\big) \\ &= P^{-1}\big(-(A^\times)^* + C^*D^{-1}CP^{-1}\big)P \\ &= P^{-1}\big(-(A^\times)^* + C^*D^{-1}CQ\big)P. \end{aligned}$$

In particular, if the eigenvalues of $A^\times - BD^{-1}B^*$ are in the open left half plane, then those of $(A^\times)^* - C^*D^{-1}CQ$ are in the open right half plane. Comparing this with (13.9), we see that in Theorem 13.2 the matrix Q is not the inverse of the matrix P.

13.2 Stabilizing solutions

The equations (13.1) and (13.7) are special cases of the general *symmetric algebraic Riccati equation*

$$-PBR^{-1}B^*P + PA + A^*P + Q = 0, \tag{13.14}$$

with R and Q selfadjoint, R invertible. Note that the Hamiltonian (see Section 12.1) corresponding to equation (13.14) is the 2×2 block matrix

$$T = \begin{bmatrix} -A^* & -Q \\ -BR^{-1}B^* & A \end{bmatrix}. \tag{13.15}$$

We shall assume throughout this section that A is an $n \times n$ matrix, B an $n \times m$ matrix, Q a selfadjoint $n \times n$ matrix, and R a positive definite $m \times m$ matrix. Thus the Hamiltonian T can be viewed as an operator on $\mathbb{C}^{2n} = \mathbb{C}^n \oplus \mathbb{C}^n$.

We shall also assume that the pair (A, B) is *stabilizable*. The latter means that there there exists an $m \times n$ matrix F such that $A - BF$ has all its eigenvalues in the open left half plane.

Equation (13.14) plays an important role in optimal control theory, where one is mainly interested in stabilizing solutions P. A solution P of (13.14) is said to be *i*$\mathbb{R}$-*stabilizing*, or simply *stabilizing* when no confusion is possible, if the matrix $A - BR^{-1}B^*P$ has all its eigenvalues in the open left half plane. In order that such a solution exists the pair (A, B) has to be stabilizable. In general, however, this condition is not sufficient. An additional condition on the eigenvalues of the Hamiltonian T is required.

Theorem 13.3. *Consider the symmetric algebraic Riccati equation* (13.14) *with R positive definite and Q selfadjoint. Then the following two statements are equivalent:*

(i) *There exists an $i\mathbb{R}$-stabilizing solution of* (13.14)*;*

(ii) *The pair (A, B) is stabilizable and the Hamiltonian T given by* (13.15) *does not have pure imaginary eigenvalues.*

Moreover, if (13.14) *has an $i\mathbb{R}$-stabilizing solution, then it is unique and Hermitian.*

The proof of the implication (i) $\Rightarrow$ (ii) and of the final statement of the theorem concerning the uniqueness of the $i\mathbb{R}$-stabilizing solution do not require R to be positive definite; selfadjointness and invertibility of R are enough.

It will be convenient first to prove a lemma using a somewhat more general setting. For this purpose we return to the general algebraic Riccati equation which was studied in Chapter 12:

$$XT_{21}X + XT_{22} - T_{11}X - T_{12} = 0. \tag{13.16}$$

Taking

$$T_{21} = -BR^{-1}B^*, \quad T_{22} = A, \quad T_{11} = -A^*, \quad T_{12} = -Q, \tag{13.17}$$

and setting $X = P$, we see that we arrive at (13.14). Note that in this case

$$T_{22} = -T_{11}^*, \qquad T_{12}^* = T_{12}, \qquad T_{21}^* = T_{21}. \tag{13.18}$$

In this symmetric case the coefficients T_{ij}, $1 \le i,\, j \le 2$, are square matrices, all of the same order, n say.

In what follows H will denote the Hamiltonian of (13.16), that is, $H = \begin{bmatrix} T_{ij} \end{bmatrix}_{i,j=1}^{2}$. Note that the identities in (13.18) hold if and only if

$$JH = -H^*J, \quad \text{where } J = \begin{bmatrix} 0 & I_n \\ -I_n & 0 \end{bmatrix}. \tag{13.19}$$

We are now ready to state the lemma.

Lemma 13.4. *Let X be a solution of* (13.16) *such that $\sigma(T_{22} + T_{21}X) \subset \mathbb{C}_{\text{left}}$. If, in addition, the coefficients of* (13.16) *satisfy the identities in* (13.18)*, then the Hamiltonian H has no pure imaginary eigenvalues and $\sigma(T_{11} - XT_{21}) \subset \mathbb{C}_{\text{right}}$.*

Proof. We shall use freely the results of Section 12.1. Let N be the angular subspace determined by X. Then N is invariant under the Hamiltonian H and the restriction $H|_N$ is similar to the matrix $T_{22}+T_{21}X$. Since the identities in (13.18) are satisfied, (13.19) holds. The symmetry relation $JH = -H^*J$ implies that the eigenvalues of H are placed symmetrically with respect to the imaginary axis (multiplicities included). Note that the dimension of the angular subspace N is equal to n, where n is the size of the matrices T_{ij}, $1 \le i, j \le 2$. Since N is invariant under H and $H|_N$ is similar to $T_{22}+T_{21}X$, the condition on the spectrum of $T_{22}+T_{21}X$, implies that $\sigma(H|_N) \subset \mathbb{C}_{\text{left}}$. It follows that H has at least n eigenvalues (multiplicities taken into account) in $\mathbb{C}_{\text{left}}$. The symmetry referred to above then gives that H also has at least n eigenvalues in $\mathbb{C}_{\text{right}}$. But the order of H is $2n$. So H has precisely n eigenvalues in $\mathbb{C}_{\text{left}}$, and also precisely n eigenvalues in $\mathbb{C}_{\text{right}}$. In particular, H has no eigenvalue on the imaginary axis.

Next, recall formula (12.5) for the present setting, that is,

$$E^{-1}HE = \begin{bmatrix} T_{11} - XT_{21} & 0 \\ T_{21} & T_{22} + T_{21}X \end{bmatrix}, \text{ where } E = \begin{bmatrix} I_n & X \\ 0 & I_n \end{bmatrix}. \tag{13.20}$$

As H and $E^{-1}HE$ have the same set of eigenvalues (multiplicities taken into account) and $\sigma(T_{22} + T_{21}X) \subset \mathbb{C}_{\text{left}}$, the result of the previous paragraph implies that $\sigma(T_{11} - XT_{21}) \subset \mathbb{C}_{\text{right}}$, which completes the proof. □

Corollary 13.5. *Assume the coefficients of the Riccati equation* (13.16) *satisfy the symmetry conditions in* (13.18). *Then equation* (13.16) *has at most one solution X such that $\sigma(T_{22}+T_{21}X) \subset \mathbb{C}_{\text{left}}$. Moreover, this solution, if it exists, is Hermitian.*

Proof. Assume X is a solution of (13.16) such that $\sigma(T_{22} + T_{21}X)$ is a subset of $\mathbb{C}_{\text{left}}$. Then, by Lemma 13.4, the Hamiltonian H has no pure imaginary eigenvalues and $\sigma(T_{11} - XT_{21}) \subset \mathbb{C}_{\text{right}}$. But then we can apply Proposition 12.1 to show that the angular subspace N determined by X is the spectral subspace of H corresponding to the eigenvalues of H in the open left half plane. In particular, N is uniquely determined and does not depend on the particular choice of the solution X. This implies that X is also uniquely determined.

Again assume that X is a solution of (13.16) such that $\sigma(T_{22} + T_{21}X)$ is a subset of $\mathbb{C}_{\text{left}}$. Then $\sigma(T_{11} - XT_{21}) \subset \mathbb{C}_{\text{right}}$. By taking adjoints and using the identities in (13.18) we see that the latter inclusion implies that $\sigma(T_{22} + T_{21}X^*)$ is a subset of $\mathbb{C}_{\text{left}}$. Furthermore, from the identities in (13.18) it also follows that X^* is a solution of (13.16). But then, by the uniqueness result of the previous paragraph, $X^* = X$. Hence X is Hermitian, as desired. □

Proof of Theorem 13.3. The implication (i) $\Rightarrow$ (ii) and the final statements of the theorem follow directly by applying Lemma 13.4 and Corollary 13.5 with the coefficients T_{ij}, $1 \le i,\ j \le 2$, being taken as in (13.17).

It remains to prove the implication (ii) $\Rightarrow$ (i). Let F be an $m \times n$ matrix such that $A-BF$ has all its eigenvalues in the open left half plane. Such a matrix exists because (A, B) is stabilizable. Introduce the rational $m \times m$ matrix function

$$V(\lambda) = R + \begin{bmatrix} B^* & -RF \end{bmatrix} (\lambda - G)^{-1} \begin{bmatrix} F^*R \\ B \end{bmatrix}, \tag{13.21}$$

where

$$G = \begin{bmatrix} -A^* + F^*B^* & -Q - F^*RF \\ 0 & A - BF \end{bmatrix}.$$

The fact that R is invertible implies that the realization (13.21) is biproper, and one verifies easily that the associate main operator is precisely the Hamiltonian T. Thus

$$V^{-1}(\lambda) = R^{-1} - \begin{bmatrix} R^{-1}B^* & -F \end{bmatrix} (\lambda - T)^{-1} \begin{bmatrix} F^* \\ BR^{-1} \end{bmatrix}.$$

Since $A - BF$ has all its eigenvalues in the open left half plane, G has no eigenvalue on the imaginary axis. By assumption the same holds true for T. Thus V has no poles or zeros on $i\mathbb{R}$. In particular, $V(\lambda)$ is invertible for each $\lambda \in i\mathbb{R}$. With J as in (13.19) we have

$$JG = -G^*J, \qquad J \begin{bmatrix} F^*R \\ B \end{bmatrix} = \begin{bmatrix} B^* & -RF \end{bmatrix}^*.$$

So, by the remark made after the proof of Theorem 9.1, the values of V on $i\mathbb{R}$ are selfadjoint matrices. Since $V(\lambda)$ is invertible for each $\lambda \in i\mathbb{R}$, it follows that the signature of the matrices $V(\lambda)$ for $\lambda \in i\mathbb{R}$, i.e., the difference between the number of positive and negative eigenvalues of the selfadjoint matrix $V(\lambda)$, is constant. As $V(\infty) = R$ is positive definite, we obtain that $V(\lambda)$ is positive definite for $\lambda \in i\mathbb{R}$. Hence we know from Theorem 9.5 that V admits a left spectral factorization with respect to $i\mathbb{R}$.

To finish the proof of (ii) $\Rightarrow$ (i), we apply (the appropriate version of) Theorem 12.5 (see the remark made between the theorem and its proof) with

$$A_{11} = -A^* + F^*B^*, \qquad A_{12} = -Q - F^*RF, \qquad A_{22} = A - BF,$$

$$B_1 = F^*R, \qquad B_2 = B, \qquad C_1 = B^*, \qquad C_2 = -RF, \qquad D = R.$$

Via these choices, equation (12.10) transforms into (13.14) with P as the unknown. Furthermore, the inclusions (12.11) and (12.12) change into

$$\sigma(-A^* + PBR^{-1}B^*) \subset \mathbb{C}_{\text{right}}, \qquad \sigma(A - BR^{-1}B^*P) \subset \mathbb{C}_{\text{left}}. \tag{13.22}$$

The conclusion is that equation (13.14) has a unique solution P satisfying the inclusions in (13.22). The second of these shows that P is a stabilizing solution of (13.14). Thus (i) is proved. □

Let P be an $i\mathbb{R}$-stabilizing solution of (13.14). Then by definition, the spectral inclusion $\sigma(A - BR^{-1}B^*P) \subset \mathbb{C}_{\mathrm{left}}$ holds. Furthermore, since P is Hermitian, also $\sigma(-A^* + PBR^{-1}B^*P) \subset \mathbb{C}_{\mathrm{right}}$; see also Lemma 13.4. So one of the spectral inclusions in (13.22) implies the other one automatically; cf., the two spectral inclusions (12.11), (12.12).

13.3 Symmetric Riccati equations and pseudo-spectral factorization

We now continue the discussion of Section 13.2. The object of study will be the algebraic Riccati equation

$$A^*P + PA + Q - (PB + S^*)R^{-1}(B^*P + S) = 0. \tag{13.23}$$

Observe that compared to (13.14) there are some additional terms. On the other hand, (13.23) can be rewritten in the more familiar form (13.14) as

$$(A^* - S^*R^{-1}B^*)P + P(A - BR^{-1}S) + (Q - S^*R^{-1}S) - PBR^{-1}B^*P = 0.$$

The Hamiltonian of this equation is given by

$$T = \begin{bmatrix} -A^* + S^*R^{-1}B^* & -Q + S^*R^{-1}S \\ -BR^{-1}B^* & A - BR^{-1}S \end{bmatrix}. \tag{13.24}$$

Also of importance is the rational matrix function

$$W(\lambda) = \begin{bmatrix} -B^*(\lambda + A^*)^{-1} & I \end{bmatrix} \begin{bmatrix} Q & S^* \\ S & R \end{bmatrix} \begin{bmatrix} (\lambda - A)^{-1}B \\ I \end{bmatrix}. \tag{13.25}$$

Note that W is selfadjoint on the imaginary axis, and admits the realization

$$W(\lambda) = R + \begin{bmatrix} B^* & S \end{bmatrix} \left(\lambda - \begin{bmatrix} -A^* & -Q \\ 0 & A \end{bmatrix} \right)^{-1} \begin{bmatrix} -S^* \\ B \end{bmatrix}. \tag{13.26}$$

For the inverse of W, one computes that

$$W(\lambda)^{-1} = R^{-1} - R^{-1} \begin{bmatrix} B^* & S \end{bmatrix} (\lambda - T)^{-1} \begin{bmatrix} -S^* \\ B \end{bmatrix} R^{-1}.$$

Letting n be the order of the matrix A and the skew-Hermitian $2n \times 2n$ matrix J as in (13.19), we have

$$J \begin{bmatrix} -A^* & -Q \\ 0 & A \end{bmatrix} = - \begin{bmatrix} -A^* & -Q \\ 0 & A \end{bmatrix}^* J, \qquad J \begin{bmatrix} -S^* \\ B \end{bmatrix} = \begin{bmatrix} B^* & S \end{bmatrix}^*,$$

and hence also $JT = -T^*J$.

The hypotheses we shall have in effect in this section are more stringent than those in Section 13.2. In fact, we shall assume A is an $n \times n$ matrix and B an $n \times m$ matrix such that (A, B) is a controllable pair (as opposed to the weaker condition of stabilizability). As in Section 13.2 we take R positive definite and Q selfadjoint.

In the next theorem we characterize when the function W introduced above is nonnegative on the imaginary axis. The characterization is given in terms of the existence of Hermitian solutions of the Riccati equation (13.23). Also we specify further the pseudo-spectral factorization result in Theorem 10.2, again in terms of Hermitian solutions of (13.23).

Theorem 13.6. *Consider the Riccati equation* (13.23) *with* (A, B) *a controllable pair,* R *positive definite and* Q *selfadjoint. Let* T *be the matrix given by* (13.24) *and let* W *be the rational matrix function defined by* (13.25). *Then the following statements are equivalent:*

(i) *Equation* (13.23) *has a Hermitian solution* P*;*

(ii) *The rational matrix function* W *is nonnegative on the imaginary axis;*

(iii) *The partial multiplicities of* T *at its pure imaginary eigenvalues are all even;*

(iv) *There exists a* T*-invariant subspace* M *such that* $J[M] = M^\perp$.

In that case, so if the equivalent conditions (i)–(iv) *hold, then, given a Hermitian solution* P *of* (13.23), *the rational matrix function* $W(\lambda)$ *factors as*

$$W(\lambda) = L(-\bar{\lambda})^* L(\lambda), \tag{13.27}$$

where

$$L(\lambda) = R^{1/2} + R^{-1/2}(B^*P + S)(\lambda I_n - A)^{-1}B. \tag{13.28}$$

Moreover, if M *is a* T*-invariant subspace such that* $J[M] = M^\perp$, *then* M *is of the form*

$$M = \operatorname{Im} \begin{bmatrix} P \\ I_n \end{bmatrix}$$

for a Hermitian solution P *of* (13.23). *In addition, if both* A *and* $T|_M$ *have all their eigenvalues in the closed left half plane, then the factorization* (13.27) *is a pseudo-spectral factorization with respect to the imaginary axis.*

Proof. (i) $\Rightarrow$ (ii) Suppose (13.23) has a Hermitian solution P. With this P, define $L(\lambda)$ by (13.28). We then have

$$\begin{aligned} L(-\bar{\lambda})^* L(\lambda) &= R - B^*(\lambda + A^*)^{-1}(S^* + PB) + (B^*P + S)(\lambda - A)^{-1}B \\ &\quad - B^*(\lambda + A^*)^{-1}(S^* + PB)R^{-1}(B^*P + S)(\lambda - A)^{-1}B. \end{aligned}$$

Using (13.23), one rewrites the last term as

$$B^*(\lambda + A^*)^{-1}\big(Q + (\lambda - A^*)P + P(A - \lambda)\big)(\lambda - A)^{-1}B$$

which, in turn, can be transformed into

$$B^*(\lambda + A^*)^{-1}Q(\lambda - A)^{-1}B + B^*P(\lambda - A)^{-1}B - B^*(\lambda + A^*)^{-1}PB.$$

Thus $L(-\bar{\lambda})^*L(\lambda) = W(\lambda)$, and (ii) holds. Moreover, the identity (13.27) is proved.

(ii) $\Rightarrow$ (iii) To prove that (ii) implies (iii) a couple of preparatory remarks are needed. Let A be an $n \times n$ matrix and B be an $n \times m$ matrix. For any $m \times n$ matrix F introduce

$$A_F = A - BF, \qquad S_F = S - RF, \qquad Q_F = Q - S^*F - F^*S + F^*RF.$$

Then,

$$\begin{bmatrix} Q_F & S_F^* \\ S_F & R \end{bmatrix} = \begin{bmatrix} I & -F^* \\ 0 & I \end{bmatrix} \begin{bmatrix} Q & S^* \\ S & R \end{bmatrix} \begin{bmatrix} I & 0 \\ -F & I \end{bmatrix}.$$

Thus

$$\begin{aligned} W(\lambda) &= \begin{bmatrix} -B^*(\lambda + A^*)^{-1} & I \end{bmatrix} \begin{bmatrix} I & F^* \\ 0 & I \end{bmatrix} \begin{bmatrix} Q_F & S_F^* \\ S_F & R \end{bmatrix} \\ &\qquad \cdot \begin{bmatrix} I & 0 \\ F & I \end{bmatrix} \begin{bmatrix} (\lambda - A)^{-1}B \\ I \end{bmatrix} \\ &= \begin{bmatrix} -B(\lambda + A^*)^{-1} & I - B^*(\lambda + A^*)^{-1}F^* \end{bmatrix} \\ &\qquad \cdot \begin{bmatrix} Q_F & S_F^* \\ S_F & R \end{bmatrix} \begin{bmatrix} (\lambda - A)^{-1}B \\ I + F(\lambda - A)^{-1}B \end{bmatrix}. \end{aligned}$$

Now introduce

$$W_F(\lambda) = \begin{bmatrix} -B^*(\lambda + A_F^*)^{-1} & I \end{bmatrix} \begin{bmatrix} Q_F & S_F^* \\ S_F & R \end{bmatrix} \begin{bmatrix} (\lambda - A_F)^{-1}B \\ I \end{bmatrix},$$

and $\Phi(\lambda) = I + F(\lambda - A)^{-1}B$. Then $\Phi(\lambda)^{-1} = I - F(\lambda - A_F)^{-1}B$. Using the fourth identity in Theorem 2.4 one sees that $(\lambda - A)^{-1}B\Phi(\lambda)^{-1} = (\lambda - A_F)^{-1}B$. Thus $W(\lambda) = \Phi(-\bar{\lambda})^*W_F(\lambda)\Phi(\lambda)$. So $W(\lambda)$ is nonnegative for $\lambda \in i\mathbb{R}$ if and only if $W_F(\lambda)$ is nonnegative for $\lambda \in i\mathbb{R}$, provided λ is not a pole of the functions involved. Next, notice that $W_F(\lambda)$ has the realization

$$W_F(\lambda) = R + \begin{bmatrix} B^* & S_F \end{bmatrix} \left(\lambda - \begin{bmatrix} -A_F^* & -Q_F \\ 0 & A_F \end{bmatrix} \right)^{-1} \begin{bmatrix} -S_F^* \\ B \end{bmatrix}.$$

One readily computes that

$$\begin{bmatrix} -A_F^* & -Q_F \\ 0 & A_F \end{bmatrix} - \begin{bmatrix} -S_F^* \\ B \end{bmatrix} R^{-1} \begin{bmatrix} B^* & S_F \end{bmatrix} = T,$$

where T is given by (13.24). So

$$W_F(\lambda)^{-1} = R^{-1} - R^{-1}\begin{bmatrix} B^* & S_F \end{bmatrix}(\lambda - T)^{-1}\begin{bmatrix} -S_F^* \\ B \end{bmatrix} R^{-1}.$$

Since the pair (A, B) is controllable, we can use the pole placement theorem from mathematical systems theory (see Theorem 19.3 in Chapter 20 below), to conclude that there exists an $m \times n$ matrix F such that all the eigenvalues of A_F are in the open left half plane. Using such an F, we see that the matrix

$$\begin{bmatrix} -A_F^* & -Q_F \\ 0 & A_F \end{bmatrix}$$

has no imaginary eigenvalues. This allows us to show (see formula (4.7) in Section 4.3 in [20]) that the matrix functions

$$\begin{bmatrix} W_F(\lambda) & 0 \\ 0 & I_{2n} \end{bmatrix} \quad \text{and} \quad \begin{bmatrix} \lambda I_{2n} - T & 0 \\ 0 & I_m \end{bmatrix}$$

are analytically equivalent on an open set containing the imaginary axis. It follows that for each $\lambda \in i\mathbb{R}$ the partial multiplicities of λ as an eigenvalue of T are equal to the partial multiplicities of λ as a zero of W_F. Since W_F is nonnegative on $i\mathbb{R}$, we know from Proposition 10.4 that the partial multiplicities of $\lambda \in i\mathbb{R}$ as a zero of W_F are even. Hence the partial multiplicities at the pure imaginary eigenvalues of T are even. Thus (ii) implies (iii).

(iii) $\Rightarrow$ (iv) This implication can be seen from Theorem 11.4 in Chapter 11 applied to $A = iT$ and $H = iJ$. Indeed, since there are no odd partial multiplicities corresponding to pure imaginary eigenvalues of T, the condition of Theorem 11.4 is satisfied. Hence there exists an A-invariant subspace M such that $H[M] = M^\perp$. This subspace then is also T-invariant and satisfies $J[M] = M^\perp$.

(iv) $\Rightarrow$ (i) Let M be T-invariant subspace such that $J[M] = M^\perp$, and write

$$M = \operatorname{Im}\begin{bmatrix} X_1 \\ X_2 \end{bmatrix},$$

for appropriate $n \times n$ matrices X_1 and X_2. It will be shown that X_2 is invertible. Once this is done, we can take $P = X_1 X_2^{-1}$. From $T[M] \subset M$ one obtains that P solves (13.23), while from $J[M] = M^\perp$ one has $P = P^*$. Hence (i) holds. We have also shown that any T-invariant J-Lagrangian subspace M is the graph of a Hermitian solution P of the Riccati equation, that is, M is of the form $M = \operatorname{Im}\begin{bmatrix} P & I \end{bmatrix}^*$ for a matrix $P = P^*$ that solves (13.23).

It remains to verify that $\operatorname{Ker} X_2 = \{0\}$. As $\dim M = n$, the null spaces $\operatorname{Ker} X_1$ and $\operatorname{Ker} X_2$ have a trivial intersection. So it is sufficient to establish that $X_2 x = 0$

implies $X_1x = 0$. Let $X_2x = 0$. Then

$$\begin{bmatrix} X_1x \\ 0 \end{bmatrix} \in M,$$

and hence

$$T\begin{bmatrix} X_1x \\ 0 \end{bmatrix} = \begin{bmatrix} -A^*X_1x + S^*R^{-1}B^*X_1x \\ -BR^{-1}B^*X_1x \end{bmatrix} \in M.$$

Now M is iJ-Lagrangian, i.e., $J[M] = M^\perp$. So

$$0 = \left\langle T\begin{bmatrix} X_1x \\ 0 \end{bmatrix}, J\begin{bmatrix} X_1x \\ 0 \end{bmatrix}\right\rangle = -\langle R^{-1}B^*X_1x, B^*X_1x\rangle.$$

As R is positive definite, we obtain $B^*X_1x = 0$. Hence

$$T\begin{bmatrix} X_1x \\ 0 \end{bmatrix} = \begin{bmatrix} -A^*X_1x \\ 0 \end{bmatrix}.$$

But this vector is in M, so it must be of the form

$$\begin{bmatrix} X_1y \\ X_2y \end{bmatrix}.$$

Thus $X_2y = 0$ and $X_1y = -A^*X_1x$. As $X_2y = 0$, we have $B^*X_1y = 0$ by the argument given above. So $B^*A^*X_1x = 0$. Now consider

$$T^2\begin{bmatrix} X_1x \\ 0 \end{bmatrix} = T\begin{bmatrix} -A^*X_1x \\ 0 \end{bmatrix} = \begin{bmatrix} A^{*2}X_1x \\ 0 \end{bmatrix}.$$

Repeating the argument we get $B^*A^{*2}X_1x = 0$. Continuing in this way we arrive at $X_1x \in \operatorname{Ker} B^*A^{*j}$ for all j. As (A, B) is a controllable pair, the pair (B^*, A^*) is observable, and thus we see that $X_1x = 0$, as desired.

It is easily seen that the eigenvalues of $T|_M$ coincide with those of the matrix $A - BR^{-1}S - BR^{-1}B^*P$. Thus, if both A and $T|_M$ have their eigenvalues in the closed left half plane, then the factorization (13.27) with L given by (13.28) is a pseudo-spectral factorization. □

Notice that the full force of the controllability condition on the pair (A, B) was only used in the last part of the proof. More precisely, the implications (i) ⇒ (ii) and (iii) ⇒ (iv) are true without any condition on (A, B), and for the implication (ii) ⇒ (iii) only stabilizability of (A, B) was used.

Notes

The connection between Riccati equations and factorizations as discussed in Section 13.1 goes back to [147] and [41]. The main result of Section 13.2 originates from [102], see also [106], Section 9.3. The results of Section 13.1 and 13.2, and similar results for the discrete time algebraic Riccati equation, play an important role in several problems in mathematical systems theory, notably, LQ-optimal control, Kalman filtering and stochastic realization (see, e.g., [84], [85], [33]). The main result of Section 13.3 appeared for the first time in [105] and [34]. See also Chapter 7 in [106]. The parametrization of solutions of the algebraic Riccati equation in terms of invariant subspaces of the matrix T, as described in Theorem 13.6, also plays a role in [135], [136].

Chapter 14

J-spectral factorization

In this chapter we continue the study of rational matrix functions that take Hermitian values on certain contours. In contrast to the previous chapters, the emphasis will not be on positive definite or nonnegative rational matrix functions, but rather on ones that have values for which the inertia is independent of the point on the contour. Such functions may still admit a symmetric canonical factorization, provided we allow for a constant Hermitian invertible matrix as a middle factor. Such a factorization is commonly known as a J-spectral factorization. We shall give necessary and sufficient conditions for its existence, and study the question when a function which admits a left J-spectral factorization also admits a right J-spectral factorization.

This chapter consists of seven sections. The first four sections and the one but last deal with J-spectral factorization with respect to the imaginary axis. Section 14.1 introduces the notion of J-spectral factorization. The next two sections provide necessary and sufficient conditions for the existence of such factorizations; in Section 14.2 these conditions are stated in terms of certain invariant subspaces and in Section 14.3 they are given in terms of Riccati equations. Two special cases are discussed in detail in Section 14.4. The fifth section (Section 14.5) deals with J-spectral factorization with respect to the unit circle and the real line. Section 14.6 concerns the topic of left versus right J-spectral factorization. In Section 14.7 an alternative approach is used to derive J -spectral factorizations with respect to the unit circle. The main result of this final section extends to a more general setting the first main result of Section 14.5.

14.1 Definition of J-spectral factorization

Throughout this chapter J is an invertible Hermitian $m \times m$ matrix. Often we shall assume additionally that $J^{-1} = J$. Thus in that case we have

$$J = J^* = J^{-1}. \tag{14.1}$$

Such a matrix is called a *signature matrix.* Up to a congruence transformation any selfadjoint invertible matrix is a signature matrix.

Suppose W is a rational $m \times m$ matrix function. A factorization

$$W(\lambda) = L(-\bar{\lambda})^* J L(\lambda) \tag{14.2}$$

is called a *right J-spectral factorization with respect to the imaginary axis* if L and L^{-1} are rational $m \times m$ matrix functions which are analytic on the closed left half plane (infinity included). In that case the function $L(-\bar{\lambda})^*$ and its inverse are analytic on the closed right half plane (including infinity). Thus a right J-spectral factorization with respect to the imaginary axis is a right canonical factorization with respect to $i\mathbb{R}$ featuring an additional symmetry property between the factors. A factorization (9.3) is called a *left J-spectral factorization with respect to the imaginary axis* if L and L^{-1} are rational $m \times m$ matrix functions which are analytic on the closed right half plane (infinity included), in which case the function $L(\bar{\lambda})^*$ and its inverse are analytic on the closed left half plane (infinity included). Such a factorization is a left canonical factorization with respect to $i\mathbb{R}$.

The existence of a right or left J-spectral factorization implies that W admits a canonical factorization with respect to the imaginary axis. In particular, in order that a right or left J-spectral factorization of W exists it is necessary that W is biproper and has no poles or zeros on the imaginary axis. Furthermore, the identity (14.2) gives that W is selfadjoint on the imaginary axis.

Contrary to spectral factorizations for positive definite rational matrix functions, J-spectral factorizations do not always exist for biproper rational matrix functions that satisfy the obvious necessary conditions mentioned in the previous paragraph. Since a J-spectral factorization is a canonical factorization, we can use Theorem 3.2 to prepare for an example of this phenomenon. Let

$$W(\lambda) = \begin{bmatrix} 0 & \dfrac{\lambda-1}{\lambda+1} \\ \dfrac{\lambda+1}{\lambda-1} & 0 \end{bmatrix}. \tag{14.3}$$

Obviously, W is biproper and its values on the imaginary axis are selfadjoint. Furthermore, W has no pole or zero on the imaginary axis. The function W has the minimal realization $W(\lambda) = D + C(\lambda - A)^{-1}B$, with

$$D = \begin{bmatrix} 0 & 1 \\ 1 & 0 \end{bmatrix}, \quad A = \begin{bmatrix} 1 & 0 \\ 0 & -1 \end{bmatrix}, \quad B = \begin{bmatrix} 1 & 0 \\ 0 & -2 \end{bmatrix}, \quad C = \begin{bmatrix} 0 & 1 \\ 2 & 0 \end{bmatrix}. \tag{14.4}$$

The associate main operator is given by

$$A^\times = A - BD^{-1}C = \begin{bmatrix} -1 & 0 \\ 0 & 1 \end{bmatrix} = -A.$$

Now for a right canonical factorization with respect to the imaginary axis to exist, we must have $\mathbb{C}^2 = M_- \dot{+} M_+^\times$, where M_- is the spectral subspace of A associated with the part of $\sigma(A)$ lying in the left half plane, and $M_+^\times$ is the spectral subspace of $A^\times$ associated with the part of $\sigma(A^\times)$ lying in the right half plane. However, since in this case $A^\times = -A$, we have $M_- = M_+^\times$. Hence a right canonical factorization of W with respect to $i\mathbb{R}$ does not exist. Analogously, a left canonical factorization does not exist either. Hence neither left nor right J-spectral factorizations of W with respect to the imaginary axis exist for any choice of $J = J^* = J^{-1}$.

To further clarify the connection between J-spectral factorization and canonical factorization we present the following proposition.

Proposition 14.1. *Let W be a biproper rational $m \times m$ matrix function that is selfadjoint on the imaginary axis and has no pole there. Then $W(\infty)$ is congruent to a signature matrix J, and for such a matrix J the function W admits a right (respectively, left) J-spectral factorization with respect to the imaginary axis if and only if it admits a right (respectively, left) canonical factorization with respect to the imaginary axis.*

Proof. Since W is selfadjoint on the imaginary axis and proper, we see that $D = W(\infty)$ is well-defined and selfadjoint. The fact that W is biproper means that D is invertible. Thus D is an invertible selfadjoint matrix, and hence congruent to a signature matrix, J say: $D = E^*JE$ for some invertible matrix E.

Let $W(\lambda) = W_-(\lambda)W_+(\lambda)$ be a right canonical factorization of W with respect to the imaginary axis. Since W, W_- and W_+ are biproper we have $D = D_-D_+$, where $D_- = W_-(\infty)$ and $D_+ = W_+(\infty)$. It follows that the factorization $W(\lambda) = W_-(\lambda)W_+(\lambda)$ can be rewritten as $W(\lambda) = V_-(\lambda)DV_+(\lambda)$, where

$$V_-(\lambda) = W_-(\lambda)D_-^{-1}, \qquad V_+(\lambda) = D_+^{-1}W_-(\lambda).$$

In particular, the values of V_- and V_+ at infinity are equal to the $m \times m$ identity matrix. Since V_+ and V_+^{-1} are analytic on the closed right half plane (infinity included) and the functions V_- and V_-^{-1} are analytic on the closed left half plane (infinity included), the factorization is unique. Now we use that D is selfadjoint and that W is selfadjoint on the imaginary axis. It follows that

$$W(\lambda) = V_+(-\bar{\lambda})^*DV_-(-\bar{\lambda})^*,$$

and in this factorization the factors have the same analyticity properties as those in $W(\lambda) = V_-(\lambda)DV_+(\lambda)$. Because of the uniqueness of the latter factorization, we conclude that $V_-(\lambda) = V_+(-\bar{\lambda})^*$. Recall that $D = E^*JE$. Put $L(\lambda) = EV_+(\lambda)$. Then $W(\lambda) = L(-\bar{\lambda})^*JL(\lambda)$, and this factorization is a left J-spectral factorization with respect to the imaginary axis. The reverse implication is trivial. □

14.2 J-spectral factorizations and invariant subspaces

In this section necessary and sufficient conditions for existence of a right or left J-spectral factorization with respect to the imaginary axis will be derived in terms of invariant subspaces. It will be assumed that the obvious necessary conditions for the existence of a J-spectral factorization are satisfied, that is, the rational $m \times m$ matrix function W for which we wish to find J-spectral factorizations with respect to $i\mathbb{R}$ is assumed to be biproper, to have no poles or zeros on $i\mathbb{R}$, and to be selfadjoint on $i\mathbb{R}$.

We begin with two lemmas which can be viewed as further refinements of Theorem 9.1(ii).

Lemma 14.2. *Let W be a biproper rational $m \times m$ matrix function that is selfadjoint on the imaginary axis and has no pole there. Then W admits a minimal realization*

$$W(\lambda) = D + B^*H^*(\lambda I_{2n} - A)^{-1}B, \tag{14.5}$$

such that $D = D^*$ is invertible, H is invertible,

$$HA = -A^*H, \qquad H^* = -H, \tag{14.6}$$

and the matrices A and H partition as

$$A = \begin{bmatrix} A_{11} & A_{12} \\ 0 & A_{22} \end{bmatrix}, \qquad H = \begin{bmatrix} 0 & -H_{21}^* \\ H_{21} & H_{22} \end{bmatrix}, \tag{14.7}$$

where A_{11} and A_{22} are $n \times n$ matrices which have all their eigenvalues in the right open half plane and left open half plane, respectively.

Proof. Since W is biproper, $D = W(\infty)$ is invertible. The fact that D is selfadjoint is covered by item (ii) in Theorem 9.1.

Next, let $W(\lambda) = D + \widehat{C}(\lambda I_p - \widehat{A})^{-1}\widehat{B}$ be a minimal realization of W. The fact that W has no poles on $i\mathbb{R}$ and the minimality of the realization imply that $\widehat{A}$ has no eigenvalue on $i\mathbb{R}$. Furthermore, using item (ii) of Theorem 9.1 again, we know that there exists a unique invertible $p \times p$ matrix $\widehat{T}$ for which we have

$$\widehat{T}\widehat{A} = -\widehat{A}^*\widehat{T}, \qquad \widehat{T}\widehat{B} = \widehat{C}^*, \qquad \widehat{T} = -\widehat{T}^*. \tag{14.8}$$

Let N_+ be the spectral subspace of $\widehat{A}$ corresponding to the eigenvalues in the open right half plane. The identity $\widehat{T}\widehat{A} = -\widehat{A}^*\widehat{T}$ yields $\widehat{T}[N_+] = N_+^\perp$. But then the invertibility of $\widehat{T}$ implies that $\dim N_+ = \dim N_+^\perp$. The latter can only happen when p is even, that is, $p = 2n$ for some nonnegative integer n. In particular, $\dim N_+ = n$. Now let $f_1, \ldots, f_n$ be an orthogonal basis of N_+, and let $f_{n+1}, \ldots, f_{2n}$ be an orthogonal basis of $N_+^\perp$. Since $\mathbb{C}^n = N_+ \oplus N_+^\perp$, the vectors $f_1, \ldots, f_{2n}$ form an orthogonal basis of $\mathbb{C}^{2n}$, and we can consider the unitary matrix U that transforms the basis $f_1, \ldots, f_{2n}$ into the standard basis $e_1, \ldots, e_{2n}$ of $\mathbb{C}^{2n}$. Define

$$A = U\widehat{A}U^{-1}, \qquad B = U\widehat{B}, \qquad C = \widehat{C}U^{-1}, \qquad H = U\widehat{T}U^*.$$

Then $W(\lambda) = D + C(\lambda I_{2n} - A)^{-1}B$ is a minimal realization of W. The fact that $U^{-1} = U^*$ together with (14.8) shows that

$$HA = -A^*H, \qquad HB = C^*, \qquad H = -H^*.$$

Thus W is of the form (14.5) and (14.6) holds.

The spectral subspace M_+ of A corresponding to the eigenvalues in the open right half plane is given by

$$M_+ = \text{span}\,\{e_1, \dots, e_n\}. \tag{14.9}$$

The first identity in (14.8) yields

$$H\big[[\text{span}\,\{e_1, \dots, e_n\}\big] = H[M_+] = M_+^\perp = \text{span}\,\{e_{n+1}, \dots, e_{2n}\}. \tag{14.10}$$

It follows that the matrices A, and H can be partitioned as in (14.7). All blocks in these representations of A and H are $n \times n$ matrices. The zero entry in A follows from the A-invariance of M_+ and the fact that this space is given by (14.9), while the zero entry in H follows from (14.10). The definition of M_+ and the identity (14.9) also imply that all the eigenvalues of A_{11} are in the open right half plane and those of A_{22} are in the open left half plane. □

Lemma 14.3. *Let W be a biproper rational $m \times m$ matrix function that is selfadjoint on the imaginary axis and has no pole there. Then W admits a minimal realization*

$$W(\lambda) = D + C(\lambda I_{2n} - A)^{-1}B, \tag{14.11}$$

such that $D = D^*$ is invertible and the matrices A, B and C can be partitioned as

$$A = \begin{bmatrix} -A_{22}^* & A_{12} \\ 0 & A_{22} \end{bmatrix}, \qquad B = \begin{bmatrix} B_1 \\ B_2 \end{bmatrix}, \qquad C = \begin{bmatrix} -B_2^* & B_1^* \end{bmatrix}, \tag{14.12}$$

where A_{12} is a selfadjoint $n \times n$ matrix, A_{22} is a $n \times n$ matrix which has all its eigenvalues in the open left half plane, and both B_1 and B_2 are $n \times m$ matrices.

Proof. From the preceding lemma we know that W admits a minimal realization

$$W(\lambda) = D + \widetilde{B}^* \widetilde{H}^* (\lambda I_{2n} - \widetilde{A})^{-1} \widetilde{B},$$

where $D = D^*$ is invertible, $\widetilde{H}$ is invertible,

$$\widetilde{H}\widetilde{A} = -\widetilde{A}^* H, \qquad \widetilde{H}^* = -\widetilde{H},$$

and the matrices $\widetilde{A}$ and $\widetilde{H}$ partition as

$$\widetilde{A} = \begin{bmatrix} \widetilde{A}_{11} & \widetilde{A}_{12} \\ 0 & \widetilde{A}_{22} \end{bmatrix}, \qquad \widetilde{H} = \begin{bmatrix} 0 & -\widetilde{H}_{21}^* \\ \widetilde{H}_{21} & \widetilde{H}_{22} \end{bmatrix}, \tag{14.13}$$

such that the eigenvalues of $\widetilde{A}_{11}$ are in the open right half plane and those of $\widetilde{A}_{22}$ are in the open left half plane.

Since $\widetilde{H}$ is invertible, it follows that $\widetilde{H}_{21}$ is invertible, and hence we can define

$$S = \begin{bmatrix} \widetilde{H}_{21}^{-1} & -\frac{1}{2}\widetilde{H}_{21}^{-1}\widetilde{H}_{22} \\ 0 & I_n \end{bmatrix}.$$

The matrix S is invertible. Put $\widetilde{C} = \widetilde{B}^*\widetilde{H}^*$, and consider the matrices

$$A = S^{-1}\widetilde{A}S, \qquad B = S^{-1}\widetilde{B}, \qquad C = \widetilde{C}S^{-1}, \qquad H = S^*\widetilde{H}S.$$

Obviously, $W(\lambda) = D + C(\lambda I_n - A)^{-1}B$ is a minimal realization of W.

It remains to prove that A, B, C can be partitioned in the desired way. A straightforward calculation shows that

$$HA = -A^*H, \qquad HB = C^*, \qquad H = \begin{bmatrix} 0 & -I_n \\ I_n & 0 \end{bmatrix}. \tag{14.14}$$

Since the matrices $\widetilde{A}$, S, and S^{-1} are all block upper triangular, the same holds true for A. The first identity in (14.14) together with the third identity in (14.14) shows that A is of the form given in (14.12) with A_{12} being selfadjoint. Furthermore, since the entry in the right lower corner of S and S^{-1} is the $n \times n$ identity matrix we see that $A_{22} = \widetilde{A}_{22}$, and hence A_{22} is an $n \times n$ matrix which has all its eigenvalues in the open left half plane. The second and third identities in (14.14) show that B and C are as in (14.12). Obviously, B_1 and B_2 are matrices of size $n \times m$. □

The external matrix D in the realizations (14.5) and (14.11) is congruent to a signature matrix J, that is, $D = E^*JE$ for some invertible matrix E. Replacing $W(\lambda)$ by $(E^*)^{-1}W(\lambda)E^{-1}$ we may assume that the external matrix is actually equal to J. In the next theorem we shall make this assumption.

Theorem 14.4. *Let W be a rational $m \times m$ matrix function that is selfadjoint on the imaginary axis and has no pole there. Suppose W is given by*

$$W(\lambda) = J + C(\lambda I_{2n} - A)^{-1}B,$$

where J is a signature matrix and

$$A = \begin{bmatrix} -A_{22}^* & A_{12} \\ 0 & A_{22} \end{bmatrix}, \qquad B = \begin{bmatrix} B_1 \\ B_2 \end{bmatrix}, \qquad C = \begin{bmatrix} -B_2^* & B_1^* \end{bmatrix},$$

such that A_{12} is a selfadjoint $n \times n$ matrix, A_{22} is an $n \times n$ matrix which has all its eigenvalues in the open left half plane, and both B_1 and B_2 are $n \times m$ matrices. Then W admits a left J-spectral factorization with respect to the imaginary axis,

$$W(\lambda) = L_-(-\overline{\lambda})^*JL_-(\lambda),$$

if and only if

$$A^\times = \begin{bmatrix} -A_{22}^* + B_1B_2^* & A_{12} - B_1B_1^* \\ B_2B_2^* & A_{22} - B_2B_1^* \end{bmatrix}$$

has no eigenvalues on the imaginary axis, and the spectral subspace of $A^\times$ corresponding to its eigenvalues in the open left half plane is of the form

$$\operatorname{Im} \begin{bmatrix} X \\ I_n \end{bmatrix}$$

for some Hermitian matrix X. In that case the unique left J-spectral factor L_- for which $L_-(\infty) = I_m$ is given by

$$L_-(\lambda) = I_m + J^{-1}(B_1^* - B_2^*X)(\lambda I_n - A_{22})^{-1}B_2.$$

In this expression (as well as in other comparable formulas below) the matrix J^{-1} can be replaced by J.

Proof. In order to prove the first part of the theorem, we have only to check when W admits a left canonical factorization with respect to the imaginary axis (see Proposition 14.1).

Let M be a spectral subspace of A corresponding to its eigenvalues in the open right half plane. Then $M = \operatorname{Im}\,[I\ \ 0]^*$. Writing $M^\times$ for the spectral subspace of $A^\times$ corresponding to its eigenvalues in the open left half plane, the matching condition

$$\mathbb{C}^n = M \dot{+} M^\times \tag{14.15}$$

is satisfied if and only if $M^\times = \operatorname{Im}\,[X^*\ \ I]^*$ for some matrix X. With H as in (14.14), the subspace $M^\times$ is iH-Lagrangian (see Section 11.1). Thus

$$\operatorname{Im} \begin{bmatrix} -I \\ X \end{bmatrix} = H[M^\times] = (M^\times)^\perp = \operatorname{Ker} \begin{bmatrix} X^* & I \end{bmatrix},$$

which implies $X = X^*$. Applying the left-version of Theorem 3.2 the first part of the theorem follows.

Next let us deal with the second part. So suppose (14.15) is satisfied and write the projection Π of $\mathbb{C}^n$ along M onto $M^\times$ in the form

$$\Pi = \begin{bmatrix} 0 & X \\ 0 & I \end{bmatrix}.$$

Then the unique right hand factor L_- in a left canonical factorization with respect to the imaginary axis of W, satisfying the additional condition that $L(\infty) = I_m$, is given by

$$\begin{aligned} L_-(\lambda) &= I + J^{-1}C\Pi(\lambda - \Pi A\Pi)^{-1}\Pi B \\ &= I + J^{-1}\begin{bmatrix} 0 & B_1^* - B_2^*X \end{bmatrix}\left(\lambda - \begin{bmatrix} 0 & XA_{22} \\ 0 & A_{22} \end{bmatrix}\right)^{-1}\begin{bmatrix} XB_2 \\ B_2 \end{bmatrix} \\ &= I + J^{-1}(B_1^* - B_2^*X)(\lambda - A_{22})^{-1}B_2, \end{aligned}$$

as was claimed. □

In Section 14.5 below we shall consider J-spectral factorization for selfadjoint rational matrix functions on the real line or on the unit circle.

14.3 J-spectral factorizations and Riccati equations

In this section, necessary and sufficient conditions for existence of a right or left J-spectral factorization with respect to the imaginary axis will be derived in terms of Riccati equations. It will be assumed that the obvious necessary conditions for the existence of a J-spectral factorization are satisfied, that is, the rational $m \times m$ matrix function W for which we wish to find J-spectral factorizations with respect to $i\mathbb{R}$ is assumed to be biproper, to have no poles or zeros on $i\mathbb{R}$, and to be selfadjoint on $i\mathbb{R}$. As in Theorem 14.4 we assume that the external matrix (that is, the value at infinity) is a signature matrix.

Theorem 14.5. *Let W be a rational $m \times m$ matrix function that is selfadjoint on the imaginary axis and has no pole there. Suppose W is given by*

$$W(\lambda) = J + C(\lambda I_{2n} - A)^{-1}B,$$

where J is a signature matrix and

$$A = \begin{bmatrix} -A_{22}^* & A_{12} \\ 0 & A_{22} \end{bmatrix}, \qquad B = \begin{bmatrix} B_1 \\ B_2 \end{bmatrix}, \qquad C = \begin{bmatrix} -B_2^* & B_1^* \end{bmatrix},$$

such that A_{12} is a selfadjoint $n \times n$ matrix, A_{22} is an $n \times n$ matrix which has all its eigenvalues in the open left half plane, and both B_1 and B_2 are $n \times m$ matrices. Then W admits a left J-spectral factorization with respect to the imaginary axis,

$$W(\lambda) = L_-(-\overline{\lambda})^* J L_-(\lambda),$$

if and only if the algebraic Riccati equation

$$\begin{aligned} XB_2J^{-1}B_2^*X + X(A_{22} - B_2J^{-1}B_1^*) + (A_{22}^* - B_1J^{-1}B_2^*)X \quad (14.16) \\ -A_{12} + B_1J^{-1}B_1^* = 0 \end{aligned}$$

has a (unique) $i\mathbb{R}$-stabilizing Hermitian solution X. In that case the unique left J-spectral factor L_- for which $L_-(\infty) = I_m$ is given by

$$L_-(\lambda) = I_m + J^{-1}(B_1^* - B_2^* X)(\lambda I_n - A_{22})^{-1} B_2. \tag{14.17}$$

In line with the definition given in the paragraph preceding Theorem 13.14, a solution of (14.16) is said to be $i\mathbb{R}$-stabilizing (or simply stabilizing) if the matrix $A_{22} - B_2 J^{-1} B_1^* + B_2 J^{-1} B_2^* X$ has its eigenvalues in the open left half plane.

Proof. In order to prove the first part of the theorem, we have only to check when W admits a left canonical factorization with respect to the imaginary axis (see Proposition 14.1).

A straightforward application of Theorem 12.5, with F_+ equal to $\mathbb{C}_{\text{left}}$ and F_- equal to $\mathbb{C}_{\text{right}}$, tells us that W admits a left canonical factorization with respect to the imaginary axis if and only if the Riccati equation (14.16) has a unique solution X satisfying the additional spectral constraints

$$\sigma\big(-A_{22}^* + (B_1 - XB_2)J^{-1}B_2^*\big) \subset \mathbb{C}_{\text{right}}, \tag{14.18}$$

$$\sigma\big(A_{22} - B_2 J^{-1}(B_1^* - B_2^* X)\big) \subset \mathbb{C}_{\text{left}}. \tag{14.19}$$

Next, note that X satisfies (14.16) and the spectral constraints (14.18) and (14.19) if and only if the same holds true for X^*. Because of uniqueness it follows that $X = X^*$. The second spectral constraint (14.19) means that X is a stabilizing solution of (14.16). This completes the proof of the first part of the theorem.

To prove the second part one applies the second part of Theorem 12.5 with $D_1 = J$ and $D_2 = I_m$. □

Theorem 14.6. *Let W be a rational $m \times m$ matrix function that is selfadjoint on the imaginary axis and has no pole there. Suppose W is given by*

$$W(\lambda) = J + B^* H^* (\lambda I_{2n} - A)^{-1} B,$$

where J is a signature matrix, H is invertible, $HA = -A^*H$ and $H^* = -H$, and the matrices A and H partition as

$$A = \begin{bmatrix} A_{11} & A_{12} \\ 0 & A_{22} \end{bmatrix}, \qquad H = \begin{bmatrix} 0 & -H_{21}^* \\ H_{21} & H_{22} \end{bmatrix},$$

where A_{11} and A_{22} are $n \times n$ matrices which have all their eigenvalues in the open right half plane and open left half plane, respectively. Put

$$\widehat{A}_{12} = \frac{1}{2}A_{22}^* H_{22} + \frac{1}{2}H_{22}A_{22} + H_{21}A_{12}, \tag{14.20}$$

$$\widehat{B}_1 = H_{21}B_1 + \frac{1}{2}H_{22}B_2. \tag{14.21}$$

Then W admits a left J-spectral factorization with respect to the imaginary axis,

$$W(\lambda) = L_-(-\overline{\lambda})^* J L_-(\lambda),$$

if and only if the algebraic Riccati equation

$$\begin{aligned} XB_2J^{-1}B_2^*X + X(A_{22} - B_2J^{-1}\widehat{B}_1^*) + (A_{22}^* - \widehat{B}_1J^{-1}B_2^*)X \\ -\widehat{A}_{12} + \widehat{B}_1J^{-1}\widehat{B}_1^* = 0. \end{aligned} \tag{14.22}$$

has a (unique) $i\mathbb{R}$-stabilizing Hermitian solution X. In that case the unique left J-spectral factor L_- for which $L_-(\infty) = I_m$ is given by

$$L_-(\lambda) = I_m + J^{-1}(\widehat{B}_1^* - B_2^*X)(\lambda I_n - A_{22})^{-1}B_2. \tag{14.23}$$

Recall that an $i\mathbb{R}$-stabilizing solution X of (14.22) is one for which the matrix $A_{22} - B_2J^{-1}\widehat{B}_1^* + B_2J^{-1}B_2^*X$ has its eigenvalues in the open left half plane.

Proof. Put $C = B^*H^*$, and consider the matrices $\widehat{A} = S^{-1}AS$, $\widehat{B} = S^{-1}B$ and $\widehat{C} = CS$, where

$$S = \begin{bmatrix} H_{21}^{-1} & -\frac{1}{2}H_{21}^{-1}H_{22} \\ 0 & I \end{bmatrix}.$$

Then $W(\lambda) = J + \widehat{C}(\lambda I_{2n} - \widehat{A})^{-1}\widehat{B}$, and from the proof of Lemma 14.3 we know that $\widehat{A}$, $\widehat{B}$ and $\widehat{C}$ partition as

$$\widehat{A} = \begin{bmatrix} -\widehat{A}_{22}^* & \widehat{A}_{12} \\ 0 & \widehat{A}_{22} \end{bmatrix}, \qquad \widehat{B} = \begin{bmatrix} \widehat{B}_1 \\ \widehat{B}_2 \end{bmatrix}, \qquad \widehat{C} = \begin{bmatrix} -\widehat{B}_2^* & \widehat{B}_1^* \end{bmatrix},$$

where $\widehat{A}_{22} = A_{22}$ and $\widehat{B}_2 = B_2$. Since

$$S^{-1} = \begin{bmatrix} H_{21} & \frac{1}{2}H_{22} \\ 0 & I \end{bmatrix},$$

one readily computes that $\widehat{A}_{12}$ and $\widehat{B}_2$ are given by (14.20) and (14.21), respectively. It follows that the realization $W(\lambda) = J + \widehat{C}(\lambda I_{2n} - \widehat{A})^{-1}\widehat{B}$ satisfies the conditions of Theorem 14.5. Note that the Riccati equation (14.16) transforms into equation (14.22) when B_1 is replaced by $\widehat{B}_1$ and the matrix A_{12} by $\widehat{A}_{12}$. Furthermore, when passing from B_1 to $\widehat{B}_1$, formula (14.17) transforms into (14.23). But then we can apply Theorem 14.5 to finish the proof. □

Note that the procedure to find the J-spectral factor, if it exists, now consists of two main steps. The first is to find a realization as in Theorem 14.6, which can be done by using an orthogonal basis transformation (see the proof of Lemma 14.2), and then to find the stabilizing solution X of (14.22) in case it exists.

With this in mind, let us return to the counterexample given in Section 14.1. Let W be the rational 2×2 matrix function given by (14.3). The realization of this function given in Section 14.1, involving the matrices featured in (14.4), can be rewritten as $W(\lambda) = J + B^*H^*(\lambda I_2 - A)^{-1}B$, where

$$J = \begin{bmatrix} 0 & 1 \\ 1 & 0 \end{bmatrix}, \qquad A = \begin{bmatrix} 1 & 0 \\ 0 & -1 \end{bmatrix}, \qquad B = \begin{bmatrix} 1 & 0 \\ 0 & -2 \end{bmatrix}, \qquad H = \begin{bmatrix} 0 & -1 \\ 1 & 0 \end{bmatrix}.$$

This realization satisfies the conditions required in the first part of Theorem 14.6. So it makes sense to check the situation with respect to the Riccati equation featured in the theorem. Note that in this case $\widehat{A}_{12} = 0$ and $\widehat{B}_1 = \begin{bmatrix} 1 & 0 \end{bmatrix}$. Since $B_2 = \begin{bmatrix} 0 & -2 \end{bmatrix}$, it follows that in the algebraic Riccati equation (14.22) both the quadratic and the constant term vanish. Hence (14.22) reduces to a linear equation, namely $2x = 0$. So $x = 0$ is the unique solution, and this solution is not stabilizing. Hence, W does not admit a J-spectral factorization with respect to the imaginary axis, which corroborates what was already observed in the paragraph preceding Proposition 14.1.

14.4 Two special cases of J-spectral factorization

In this section we consider two special cases. The first concerns the situation where the rational matrix function appears already as a product

$$W(\lambda) = V(-\bar{\lambda})^* J' V(\lambda) \tag{14.24}$$

where J' is a signature matrix and V has all its poles in the open left half plane. This situation is encountered in several problems in mathematical systems theory, notably in the theory of H_∞-control (see Chapter 20 below).

Let W be the rational $m \times m$ matrix function given by the product (14.24), where $V(\lambda) = D + C(\lambda I_n - A)^{-1}B$. Observe that W is selfadjoint on the imaginary axis. We assume that A has all its eigenvalues in the open left half plane and that the (possibly non-square) matrix D is of full column rank (that is, $\operatorname{Ker} D = \{0\}$). The latter implies that $D^*J'D$ is selfadjoint and invertible, and hence $D^*J'D$ is congruent to some signature matrix, J say. We are looking for a J-spectral factorization of D.

Theorem 14.7. *Let $V(\lambda) = D + C(\lambda I_n - A)^{-1}B$ be a given rational $p \times m$ matrix function. Assume A has all its eigenvalues in the open left half plane and the $p \times m$ matrix D has full column rank. Let J' be a $p \times p$ signature matrix, and let E be an invertible $m \times m$ matrix such that $J = E^*D^*J'DE$ is an $m \times m$ signature* *matrix. Then the rational* $m \times m$ *matrix function* $W(\lambda) = V(-\bar{\lambda})^* J' V(\lambda)$ *has a left* J*-spectral factorization with respect to the imaginary axis,*

$$W(\lambda) = L_-(-\overline{\lambda})^* J L_-(\lambda),$$

if and only if the algebraic Riccati equation

$$\begin{aligned} XBJ^{-1}B^*X + X(A - BJ^{-1}D^*J'C) + (A^* - C^*J'DJ^{-1}B^*)X \quad (14.25) \\ + C^*J'DJD^*J'C - C^*J'C = 0 \end{aligned}$$

has a (unique) $i\mathbb{R}$*-stabilizing Hermitian solution* X*. In that case, the corresponding left* J*-spectral factor of* W *is given by*

$$L_-(\lambda) = E^{-1} + JE^*(D^*J'C - B^*X)(\lambda I_n - A)^{-1}B.$$

Recall that an $i\mathbb{R}$-stabilizing solution X of (14.25) is one such that the matrix $A - BJ^{-1}D^*J'C + BJ^{-1}B^*X$ has its eigenvalues in the open left half plane.

Proof. Put $\widehat{D} = DE$, $\widehat{B} = BE$, and consider the rational $m \times m$ matrix function

$$\widehat{W}(\lambda) = E^*W(\lambda)E = \widehat{V}(-\bar{\lambda})^* J' \widehat{V}(\lambda),$$

where $\widehat{V}(\lambda) = V(\lambda)E = DE + C(\lambda I_n - A)^{-1}BE$. Using the product rule for realizations, we see that W admits the realization $W(\lambda) = J + \widehat{C}(\lambda I_{2n} - \widehat{A})^{-1}\widehat{B}$, where

$$\widehat{A} = \begin{bmatrix} -A^* & C^*J'C \\ 0 & A \end{bmatrix}, \qquad \widehat{B} = \begin{bmatrix} C^*J'DE \\ BE \end{bmatrix}, \qquad \widehat{C} = \begin{bmatrix} -E^*B^* & E^*D^*J'C \end{bmatrix}.$$

Obviously, $\widehat{W}$ is selfadjoint on the imaginary axis. Furthermore, $\widehat{W}$ is biproper. Since A has all its eigenvalues in the open left half plane, we know that $\widehat{A}$ has no eigenvalue on $i\mathbb{R}$, and hence $\widehat{W}$ has no pole on $i\mathbb{R}$. We conclude that the realization $\widehat{W}(\lambda) = J + \widehat{C}(\lambda I_{2n} - \widehat{A})^{-1}\widehat{B}$ meets all the requirements of the first part of Theorem 14.5. It follows that $\widehat{W}$ admits a left J-spectral factorization with respect to the imaginary axis if and only if the Riccati equation (14.25) has a unique stabilizing Hermitian solution X. Moreover, in that case a left J-spectral factorization $\widehat{W}(\lambda) = K_-(-\bar{\lambda})^* J K_-(\lambda)$ of $\widehat{W}$ with respect to the imaginary axis is obtained by taking

$$K_-(\lambda) = I_m + J^{-1}E^*(D^*J'C - B^*X)(\lambda I_n - A)^{-1}BE.$$

Recall that $W(\lambda) = E^{-*}\widehat{W}(\lambda)E^{-1}$. It follows that W admits a left J-spectral factorization with respect to the imaginary axis if and only if so does $\widetilde{W}$. Thus the result of the preceding paragraph shows that W admits a left J-spectral factorization with respect to the imaginary axis if and only if the Riccati equation (14.25) has a unique stabilizing Hermitian solution X. Moreover, in that case a left J-spectral factorization $W(\lambda) = L_-(-\bar{\lambda})^* J L_-(\lambda)$ of W with respect to the imaginary axis is obtained by taking $L_-(\lambda) = K_-(\lambda)E^{-1}$. □

In our second example we assume that the rational $m \times m$ matrix function is given in the following manner (cf., the paragraph preceding Theorem 13.2):

$$W(\lambda) = J + C(\lambda I_n - A)^{-1}B - B^*(\lambda I_n + A^*)^{-1}C^*, \tag{14.26}$$

where A has only eigenvalues in the open left plane and J is a signature matrix. The function W admits a realization

$$W(\lambda) = J + \begin{bmatrix} -B^* & C \end{bmatrix} \left(\lambda I_{2n} - \begin{bmatrix} -A^* & 0 \\ 0 & A \end{bmatrix} \right)^{-1} \begin{bmatrix} C^* \\ B \end{bmatrix}. \tag{14.27}$$

This realization satisfies all the requirements of the first part of Theorem 14.5, which yields immediately the following result.

Theorem 14.8. *Let the rational $m \times m$ matrix function W be given by (14.26), where J is a signature matrix and A has its eigenvalues in the open left half plane. Then W admits a left J-spectral factorization with respect to the imaginary axis,*

$$W(\lambda) = L_-(-\bar{\lambda})^* J L_-(\lambda),$$

if and only if the algebraic Riccati equation

$$XBJB^*X + X(A - BJC) + (A^* - C^*JB^*)X + C^*JC = 0$$

has a (unique) Hermitian solution X such that the matrix $A - BJC + BJB^*X$ has all its eigenvalues in the open left half plane (so X is $i\mathbb{R}$-stabilizing). In that case the unique left J-spectral factor L_- for which $L_-(\infty) = I_m$ and its inverse L_-^{-1} are given by

$$L_-(\lambda) = I_m + J(C - B^*X)(\lambda I_n - A)^{-1}B,$$

So far we have mainly concentrated on left J-spectral factorizations. The analogous results for right J-spectral factorization of W can be obtained by simply applying the left factorization results to $V(\lambda) = W(-\lambda)$. Indeed, a left J-spectral factorization,

$$V(\lambda) = K_-(-\bar{\lambda})^* J K_-(\lambda),$$

of V with respect to $i\mathbb{R}$ yields a right J-spectral factorization,

$$W(\lambda) = L_+(-\bar{\lambda})^* J L_+(\lambda),$$

of W with respect to $i\mathbb{R}$ by taking $L_+(\lambda) = K_-(-\lambda)$.

Let us apply this observation to W given by the realization (14.27). Note that

$$V(\lambda) = W(-\lambda) = J + \begin{bmatrix} -C & B^* \end{bmatrix} \left(\lambda I_{2n} - \begin{bmatrix} -A & 0 \\ 0 & A^* \end{bmatrix} \right)^{-1} \begin{bmatrix} B \\ C^* \end{bmatrix}.$$

Since A has all its eigenvalues in the open left half plane, the same holds true for A^*. Thus we can apply Theorem 14.8 together with the above scheme to get the following right J-spectral factorization result.

Theorem 14.9. *Let the rational matrix function W be given by (14.26), where J is a signature matrix and A has its eigenvalues in the open left half plane. Then W admits a right J-spectral factorization with respect to the imaginary axis,*

$$W(\lambda) = L_+(-\bar{\lambda})^* J L_+(\lambda),$$

if and only if the algebraic Riccati equation

$$YC^*JCY + Y(A^* - C^*JB^*) + (A - BJC)Y + BJB^* = 0 \qquad (14.28)$$

has a (unique) Hermitian solution Y such that $A^* - C^*JB + C^*JCY$ has all its eigenvalues in the open left half plane (so X is $i\mathbb{R}$-stabilizing). In that case the unique right J-spectral factor L_+ for which $L_+(\infty) = I_m$ and its inverse L_+^{-1} are given by

$$L_+(\lambda) = I_m + J(CY - B^*)(\lambda I_n + A^*)^{-1}C^*.$$

14.5 J-spectral factorization with respect to other contours

In this section we consider J-spectral factorizations with respect to the real line $\mathbb{R}$ and to the unit circle $\mathbb{T}$ featuring an additional symmetry property between the factors. Here, as before, J is is an invertible Hermitian $m \times m$ matrix. We begin by considering the case of the unit circle.

Suppose W is a rational $m \times m$ matrix function. A factorization

$$W(\lambda) = L(\bar{\lambda}^{-1})^* J L(\lambda) \qquad (14.29)$$

is called a *right J-spectral factorization with respect to the unit circle* if L and L^{-1} are rational $m \times m$ matrix functions which are analytic on the closed unit disc. In that case the function $L(\bar{\lambda}^{-1})^*$ and its inverse are analytic on the closure of $\mathbb{D}_{\text{ext}}$ (infinity included). Thus a right J-spectral factorization with respect to the unit circle is a right canonical factorization with respect to $\mathbb{T}$ featuring an additional symmetry property between the factors. A factorization (14.29) is called a *left* J-spectral *spectral factorization with respect to the unit circle* if L and L^{-1} are rational $m \times m$ matrix functions which are analytic on the closure of $\mathbb{D}_{\text{ext}}$ (infinity included), in which case the function $L(\bar{\lambda}^{-1})^*$ and its inverse are analytic on the closed unit disc. Such a factorization is a left canonical factorization with respect to $\mathbb{T}$.

The case of J-spectral factorization with respect to the unit circle is somewhat more complicated than that of J-spectral factorization with respect to the imaginary axis. The first result is an analogue of Proposition 14.1.

Proposition 14.10. *Let W be a rational $m \times m$ matrix function that is selfadjoint on the unit circle and has neither poles nor zeros there. Then there exists a signature matrix J such for each $\lambda \in \mathbb{T}$ the matrix $W(\lambda)$ is congruent to J. For such a matrix*

J, the function W admits a right (respectively, left) J-spectral factorization with respect to the unit circle if and only if it admits a right (respectively, left) canonical factorization with respect to the unit circle.

We can use a Möbius transform to reduce the case of the unit circle to the case of the imaginary axis. To be precise, let $V(\lambda) = W\big((\lambda - i)/(\lambda + i)\big)$. Then V is a rational $m \times m$ matrix function that has neither poles nor zeros on the imaginary axis, and has selfadjoint values there. Moreover, $V(\infty) = W(1)$, and thus V is biproper. Also, right and left J-spectral factorizations of W, and right and left canonical factorization of W can easily be obtained from the corresponding factorizations of V. Thus the proposition above actually follows from Proposition 14.1. For the sake of completeness we shall give a direct proof.

Proof. By assumption, $W(\lambda)$ is invertible and selfadjoint for each $\lambda \in \mathbb{T}$. Thus the number of eigenvalues of $W(\lambda)$ in the open unit disc does not depend on the particular choice of $\lambda \in \mathbb{T}$. In other words $W(\lambda)$ has constant signature on $\mathbb{T}$. Now let J be a signature matrix the signature of which is equal to this constant signature. Then for each $\lambda \in \mathbb{T}$ the matrix $W(\lambda)$ is congruent to J.

Let $W(\lambda) = W_-(\lambda)W_+(\lambda)$ be a right canonical factorization of W with respect to $\mathbb{T}$. Consider

$$\widetilde{W}_+(\lambda) = W_+(\bar{\lambda}^{-1})^*, \qquad \widetilde{W}_-(\lambda) = W_-(\bar{\lambda}^{-1})^*.$$

Then $W(\lambda) = \widetilde{W}_+(\lambda)\widetilde{W}_-(\lambda)$ is again a right canonical factorization of W with respect to $\mathbb{T}$. It follows that $\widetilde{W}_+(\lambda)^{-1}W_-(\lambda)$ is a constant matrix, F say. This shows that $W(\lambda) = W_+(\bar{\lambda}^{-1})^* F W_+(\lambda)$. Since $W(\lambda)$ is selfadjoint for $\lambda \in \mathbb{T}$, it follows that F is congruent to the signature matrix J introduced in the first paragraph of the proof. Thus $F = E^*JE$ for some invertible matrix E. Put $L_+(\lambda) = EW_+(\lambda)$. Then $W(\lambda) = L_+(\bar{\lambda}^{-1})^* J L_+(\lambda)$ is a left J-spectral factorization of W with respect to the unit circle. The reverse implication is trivial. □

In what follows we assume that W is a biproper rational $m \times m$ matrix function which is selfadjoint on the unit circle and has no pole there. Such a function can be represented in the form

$$W(\lambda) = D_0 + C(\lambda I_n - A)^{-1}B + B^*(\lambda^{-1}I_n - A^*)^{-1}C^*,$$

where A has all its eigenvalues in the open unit disc. The fact that W is proper implies that W is analytic at zero. We shall assume additionally that A is invertible. Note that the invertibility of A follows from the analyticity at zero whenever the realization $C(\lambda - A)^{-1}B$ is minimal.

The invertibility assumption on A allows us to write

$$W(\lambda) = D_0 - B^*A^{-*}C^* + C(\lambda - A)^{-1}B - B^*A^{-*}(\lambda - A^{-*})^{-1}A^{-*}C^*.$$

Since $W(\infty) = D_0 - B^*A^{-*}C^* = W(0)^*$ one has

$$D_0 - B^*A^{-*}C^* = (D_0 - CA^{-1}B)^*.$$

Hence D_0 is selfadjoint. We shall assume additionally that $D_0 = J_0$ for some signature matrix J_0. Thus W is of the form

$$W(\lambda) = J_0 - B^*A^{-*}C^* + C(\lambda - A)^{-1}B - B^*A^{-*}(\lambda - A^{-*})^{-1}A^{-*}C^*. \quad (14.30)$$

We shall prove the following factorization result.

Theorem 14.11. *Let W be a biproper rational $m \times m$ matrix function given by (14.30), where J_0 is a signature matrix and A is an invertible $n \times n$ matrix having all its eigenvalues in the open unit disc. In order that, for some signature matrix J the function W admits a left J-spectral factorization with respect to the unit circle, it is necessary and sufficient that there exists a Hermitian $n \times n$ matrix Y such that $J_0 + B^*YB$ is invertible and Y is a solution of the equation*

$$Y = A^*YA - (C^* + A^*YB)(J_0 + B^*YB)^{-1}(C + B^*YA) \quad (14.31)$$

with $A - B(J_0 + B^*YB)^{-1}(C + B^*YA)$ having all its eigenvalues in the open unit disc. In that case Y is unique and for J one can take any signature matrix J determined by

$$J_0 + B^*YB = E^*JE, \quad (14.32)$$

where E is some invertible matrix. Furthermore, if Y is a Hermitian matrix with the properties mentioned above, then for a signature matrix J determined by the expression (14.32), a left J-spectral factorization $W(\lambda) = L_-(\bar{\lambda}^{-1})^*JL_-(\lambda)$ of W with respect to the unit circle is obtained by taking

$$L_-(\lambda) = E + E(J_0 + B^*YB)^{-1}(C + B^*YA)(\lambda I_n - A)^{-1}B. \quad (14.33)$$

Equation (14.31) is a particular case of the so-called *discrete algebraic Riccati equation*. A solution Y of equation (14.31) is called $\mathbb{T}$-*stabilizing*, or simply *stabilizing* when no confusion can arise, if $J_0 + B^*YB$ is invertible and the matrix $A - B(J_0 + B^*YB)^{-1}(C + B^*YA)$ has all its eigenvalues in the open unit disc. In the above theorem, the existence of such a solution is required.

Proof. We split the proof into six parts.

Part 1. Since W is biproper and given by (14.30), we can write a realization for W. In fact $W(\lambda) = D + \widetilde{C}(\lambda - \widetilde{A})^{-1}\widetilde{B}$, where $D = W(\infty) = J_0 - B^*A^{-*}C^*$ and

$$\widetilde{A} = \begin{bmatrix} A^{-*} & 0 \\ 0 & A \end{bmatrix}, \quad \widetilde{B} = \begin{bmatrix} -A^{-*}C^* \\ B \end{bmatrix}, \quad \widetilde{C} = \begin{bmatrix} B^*A^{-*} & C \end{bmatrix}. \quad (14.34)$$

Recall that the matrix A is invertible and has all its eigenvalues in the open unit disc $\mathbb{D}$. Hence A^{-*} has all its eigenvalues in $\mathbb{D}_{\text{ext}}$. This allows us to apply Theorem 12.5 with $F_- = \mathbb{D}$ and $F_+ = \mathbb{D}_{\text{ext}}$. It follows that W admits a left canonical factorization with respect to $\mathbb{T}$ if and only if the equation

$$\begin{aligned} YBD^{-1}B^*A^{-*}Y - Y(A - BD^{-1}C) \\ + (A^{-*} + A^{-*}C^*D^{-1}B^*A^{-*})Y + A^{-*}C^*D^{-1}C = 0 \end{aligned} \quad (14.35)$$

has a unique solution Y satisfying the following additional spectral constraints:

$$\sigma\big(A^{-*} + (A^{-*}C^* + YB)D^{-1}B^*A^{-*}\big) \quad \subset \quad \mathbb{D}_{\text{ext}}, \tag{14.36}$$

$$\sigma\big(A - BD^{-1}(B^*A^{-*}Y + C)\big) \quad \subset \quad \mathbb{D}. \tag{14.37}$$

Furthermore, if Y is such a solution of (14.35), then a left canonical factorization $W(\lambda) = W_1(\lambda)W_2(\lambda)$ of W with respect to $\mathbb{T}$ is obtained by taking

$$W_1(\lambda) = D - B^*A^{-*}(\lambda - A^{-*})^{-1}(A^{-*}C^* + YB), \tag{14.38}$$

$$W_2(\lambda) = I + D^{-1}(B^*A^{-*}Y + C)(\lambda - A)^{-1}B. \tag{14.39}$$

Let Y be the solution of (14.35) satisfying (14.36) and (14.37). We claim that $J_0 + B^*YB$ is invertible. To prove this it will be convenient to rewrite W_1 as a function of λ^{-1}. This can be done as follows:

$$\begin{aligned} W_1(\lambda) &= D - B^*(\lambda A^* - I)^{-1}(A^{-*}C^* + YB) \\ &= D + B^*\lambda^{-1}(\lambda^{-1} - A^*)^{-1}(A^{-*}C^* + YB) \\ &= D + B^*(\lambda^{-1} - A^* + A^*)(\lambda^{-1} - A^*)^{-1}(A^{-*}C^* + YB) \\ &= D + B^*A^{-*}C^* + B^*YB + B^*(\lambda^{-1} - A^*)^{-1}(C^* + A^*YB). \end{aligned}$$

Recall that $D = J_0 - B^*A^{-*}C^*$. Thus

$$W_1(\lambda) = J_0 + B^*YB + B^*(\lambda^{-1} - A^*)^{-1}(C^* + A^*YB). \tag{14.40}$$

Since A is invertible, both W and W_2 are analytic at zero. From the above formula for W_1 we see that W_1 is also analytic at zero. Hence $W(0) = W_1(0)W_2(0)$. But $W(0)$ is invertible. Thus $W_1(0) = J_0 + B^*YB$ is invertible too.

Part 2. In this part Y stands for a solution of (14.35) such that $J_0 + B^*YB$ is invertible. We prove that in this case Y is also a solution of (14.31). Furthermore, we show that

$$D^{-1}(C + B^*A^{-*}Y) = (J_0 + B^*YB)^{-1}(C + B^*YA). \tag{14.41}$$

Multiplying (14.35) on the left by A^* and regrouping terms one obtains

$$A^*YA - Y - (A^*YB + C^*)D^{-1}(C + B^*A^{-*}Y) = 0. \tag{14.42}$$

So $Y = A^*YA - (A^*YB + C^*)D^{-1}(C + B^*A^{-*}Y)$. Multiplying the latter identity on the left with B^*A^{-*} and adding C to both sides gives

$$C + B^*A^{-*}Y = C + B^*YA - (B^*YB + B^*A^{-*}C^*)D^{-1}(C + B^*A^{-*}Y).$$

It follows that

$$\begin{aligned} C + B^*YA &= \left(I + (B^*YB + B^*A^{-*}C^*)D^{-1}\right)(C + B^*A^{-*}Y) \\ &= \left(D + B^*A^{-*}C^* + B^*YB\right) D^{-1}(C + B^*A^{-*}Y) \\ &= (J_0 + B^*YB)D^{-1}(C + B^*A^{-*}Y). \end{aligned}$$

Since $J_0 + B^*YB$ is invertible, we see that (14.41) holds. Using (14.41) in (14.42) gives that Y is a solution of (14.31).

Part 3. In this part we show that Y^* is a solution of (14.35) whenever so is Y. For this purpose we consider the Hamiltonian T of (14.35), that is,

$$T = \begin{bmatrix} -A^{-*} - A^{-*}C^*D^{-1}B^*A^{-*} & -A^{-*}C^*D^{-1}C \\ BD^{-1}B^*A^{-*} & -(A - BD^{-1}C) \end{bmatrix}.$$

Note that $T = -(\widetilde{A} - \widetilde{B}D^{-1}\widetilde{C})$, where $\widetilde{A}$, $\widetilde{B}$ and $\widetilde{C}$ are given by (14.34). Put

$$H = \begin{bmatrix} 0 & I \\ -I & 0 \end{bmatrix}.$$

Then $H\widetilde{A} = \widetilde{A}^{-*}H$, $H\widetilde{B} = \widetilde{A}^{-*}\widetilde{C^*}$ and $H = -H^*$. Next we carry out the following computation:

$$\begin{aligned} D - \widetilde{C}\widetilde{A}^{-1}\widetilde{B} &= D - \begin{bmatrix} B^*A^{-*} & C \end{bmatrix} \begin{bmatrix} A^* & 0 \\ 0 & A^{-1} \end{bmatrix} \begin{bmatrix} -A^{-*}C^* \\ B \end{bmatrix} \\ &= D - \begin{bmatrix} B^* & CA^{-1} \end{bmatrix} \begin{bmatrix} -A^{-*}C^* \\ B \end{bmatrix} \\ &= D + B^*A^{-*}C^* - CA^{-1}B = J_0 - CA^{-1}B = D^*. \end{aligned}$$

Thus $D - \widetilde{C}\widetilde{A}^{-1}\widetilde{B} = D^*$ and we can apply item (iii) in Proposition 9.2 to show that T is invertible and $HT = T^*H$.

Taking adjoints in (14.35) we obtain the equation

$$\begin{aligned} Y^*A^{-1}BD^{-*}B^*Y^* + Y^*(A^{-1} + A^{-1}BD^{-*}CA^{-1}) \\ -(A^* - C^*D^{-*}B^*)Y^* + C^*D^{-*}CA^{-1} = 0, \end{aligned} \tag{14.43}$$

where Y^* is the unknown. The Hamiltonian T_* of this equation is given by

$$T_* = \begin{bmatrix} A^* - C^*D^{-*}B^* & -C^*D^{-*}CA^{-1} \\ A^{-1}BD^{-*}B^* & A^{-1} + A^{-1}BD^{-*}CA^{-1} \end{bmatrix}.$$

It follows that $T_* = HT^*H$. This together with the result of the previous paragraph shows that $T_* = T^{-1}$.

Now let Y be a solution of (14.35). It follows that Y^* is a solution of (14.43). Using the general theory of Riccati equations (see Section 12.1), this implies that the space

$$N_* = \operatorname{Im} \begin{bmatrix} Y^* \\ I \end{bmatrix}$$

is invariant under T_*. But $T_* = T^{-1}$. Thus the finite dimensional space N_* is invariant under the Hamiltonian T of (14.35). But then (again see Section 12.1) we may conclude that Y^* is a solution of (14.35) too.

Part 4. Let Y be a solution of (14.35) satisfying the additional spectral constraints (14.36) and (14.37). In this part we show that Y must be Hermitian. Now Y is uniquely determined by the given properties. Since, by the result of the previous part of the proof, Y^* a solution of (14.35), it thus suffices to show that the conditions (14.36) and (14.37) hold with Y^* in place of the matrix Y.

From the first part of the proof we know that $J_0 + B^*YB$ is invertible. Hence the identity (14.41) holds. Using this identity, we can rewrite (14.37) as

$$\sigma\big(A - B(J_0 + B^*YB)^{-1}(C + B^*YA)\big) \subset \mathbb{D}.$$

Taking adjoints, we arrive at $\sigma\big((A^* - (A^*Y^*B + C^*)(J_0 + B^*YB)^{-1}B^*\big) \subset \mathbb{D}$. Next, note that

$$\begin{aligned}
\big(A^* - (A^*Y^*B + C^*)&(J_0 + B^*YB)^{-1}B^*\big)^{-1} \\
&= \big(I - (Y^*B + A^{-*}C^*)(J_0 + B^*YB)^{-1}B^*\big)^{-1}A^{-*} \\
&= \Big(I + (Y^*B + A^{-*}C^*)\big(J_0 + B^*YB \\
&\qquad\qquad -B^*(Y^*B + A^{-*}C^*)\big)^{-1}B^*\Big)A^{-*} \\
&= A^{-*} + (Y^*B + A^{-*}C^*)D^{-1}B^*A^{-*}.
\end{aligned}$$

Here we used that $D = J_0 - B^*A^{-*}C^*$. We conclude that

$$\sigma(A^{-*} + (Y^*B + A^{-*}C^*)D^{-1}B^*A^{-*}) \subset \mathbb{D}_{\text{ext}},$$

which is (14.36) with Y^* in place of Y.

In Part 3 of the proof we saw that Y^* is a solution of (14.35). Furthermore, $J_0 + B^*Y^*B = (J_0 + B^*Y^*B)^*$ is invertible. Thus we know that (14.41) holds with Y^* in place of Y, that is,

$$D^{-1}(C + B^*A^{-*}Y^*) = (J_0 + B^*Y^*B)^{-1}(C + B^*Y^*A). \tag{14.44}$$

Using this we show that (14.37) holds with Y^* in place of Y. Indeed, taking adjoints in (14.36) we get $\sigma(A^{-1} + A^{-1}BD^{-*}(B^*Y^* + CA^{-1}) \subset \mathbb{D}\text{ext}$. Now

$$\begin{aligned}(A^{-1} + A^{-1}BD^{-*}(B^*Y^* + CA^{-1})^{-1} \\ &= \left(I + BD^{-*}(B^*Y^* + CA^{-1})\right)^{-1}A \\ &= \left(I - B(D^* + B^*Y^*B + CA^{-1}B)^{-1}(B^*Y^* + CA^{-1})\right)^{-1}A \\ &= A - B(J_0 + B^*Y^*B)^{-1}(B^*Y^*A + C).\end{aligned}$$

Here we used that $D^* = J_0 - CA^{-1}B$. Now apply the identity (14.44). It follows that $\sigma(A - BD^{-1}(C + B^*A^{-*}Y^*)) \subset \mathbb{D}$, which is (14.37) with Y^* in place of Y.

Part 5. Let Y be a Hermitian matrix such that $J_0 + B^*YB$ is invertible and Y is a stabilizing solution of (14.31). In this part we show that in that case Y is a solution of (14.35) and that Y satisfies the spectral constraints (14.36) and (14.37).

As a first step let us prove that under the above conditions on Y again (14.41) holds. Indeed, multiplying (14.31) from the left by B^*A^{-*} and adding C to both sides we get

$$\begin{aligned}C + B^*A^{-*}Y &= C + B^*YA - (B^*A^{-*}C^* + B^*YB) \\ &\qquad \cdot(J_0 + B^*YB)^{-1}(C + B^*YA) \\ &= ((J_0 + B^*YB) - (B^*A^{-*}C^* + B^*YB)) \\ &\qquad \cdot(J_0 + B^*YB)^{-1}(C + B^*YA) \\ &= (J_0 - B^*YB)(J_0 + B^*YB)^{-1}(C + B^*YA) \\ &= D(J_0 + B^*YB)^{-1}(C + B^*YA).\end{aligned}$$

Hence (14.41) holds indeed. Using this we can rewrite (14.31) as

$$A^*YA - Y - (A^*YB + C^*)D^{-1}(C + B^*A^{-*}Y) = 0.$$

Multiplying the latter on the left by A^{-*} and regrouping terms we see that Y satisfies (14.35).

Since Y is a stabilizing solution of (14.31) and (14.41) holds, the spectral constraint (14.37) is satisfied too. It remains to prove (14.36) To do this we first note that

$$
\begin{aligned}
&\left(A^{-1} + A^{-1}BD^{-*}(B^*Y + CA^{-1})\right)^{-1} \\
&\qquad = \left(I + BD^{-*}(B^*Y + CA^{-1})\right)^{-1} A \\
&\qquad = \left(I - B(D^* + B^*Y^*B + CA^{-1}B)^{-1}(B^*Y^* + CA^{-1})\right)^{-1} A \\
&\qquad = A - B(J_0 + B^*Y^*B)^{-1}(B^*Y^*A + C) \\
&\qquad = A - BD^{-1}(C + B^*A^{-*}Y).
\end{aligned}
$$

Thus, since Y is Hermitian, we see that (14.36) follows from (14.37) by taking adjoints and an inverse.

Because of the uniqueness of the solution Y in the first part of the proof, the result of the present part also shows that the Hermitian stabilizing solution of (14.31), if it exists, is unique

Part 6. In this final part we complete the argument. Assume that for some J the function W admits a left J-spectral factorization with respect to the unit circle. Then by the first part of the proof, equation (14.35) has a solution Y satisfying (14.36) and (14.37). Moreover for this Y we have that $J_0 + B^*YB$ is invertible. Part 4 of the proof tells us that Y is Hermitian. From Part 2 we know that Y is a solution of (14.31) which, according to (14.37) and (14.41), is stabilizing.

Conversely, if Y is a Hermitian matrix such that $J_0 + B^*YB$ is invertible and Y is a stabilizing solution of (14.31), then Y is a solution of (14.35) and Y satisfies (14.36) and (14.37). Hence W admits a left canonical factorization with respect to the unit circle, and thus, by Proposition 14.10, also a left J-spectral factorization with respect to the unit circle.

Finally, take a signature matrix J such that (14.32) holds. It remains to establish the formula for the left spectral factor L_-. To do this we use the left canonical factorization $W(\lambda) = W_1(\lambda)W_2(\lambda)$ obtained in Part 1. Combining (14.39) and (14.41) we get $W_2(\lambda) = I + (J_0 + B^*YB)^{-1}(C + B^*YA)(\lambda - A)^{-1}B$. Thus, using the expression (14.40) for $W_1(\lambda)$,

$$
\begin{aligned}
W_2(\bar{\lambda}^{-1})^* &= I + B^*(\lambda^{-1} - A^*)^{-1}(C^* + A^*YB)(J_0 + B^*YB)^{-1} \\
&= \left(J_0 + B^*YB + B^*(\lambda^{-1} - A^*)^{-1}(C^* + A^*YB)\right)(J_0 + B^*YB)^{-1} \\
&= W_1(\lambda)(J_0 + B^*YB)^{-1},
\end{aligned}
$$

and it follows that $W(\lambda) = W_2(\bar{\lambda}^{-1})^*(J_0 + B^*YB)W_2(\lambda)$. Now let J be a signature matrix such that (14.32) holds. Then we see that $W(\lambda) = L_-(\bar{\lambda}^{-1})^*JL_-(\lambda)$, with L_- given by (14.33), is a left J-spectral factorization with respect to the unit circle. □

We now turn to a situation arising from linear-quadratic optimal control theory. It concerns the following version of the *discrete algebraic Riccati equation*

$$X = A^*XA + Q - A^*XB(R + B^*XB)^{-1}B^*XA. \tag{14.45}$$

Here A, B, Q and R are given matrices of sizes $n \times n$, $n \times m$, $n \times n$ and $m \times m$, respectively. We will consider the case when A has all its eigenvalues in the open unit circle, R and Q are Hermitian, and R is invertible. Of special interest are the stabilizing solutions of (14.45). A solution X of (14.45) is said to be $\mathbb{T}$-*stabilizing*, or simply *stabilizing* when there is no danger of confusion, if $R+B^*XB$ is invertible and $A - B(R + B^*XB)^{-1}B^*XA$ has all its eigenvalues in the open unit disc. In connection with (14.45) we consider the rational matrix function

$$W(\lambda) = R + B^*(\lambda^{-1}I_n - A^*)^{-1}Q(\lambda I_n - A)^{-1}B. \tag{14.46}$$

Note that this function is Hermitian on the unit circle.

Proposition 14.12. *Let A, B, Q and R be as above, so A is an $n \times n$ matrix having its eigenvalues in the open unit disc, B is an $m \times m$ matrix, R is an invertible Hermitian $m \times m$ matrix, and Q is a Hermitian $n \times n$ matrix. Assume in addition that A is invertible. The following two statements are equivalent:*

(i) *The Riccati equation* (14.45) *has a (unique) Hermitian $\mathbb{T}$-stabilizing solution;*

(ii) *For some Hermitian matrix J, the rational matrix function* (14.46) *admits a left J-spectral factorization with respect to the unit circle.*

In that case J is congruent to $R+B^*XB$. Also, if X is the Hermitian $\mathbb{T}$-stabilizing solution of (14.45), *then*

$$W(\lambda) = L_-(\bar{\lambda}^{-1})^*(R + B^*XB)L_-(\lambda),$$

with

$$L_-(\lambda) = I_m + (R + B^*XB)^{-1}B^*XA(\lambda I_n - A)^{-1}B,$$

is a left $(R + B^*XB)$-spectral factorization with respect to the unit disc. The function L_- is the unique left $(R + B^*XB)$-spectral factor with $L_-(\infty) = I_m$.

The additional assumption that A is invertible plays an essential role in the proof as we give it below. Indeed, the argument involves a reduction to earlier results, in particular to Theorem 14.11. However, instead of Theorem 14.11 one can employ Theorem 14.15 below which does not feature the hypothesis that A is invertible.

Before we prove the proposition, let us remark that in the case of the linear quadratic optimal control problem of mathematical systems theory, one has that R is positive definite and Q is positive semidefinite. Hence the function (14.46) is positive definite on the unit circle, and as A has is eigenvalues in the open unit disc, it has no poles on the unit circle. Thus, in that case, the function does admit a right spectral factorization with $J = I$, and hence there is a stabilizing solution X to the discrete algebraic Riccati equation. In addition, for that solution the matrix $R + B^*XB$ is positive definite.

Proof. We shall deduce Proposition 14.12 from Theorem 14.11. First, a realization for (14.46) is given as

$$W(\lambda) = R + \begin{bmatrix} -B^*A^{-*}Q & B^*A^{-*} \end{bmatrix} \left(\lambda - \begin{bmatrix} A & 0 \\ -A^{-*}Q & A^{-*} \end{bmatrix} \right)^{-1} \begin{bmatrix} B \\ 0 \end{bmatrix}.$$

Since A is has all its eigenvalues in the open unit disc, there is a unique solution to the equation

$$X_0 - A^*X_0A = Q. \tag{14.47}$$

Taking as a similarity transformation the matrix

$$\begin{bmatrix} I & 0 \\ X_0 & I \end{bmatrix},$$

and using $Q - X_0 = -A^*X_0A$, the realization above can be rewritten as:

$$\begin{aligned} W(\lambda) &= R + \begin{bmatrix} -B^*A^{-*}(Q - X_0) & B^*A^{-*} \end{bmatrix} \left(\lambda - \begin{bmatrix} A & 0 \\ 0 & A^{-*} \end{bmatrix} \right)^{-1} \begin{bmatrix} B \\ X_0B \end{bmatrix} \\ &= R + B^*X_0A(\lambda - A)^{-1} + B^*A^{-*}(\lambda - A^{-*})^{-1}X_0B. \end{aligned}$$

The latter expression is of the form (14.30), with $C = B^*X_0A$ and with $J_0 = R + B^*X_0B$. So, we can apply Theorem 14.11, with (14.31) suitably modified, to conclude that W admits a left J-spectral factorization if and only if there is a solution Y, satisfying additional constraints, of the equation

$$Y = A^*YA - (A^*X_0B + A^*YB)(R + B^*X_0B + B^*YB)^{-1}(B^*YA + B^*X_0A).$$

Putting $X = X_0 + Y$ and taking into account (14.47), we see that the above equation becomes (14.45) for X. The additional constraints referred to above are: in the first place, invertibility of $R + B^*X_0B + B^*YB = R + B^*XB$, which we also required for the solution of (14.45), and, secondly, the condition that the eigenvalues of

$$A - B(R + B^*(X_0 + Y)B)^{-1}B^*(X_0 + Y)A = A - B(R + B^*XB)^{-1}B^*XA$$

are in the open unit disc. But this is exactly what is required for the stabilizing solution of the equation (14.45).

The expressions for the factorization also follow directly from the formulas in Theorem 14.11. □

We conclude this section by considering J-spectral factorization of a selfadjoint function on the real line. As before J is an invertible Hermitian $m \times m$ matrix.

Suppose W is a rational $m \times m$ matrix function. A factorization

$$W(\lambda) = L(\bar{\lambda})^* J L(\lambda) \tag{14.48}$$

is called a *right J-spectral factorization with respect to the real line* if L and L^{-1} are rational $m \times m$ matrix functions which are analytic on the closed upper half plane (infinity included). In that case the function $L(\bar{\lambda})^*$ and its inverse are analytic on the closed lower half plane (infinity included). Thus a right J-spectral factorization with respect to the real line is a right canonical factorization with respect to $\mathbb{R}$ featuring an additional symmetry property between the factors. A factorization (14.48) is called a *left J-spectral factorization with respect to the real line* if L and L^{-1} are rational $m \times m$ matrix functions which are analytic on the closed lower half plane (infinity included), in which case the function $L(\bar{\lambda})^*$ and its inverse are analytic on the closed upper half plane (infinity included). Such a factorization is a left canonical factorization with respect to $\mathbb{R}$.

Results for this type of factorization can be derived in a straightforward manner from J-spectral factorization theorems with respect to the imaginary axis. Indeed, if W is selfadjoint on the real line, then V given by $V(\lambda) = W(-i\lambda)$ is selfadjoint on the imaginary axis. Also $W(\lambda) = L_+(\bar{\lambda}) J L_+(\lambda)$ is a right J-spectral factorization of W with respect to the real line if and only $V(\lambda) = K_+(-\bar{\lambda}) J K_+(\lambda)$, with $K_+(\lambda) = L_+(-i\lambda)$, is a right J-spectral factorization of V with respect to the imaginary axis. As an illustration we show how one can derive the following result as a corollary from Theorem 14.9.

Theorem 14.13. *Let the rational $m \times m$ matrix function W be given by*

$$W(\lambda) = J + C(\lambda I_n - A)^{-1} B + B^*(\lambda I_n - A^*)^{-1} C^*,$$

where J is an $m \times m$ signature matrix and A is an $n \times n$ matrix having all its eigenvalues in the open upper half plane. Then W admits a right J-spectral factorization with respect to the real line,

$$W(\lambda) = L_+(\bar{\lambda})^* J L_+(\lambda),$$

if and only if the algebraic Riccati equation

$$YC^*JCY - Y(A^* - C^*JB^*) + (A - BJC)Y - BJB^* = 0 \tag{14.49}$$

has a (unique) skew-Hermitian solution Y such that $A^* - C^*JB^* - C^*JCY$ has all its eigenvalues in the open lower half plane. In that case, the unique right J-spectral factor L_+ for which $L_+(\infty) = I_m$ is given by

$$L_+(\lambda) = I_m + J(CY + B^*)(\lambda I_n - A^*)^{-1} C^*.$$

A solution Y of the Riccati equation (14.49) is called $\mathbb{R}$-*stabilizing*, or simply *stabilizing* when confusion is not possible, if $A^* - C^*JB^* - C^*JCY$ has all its eigenvalues in the open lower half plane. In the above theorem, the existence of such a solution is required.

Proof. Write $V(\lambda) = W(-i\lambda)$. Then

$$\begin{aligned} V(\lambda) &= J + C(-i\lambda - A)^{-1}B + B^*(-i\lambda - A^*)^{-1}C \\ &= J + (iC)\big(\lambda - (iA)\big)^{-1}B + B^*\big(\lambda + (iA)^*\big)^{-1}(iC). \end{aligned}$$

Notice that iA has all its eigenvalues in the open left half plane. By Theorem 14.9 the function V admits a right J-spectral factorization with respect to the imaginary axis if and only if the equation

$$\begin{aligned} X(iC)^*J(iC)X + X\big((iA)^* - (iC)^*JB^*\big) \\ +\big(iA - BJ(iC)\big)X + BJB^* = 0 \end{aligned} \tag{14.50}$$

has a Hermitian solution X such that the matrix $(iA)^* - (iC)^*JB^* + (iC)^*J(iC)X$ has all its eigenvalues in the open left half plane. In that case, a right J-spectral factorization $V(\lambda) = K_+(-\bar{\lambda})^*JK_+(\lambda)$ of V with respect to the imaginary axis is obtained by taking $K_+(\lambda) = I + J(iCX - B^*)\big(\lambda + (iA)^*\big)^{-1}(iC)^*$. Next we replace X by iY and multiply equation (14.50) by -1. In this way (14.50) is shown to be equivalent to (14.49). Furthermore Y is skew-Hermitian if and only if X is Hermitian, and $A^* - C^*JB^* - C^*JCY = i\big((iA)^* - (iC)^*JB^* + (iC)^*J(iC)X\big)$. Finally, put $L_+(\lambda) = K_+(i\lambda)$. Then

$$\begin{aligned} L_+(\lambda) &= I + J(iCX - B^*)\big(\lambda + (iA)^*\big)^{-1}(iC)^* \\ &= I + J(-CY - B^*)(i\lambda - iA^*)^{-1}(-i)C^* \\ &= I + J(CY + B^*)(\lambda - A^*)^{-1}C^*. \end{aligned}$$

Using these formulas it is now straightforward to complete the argument. □

14.6 Left versus right J-spectral factorization

The existence of a left canonical factorization does not always imply the existence of a right canonical factorization. The same is true for J-spectral factorization. In this section we answer the following question: if a rational matrix function W admits a left J-spectral factorization, under what conditions does it also have a right J-spectral factorization? And, if so, how can the right factorization be obtained from the left one? The main result can be viewed as a symmetric version of Theorem 12.6. We restrict our attention to factorization with respect to the imaginary axis.

For later purposes it will be convenient to only assume that J is an invertible Hermitian matrix. We do not stipulate it to be a signature matrix here.

Theorem 14.14. *Let J be an invertible Hermitian $m \times m$ matrix, and let W be a rational $m \times m$ matrix function. Suppose*

$$W(\lambda) = L_-(-\bar{\lambda})^* J L_-(\lambda)$$

is a left J-spectral factorization with respect to the imaginary axis, and L_- admits the realization

$$L_-(\lambda) = I_m + C(\lambda I_n - A)^{-1}B \tag{14.51}$$

with A and $A^\times = A - BC$ having their eigenvalues in the open left half plane. Let Q and P be the unique (Hermitian) solutions of the Lyapunov equations

$$QA + A^*Q = C^*JC. \tag{14.52}$$

$$A^\times P + P(A^\times)^* = -BJ^{-1}B^*. \tag{14.53}$$

Then W admits a right J-spectral factorization with respect to the imaginary axis if and only if $I - QP$ is invertible, or, which amounts to the same, $I - PQ$ is invertible. In that case, a right J-spectral factorization of W with respect to the imaginary axis is given by

$$W(\lambda) = L_+(-\bar{\lambda})^* J L_+(\lambda), \tag{14.54}$$

where $L_+(\lambda)$ and its inverse are given by

$$L_+(\lambda) = I_m + (CP - J^{-1}B^*)(I - QP)^{-1} \cdot(\lambda I_n + A^*)^{-1}(C^*J - QB), \tag{14.55}$$

$$L_+^{-1}(\lambda) = I_m - (CP - J^{-1}B^*)\big(\lambda I_n + (A^\times)^*\big)^{-1} \cdot(I - QP)^{-1}(C^*J - QB). \tag{14.56}$$

Proof. We bring ourselves in the situation of Section 12.4 by introducing

$$Y_+(\lambda) = L_-(-\bar{\lambda})^* = I_m - B^*(\lambda I_n + A^*)^{-1}C^*,$$

$$Y_-(\lambda) = JL_-(\lambda) = J + JC(\lambda I_n - A)^{-1}B.$$

Then $W(\lambda) = Y_+(\lambda)Y_-(\lambda)$ is a left canonical factorization, here taken with respect to the imaginary axis (cf., the remark made after the proof of Theorem 12.6). In terms of the notation employed in Section 12.4,

$$Y_+(\lambda) = D_+ + C_+(\lambda - A_+)^{-1}B_+,$$

$$Y_-(\lambda) = D_- + C_-(\lambda - A_-)^{-1}B_-,$$

with

$$D_+ = I_m, \qquad A_+ = -A^*, \qquad B_+ = C^*, \qquad C_+ = -B^*,$$
$$D_- = J, \qquad A_- = A, \qquad B_- = B, \qquad C_- = JC.$$

For the associate main matrices we have $A_+^\times = -(A^\times)^*$ and $A_-^\times = A^\times$. Thus the Lyapunov equations (12.19) reduce to the equations (14.53) and (14.52). Application of Theorem 12.6 now shows that W admits a right canonical factorization with respect to the imaginary axis if and only if $I - QP$ is invertible, or, which amounts to the same, $I - PQ$ is invertible.

Assume this is the case. Then, again by virtue of Theorem 12.6, we have the right canonical factorization $W(\lambda) = W_-(\lambda)W_+(\lambda)$, where

$$\begin{aligned} W_-(\lambda) &= D_+ + (D_+C_- + C_+Q)(\lambda I_{X_-} - A_-)^{-1} \\ &\qquad \cdot (I_{X_-} - PQ)^{-1}(B_-D_-^{-1} - PB_+), \\ W_+(\lambda) &= D_- + (D_+^{-1}C_+ + C_-P)(I_{X_+} - QP)^{-1} \\ &\qquad \cdot (\lambda I_{X_+} - A_+)^{-1}(B_+D_- - QB_-). \end{aligned}$$

Making the appropriate substitutions, we get

$$\begin{aligned} W_-(\lambda) &= I + (JC - B^*Q)(\lambda - A)^{-1}(I - PQ)^{-1}(BJ^{-1} - PC^*), \\ W_+(\lambda) &= J + (JCP - B^*)(I - QP)^{-1}(\lambda + A^*)^{-1}(C^*J - QB). \end{aligned}$$

Put $L_+(\lambda) = J^{-1}W_+(\lambda)$. Then $L_+(\lambda)$ is given by (14.55). Taking into account the selfadjointness of Q and P, one sees that $L_+(-\bar{\lambda})^*$ is precisely $W_-(\lambda)$. It follows that $W(\lambda) = L_+(-\bar{\lambda})^* J L_+(\lambda)$, and this factorization is a right J-spectral factorization of W with respect to the imaginary axis. Finally, $L_+^{-1}(\lambda) = W_+^{-1}(\lambda)J$, and according to Theorem 12.6,

$$\begin{aligned} W_+^{-1}(\lambda) &= D_-^{-1} - D_-^{-1}(D_+^{-1}C_+ + C_-P)(\lambda I_{X_+} - A_+^\times)^{-1} \\ &\qquad \cdot (I_{X_+} - QP)^{-1}(B_+D_- - QB_-)D_-^{-1}. \end{aligned}$$

Via the appropriate substitutions this becomes

$$W_+^{-1}(\lambda) = J^{-1} - (CP - J^{-1}B^*)\big(\lambda + (A^\times)^*\big)^{-1}(I - QP)^{-1}(C^*J - QB)J^{-1}.$$

Multiplying the latter identity from the right by J gives (14.56). □

For the case when J is a signature matrix (that is, $J = J^* = J^{-1}$) it is also possible to derive the previous result from Theorem 14.9. Indeed, let Q be the solution of (14.52), and introduce

$$T = \begin{bmatrix} I & 0 \\ Q & I \end{bmatrix}.$$

Then one has (via the product rule for realizations)

$$\begin{aligned} W(\lambda) &= L_-(-\bar{\lambda})^* J L_-(\lambda) \\ &= J + \begin{bmatrix} JC & -B^* \end{bmatrix} T \left(\lambda - T^{-1} \begin{bmatrix} A & 0 \\ C^*JC & -A^* \end{bmatrix} T \right)^{-1} T^{-1} \begin{bmatrix} B \\ C^*J \end{bmatrix} \\ &= J + (JC - B^*Q)(\lambda - A)^{-1}B - B^*(\lambda + A^*)^{-1}(C^*J - QB). \end{aligned}$$

Clearly, one can now apply Theorem 14.9. The stabilizing solution of equation (14.28), taken for this particular situation, and the solution P of (14.53) are related as follows: if Y is the stabilizing solution, then $I+QY$ is invertible, the matrix $P = Y(I+QY)^{-1}$ solves (14.53), and $I - QP = (I+QY)^{-1}$ is invertible. Conversely, if P is the solution of (14.53) and $I - QP$ is invertible, then $Y = P(I-QP)^{-1}$ is Hermitian and it is the desired stabilizing solution.

Finally, for the case where $J = I$, and so W is positive definite on the imaginary line, the condition that $I - QP$ is invertible should be automatically fulfilled on account of Theorem 9.4. That this is indeed the case can be seen as follows. First recall that A has all its eigenvalues in the open left half plane. This implies that P is positive semidefinite and Q is negative semidefinite. Since $J = I$ we get from (14.53) that $\operatorname{Ker} P$ is invariant under A^*. Now write P, Q, A and C with respect to the decomposition $\mathbb{C}^n = \operatorname{Ker} P \dot{+} \operatorname{Im} P$ as

$$P = \begin{bmatrix} 0 & 0 \\ 0 & P_{22} \end{bmatrix}, \quad Q = \begin{bmatrix} Q_{11} & Q_{12} \\ Q_{21} & Q_{22} \end{bmatrix}, \quad A = \begin{bmatrix} A_{11} & 0 \\ A_{21} & A_{22} \end{bmatrix}, \quad C = \begin{bmatrix} C_1 & C_2 \end{bmatrix}.$$

Then Q_{22} is negative semidefinite and P_{22} is positive definite. Finally, $I - QP$ is invertible if and only if $I - Q_{22}P_{22}$ is invertible as a map from $\operatorname{Im} P$ to itself. Since $I - Q_{22}P_{22}$ is similar to $I - P_{22}^{1/2} Q_{22} P_{22}^{1/2}$, and the latter is positive definite, we see that invertibility of $I - QP$ is indeed automatically satisfied.

14.7 J-spectral factorization relative to the unit circle revisited

In this section we present a somewhat more general form of Theorem 14.11, using an alternative approach. As in the first part of Section 14.5, the function W is a rational $m \times m$ matrix function which is selfadjoint on the unit circle and has no pole there. Such a function can be represented in the form

$$W(\lambda) = D_0 + C(\lambda I_n - A)^{-1}B + B^*(\lambda^{-1}I_n - A^*)^{-1}C^*, \tag{14.57}$$

where D_0 is a Hermitian $m \times m$ matrix and A is an $n \times n$ matrix having all its eigenvalues in the open unit disc. In contrast to the situation considered in

Section 14.5 we do not assume that A is invertible, and hence the representation (14.30) is not available in the present context.

Similar to what was done in Theorem 14.11, we associate with the representation (14.57) the Riccati equation

$$Y = A^*YA - (C^* + A^*YB)(D_0 + B^*YB)^{-1}(C + B^*YA). \tag{14.58}$$

Recall from the paragraph directly following Theorem 14.11 that a solution Y to this Riccati equation is called $\mathbb{T}$-stabilizing (or simply stabilizing) if $D_0 + B^*YB$ is invertible and the matrix

$$A - B(D_0 + B^*YB)^{-1}(C + B^*YA) \tag{14.59}$$

has all its eigenvalues in the open unit disc. The following theorem is the main result of this section.

Theorem 14.15. *Let W be a rational $m \times m$ matrix function given by* (14.57), *where D_0 is a Hermitian matrix and A is an $n \times n$ matrix having all its eigenvalues in the open unit disc. In order that, for some signature matrix J the function W admits a left J-spectral factorization with respect to the unit circle, it is necessary and sufficient that the Riccati equation* (14.58) *has a Hermitian $\mathbb{T}$-stabilizing solution Y. In that case Y is unique, and for J one can take any signature matrix J determined by*

$$D_0 + B^*YB = E^*JE, \tag{14.60}$$

where E is some invertible matrix. Furthermore, if Y is the Hermitian $\mathbb{T}$-stabilizing solution to (14.58), *then for a signature matrix J determined by* (14.60), *a left J-spectral factorization $W(\lambda) = L_-(\bar{\lambda}^{-1})^*JL_-(\lambda)$ of W with respect to the unit circle can be obtained by taking*

$$L_-(\lambda) = E + E(D_0 + B^*YB)^{-1}(C + B^*YA)(\lambda I_n - A)^{-1}B. \tag{14.61}$$

To prove the above theorem we cannot use the method employed in Section 14.5. Instead we shall use the connection between canonical factorization and invertibility of Toeplitz operators described in Section 1.2. For this purpose we need the block Toeplitz operator T on ℓ_2^m defined by the rational $m \times m$ matrix function $W(\lambda^{-1})$. Recall (see Section 1.2) that $\ell_2^m = \ell_2(\mathbb{C}^m)$ stands for the Hilbert space of all square summable sequences $(x_0, x_1, x_2, \ldots)$ with entries in $\mathbb{C}^m$. Furthermore, by definition, T is the operator on ℓ_2^m given by the block matrix representation

$$T = \begin{bmatrix} R_0 & R_{-1} & R_{-2} & \cdots \\ R_1 & R_0 & R_{-1} & \cdots \\ R_2 & R_1 & R_0 & \cdots \\ \vdots & \vdots & \vdots & \ddots \end{bmatrix}, \tag{14.62}$$

where $\ldots, R_{-1}, R_0, R_1, \ldots$ are the coefficients in the Laurent expansion

$$W(\lambda^{-1}) = \sum_{j=-\infty}^{\infty} \lambda^j R_j$$

of the function $W(\lambda^{-1})$ on the unit circle. When W is given by (14.57), we have

$$R_0 = D_0, \qquad R_j = R_{-j}^* = CA^{j-1}B, \quad j = 1, 2, \ldots . \tag{14.63}$$

The following lemma provides one of the main steps in the proof of Theorem 14.15. As always in this section, J stands for a signature matrix.

Lemma 14.16. *Let W be a rational $m \times m$ matrix function given by* (14.57), *where D_0 is a Hermitian matrix and A is an $n \times n$ matrix having all its eigenvalues in the open unit disc. Assume W admits a left J-spectral factorization with respect to the unit circle. Then the block Toeplitz operator T on ℓ_2^m defined by the rational $m \times m$ matrix function $W(\lambda^{-1})$ is invertible, and the $n \times n$ matrix Y given by*

$$Y = -\begin{bmatrix} C^* & A^*C^* & A^{*2}C^* & \cdots \end{bmatrix} T^{-1} \begin{bmatrix} C \\ CA \\ CA^2 \\ \vdots \end{bmatrix} \tag{14.64}$$

is a Hermitian stabilizing solution to the Riccati equation (14.58).

Proof. A left J-spectral factorization with respect to the unit circle is, in particular, a left canonical factorization with respect to the unit circle. But then the function $W(\lambda^{-1})$ admits a right canonical factorization with respect to the unit circle, and Theorem 1.2 tells us that the block Toeplitz operator T is invertible. This, together with the fact that A has all its eigenvalues in the open unit disc, gives that the matrix Y is well-defined by (14.64). Note that T is selfadjoint because $W(\lambda^{-1})$ has Hermitian values on the unit circle. But then T^{-1} is selfadjoint too, and (14.64) shows that Y is Hermitian

Note that ℓ_2^m can be identified with the Hilbert space direct sum $\mathbb{C}^m \oplus \ell_2^m$. Via this identification the operator T partitions as

$$T = \begin{bmatrix} R_0 & \Lambda^* \\ \Lambda & T \end{bmatrix}, \quad \text{where } \Lambda = \begin{bmatrix} R_1 \\ R_2 \\ R_3 \\ \vdots \end{bmatrix} : \mathbb{C}^m \to \ell_2^m. \tag{14.65}$$

Put $\Delta = R_0 - \Lambda^* T^{-1} \Lambda$. Since the 2×2 operator matrix in (14.65) and the operator in its right lower corner are both invertible, a standard Schur complement argument (see [19] or the second proof of Theorem 2.1 in [20]) tells us that Δ is invertible as well. Furthermore, relative to the Hilbert space direct sum decomposition $\mathbb{C}^m \oplus \ell_2^m$ the inverse of T admits the block matrix representation

$$T^{-1} = \begin{bmatrix} \Delta^{-1} & -\Delta^{-1}\Lambda^* T^{-1} \\ -T^{-1}\Lambda\Delta^{-1} & T^{-1} + T^{-1}\Lambda\Delta^{-1}\Lambda^* T^{-1} \end{bmatrix}. \tag{14.66}$$

Recall from (14.63) that $R_0 = D_0$. Combining the second part of (14.63) with (14.64) we obtain that $B^*YB = -\Lambda^* T^{-1}\Lambda$. It follows that $D_0 + B^*YB = \Delta$, and hence $D_0 + B^*YB$ is invertible, as desired.

To prove that Y satisfies the Riccati equation (14.58) we first consider the operator $T^{-1} - ST^{-1}S^*$, where S is the (block) forward shift on ℓ_2^m. Thus the actions of S and S^* on ℓ_2^m are given by

$$S(x_0, x_1, x_2, \ldots) = (0, x_0, x_1, \ldots), \qquad S^*(x_0, x_1, x_2, \ldots) = (x_1, x_2, x_3, \ldots).$$

A straightforward computation shows that the partitioning of $ST^{-1}S^*$ relative to the Hilbert space direct sum $\mathbb{C}^m \oplus \ell_2^m$ is given by

$$ST^{-1}S^* = \begin{bmatrix} 0 & 0 \\ 0 & T^{-1} \end{bmatrix}.$$

This identity, together with the identity (14.66), yields

$$\begin{aligned} T^{-1} - ST^{-1}S^* &= \begin{bmatrix} \Delta^{-1} & -\Delta^{-1}\Lambda^* T^{-1} \\ -T^{-1}\Lambda\Delta^{-1} & T^{-1}\Lambda\Delta^{-1}\Lambda^* T^{-1} \end{bmatrix} \\ &= \begin{bmatrix} I \\ -T^{-1}\Lambda \end{bmatrix} \Delta^{-1} \begin{bmatrix} I & -\Lambda^* T^{-1} \end{bmatrix}. \end{aligned} \tag{14.67}$$

Next, let Γ be the operator from $\mathbb{C}^n$ to ℓ_2^m given by

$$\Gamma = \begin{bmatrix} C \\ CA \\ CA^2 \\ \vdots \end{bmatrix}. \tag{14.68}$$

Note that this operator Γ is well-defined because the matrix A has all its eigenvalues in the open unit disc. As is easily checked

$$\Gamma A = S^*\Gamma, \qquad \Gamma B = \Lambda, \qquad Y = -\Gamma^* T^{-1}\Gamma. \tag{14.69}$$

From these identities and (14.67) it follows that

$$\begin{aligned} Y - A^*YA &= -\Gamma^*T^{-1}\Gamma + A^*\Gamma^*T^{-1}\Gamma A \\ &= -\Gamma^*T^{-1}\Gamma + \Gamma^*ST^{-1}S^*\Gamma \\ &= -\Gamma^*(T^{-1} - ST^{-1}S^*)\Gamma \\ &= -\Gamma^*\begin{bmatrix} I \\ -T^{-1}\Lambda \end{bmatrix}\Delta^{-1}\begin{bmatrix} I & -\Lambda^*T^{-1} \end{bmatrix}\Gamma. \end{aligned}$$

Furthermore

$$[I \quad -\Lambda^*T^{-1}]\Gamma = C - \Lambda^*T^{-1}S^*\Gamma = C - B^*\Gamma^*T^{-1}\Gamma = C + B^*YA. \qquad (14.70)$$

Summarizing (and using that Y is Hermitian) we have

$$Y - A^*YA = -(C + B^*YA)^*\Delta^{-1}(C + B^*YA).$$

Since $\Delta = D_0 + B^*YB$, this identity shows that Y satisfies the Riccati equation (14.58).

Write $A^\times$ for the matrix (14.59). We need to show that for Y given by (14.64), all eigenvalues of $A^\times$ are in the open unit disc. Using (14.67), the fact that S^*S is the identity operator on ℓ_2^m, and the identities in (14.69) and (14.70), we see that

$$\begin{aligned} S^*T^{-1}\Gamma &= T^{-1}S^*\Gamma + S^*\begin{bmatrix} I \\ -T^{-1}\Lambda \end{bmatrix}\Delta^{-1}\begin{bmatrix} I & -\Lambda^*T^{-1} \end{bmatrix}\Gamma \\ &= T^{-1}\Gamma A - T^{-1}\Lambda\Delta^{-1}(C + B^*YA) \\ &= T^{-1}\Gamma(A - B\Delta^{-1}(C + B^*YA)) = T^{-1}\Gamma A^\times. \end{aligned}$$

Thus $S^*T^{-1}\Gamma = T^{-1}\Gamma A^\times$. It follows that

$$(S^*)^kT^{-1}\Gamma = T^{-1}\Gamma(A^\times)^k, \qquad k = 1, 2, \ldots.$$

But then the fact that S^{*n} converges to zero in the strong operator topology yields

$$\lim_{k\to\infty} T^{-1}\Gamma(A^\times)^kx = \lim_{k\to\infty} (S^*)^kT^{-1}\Gamma x = 0, \qquad x \in \mathbb{C}^n. \qquad (14.71)$$

We shall use (14.71) to prove that $A^\times$ has all its eigenvalues in the open unit disc. To do this we first decompose $\mathbb{C}^n$ as $\mathbb{C}^n = \mathcal{X}_1 \oplus \mathcal{X}_2$, where $\mathcal{X}_2 = \operatorname{Ker}\Gamma$ and $\mathcal{X}_1 = (\operatorname{Ker}\Gamma)^\perp$. Notice that $\mathcal{X}_2$ is an invariant subspace for A, and $C[\mathcal{X}_2] = \{0\}$. We also have $YA[\mathcal{X}_2] = \{0\}$. Indeed

$$Y[A\mathcal{X}_2] \subset Y[\mathcal{X}_2] = -\Gamma^*T^{-1}\Gamma[\mathcal{X}_2] = \{0\}.$$

Using $C[\mathcal{X}_2] = \{0\}$ and $YA[\mathcal{X}_2] = \{0\}$ in (14.59), we see that $A^\times|_{\mathcal{X}_2} = A|_{\mathcal{X}_2}$, and $\mathcal{X}_2$ is an invariant subspace for $A^\times$ too. In other words, $A^\times$ admits a matrix representation of the form

$$A^\times = \begin{bmatrix} A_{11}^\times & 0 \\ A_{21}^\times & A_{22}^\times \end{bmatrix} : \mathcal{X}_1 \oplus \mathcal{X}_2 \to \mathcal{X}_1 \oplus \mathcal{X}_2, \qquad (14.72)$$

where $A_{22}^\times = A|_{\mathcal{X}_2} : \mathcal{X}_2 \to \mathcal{X}_2$. Since $\mathcal{X}_2$ is an invariant subspace for A and A has all its eigenvalues in the open unit disc, A_{22} has all its eigenvalues in the open unit disc too. Hence, in order to prove that $A^\times$ has all its eigenvalues in the open unit disc, it now suffices to prove that $A_{11}^\times$ has this property. Let τ_1 be the canonical embedding of $\mathcal{X}_1$ into $\mathbb{C}^n = \mathcal{X}_1 \oplus \mathcal{X}_2$, and let Γ_1 be the one-to-one operator from $\mathcal{X}_1$ into ℓ_2^m defined by $\Gamma_1 = \Gamma\tau_1$. Take $x \in \mathcal{X}_1$. Since Γ is equal to zero on $\mathcal{X}_2$, we see from (14.72) that $T^{-1}\Gamma(A^\times)^k x = T^{-1}\Gamma_1(A_{11}^\times)^k x$. But then (14.71) tells us that $\lim_{k\to\infty} T^{-1}\Gamma_1(A_{11}^\times)^k x = 0$. Observe that $T^{-1}\Gamma_1$ is one-to-one and has a closed (finite dimensional) range, that is, $T^{-1}\Gamma_1$ is left invertible. Hence $\lim_{k\to\infty} T^{-1}\Gamma_1(A_{11}^\times)^k x = 0$ implies that $\lim_{k\to\infty}(A_{11}^\times)^k x = 0$. Since x is an arbitrary element of $\mathcal{X}_1$, the latter holds if and only if the eigenvalues of $A_{11}^\times$ are in the open unit disc. □

Lemma 14.16 proves the necessity part of Theorem 14.15. The sufficiency part, the formula for the J-spectral factorization, and the uniqueness statement are covered by the next two lemmas.

Lemma 14.17. *Let W be a rational $m \times m$ matrix function given by* (14.57), *where D_0 is a Hermitian matrix and A is an $n \times n$ matrix having all its eigenvalues in the open unit disc. Assume Y is a Hermitian stabilizing solution of the Riccati equation* (14.58). *Then W admits a left J-spectral factorization with respect to the unit circle. Such a factorization can be obtained as follows. Choose an $m \times m$ signature matrix J such that $D_0 + B^*YB = E^*JE$, where E is some invertible matrix, and define L_- by* (14.61), *i.e.,*

$$L_-(\lambda) = E + E(D_0 + B^*YB)^{-1}(C + B^*YA)(\lambda I_n - A)^{-1}B.$$

Then $W(\lambda) = L_-(\bar{\lambda}^{-1})^* J L_-(\lambda)$ is a left J-spectral factorization of W with respect to the unit circle.

Proof. Put $\Delta = D_0 + B^*YB$, $C_0 = C + B^*YA$, and set

$$\Psi(\lambda) = \Delta + C_0(\lambda - A)^{-1}B. \qquad (14.73)$$

Note that $A - B\Delta^{-1}C_0$ is equal to the matrix $A^\times$ defined by (14.59). Thus

$$\Psi(\lambda)^{-1} = \Delta^{-1} - \Delta^{-1}C_0(\lambda - A^\times)^{-1}B\Delta^{-1}. \qquad (14.74)$$

The fact that A and $A^\times$ have all their eigenvalues in the open unit disc implies that $\Psi(\lambda)$ and $\Psi(\lambda)^{-1}$ are both analytic on the closure of the exterior of the unit disc, infinity included. Since $L_-(\lambda) = E\Delta^{-1}\Psi(\lambda)$, the same holds true for $L_-(\lambda)$ and $L_-(\lambda)^{-1}$. It follows that $L_-(\bar\lambda^{-1})^*JL_-(\lambda)$ is a left spectral factorization with respect to the unit circle. It remains to show that

$$W(\lambda) = L_-(\bar\lambda^{-1})^*JL_-(\lambda). \tag{14.75}$$

From $L_-(\lambda) = E\Delta^{-1}\Psi(\lambda)$ and $\Delta = D_0 + B^*YB = E^*JE$ we see that

$$L_-(\bar\lambda^{-1})^*JL_-(\lambda) = \Psi(\bar\lambda^{-1})^*\Delta^{-1}\Psi(\lambda).$$

Using the definitions of Δ and C_0, the Riccati equation (14.58) can be rewritten as $Y - A^*YA = -C_0^*\Delta^{-1}C_0$. It follows that

$$\lambda C_0^*\Delta^{-1}C_0 = -Y(\lambda - A) + (I - \lambda A^*)Y(\lambda - A) - \lambda(I - \lambda A^*)Y.$$

Using this identity we obtain

$$\begin{aligned} B^*(I - \lambda A^*)^{-1}&(\lambda C_0^*\Delta^{-1}C_0)(\lambda - A)^{-1}B \\ &= -B^*(I - \lambda A^*)^{-1}YB + B^*YB - \lambda B^*Y(\lambda - A)^{-1}B \\ &= -\lambda B^*(I - \lambda A^*)^{-1}A^*YB - B^*YB - B^*YA(\lambda - A)^{-1}B. \end{aligned}$$

Hence

$$\begin{aligned} \Psi(\bar\lambda^{-1})^*&\Delta^{-1}\Psi(\lambda) \\ &= (\Delta + \lambda B^*(I - \lambda A^*)^{-1}C_0^*)\Delta^{-1}(\Delta + C_0(\lambda - A)^{-1}B) \\ &= \Delta + \lambda B^*(I - \lambda A^*)^{-1}C_0^* + C_0(\lambda - A)^{-1}B \\ &\qquad + B^*(I - \lambda A^*)^{-1}(\lambda C_0^*\Delta^{-1}C_0)(\lambda - A)^{-1}B. \end{aligned}$$

From the definitions of Δ and C_0 given in the beginning of the proof we see that $\Delta - B^*YB = D_0$ and $C_0 - B^*YA = C$. Thus the calculations above yield

$$\Psi(\bar\lambda^{-1})^*\Delta^{-1}\Psi(\lambda) = D_0 + \lambda C(I - \lambda A)^{-1} + B^*(\lambda - A^*)^{-1}C^*.$$

According to (14.57) the right-hand side in the previous identity is equal to $W(\lambda)$. Thus $\Psi(\bar\lambda^{-1})^*\Delta^{-1}\Psi(\lambda) = W(\lambda)$, as desired. □

Lemma 14.18. *Let W be a rational $m \times m$ matrix function given by* (14.57)*, where D_0 is a Hermitian matrix and A is an $n \times n$ matrix having all its eigenvalues in the open unit disc. Assume Y is a Hermitian stabilizing solution of the Riccati equation* (14.58)*. Then the block Toeplitz operator T on ℓ_2^m defined by the rational $m \times m$ matrix function $W(\lambda^{-1})$ is invertible and Y is uniquely determined by the expression* (14.64).

Proof. As in the proof of the preceding lemma, we set $\Delta = D_0 + B^*YB$ and $C_0 = C + B^*YA$. Furthermore, $\Psi(\lambda)$ is the rational $m \times m$ matrix function defined by (14.73). Put $\Theta(\lambda) = \Psi(\lambda^{-1})$. The proof of the preceding lemma tells us that

$$W(\lambda^{-1}) = \Theta(\bar{\lambda}^{-1})^* \Delta^{-1} \Theta(\lambda).$$

Hence the block Toeplitz operator T on ℓ_2^m defined by $W(\lambda^{-1})$ admits the factorization $T = (T_\Theta)^* \, \Xi \, T_\Theta$, where T_Θ is the block Toeplitz operator on ℓ_2^m defined by Θ, and Ξ is the block diagonal operator on ℓ_2^m given by

$$\Xi = \operatorname{diag}(\Delta^{-1}, \Delta^{-1}, \Delta^{-1}, \ldots).$$

From (14.73), (14.74) and $\Theta(\lambda) = \Psi(\lambda^{-1})$ we know that

$$\Theta(\lambda) = \Delta + \lambda C_0 (I - \lambda A)^{-1} B, \tag{14.76}$$

$$\Theta(\lambda)^{-1} = \Delta^{-1} - \lambda \Delta^{-1} C_0 (I - \lambda A^\times)^{-1} B \Delta^{-1}, \tag{14.77}$$

where $A^\times$ is given by (14.59). From (14.76), (14.77), and the fact that both A and $A^\times$ have all their eigenvalues in the open unit disc we see that T_Θ is invertible and T_Θ^{-1} is given by

$$T_\Theta^{-1} = \begin{bmatrix} \Theta_0^\times & 0 & 0 & \cdots \\ \Theta_1^\times & \Theta_0^\times & 0 & \cdots \\ \Theta_2^\times & \Theta_1^\times & \Theta_0^\times & \\ \vdots & \vdots & & \ddots \end{bmatrix}, \tag{14.78}$$

where $\Theta_0^\times, \Theta_1^\times, \Theta_2^\times, \ldots$ are the Taylor coefficients of $\Theta(\lambda)^{-1}$ at zero. Furthermore, (14.76) yields

$$\Theta_0^\times = \Delta^{-1}, \qquad \Theta_j^\times = -\Delta^{-1} C_0 (A^\times)^{j-1} B \Delta^{-1}, \quad j = 1, 2, \ldots. \tag{14.79}$$

Let Γ be the operator from $\mathbb{C}^n$ into ℓ_2^m defined by (14.68). Using the identities in (14.78) and (14.79) we compute that

$$\Gamma^* T_\Theta^{-1} = \begin{bmatrix} \widetilde{\beta} & A^* \widetilde{\beta} & (A^*)^2 \widetilde{\beta} & \cdots \end{bmatrix}, \tag{14.80}$$

with $\widetilde{\beta}$ given by

$$\widetilde{\beta} = C^* \Delta^{-1} - A^* \Big(\sum_{j=0}^{\infty} (A^*)^j C^* \Delta^{-1} C_0 (A^\times)^j \Big) B \Delta^{-1}. \tag{14.81}$$

As $T = (T_\Theta)^* \Xi T_\Theta$ and T_Θ is invertible, we conclude that T is invertible. Moreover, using (14.80), we have

$$\Gamma^* T^{-1} \Gamma = (\Gamma^* T_\Theta^{-1}) \, \Xi^{-1} (\Gamma^* T_\Theta^{-1})^* = \sum_{j=0}^{\infty} (A^*)^j \widetilde{\beta} \Delta \widetilde{\beta}^* A^j. \tag{14.82}$$

We proceed by showing that $\widetilde{\beta} = (C^* + A^*YB)\Delta^{-1}$, where $A^\times$ is given by (14.59). To prove this we use the fact that Y satisfies the Riccati equation (14.58). A straightforward computation gives

$$\begin{aligned} Y &= A^*YA - (C^* + A^*YB)\Delta^{-1}(C + B^*YA) \\ &= A^*Y\Big(A - B\Delta^{-1}(C + B^*YA)\Big) - C^*\Delta^{-1}(C + B^*YA) \\ &= A^*YA^\times - C^*\Delta^{-1}C_0. \end{aligned}$$

We conclude that $Y - A^*YA^\times = -C^*\Delta^{-1}C_0$. Since both A and $A^\times$ have all their eigenvalues in the open unit disc, we obtain

$$Y = -\sum_{j=0}^{\infty}(A^*)^jC^*\Delta^{-1}C_0(A^\times)^j.$$

Using the latter identity in (14.81) we arrive at

$$\widetilde{\beta} = C^*\Delta^{-1} + A^*YB\Delta^{-1} = (C^* + A^*YB)\Delta^{-1}.$$

Finally, the identity $\widetilde{\beta} = (C^* + A^*YB)\Delta^{-1}$ and the fact that Y satisfies the Riccati equation yield

$$Y - A^*YA = -(C^* + A^*YB)\Delta^{-1}(C + B^*YA) = -\widetilde{\beta}\Delta\widetilde{\beta}^*. \tag{14.83}$$

But then $Y = -\sum_{j=0}^{\infty}(A^*)^j\widetilde{\beta}\Delta\widetilde{\beta}^*A^j$ because A has all its eigenvalues in the open unit disc. Comparing the latter expression for Y with (14.82) we see that $Y = -\Gamma^*T^{-1}\Gamma$. Thus Y is given by (14.64), as desired. □

In Theorem 14.15 we restricted the attention to stabilizing solutions of the Riccati equation (14.58) that are required to be Hermitian. This requirement is not essential: Theorem 14.15 remains true if Y is just an arbitrary stabilizing solution of (14.58). The reason is that a stabilizing solution of (14.58) is always Hermitian. This result is the contents of the following proposition.

Proposition 14.19. *If Y is a $\mathbb{T}$-stabilizing solution of the Riccati equation* (14.58), *then Y is Hermitian.*

Proof. Let Y be a stabilizing solution of (14.58), and put $\Delta = D_0 + B^*YB$. Then Δ is invertible, and

$$\sigma\big(A - B\Delta^{-1}(C + B^*YA)\big) \subset \mathbb{D}. \tag{14.84}$$

Consider the $m \times m$ rational matrix functions

$$W_-(\lambda) = I_m + \Delta^{-1}(C + B^*YA)(\lambda I_n - A)^{-1}B, \tag{14.85}$$

$$W_+(\lambda) = \Delta + B^*(\lambda^{-1}I_n - A^*)^{-1}(C^* + A^*YB). \tag{14.86}$$

The first part of the proof consists of showing that $W(\lambda) = W_+(\lambda)W_-(\lambda)$ and that this factorization is a left canonical one with respect to the unit circle.

Part 1. To prove that $W(\lambda) = W_+(\lambda)W_-(\lambda)$, we use a modification of the argument used to prove (14.75). Put

$$C_0 = C + B^*YA, \qquad B_0 = C^* + A^*YB. \tag{14.87}$$

Then equation (14.58) can be rewritten as $Y - A^*YA = -C_0\Delta^{-1}B_0$, and hence

$$\lambda B_0\Delta^{-1}C_0 = -Y(\lambda - A) + (I - \lambda A^*)Y(\lambda - A) - \lambda(I - \lambda A^*)Y.$$

It then follows that

$$\begin{aligned} &B^*(\lambda^{-1} - A^*)^{-1}B_0\Delta^{-1}C_0(\lambda I_n - A)^{-1}B \\ &\qquad = -B^*(\lambda^{-1} - A^*)^{-1}A^*YB - B^*YB - B^*YA)(\lambda I_n - A)^{-1}B. \end{aligned}$$

This yields

$$\begin{aligned} W_+(\lambda)W_-(\lambda) &= \Delta + B^*(\lambda^{-1} - A^*)^{-1}B_0 + C_0(\lambda I_n - A)^{-1}B \\ &\qquad + B^*(\lambda^{-1} - A^*)^{-1}B_0\Delta^{-1}(\lambda I_n - A)^{-1}B \\ &= D_0 + B^*(\lambda^{-1} - A^*)^{-1}C^* + C(\lambda I_n - A)^{-1}B = W(\lambda). \end{aligned}$$

Next we prove that $W(\lambda) = W_+(\lambda)W_-(\lambda)$ is a left canonical factorization with respect to the unit circle. To do this, using (14.85), we first note that

$$W_-(\lambda)^{-1} = I_m - \Delta^{-1}(C + B^*YA)(\lambda I_n - A^\times)^{-1}B, \tag{14.88}$$

where $A^\times = A - B\Delta^{-1}(C + B^*YA)$. From (14.84) we know that $A^\times$ has all its eigenvalues in $\mathbb{D}$. By assumption the same holds true for the matrix A. Thus (14.85) and (14.88) tell us that both W_- and W_-^{-1} are analytic on the complement of $\mathbb{D}$, infinity included. Thus the factor W_- has the desired properties.

As A has all its eigenvalues in $\mathbb{D}$, the same holds true for A^*. Thus (14.86) tells us that W_+ is analytic on the closed unit disc $\overline{\mathbb{D}}$. We have to show that W_+^{-1} also is analytic on $\overline{\mathbb{D}}$. To do this, put

$$V_+(\lambda) = W_-(\bar{\lambda}^{-1})^*, \qquad V_-(\lambda) = W_+(\bar{\lambda}^{-1})^*.$$

Using the properties of W_- derived in the previous paragraph, we see that V_+ and V_+^{-1} are analytic on $\overline{\mathbb{D}}$. Furthermore, V_- is analytic on $|\lambda| \geq 1$, infinity included. Now, recall that W is selfadjoint on the unit circle. Hence $W(\lambda) = W(\bar{\lambda}^{-1})^*$, and thus $W(\lambda) = W_+(\lambda)W_-(\lambda) = V_+(\lambda)V_-(\lambda)$. But then

$$V_-(\lambda)W_-(\lambda)^{-1} = V_+(\lambda)^{-1}W_+(\lambda). \tag{14.89}$$

The left-hand side of (14.89) is analytic on $|\lambda| \geq 1$ with infinity included, and the right-hand side of (14.89) is analytic on $\overline{\mathbb{D}}$. By Liouville's theorem, there exists a constant matrix K such that

$$V_-(\lambda) = KW_-(\lambda), \qquad W_+(\lambda) = V_+(\lambda)K. \tag{14.90}$$

As $\det W_+(\lambda)$ does not vanish identically, K is invertible. Hence the second identity in (14.90) tells us that $W_+(\lambda) = K^{-1}V_+(\lambda)^{-1}$ is analytic on $\overline{\mathbb{D}}$. Thus W_+ and W_+^{-1} are analytic on $\overline{\mathbb{D}}$, as desired. We conclude that $W(\lambda) = W_+(\lambda)W_-(\lambda)$ is a left canonical factorization with respect to the unit circle.

Part 2. In this part we establish the inclusion

$$\sigma\big(A^* - (C^* + A^*YB)\Delta^{-1}B^*\big) \subset \mathbb{D}. \tag{14.91}$$

Put $\Phi(\lambda) = W_+(\lambda^{-1})$. Then, with $\Omega = A^*$ and $\Omega^\times = A^* - (C^* + A^*YB)\Delta^{-1}B^*$,

$$\Phi(\lambda) \quad = \quad \Delta + B^*(\lambda I_n - \Omega)^{-1}(C^* + A^*YB), \tag{14.92}$$

$$\Phi^{-1}(\lambda) \quad = \quad \Delta^{-1} - \Delta^{-1}B^*(\lambda I_n - \Omega^\times)^{-1}(C^* + A^*YB)\Delta^{-1}. \tag{14.93}$$

We want to prove that $\sigma(\Omega^\times) \subset \mathbb{D}$. Take $|\lambda_0| \geq 1$. As $\sigma(\Omega) \subset \mathbb{D}$, we have $\lambda_0 \notin \sigma(\Omega)$, and hence $\lambda_0 \notin \sigma(\Omega) \cap \sigma(\Omega^\times)$. From (14.92) and (14.93) we see that $\Omega^\times$ is the associate main matrix of the realization (14.92). But then $\lambda_0 \notin \sigma(\Omega) \cap \sigma(\Omega^\times)$ implies that the realization in (14.92) is locally minimal at λ_0. Since W_+ and W_+^{-1} are analytic on $\overline{\mathbb{D}}$, the rational matrix function Φ has no poles or zeros on $|\lambda| \geq 1$. But then the local minimality at λ_0 implies that λ_0 is not an eigenvalue of $\Omega^\times$. Recall that λ_0 is an arbitary complex number with $|\lambda_0| \geq 1$. We conclude that $\sigma(\Omega^\times)$ is contained in $\mathbb{D}$, that is, (14.91) is proved.

Part 3. Let T be the block Toeplitz operator on ℓ_2^m determined by $W(\lambda^{-1})$. Since W admits a left canonical factorization with respect to the unit circle, the function $W(\lambda^{-1})$ admits a right canonical factorization with respect to the unit circle, and hence T is invertible. We claim that

$$Y = -\begin{bmatrix} C^* & A^*C^* & A^{*2}C^* & \cdots \end{bmatrix} T^{-1} \begin{bmatrix} C \\ CA \\ CA^2 \\ \vdots \end{bmatrix}. \tag{14.94}$$

Since the values of $W(\lambda^{-1})$ on the unit circle are Hermitian, the operator T is selfadjoint, and hence the same holds true for T^{-1}. But then the identity (14.94) shows that Y is Hermitian. Thus it remains to prove (14.94).

To prove (14.94) we follow the same line of reasoning as in the proof of Lemma 14.18. Put

$$\Theta(\lambda) = \Delta W_-(\lambda^{-1}), \qquad \Phi(\lambda) = W_+(\lambda^{-1}). \tag{14.95}$$

Here W_+ and W_- are as in Part 1 of the proof; see (14.85) and (14.86). By the result of Part 1 we have that $W_-(\lambda^{-1}) = \Phi(\lambda)\Delta^{-1}\Theta(\lambda)$. Moreover, Θ and Θ^{-1} are analytic on $\overline{\mathbb{D}}$, and Φ and Φ^{-1} are analytic on $|\lambda| \geq 1$ with infinity included.

Let T_Θ and T_Φ be the block Toeplitz operators on ℓ_2^m determined by Θ and Φ, respectively. By the results mentioned in the previous paragraph, the operators T_Θ and T_Φ are invertible, $T_\Theta^{-1} = T_{\Theta^{-1}}$ and $T_\Phi^{-1} = T_{\Phi^{-1}}$. Furthermore, $T^{-1} = T_{\Theta^{-1}}\Xi^{-1}T_{\Phi^{-1}}$, where, as in the proof of Lemma 14.18, the operator Ξ is the block diagonal operator on ℓ_2^m given by

$$\Xi = \operatorname{diag}(\Delta^{-1}, \Delta^{-1}, \Delta^{-1}, \ldots).$$

Note that

$$\begin{aligned}\Theta^{-1}(\lambda) &= \Delta^{-1} - \Delta^{-1}(C + B^*YA)(\lambda^{-1}I_n - A^\times)^{-1}B\Delta^{-1},\\ \Phi^{-1}(\lambda) &= \Delta^{-1} - \Delta^{-1}B^*(\lambda I_n - \Omega^\times)^{-1}(C^* + A^*YB)\Delta^{-1}.\end{aligned}$$

Here

$$A^\times = A - B\Delta^{-1}(C + B^*YA), \qquad \Omega^\times = A^* - (C^* + A^*YB)\Delta^{-1}B^*,$$

and the eigenvalues of these two matrices are all in the open unit disc.

Let Γ be the operator defined by (14.68). We now repeat the arguments used in the proof of Lemma 14.18, more specifically appearing in the paragraphs after (14.79). This together with a duality argument yields

$$\Gamma^*T^{-1}\Gamma = (\Gamma^*T_\Theta^{-1})\Xi^{-1}(T_\Phi^{-1}\Gamma) = \sum_{j=0}^{\infty}(A^*)^j\widetilde{\beta}\Delta^{-1}\widetilde{\gamma}A^j. \tag{14.96}$$

Here

$$\begin{aligned}\widetilde{\beta} &= C^*\Delta^{-1} - A^*\Big(\sum_{j=0}^{\infty}(A^*)^jC^*\Delta^{-1}(C + B^*YA)(A^\times)^j\Big)B\Delta^{-1},\\ \widetilde{\gamma} &= \Delta^{-1}C - \Delta^{-1}B^*\Big(\sum_{j=0}(\Omega^\times)^j(C^* + A^*YB)\Delta^{-1}CA^J\Big)A.\end{aligned}$$

Note that the Riccati equation (14.58) can be rewritten in the following two equivalent forms

$$\begin{aligned}Y - A^*YA^\times &= -C^*\Delta^{-1}(C + B^*YA),\\ Y - \Omega^\times YA &= -(C^* + A^*YB)\Delta^{-1}C.\end{aligned}$$

Since the eigenvalues of the matrices A, A^*, $A^\times$ and $\Omega^\times$ are all in the open unit disc, we see that the formulas for $\widetilde{\beta}$ and $\widetilde{\gamma}$ can be transformed into

$$\widetilde{\beta} = (C^* + A^*YB)\Delta^{-1}, \qquad \widetilde{\gamma} = \Delta^{-1}(C + B^*YA).$$

This allows us to rewrite (14.58) as $Y - A^*YA = -\widetilde{\beta}\Delta\widetilde{\gamma}$, and we see from (14.96) that (14.94) holds. $\square$

Notes

As noted J-spectral factorization is a special form of canonical factorization, reflecting the symmetry condition on the given function. This chapter develops this theme in a systematic way for rational matrix functions. Sections 14.2 and 14.3 are based on [121]. For Section 14.4 we refer to [76], see also [112] and [83]. A good source for Section 14.5 is [98], see also [97]. The linear quadratic optimal control problem for discrete time systems, mentioned in Section 14.5 in the paragraph before Proposition 14.12, can be found in many books on mathematical systems theory, see, e.g., [85]. The connection with the algebraic Riccati equation of the form (14.45) is also shown in the latter book. Much more information on this equation, including its connection to factorization in more general setting than the one exhibited in Proposition 14.12, can be found in Part III of [106]. Section 14.6 is based on [9], see also [8]. The final section is inspired by [44]. In fact, Theorem 14.15 is just the symmetric version of Theorem 1.1 in [44].

The notion of J-spectral factorization plays an important role in control theory; see, e.g., the books [43], [85], [150], the papers [76], [145] and the references in these papers. The final part of this book is devoted to this connection, with an emphasis on H_∞-problems.

Part VI
Factorizations and symmetries

In this part we study rational matrix functions that are unitary or of the form identity matrix plus contractions, and rational matrix functions that have a positive real part. Because of the state space similarity theorem, these additional symmetries can be restated in terms of special properties of the minimal realizations of the rational matrix functions considered. These reformulations involve an algebraic Riccati equation. The results are known in systems theory as the bounded real lemma and the positive real lemma, respectively.

This part consists of three chapters. In the first chapter (Chapter 15) we study rational matrix functions that have a positive definite real part or a non-negative real part on the real line, and we present canonical and pseudo-canonical factorization theorems for such functions in state space form. In the second chapter (Chapter 16) realizations are used to study rational matrix functions of which the values on the imaginary axis (or on the real line) are contractive matrices. Included are solutions to spectral and canonical factorization problems for functions V of the form

$$V(\lambda) = I - W(-\bar{\lambda})^* W(\lambda), \qquad V(\lambda) = I + W(\lambda),$$

where W has contractive values on the imaginary axis (or on the real line) and is strictly contractive at infinity. In the third chapter (Chapter 17) realizations are used to study rational matrix functions of which the values on the imaginary axis are J-unitary matrices. Solutions to various factorization problems are given. Special attention is paid to factorization of J-unitary rational matrix functions into J-unitary factors. In this chapter we also discuss problems of embedding a contractive rational matrix function into a unitary rational matrix function of larger size.

Chapter 15

Factorization of positive real rational matrix functions

This chapter is concerned with canonical factorization (with respect to the real line) of rational matrix functions with a positive definite real part on the real line. Also the generalization to pseudo-canonical factorization for functions that have a nonnegative real part is developed. All factorizations are obtained explicitly using state space realizations of the functions involved. In Section 15.1 rational matrix functions that have a positive definite real part or a nonnegative real part on the real line are characterized in terms of realizations. Section 15.2 deals with canonical factorization, and Section 15.3, the final section of the chapter, with pseudo-canonical factorization.

15.1 Rational matrix functions with a positive definite real part

In this section we consider rational $m \times m$ matrix functions W which have the property that

$$W(\lambda) + W(\lambda)^* \geq 0, \qquad \lambda \in \mathbb{R}, \ \lambda \text{ not a pole of } W. \tag{15.1}$$

In this case we say that W has a *nonnegative real part on the real line*. If in (15.1) the inequality is strict, that is,

$$W(\lambda) + W(\lambda)^* > 0, \qquad \lambda \in \mathbb{R}, \ \lambda \text{ not a pole of } W. \tag{15.2}$$

we say that W has a *positive definite real part on the real line*. The following two theorems characterize these properties in terms of realizations of W.

Theorem 15.1. *Let $W(\lambda) = D + C(\lambda I_n - A)^{-1}B$ be a rational $m \times m$ matrix function, and let (A, B) be controllable. Write $G = D + D^*$ and assume G is positive definite. Then W has a nonnegative real part on the real line if and only if there is a Hermitian solution X of the equation*

$$-iA^*X + iXA - (XB - iC^*)G^{-1}(B^*X + iC) = 0. \tag{15.3}$$

Furthermore, for any Hermitian solution X of (15.3) one has

$$W(\lambda) + W(\bar{\lambda})^* = K(\bar{\lambda})^*K(\lambda), \tag{15.4}$$

where

$$K(\lambda) = G^{1/2} + G^{-1/2}(C - iB^*X)(\lambda I_n - A)^{-1}B. \tag{15.5}$$

Finally, if, in addition, the pair (C, A) is observable, then each solution X of (15.3) is invertible.

For later use we note that equation (15.3) can be rewritten as

$$-(iA^* - iC^*G^{-1}B^*)X + X(iA - iBG^{-1}C) - C^*G^{-1}C - XBG^{-1}B^*X = 0. \tag{15.6}$$

Proof. Put $V(\lambda) = W(-i\lambda) + W(i\bar{\lambda})^*$. Then W has a nonnegative real part on $\mathbb{R}$ if and only if V is nonnegative on the imaginary axis. Using the given realization of W we have

$$\begin{aligned} V(\lambda) &= D + C(-i\lambda I_n - A)^{-1}B + D^* + B^*(-i\lambda I_n - A^*)^{-1}C^* \\ &= G + (iC)\big(\lambda I_n - (iA)\big)^{-1}B - B^*\big(\lambda I_n + (iA)^*\big)^{-1}(iC)^* \\ &= \begin{bmatrix} -B^*\big(\lambda + (iA)^*\big)^{-1} & I \end{bmatrix} \begin{bmatrix} 0 & (iC)^* \\ iC & G \end{bmatrix} \begin{bmatrix} (\lambda - iA)^{-1}B \\ I \end{bmatrix}. \end{aligned}$$

Thus we can apply Theorem 13.6, with $R = G$, $Q = 0$, $S = iC$ and iA instead of A, to show that W has a nonnegative real part on $\mathbb{R}$ if and only equation (15.3) has a Hermitian solution.

Next, let X be a Hermitian solution of (15.3). By the second part of Theorem 13.6, the function V admits a factorization $V(\lambda) = L(-\bar{\lambda})^*L(\lambda)$, where

$$L(\lambda) = G^{1/2} + G^{-1/2}(B^*X + iC)(\lambda - iA)^{-1}B.$$

As $W(\lambda) + W(\bar{\lambda})^* = V(i\lambda)$, we see that (15.4) holds with K being given by (15.5)

To prove the final part, assume additionally that the pair (C, A) is observable, and let X be a Hermitian solution of (15.3). We have to show that X is invertible. Since X is square it suffices to prove that $\operatorname{Ker} X = \{0\}$. Assume $Xx = 0$. Then $x^*X = 0$ because X is Hermitian, and by (15.3) we have $0 = -\langle C^*G^{-1}Cx, x\rangle$. As $G > 0$, this gives $Cx = 0$. Multiplying (15.3) on the right by x we then obtain $iXAx = 0$. So $\operatorname{Ker} X$ is A-invariant and contained in $\operatorname{Ker} C$. Therefore $\operatorname{Ker} X = \{0\}$ and X is invertible. □

Theorem 15.2. *Let $W(\lambda) = D + C(\lambda I_n - A)^{-1}B$ be a rational $m \times m$ matrix function, and let (A, B) be controllable. Write $G = D + D^*$ and assume G is positive definite. If, in addition, A has no real eigenvalues, then the following statements are equivalent:*

(i) *The function W has a positive definite real part on the real line;*

(ii) *Equation* (15.3) *has a Hermitian solution X such that the matrix*

$$A - BG^{-1}C + iBG^{-1}B^*X \tag{15.7}$$

has no real eigenvalues;

(iii) *The matrix*

$$H = \begin{bmatrix} iA^* - iC^*G^{-1}B^* & C^*G^{-1}C \\ -BG^{-1}B^* & iA - iBG^{-1}C \end{bmatrix}$$

has no pure imaginary eigenvalues.

Moreover, in that case equation (15.3) *has a unique Hermitian solution X such that the matrix* (15.7) *has its eigenvalues in the open upper half plane.*

Proof. As in the proof of the previous theorem, we consider the rational $m \times m$ matrix function $V(\lambda) = W(-i\lambda) + W(i\bar{\lambda})^*$. Using the given realization of W we see (see (13.6) and the second part of the proof of Theorem 13.2) that V admits the realization $V(\lambda) = G + \widehat{C}(\lambda I_{2n} - \widehat{A})^{-1}\widehat{B}$, where

$$\widehat{A} = \begin{bmatrix} iA^* & 0 \\ 0 & iA \end{bmatrix}, \qquad \widehat{B} = \begin{bmatrix} iC^* \\ B \end{bmatrix}, \qquad \widehat{C} = \begin{bmatrix} B^* & iC \end{bmatrix}.$$

It follows that $\widehat{A}^\times = \widehat{A} - \widehat{B}G^{-1}\widehat{C}$ is precisely equal to the block matrix H appearing in item (c). Since A has no real eigenvalue, the matrix $\widehat{A}$ has no pure imaginary eigenvalue. Thus V has no pole on the imaginary axis. Hence (cf., Section 8.1) the realization $V^{-1}(\lambda) = G^{-1} - G^{-1}\widehat{C}(\lambda I_{2n} - \widehat{A}^\times)^{-1}\widehat{B}G^{-1}$ is minimal at each point of the imaginary axis. But then V^{-1} has no pole on the imaginary axis if and only if $\widehat{A}^\times$ has no pure imaginary eigenvalue. As $\widehat{A}^\times = H$, we conclude that condition (iii) is equivalent to the requirement that $V(\lambda)$ is invertible for each $\lambda \in i\mathbb{R}$.

(i) $\Rightarrow$ (iii) If (i) is satisfied, then $V(\lambda)$ is positive definite for each $\lambda \in i\mathbb{R}$. In particular, $V(\lambda)$ is invertible for each $\lambda \in i\mathbb{R}$, and hence, by the result of the previous paragraph, (iii) holds.

(iii) $\Rightarrow$ (i) Conversely, assume (iii) is satisfied. Recall that V has no pole on the imaginary axis. Furthermore, $V(\lambda)$ is selfadjoint for $\lambda \in i\mathbb{R}$. Since $V(\lambda)$ is invertible for each $\lambda \in i\mathbb{R}$, it follows that for imaginary λ the signature of the matrix $V(\lambda)$ does not depend on λ. Next, observe that the rational matrix function V is biproper and that its value at infinity is equal to G. Hence the value of V

at infinity is positive definite. We obtain that $V(\lambda)$ is positive definite for each $\lambda \in i\mathbb{R}$. Thus (i) holds.

(i) $\Rightarrow$ (ii) Assume W has a positive definite real part on $\mathbb{R}$. Theorem 15.1 implies that equation (15.3) has a Hermitian solution X. Hence we have the factorization $W(\lambda) + W(\bar{\lambda})^* = K(\bar{\lambda})^* K(\lambda)$ with $K(\lambda)$ being given by (15.5). Since A has no eigenvalue on $\mathbb{R}$, the functions W and K have no pole on $\mathbb{R}$. The fact that W has a positive definite real part on $\mathbb{R}$ and the fact that W has no pole on $\mathbb{R}$ together imply that $W(\lambda) + W(\bar{\lambda})^*$ is invertible for each $\lambda \in \mathbb{R}$. Hence $K(\lambda)$ is also invertible for each $\lambda \in \mathbb{R}$. Thus $K(\lambda)^{-1}$ has no pole on $\mathbb{R}$. Notice that

$$K(\lambda)^{-1} = G^{-1/2} - G^{-1}(C - iB^*X)(\lambda - Z)^{-1}B, \tag{15.8}$$

where $Z = A - BG^{-1}(C - iB^*)$. Let $\lambda_0 \in \mathbb{R}$. Then λ_0 is not a common eigenvalue of A and Z. Thus we can apply the material presented in Section 8.1 to show that the realization given by the right-hand side of (15.8) is minimal at λ_0. But then the fact that $K(\lambda)^{-1}$ has no pole on $\mathbb{R}$ implies that λ_0 is not an eigenvalue of Z. Thus $Z = A - BG^{-1}C + iBG^{-1}B^*X$ has no real eigenvalue. This proves (ii).

(ii) $\Rightarrow$ (i) Let X be as in (ii). Then $W(\lambda) + W(\bar{\lambda})^* = K(\bar{\lambda})^* K(\lambda)$ with $K(\lambda)$ being given by (15.5). Observe that $K(\lambda)^{-1}$ is given by (15.8), where Z is as above. According to our hypothesis Z has no real eigenvalue. Hence $K(\bar{\lambda})^* K(\lambda)$ is positive definite for each $\lambda \in \mathbb{R}$. Thus (i) holds.

To prove the second part of the theorem, we apply Theorem 13.3. Recall that equation (15.3) can be rewritten into the algebraic Riccati equation (15.6). The Hamiltonian of this Riccati equation is precisely the block matrix H defined in item (iii). According to our hypotheses (A, B) is controllable. This implies that the pair $(iA - iBG^{-1}C, B)$ is also controllable. But controllability implies stabilizability. Thus the pair $(iA - iBG^{-1}C, B)$ is stabilizable. But then Theorem 13.3 tells us that condition (iii) implies that equation (15.3) has a unique Hermitian solution X such that the eigenvalues of $iA - iBG^{-1}C - BG^{-1}B^*X$ are in the open left half plane. Multiplication by $-i$ then gives the desired result. □

15.2 Canonical factorization of functions with a positive definite real part

In this section we consider canonical factorization of functions with a positive definite real part on the real line. Using state space realizations we shall prove the following result.

Theorem 15.3. *Let W be a proper rational matrix function having no real poles and such that $D = W(\infty)$ satisfies $D + D^* > 0$. Assume that W has a positive definite real part on the real line. Then W admits both a right and a left canonical factorization with respect to the real line.*

We start with some preparations that are of independent interest and will be useful in the next section too. Let T be a square matrix. If the real part of T

is positive definite, then T is injective, hence invertible. Indeed, for non-zero x we have $2\Re(\langle Tx, x\rangle) = \langle (T + T^*)x, x\rangle > 0$. Also, if T is invertible, then T^{-1} has a positive definite real part if and only if this is the case for T. This is immediate from either of the identities

$$T^{-1} + T^{-*} = T^{-1}(T + T^*)T^{-*}, \qquad T^{-1} + T^{-*} = T^{-*}(T + T^*)T^{-1}.$$

Now let $W(\lambda) = D + C(\lambda I_n - A)^{-1}B$ be a rational $m \times m$ matrix function with $G = D + D^*$ positive definite, and assume W has a nonnegative real part on $\mathbb{R}$. Then D is invertible, $G^\times$ defined by $G^\times = D^{-1} + D^{-*}$ is positive definite, $G^\times = D^{-1}G\,D^{-*}$, and W^{-1} has a nonnegative real part on $\mathbb{R}$. For W^{-1} we have the realization

$$W^{-1}(\lambda) = D^{-1} - D^{-1}C(\lambda I_n - A^\times)^{-1}BD^{-1}, \tag{15.9}$$

where, as usual, $A^\times = A - BD^{-1}C$. This gives rise to the following analogue of equation (15.3):

$$-i(A^\times)^*X + iXA^\times - (XBD^{-1} + iC^*D^{-*})(G^\times)^{-1}(D^{-*}B^*X - iD^{-1}C) = 0, \tag{15.10}$$

which can also be written as an algebraic Riccati equation

$$\begin{aligned} -\big(i(A^\times)^* + iC^*D^{-*}(G^\times)^{-1}D^{-*}B^*\big)X& \\ +X\big(iA^\times + iBD^{-1}(G^\times)^{-1}D^{-1}C\big)& \\ -C^*D^{-*}(G^\times)^{-1}D^{-1}C - XBD^{-1}(G^\times)^{-1}D^{-*}B^*X &= 0. \end{aligned} \tag{15.11}$$

Now let us look at the right coefficient of X in this expression. Using the identity $(G^\times)^{-1} = DG^{-1}D^*$, we get

$$\begin{aligned} iA^\times + iBD^{-1}(G^\times)^{-1}D^{-1}C &= iA - iBD^{-1}C + iBD^{-1}(DG^{-1}D^*)D^{-1}C \\ &= iA - iBD^{-1}C + iBG^{-1}D^*D^{-1}C \\ &= iA - iBG^{-1}(G - D^*)D^{-1}C \\ &= iA - iBG^{-1}DD^{-1}C = iA - iBG^{-1}C. \end{aligned}$$

Thus the right coefficient of X in (15.11) is equal to the right coefficient of X in (15.6). The left coefficient of X in (15.11) is the adjoint of the right coefficient of X in (15.11), and the same is true with (15.11) replaced by (15.6). Hence the left coefficient of X in (15.11) is equal to the left coefficient of X in (15.6). For the constant term in (15.11), we have

$$-C^*D^{-*}(G^\times)^{-1}D^{-1}C = -C^*D^{-*}(D^*G^{-1}D)D^{-1}C = -C^*G^{-1}C,$$

and the latter is the constant term in (15.11). Finally, the identities

$$-BD^{-1}(G^\times)^{-1}D^{-*}B^* = -BD^{-1}(DG^{-1}D^*)D^{-*}B^* = -BG^{-1}B^*$$

show that the coefficients of the quadratic terms in (15.11) and (15.6) coincide too. We conclude that the equations (15.3), (15.6), (15.10) and (15.11) all amount to the same.

Lemma 15.4. *Let $W(\lambda) = D + C(\lambda I_n - A)^{-1}B$ be a rational $m \times m$ matrix function such that $G = D + D^* > 0$. Assume X is an invertible Hermitian matrix satisfying* (15.3). *Then*

$$\begin{aligned}\frac{1}{2i}(XA - A^*X) &= -\frac{1}{2}(B^*X + iC)^*G^{-1}(B^*X + iC),\\ \frac{1}{2i}(XA^\times - (A^\times)^*X) &= -\frac{1}{2}(DD^{-*}B^*X - iC)^*G^{-1}(DD^{-*}B^*X - iC).\end{aligned}$$

In particular both A and $A^\times$ are $(-X)$-dissipative.

Proof. The first identity is just a restatement of (15.3). Recall that (15.3) and (15.10) amount to the same. Hence X also satisfies (15.10). Now note that the second identity in the lemma is just another way of writing (15.10). Here we use that $(G^\times)^{-1} = D^*G^{-1}D$. □

Before turning to the proof of Theorem 15.3 we present another lemma.

Lemma 15.5. *Let $W(\lambda) = D + C(\lambda I_n - A)^{-1}B$ be a rational $m \times m$ matrix function such that $G = D + D^* > 0$ and the pair (C, A) is observable. Assume X is an invertible Hermitian matrix satisfying* (15.3). *Let $N_1, N_1^\times$ be maximal X-nonpositive subspaces and $N_2, N_2^\times$ be maximal X-nonnegative subspaces such that N_1, N_2 are invariant under A and $N_1^\times, N_2^\times$ are invariant under $A^\times$. Then*

$$\mathbb{C}^n = N_1 \dot{+} N_2^\times, \qquad \mathbb{C}^n = N_2 \dot{+} N_1^\times. \tag{15.12}$$

Proof. Applying Proposition 11.1 we obtain

$$\dim N_1 + \dim N_2^\times = n, \qquad \dim N_2 + \dim N_1^\times = n.$$

Therefore in order to prove that (15.12) holds, it suffices to show that the intersections $N_1 \cap N_2^\times$ and $N_2 \cap N_1^\times$ are both trivial. Take $x \in N_1 \cap N_2^\times$. Then $\langle Xx, x\rangle = 0$. Now the Cauchy-Schwartz inequality holds on N_1. Thus

$$\begin{aligned}|\langle XAx, x\rangle|^2 &\le \langle XAx, Ax\rangle\langle Xx, x\rangle = 0,\\ |\langle Xx, Ax\rangle|^2 &\le \langle XAx, Ax\rangle\langle Xx, x\rangle = 0.\end{aligned}$$

Using this together with the first identity in Lemma 15.4, we get

$$0 = \Im\langle XAx, x\rangle = -\frac{1}{2}\|G^{-1/2}(B^*X + iC)x\|^2.$$

Similarly, employing the Cauchy-Schwartz inequality on $N_2^\times$ and the second identity in Lemma 15.4, we get

$$0 = \Im\langle XA^\times x, x\rangle = -\frac{1}{2}\|G^{-1/2}(DD^{-*}B^*X - iC)x\|^2.$$

Thus $(B^*X + iC)x = 0$ and $(DD^{-*}B^*X - iC)x = 0$. Adding these two identities we arrive at $0 = (I + DD^{-*})B^*Xx = GD^{-*}B^*Xx$. Hence $B^*Xx = 0$, and it also follows that $Cx = 0$. Thus $Ax = A^\times x$ for $x \in N_1 \cap N_2^\times$. Hence $N_1 \cap N_2^\times$ is an A-invariant subspace contained in $\operatorname{Ker} C$. Given the observability of the pair (C, A), this yields $N_1 \cap N_2^\times = \{0\}$.

The proof of $N_2 \cap N_1^\times = \{0\}$ is analogous. It can also be obtained by applying the result of the previous paragraph to the rational matrix function W^{-1}. □

Proof of Theorem 15.3. Let $W(\lambda) = D + C(\lambda I_n - A)^{-1}B$ be a minimal realization of W. Since W has no poles on $\mathbb{R}$, the minimality of the realization guarantees that A has no eigenvalues on $\mathbb{R}$. As we have seen (in the first paragraph after Theorem 15.3), the positive definiteness of $D + D^*$ implies that D is invertible. Similarly we conclude that W takes invertible values on $\mathbb{R}$. Hence we know from Theorem 2.4 that $A^\times = A - BD^{-1}C$ has no real eigenvalues either.

Since W has a positive definite real part, we can use Theorem 15.1 to deduce that equation (15.3) has an invertible Hermitian solution X, say. Lemma 15.4 now gives that both A and $A^\times$ are $(-X)$-dissipative.

Let M_+ and $M_+^\times$ be the spectral subspaces of A and $A^\times$, respectively, corresponding to the open upper half plane, and let M_- and $M_-^\times$, be the spectral subspaces of A and $A^\times$, respectively, corresponding to the open lower half plane. As A and $A^\times$ are $(-X)$-dissipative, we have that M_+ and $M_+^\times$ are maximal X-nonpositive. Similarly, the spaces M_- and $M_-^\times$ are maximal X-nonnegative. Using Lemma 15.5 we may conclude that $\mathbb{C}^n = M_+ \dot{+} M_-^\times$ and $\mathbb{C}^n = M_- \dot{+} M_+^\times$. But then Theorem 3.2 guarantees that W admits the desired canonical factorizations. □

15.3 Generalization to pseudo-canonical factorization

In this section the results of the previous section concerning canonical factorizations will be generalized to pseudo-canonical factorizations.

Theorem 15.6. *Let W be a proper rational $m \times m$ matrix function having no real poles such that $D = W(\infty)$ satisfies $D + D^* > 0$. Assume that W has a nonnegative real part on the real line. Then, with respect to the real line, W admits both right and left pseudo-canonical factorization. Such factorizations can be obtained in the following manner. Let $W(\lambda) = D + C(\lambda I_n - A)^{-1}B$ be a minimal realization, and put $G = D + D^*$. Then there exists an invertible Hermitian matrix X satisfying*

$$-iA^*X + iXA - (XB - iC^*)\,G^{-1}(B^*X + iC) = 0. \tag{15.13}$$

Also there are A-invariant subspaces M_+ and M_-, and $A^\times$-invariant subspaces $M_+^\times$ and $M_-^\times$, such that

(i) *M_+ is maximal X-nonpositive, M_+ contains the spectral subspace of A associated with the part of $\sigma(A)$ lying in the open upper half plane, and $\sigma(A|_{M_+}) \subset \{\lambda \mid \Im\lambda \geq 0\}$,*

(ii) M_- *is maximal* X*-nonnegative,* M_- *contains the spectral subspace of* A *associated with the part of* $\sigma(A)$ *lying in the open lower half plane, and* $\sigma(A|_{M_-}) \subset \{\lambda \mid \Im\lambda \leq 0\}$,

(iii) $M_+^\times$ *is maximal* X*-nonpositive,* $M_+^\times$ *contains the spectral subspace of* $A^\times$ *associated with the part of* $\sigma(A^\times)$ *lying in the open upper half plane, and* $\sigma(A^\times|_{M_+^\times}) \subset \{\lambda \mid \Im\lambda \geq 0\}$,

(iv) $M_-^\times$ *is maximal* X*-nonnegative,* $M_-^\times$ *contains the spectral subspace of* $A^\times$ *associated with the part of* $\sigma(A^\times)$ *lying in the open lower half plane, and* $\sigma(A^\times|_{M_-^\times}) \subset \{\lambda \mid \Im\lambda \leq 0\}$.

For such subspaces the matching conditions

$$\mathbb{C}^n = M_+ \dot{+} M_-^\times, \qquad \mathbb{C}^n = M_- \dot{+} M_+^\times \tag{15.14}$$

are satisfied. Write Π_r *for the projection along* M_+ *onto* $M_-^\times$ *and* Π_l *for the projection along* M_- *onto* $M_+^\times$*. Further put*

$$\begin{aligned}
\widetilde{W}_-(\lambda) &= D + C(\lambda I_n - A)^{-1}(I_n - \Pi_r)B,\\
\widetilde{W}_+(\lambda) &= I_n + D^{-1}C\Pi_r(\lambda I_n - A)^{-1}B,\\
\widehat{W}_+(\lambda) &= D + C(\lambda I_n - A)^{-1}(I_n - \Pi_l)B,\\
\widehat{W}_-(\lambda) &= I_n + D^{-1}C\Pi_l(\lambda I_n - A)^{-1}B.
\end{aligned}$$

Then $W(\lambda) = \widetilde{W}_-(\lambda)\widetilde{W}_+(\lambda)$ *and* $W(\lambda) = \widehat{W}_+(\lambda)\widehat{W}_-(\lambda)$ *are a right and a left pseudo-canonical factorization with respect to the real line, respectively.*

Proof. In view of the minimality of the given realization we can employ Theorem 15.1 to show that there is an invertible Hermitian matrix X such that (15.13), which is identical to (15.3), holds. By Lemma 15.4 the matrices A and $A^\times$ are $(-X)$-dissipative. The existence of subspaces $M_+, M_-, M_+^\times$ and $M_-^\times$ with the properties mentioned above is now guaranteed by Theorem 11.6. Lemma 15.5 gives the direct sums (15.14), and the conclusion of the theorem is straightforward by Theorem 8.6. □

As a further application of Lemma 15.5 we prove the following result on skew selfadjoint matrix functions. A rational $m \times m$ matrix function W is called *skew-Hermitian on the real line* if $W(\lambda)$ is skew-Hermitian for all λ in $\mathbb{R}$, λ not a pole of W.

Proposition 15.7. *Let* $W(\lambda) = D + V(\lambda)$*, where* V *is a strictly proper rational* $m \times m$ *matrix function that has no real poles, is skew-Hermitian on the real line and vanishes at infinity. Assume* $D + D^* > 0$*. The following statements are true.*

(i) *W admits a minimal factorization $W(\lambda) = \widetilde{W}_1(\lambda)\widetilde{W}_2(\lambda)$ where $\widetilde{W}_1$ has all its poles, respectively zeros, in the open upper, respectively lower, half plane, and $\widetilde{W}_2$ has all its poles, respectively zeros, in the open lower, respectively upper, half plane.*

(ii) *W admits a minimal factorization $W(\lambda) = \widehat{W}_1(\lambda)\widehat{W}_2(\lambda)$ where $\widehat{W}_1$ has all its poles, respectively zeros, in the open lower, respectively upper, half plane, and $\widehat{W}_2$ has all its poles, respectively zeros, in the open upper, respectively lower, half plane.*

Proof. Recall that $D + D^* > 0$ implies that D is invertible. Since V is skew-Hermitian on the real line, we see that $W(\lambda) + W(\lambda)^* = D + D^* > 0$ for $\lambda \in \mathbb{R}$. From the latter it follows that $W(\lambda)$ is invertible for each $\lambda \in \mathbb{R}$. Now let $W(\lambda) = D + C(\lambda I_n - A)^{-1}B$ be a minimal realization of W. Then both A and $A^\times$ have no eigenvalues on the real line.

From $C(\lambda I_n - A)^{-1}B = -(C(\lambda I_n - A)^{-1}B)^*$ for $\lambda \in \mathbb{R}$ and the minimality of the realization we may conclude (by the state space similarity theorem) that there is a unique invertible matrix Y such that

$$YA = A^*Y, \qquad YB = C^*, \qquad C = -B^*Y.$$

Taking adjoints in the above equations, and using the uniqueness of Y, one deduces that $Y = -Y^*$. Put $X = -iY$. Then X is selfadjoint. As $XA = A^*X$, the matrix A is X-selfadjoint. Furthermore, from $-iA^*X + iXA = 0$ and $XB - iC^* = 0$, we see that X is an invertible Hermitian solution of (15.13). But then we can use Lemma 15.4 to show that $A^\times$ is $(-X)$-dissipative.

Let M_u and M_l be the spectral subspaces of A associated with the part of $\sigma(A)$ lying in the open upper and open lower half plane, respectively. Also let $M_u^\times$ and $M_l^\times$ be the spectral subspaces of $A^\times$ associated with the part of $\sigma(A^\times)$ lying in the open upper and open lower half plane, respectively. Since the matrix A is X-selfadjoint and has no real eigenvalues, we know (see Theorem 11.5) that the spaces M_u and M_l are X-Lagrangian. In particular, these spaces are both maximal X-nonpositive and maximal X-nonnegative. The fact that $A^\times$ is $(-X)$-dissipative and has no real eigenvalues either, gives that the same conclusion holds for $M_u^\times$ and $M_l^\times$. But then Lemma 15.5 gives $\mathbb{C}^n = M_u \dot{+} M_u^\times$ as well as $\mathbb{C}^n = M_l \dot{+} M_l^\times$.

Let Π bethe projection of $\mathbb{C}^n$ along M_u onto $M_u^\times$. Then Π is a supporting projection of the minimal realization $W(\lambda) = D + C(\lambda I - A)^{-1}B$. Hence W admits a minimal factorization $W(\lambda) = \widetilde{W}_1(\lambda)\widetilde{W}_2(\lambda)$ such that (see Chapter 8) the following holds: the poles of $\widetilde{W}_1$ and $\widetilde{W}_2$ coincide with the eigenvalues of $A|_{M_u}$ and $A|_{M_u^\times}$, respectively, and the zeros of $\widetilde{W}_1$ and $\widetilde{W}_2$ coincide with the eigenvalues of $A^\times|_{M_u}$ and $A^\times|_{M_u^\times}$, respectively. Since A and $A^\times$ have no real eigenvalues and $M_u \dot{+} M_u^\times = \mathbb{C}^n$, we have $\sigma(A|_{M_u^\times}) = \sigma(A|_{M_l})$ and $\sigma(A^\times|_{M_u}) = \sigma(A^\times|_{M_l^\times})$. From these remarks it is clear that the factorization $W(\lambda) = \widetilde{W}_1(\lambda)\widetilde{W}_2(\lambda)$ has the desired properties. The factorization $W(\lambda) = \widehat{W}_1(\lambda)\widehat{W}_2(\lambda)$ is obtained in a similar way using the other direct sum decomposition $\mathbb{C}^n = M_l \dot{+} M_l^\times$. □

Notes

This chapter is based on [126], see also [129] and [128]. Rational matrix functions with a positive definite real part play a role in circuit and systems theory. In particular, Theorem 15.2 is a version of what is known as the positive real lemma. There are several variants of this result, see, for instance, Section 5.2 in [4], where also the connection with spectral factorization and Riccati equations is discussed. Another version in terms of Riccati inequalities is given in Section 12.6.3 in [83]. An infinite dimensional version may be found as Exercise 6.28 in [35].

Chapter 16

Contractive rational matrix functions

In this chapter rational matrix functions are studied of which the values on the imaginary axis or on the real line are contractive matrices. Included are solutions to spectral or canonical factorization problems for functions V of the form

$$V(\lambda) = I - W(-\bar{\lambda})^* W(\lambda) \quad \text{or} \quad V(\lambda) = I + W(\lambda),$$

where W is a rational matrix function which has contractive values on the imaginary axis or on the real line and, in addition, has a strictly contractive value at infinity.

This chapter consists of five sections. Sections 16.1 and 16.2 present a state space analysis (involving algebraic Riccati equations) of rational matrix functions that are contractive or strictly contractive on the imaginary axis. In Section 16.3 a state space formula is derived for the spectral factor in a spectral factorization of a rational matrix function of the form $V(\lambda) = I - W(-\bar{\lambda})^* W(\lambda)$, where W is strictly proper and strictly contractive on the imaginary axis. The final two sections of the chapter deal with canonical and pseudo-canonical factorization, respectively, for functions of the form $V(\lambda) = I + W(\lambda)$, where $W(\lambda)$ is strictly proper and strictly contractive for real λ (Section 16.4) or just contractive (Section 16.5).

16.1 State space analysis of contractive rational matrix functions

A rational $p \times m$ matrix function W is called *contractive on the imaginary axis* if the values that W takes on the imaginary axis are contractive matrices. Such a function does not have a pole on the imaginary axis. Moreover, it is proper and the value at infinity is again contractive. Of special interest is the subclass consisting of the contractive rational matrix functions W that are *strictly contractive at infinity*, i.e., the value of W at ∞ has norm smaller than 1. The first main result of this section is a characterization of this subclass in terms of realizations.

Theorem 16.1. *Let $W(\lambda) = D + C(\lambda I_n - A)^{-1}B$ be a realization of a $p \times m$ rational matrix function, and assume D is a strict contraction. Then the following two assertions hold:*

(i) *Assume (C, A) is an observable pair. Then W is contractive on the imaginary axis if and only if the algebraic Riccati equation*

$$-AP - PA^* - BB^* - (PC^* + BD^*)(I - DD^*)^{-1}(CP + DB^*) = 0 \quad (16.1)$$

has a Hermitian solution P.

(ii) *Assume (A, B) is a controllable pair. Then W is contractive on the imaginary axis if and only if the algebraic Riccati equation*

$$A^*P + PA - C^*C - (PB - C^*D)(I - D^*D)^{-1}(B^*P - D^*C) = 0 \quad (16.2)$$

has a Hermitian solution P.

Proof. Put $V(\lambda) = I - W(\lambda)W(-\bar{\lambda})^*$. Since W is proper, the same holds true for V. Moreover, $V(\infty) = I - DD^*$, and hence $V(\infty)$ is positive definite, because D is assumed to be a strict contraction. Note that W is contractive on $i\mathbb{R}$ if and only if V is nonnegative on $i\mathbb{R}$. Using the given realization for W we have

$$\begin{aligned} V(\lambda) &= I - DD^* + \begin{bmatrix} C & DB^* \end{bmatrix} \left(\lambda - \begin{bmatrix} A & BB^* \\ 0 & -A^* \end{bmatrix} \right)^{-1} \begin{bmatrix} -BD^* \\ C^* \end{bmatrix} \\ &= \begin{bmatrix} -C(\lambda - A)^{-1} & I \end{bmatrix} \begin{bmatrix} -BB^* & BD^* \\ DB^* & I - DD^* \end{bmatrix} \begin{bmatrix} (\lambda + A^*)^{-1}C^* \\ I \end{bmatrix}. \end{aligned}$$

The latter expression is of the form (13.25) and we see that (i) is an immediate consequence of the equivalence of statements (i) and (ii) in Theorem 13.6.

To prove assertion (ii) we use a duality argument. First note that a matrix X is a (strict) contraction if and only if X^* is a (strict) contraction. So W is contractive on $i\mathbb{R}$ if and only this is the case for the function $W(-\bar{\lambda})^*$. The latter has the realization $W(-\bar{\lambda})^* = D^* - B^*(\lambda + A^*)^{-1}C^*$. Also, the controllability of the pair (A, B) implies the observability of $(B^*, -A^*)$. Finally, D^* is a strict contraction. Thus assertion (ii) follows from part (i) by taking adjoints. □

Suppose D is a strict contraction. If the pair (C, A) is observable, then each Hermitian solution P of (16.2) is invertible. To see this, we argue as follows. Assume $Px = 0$. Multiplying (16.2) from the left by x^* and from the right by x yields $x^*C^*Cx + x^*C^*D(I - D^*D)^{-1}DC^*x = 0$. Now C^*C and $I - D^*D$ are nonnegative (in fact even $I - DD^* > 0$), and it follows that $x^*C^*Cx = 0$. Hence $Cx = 0$. But then, multiplying (16.2) on the right by x, we get $PAx = 0$. So $\operatorname{Ker} P$ is A-invariant and contained in $\operatorname{Ker} C$. As (C, A) is an observable pair, it follows that $\operatorname{Ker} P = \{0\}$. Since P is a square matrix, this yields the invertibility of P. In a similar fashion one proves that each solution of (16.1) is invertible provided that the pair (A, B) is controllable, or, which amounts to the same, the pair $B^*, A^*)$ is observable. Finally we note that P is an invertible solution of (16.2) if and only if $-P^1$ is an invertible solution of (16.1). Indeed, replacing P by $-P^{-1}$ in (16.1) and multiplying from the left and the right with P, one gets (16.2). In working out the details, identities of the type $D(I - D^*D)^{-1} = (I - DD^*)^{-1}D$ and $I + D(I - D^*D)^{-1}D^* = (I - DD^*)^{-1}$ play a role.

Theorem 16.2. *Let $W(\lambda) = D + C(\lambda I_n - A)^{-1}B$ be a realization of a $p \times m$ rational matrix function, and let D be a strict contraction. Assume, in addition, that the pair (C, A) is observable. Then W is contractive on the imaginary axis if and only if the matrix*

$$T = \begin{bmatrix} A + BD^*(I - DD^*)^{-1}C & B(I - D^*D)^{-1}B^* \\ -C^*(I - DD^*)^{-1}C & -A^* - C^*(I - DD^*)^{-1}DB^* \end{bmatrix} \tag{16.3}$$

has only even partial multiplicities at its pure imaginary eigenvalues.

Proof. Let V be as in the proof of Theorem 16.1, and recall that W is contractive on $i\mathbb{R}$ if and only if V is nonnegative on $i\mathbb{R}$. The desired result is now immediate from the equivalence of statements (ii) and (iii) in Theorem 13.6 combined with the fact that (16.3) is the Hamiltonian of the equation (16.1). □

Theorem 16.2 has a counterpart in which (16.3) is replaced by the Hamiltonian of (16.2).

As a special case of Theorem 16.1 let us consider rational matrix functions which are contractive not only on the imaginary axis but on the full closed right half plane.

Theorem 16.3. *Let $W(\lambda) = D + C(\lambda I_n - A)^{-1}B$ be a minimal realization of a rational $p \times m$ matrix function. Assume that W is contractive on the imaginary axis, and let D be a strict contraction. Then the following statements are equivalent:*

(i) *For each λ in the closed right half plane, λ not a pole of W, the matrix $W(\lambda)$ is a contraction;*

(ii) *The matrix A has all its eigenvalues in the open left half plane;*

(iii) *There is a positive definite solution of* (16.1).

Proof. Suppose A has all its eigenvalues in the open left half plane. Then $W(\lambda)$ is analytic in the closed right half plane. As $W(\lambda)$ is a contraction for each $\lambda \in i\mathbb{R}$ and at infinity, the maximum modulus theorem implies $W(\lambda)$ is contractive for all

λ in the open right half plane as well. Thus (ii) implies (i). Conversely, suppose (i) holds. Then W must be analytic in the closed right half plane (infinity included), and by minimality of the realization the matrix A has all its eigenvalues in the open left half plane. The equivalence of (ii) and (iii) follows by rewriting (16.1) as

$$AP + PA^* = RR^*,$$

where $Q = -P$ and $R = \begin{bmatrix} B & (PC^* + BD^*)(I - DD^*)^{-1/2} \end{bmatrix}$. Since the realization is minimal, (A, B) is a controllable pair, and hence the same holds true for the pair (A, R). But then we can apply a well-known inertia theorem (see, e.g., Theorem 13.1.4 in [107]) to show that (ii) and (iii) are equivalent. □

16.2 Strictly contractive rational matrix functions

In this section we specify further the results of the previous section for the case of rational matrix functions W that are *strictly contractive on the imaginary axis*. By this we mean that $\|W(\lambda)\| < 1$ for $\lambda \in i\mathbb{R}$. Such a function does not have a pole on $i\mathbb{R}$ and is proper.

Theorem 16.4. *Let $W(\lambda) = D + C(\lambda I_n - A)^{-1}B$ be a realization of a $p \times m$ rational matrix function W. Assume A has no pure imaginary eigenvalues, D is a strict contraction, and the pair (C, A) is observable. Then the following statements are equivalent:*

(i) *The function W is strictly contractive on the imaginary axis;*

(ii) *Equation (16.1) has an $i\mathbb{R}$-stabilizing solution P, that is, it has a solution P such that $-A^* - C^*(I - DD^*)^{-1}DB^* - C^*(I - DD^*)^{-1}CP$ has all its eigenvalues in the open left half plane;*

(iii) *The matrix T given by (16.3) has no pure imaginary eigenvalues.*

Moreover, if one of the above conditions is satisfied, then the $i\mathbb{R}$-stabilizing solution P in (ii) is unique and Hermitian.

Proof. Suppose W is strictly contractive on $i\mathbb{R}$. Then the rational $m \times m$ matrix function $V(\lambda) = I - W(\lambda)W(-\bar{\lambda})^*$ is positive definite on $i\mathbb{R}$ and $V(\infty) = I - DD^*$ is positive definite too. In particular, V is biproper and V has no pole or zero on $i\mathbb{R}$. Recall (see the proof of Theorem 16.1) that

$$V(\lambda) = I - DD^* + \begin{bmatrix} C & DB^* \end{bmatrix} \left(\lambda - \begin{bmatrix} A & BB^* \\ 0 & -A^* \end{bmatrix} \right)^{-1} \begin{bmatrix} -BD^* \\ C^* \end{bmatrix}.$$

The associate main matrix of this realization is T given by (16.3). It follows that T has no eigenvalues on $i\mathbb{R}$. So (iii) holds. Conversely, if T has no pure imaginary eigenvalues, then V has no poles or zeros on $i\mathbb{R}$. As $V(\infty)$ is positive definite, it follows that $V(\lambda)$ is positive definite for $\lambda \in i\mathbb{R}$. Hence $W(\infty)$ is strictly contractive for $\lambda \in i\mathbb{R}$. We have now proved the equivalence of (i) and (iii). The equivalence of (ii) and (iii) is a direct consequence of Theorem 13.3. Note here that the observability of the pair (C, A) is equivalent to the controllability of (A^*, C^*), and the latter implies the stabilizabilty of (A^*, C^*). The final statement of the theorem is covered by Theorem 13.3 as well. □

Corollary 16.5. *Let (C, A) be an observable pair, and assume that A has no pure imaginary eigenvalue. Then the Riccati equation*

$$YC^*CY - YA^* - AY = 0$$

has a unique Hermitian solution Y such that $A - YC^*C$ has all its eigenvalues in the open right half plane.

Proof. Apply Theorem 16.4 with $D = 0$ and $B = 0$. Then D is a strict contraction and W is identically equal to zero. In particular, (i) in Theorem 16.4 is satisfied. Next, note that with $D = 0$ and $B = 0$ equation (16.1) reduces to

$$-AP - PA^* - PC^*CP = 0,$$

and by Theorem 16.4, with $D = 0$ and $B = 0$, this equation has a unique Hermitian solution P such that $-A^* - C^*CP$ has all its eigenvalues in the open left half plane. But then $A - YC^*C$ has all its eigenvalues in the open right half plane. Now put $Y = -P$, then we see that Y has all the desired properties. □

16.3 An application to spectral factorization

In this section we consider functions of the form

$$V(\lambda) = I - W(-\bar{\lambda})^* W(\lambda), \tag{16.4}$$

where W is a proper rational $p \times m$ matrix function which is strictly contractive on the imaginary axis. In fact we shall assume that W is *strictly proper*, that is W vanishes at infinity. Thus V is positive definite on the imaginary axis and has a positive definite value at infinity (namely I_m). Hence W admits a right spectral factorization. Using a minimal realization of W, such a factorization is constructed in the following theorem.

Theorem 16.6. *Let $W(\lambda) = C(\lambda I_n - A)^{-1}B$ be a minimal realization of the $p \times m$ rational matrix function W which is strictly contractive on the imaginary axis. Then the Riccati equations*

$$XBB^*X - XA - A^*X + C^*C = 0, \tag{16.5}$$

$$YC^*CY - YA^* - AY = 0, \tag{16.6}$$

have Hermitian solutions X and Y, respectively, such that the matrices $A - BB^*X$ and $A - YC^*C$ have all their eigenvalues in the open right half plane, and $I_n - XY$

is invertible (or, which amounts to the same, $I_n - YX$ is invertible). Furthermore, with respect to the imaginary axis, the function $V(\lambda) = I_m - W(-\bar\lambda)^* W(\lambda)$ admits the right spectral factorization $V(\lambda) = L_+(-\bar\lambda)^* L_+(\lambda)$ with L_+ and its inverse L_+^{-1} being given by

$$L_+(\lambda) = I + B^* X(I_n - YX)^{-1}(\lambda I_n - A + YC^*C)^{-1}B, \tag{16.7}$$

$$L_+^{-1}(\lambda) = I - B^* X(\lambda I_n - A + BB^*X)^{-1}(I_n - YX)^{-1}B. \tag{16.8}$$

Proof. By Corollary 16.5, the equation (16.6) has a Hermitian solution Y such that $A - YC^*C$ has all its eigenvalues in the open right half plane. Next, we apply Theorem 16.4 to

$$\widetilde{W}(\lambda) = W(-\bar\lambda)^* = -B^*(\lambda + A^*)^{-1}C^*.$$

Notice that $\widetilde{W}(\lambda) = -B^*(\lambda + A^*)^{-1}C^*$ satisfies the general hypotheses of Theorem 16.4. Furthermore, $\widetilde{W}$ is strictly contractive on $i\mathbb{R}$ and its value at infinity is zero. In particular, item (i) in Theorem 16.4 is satisfied. Hence item (iii) is satisfied as well, i.e., the matrix T of (16.3) has no pure-imaginary eigenvalues.

Now consider the function $V(\lambda) = I - W(-\bar\lambda)^* W(\lambda)$ which has the realization

$$V(\lambda) = I + \begin{bmatrix} B^* & 0 \end{bmatrix} \left(\lambda - \begin{bmatrix} -A^* & C^*C \\ 0 & A \end{bmatrix}\right)^{-1} \begin{bmatrix} 0 \\ B \end{bmatrix}. \tag{16.9}$$

Put

$$\widehat{A} = \begin{bmatrix} -A^* & C^*C \\ 0 & A \end{bmatrix}, \qquad \widehat{A}^\times = \begin{bmatrix} -A^* & C^*C \\ -B^*B & A \end{bmatrix}.$$

Since W is contractive on the imaginary axis, the function W has no pure imaginary poles. According to our assumptions the given realization of W is minimal. This implies that A has no eigenvalue on $i\mathbb{R}$. But then we can use the triangular form of $\widehat{A}$ to show that the same holds true for the matrix $\widehat{A}$. Since V is positive definite on the imaginary axis, we know that $V(\lambda)$ is invertible for each $\lambda \in i\mathbb{R}$. But then Theorem 2.4 gives that $\widehat{A}^\times$ has no pure imaginary eigenvalues either. (Alternatively, this may be seen from the fact that T and $\widehat{A}^\times$ are similar.)

Let M_- be the spectral subspace of the matrix $\widehat{A}$ with respect to the open left half plane, and let $M_+^\times$ be the spectral subspace of $\widehat{A}^\times$ with respect to the open right half plane. Observe that V is positive definite on the imaginary axis and has a positive definite value at infinity, namely I_m. This suggests the use of Theorem 9.4 to show that $\mathbb{C}^{2n} = M_- \dot{+} M_+^\times$. For this a skew-Hermitian H must be identified with the properties required in Theorem 9.4. This can be done along the lines indicated in the proof of Theorem 13.1. So indeed $\mathbb{C}^{2n} = M_- \dot{+} M_+^\times$. The fact that Y is Hermitian and the eigenvalues of $A^* - C^*CY$ are in the open right half plane implies that $\sigma(A - YC^*C) \cap \sigma(-A^* + C^*CY) = \emptyset$. Hence Proposition 12.1 gives that the spectral subspace M_- is given by $M_- = \operatorname{Im} \begin{bmatrix} I & Y \end{bmatrix}^*$.

Now $M_+^\times$ is an H-Lagrangian invariant subspace for $\widehat{A}^\times$ by Theorem 11.5. From Theorem 13.6 we see that there is a Hermitian solution X of (16.5) such that $M_+^\times = \operatorname{Im} \begin{bmatrix} X & I \end{bmatrix}^*$. Moreover, $A - BB^*X$ and the restriction of $\widehat{A}^\times$ to $M_+^\times$ have the same eigenvalues. Thus, the eigenvalues of $A - BB^*X$ are in the open right half plane. As $\mathbb{C}^{2n} = M_- \dot{+} M_+^\times$, the invertibility of $I - XY$ follows from Lemma 12.4.

Finally, we apply Theorem 12.3 to show that V admits the factorization $V(\lambda) = V_1(\lambda)V_2(\lambda)$, where

$$\begin{aligned}
V_1(\lambda) &= I - B^*(\lambda + A^* - C^*CY)^{-1}(I - XY)^{-1}XB^*,\\
V_2(\lambda) &= I + B^*X(I - YX)^{-1}(\lambda - A + YC^*C)^{-1}B,\\
V_1^{-1}(\lambda) &= I + B^*(I - XY)^{-1}(\lambda + A^* - XBB^*)^{-1}XB,\\
V_2^{-1}(\lambda) &= I - B^*X(\lambda - A + BB^*X)^{-1}(I - YX)^{-1}B.
\end{aligned}$$

Clearly, $V_2 = L_+$ and $V_2^{-1} = L_+^{-1}$ with L_+ and L_+^{-1} being given by (16.7) and (16.7), respectively. Furthermore, taking into account that X and Y are Hermitian,

$$\begin{aligned}
L_+(-\bar{\lambda})^* &= V_2(-\bar{\lambda})^*\\
&= I + B(-\lambda - A^* + CC^*Y)^{-1}(I - XY)^{-1}XB = V_1(\lambda).
\end{aligned}$$

Thus we have $V(\lambda) = L_+(-\bar{\lambda})^*L_+(\lambda)$, and from the location of the eigenvalues of $A - YC^*C$ and $A - BB^*X$ we see that this is a right spectral factorization. □

16.4 An application to canonical factorization

Consider a function of the form

$$V(\lambda) = I_m + C(\lambda I_n - A)^{-1}B, \tag{16.10}$$

where $W(\lambda) = C(\lambda I_n - A)^{-1}B$ is *strictly contractive on the real line*. By this we mean that the values of W on $\mathbb{R}$ are strict contractions, and this implies that W has no pole on the real line. Hence the latter holds true for V too. It follows also that V takes invertible values on the real line, i.e., V has no zero there.

Now assume for the moment that (16.10) is a minimal realization for W. Since V has neither a pole nor a zero on the real line, the minimality of the realization implies that the matrices A and $A^\times = A - BC$ have no real eigenvalues. Furthermore, since the function $\widetilde{W}(\lambda) = C(i\lambda I_n - A)^{-1}B$ is strictly contractive on the imaginary axis, we can apply Theorem 16.1(ii) to establish the existence of a Hermitian matrix X for which

$$iXA - iA^*X + XBB^*X + C^*C = 0. \tag{16.11}$$

Finally, because of the minimality (see the remark in the paragraph after the proof of Theorem 16.1), such a matrix X is invertible.

Summarizing, if (16.10) is a minimal realization and the matrix function $W(\lambda) = C(\lambda I_n - A)^{-1}B$ is strictly contractive for real λ, then both A and $A^\times$ have no real eigenvalues and there exists a Hermitian invertible matrix X solving (16.11). The next theorem describes canonical factorizations of a function of the form (16.10) in terms of a realization having the properties just described.

Theorem 16.7. *Let $V(\lambda) = I_m + C(\lambda I_n - A)^{-1}B$ be a realization of an $m \times m$ rational matrix function such that A and $A^\times = A - BC$ have no real eigenvalues, and assume that there exists a Hermitian invertible X satisfying* (16.11), *i.e.,*

$$iXA - iA^*X + XBB^*X + C^*C = 0.$$

Let M_- and M_+ be the spectral subspaces of A associated with the parts of $\sigma(A)$ lying in the open lower and open upper half plane, respectively, and let $M_-^\times$ and $M_+^\times$ be the spectral subspaces of $A^\times$ associated with the parts of $\sigma(A^\times)$ lying in the open lower and open upper half plane, respectively. Then

$$\mathbb{C}^n = M_- \dot{+} M_+^\times, \qquad \mathbb{C}^n = M_+ \dot{+} M_-^\times. \tag{16.12}$$

Moreover, V admits both a left and a right canonical factorization with respect to the real line,

$$V(\lambda) = \widetilde{V}_+(\lambda)\widetilde{V}_-(\lambda), \qquad V(\lambda) = \widehat{V}_-(\lambda)\widehat{V}_+(\lambda),$$

with the factors being given by

$$\begin{aligned}
\widetilde{V}_+(\lambda) &= I_m + C(\lambda I_n - A)^{-1}(I_n - \Pi_l)B,\\
\widetilde{V}_-(\lambda) &= I_m + C\Pi_l(\lambda I_n - A)^{-1}B,\\
\widehat{V}_-(\lambda) &= I_n + C(\lambda I_n - A)^{-1}(I_n - \Pi_r)B,\\
\widehat{V}_+(\lambda) &= I_m + C\Pi_r(\lambda I_n - A)^{-1}B.
\end{aligned}$$

Here Π_l is the projection along M_- onto $M_+^\times$, and Π_r is the projection along M_+ onto $M_-^\times$.

Proof. In view of Theorem 3.2, only (16.12) needs to be proved. We begin the verification of (16.12) by observing that (16.11) implies

$$\begin{aligned}
\frac{1}{2i}(XA - A^*X) &= \frac{1}{2}(XBB^*X + C^*C), &(16.13)\\
\frac{1}{2i}(XA^\times - (A^\times)^*X) &= \frac{1}{2}(iXB + C^*)(C - iB^*X). &(16.14)
\end{aligned}$$

These two identities imply that $\Im\langle XAx, x\rangle$ and $\Im\langle XA^\times x, x\rangle$ are nonnegative for all $x \in \mathbb{C}^n$. In other words, both A and $A^\times$ are X-dissipative, that is, they are dissipative in the indefinite inner product given by X (cf., Section 11.3). Because of this property, it follows that M_+ and $M_+^\times$ are maximal X-nonnegative, while M_- and $M_-^\times$ are maximal X-nonpositive (see Section 11.3). Using Proposition 11.1 it follows that $\dim M_+ + \dim M_-^\times = n$ and $\dim M_- + \dim M_+^\times = n$. Thus (16.12) is obtained via a dimension argument as soon as we have shown that $M_+ \cap M_-^\times = M_- \cap M_+^\times = \{0\}$.

Take $x \in M_+ \cap M_-^\times$. Then $\langle Xx, x\rangle = 0$, as x belongs to both an X-nonnegative subspace and an X-nonpositive subspace. Now the Cauchy-Schwartz inequality holds on M_+. Thus

$$|\langle XAx, x\rangle|^2 \le \langle XAx, Ax\rangle\langle Xx, x\rangle = 0,$$

and

$$|\langle Xx, Ax\rangle|^2 \le \langle XAx, Ax\rangle\langle Xx, x\rangle = 0.$$

From (16.13) we get

$$0 = \frac{1}{2i}\langle (XA - A^*X)x, x\rangle = \frac{1}{2}(\|Cx\|^2 + \|B^*Xx\|^2).$$

Hence $Cx = 0$, and so for $x \in M_+ \cap M_-^\times$ we have $A^\times x = (A - BC)x = Ax$. Consequently $M_+ \cap M_-^\times$ is both A-invariant and $A^\times$-invariant. As

$$\sigma(A|_{M_+\cap M_-^\times}) \subset \sigma(A|_{M_+}) \subset \{\lambda \mid \Im\lambda > 0\},$$

$$\sigma(A^\times|_{M_+\cap M_-^\times}) \subset \sigma(A^\times|_{M_-^\times}) \subset \{\lambda \mid \Im\lambda < 0\},$$

and $A|_{M_+\cap M_-^\times} = A^\times|_{M_+\cap M_-^\times}$, we have that $M_+\cap M_-^\times = \{0\}$. In a similar way one shows that $M_- \cap M_+^\times = \{0\}$. □

Note that the above theorem together with the arguments given in the first two paragraphs of this section yield the following corollary.

Corollary 16.8. *Let $V(\lambda) = I_m + W(\lambda)$, where W is a strictly proper rational matrix function which is strictly contractive on the real line. Then V admits both a right and a left canonical factorization with respect to the real line.*

16.5 A generalization to pseudo-canonical factorization

In this section the result of the previous section is generalized to pseudo-canonical factorizations. As a preparation we recall from Theorem 11.6 the following facts. Let X be an $n \times n$ invertible Hermitian matrix and let A be an $n \times n$ matrix which is X-dissipative. Then there exist A-invariant subspaces M_+ and M_- such that M_+ is maximal X-nonnegative and M_- is maximal X-nonpositive,

$$\sigma(A|_{M_+}) \subset \{\lambda \mid \Im\lambda \ge 0\}, \qquad \sigma(A|_{M_-}) \subset \{\lambda \mid \Im\lambda \le 0\},$$

M_+ contains the spectral subspace of A corresponding to the eigenvalues of A in the open upper half plane, and M_- contains the spectral subspace of A corresponding to the eigenvalues of A in the open lower half plane.

These facts allow us to deal with rational matrix functions that are contractive on the real line. A rational matrix function W is called *contractive on the real line* if the values that W takes on $\mathbb{R}$ are contractive matrices. Such a function does not have a pole on the real line.

Theorem 16.9. *Let W be a strictly proper rational $m \times m$ matrix function which is contractive on the real line. Then $V(\lambda) = I_m + W(\lambda)$ admits both a right and a left pseudo-canonical factorization with respect to the real line. Such factorizations can be obtained as follows. Let $W(\lambda) = C(\lambda I_n - A)^{-1}B$ be a minimal realization. Then there exists an invertible Hermitian matrix X satisfying*

$$iXA - iA^*X + XBB^*X + C^*C = 0. \tag{16.15}$$

Let M_- and $M_-^\times$ be maximal X-nonpositive subspaces that are invariant under A and $A^\times$, respectively, such that

$$\sigma(A|_{M_-}) \subset \{\lambda \mid \Im\lambda \leq 0\}, \qquad \sigma(A^\times|_{M_-^\times}) \subset \{\lambda \mid \Im\lambda \leq 0\},$$

and let M_+ and $M_+^\times$ be maximal X-nonnegative subspaces that are invariant under A and $A^\times$, respectively, such that

$$\sigma(A|_{M_+}) \subset \{\lambda \mid \Im\lambda \geq 0\}, \qquad \sigma(A^\times|_{M_+^\times}) \subset \{\lambda \mid \Im\lambda \geq 0\}.$$

Then (16.12) holds, that is $\mathbb{C}^n = M_- \dot{+} M_+^\times$ and $\mathbb{C}^n = M_+ \dot{+} M_-^\times$. Let Π_l be the projection along M_- onto $M_+^\times$, and put

$$\begin{aligned} \widetilde{V}_+(\lambda) &= I_m + C(\lambda I_n - A)^{-1}(I_n - \Pi_l)B, \\ \widetilde{V}_-(\lambda) &= I_m + C\Pi_l(\lambda I_n - A)^{-1}B. \end{aligned}$$

Then $V(\lambda) = \widetilde{V}_+(\lambda)\widetilde{V}_-(\lambda)$ is a left pseudo-canonical factorization with respect to the real line. Write Π_r for the projection along M_+ onto $M_-^\times$, and set

$$\begin{aligned} \widehat{V}_-(\lambda) &= I_m + C(\lambda - A)^{-1}(I_n - \Pi_r)B, \\ \widehat{V}_+(\lambda) &= I_m + C\Pi_r(\lambda I_n - A)^{-1}B. \end{aligned}$$

Then $V(\lambda) = \widehat{V}_-(\lambda)\widehat{V}_+(\lambda)$ is a right pseudo-canonical factorization with respect to the real line .

Proof. By applying Theorem 16.1 (ii) to $W(\lambda) = C(i\lambda I_n - A)^{-1}B$ we see that there is an invertible Hermitian X such that (16.15) holds. Once (16.12) is proved the rest of the theorem is a consequence of Theorem 8.5. Of the two equalities in (16.12) only the first will be proved, the second can be established analogously.

As M_+ is maximal X-nonnegative and $M_-^\times$ is maximal X-nonpositive we have $\dim M_+ + \dim M_-^\times = n$, by Proposition 11.1. So it remains to show that $M_+ \cap M_-^\times = \{0\}$. Take $x \in M_+ \cap M_-^\times$. As in the proof of Theorem 16.7, one shows that $Cx = 0$, and thus $Ax = A^\times x$. Obviously, it follows from this that $M_+ \cap M_-^\times$ is A-invariant and contained in $\operatorname{Ker} C$. Because of the minimality, we can conclude that $M_+ \cap M_-^\times = \{0\}$. □

Note that the location of the spectra of the operators $A|_{M_-}, A^\times|_{M_-^\times}, A|_{M_+}$ and $A^\times|_{M_+^\times}$ do not play a role in the proof of the identities in (16.12). Thus we also have the following result.

Proposition 16.10. *Let $V(\lambda) = I_m + C(\lambda I_n - A)^{-1}B$ be a minimal realization, and let X be an invertible Hermitian solution of (16.15). Let M be any A-invariant maximal X-nonnegative subspace, and let $M^\times$ be any $A^\times$-invariant maximal X-nonpositive subspace. Then $\mathbb{C}^n = M \dot{+} M^\times$. Let Π be the projection along M onto $M^\times$, and write*

$$\begin{aligned} V_1(\lambda) &= I_m + C(\lambda I_n - A)^{-1}(I - \Pi)B, \\ V_2(\lambda) &= I_m + C\Pi(\lambda I_n - A)^{-1}B. \end{aligned}$$

Then $V(\lambda) = V_1(\lambda)V_2(\lambda)$ is a minimal factorization.

A similar result holds for any A-invariant maximal X-nonpositive subspace M and any $A^\times$-invariant maximal X-nonnegative subspace $M^\times$.

Notice that there are various similarities between the proofs of Theorems 16.7 and 16.9 on the one hand and those of Theorems 15.3 and 15.6 on the other hand. These similarities are not surprising. In fact, the main results of the previous two sections are closely related to those in Sections 15.2 and 15.3 of the previous chapter. To see this we use the Cayley transformation

$$F(\lambda) = \big(I - W(\lambda)\big)\big(I + W(\lambda)\big)^{-1}. \tag{16.16}$$

Here are the details.

Let W be a strictly proper rational $m \times m$ matrix function, and let F be the rational $m \times m$ matrix function given by (16.16). Since W is strictly proper, $I + W(\lambda)$ is biproper, and hence F is well-defined. Furthermore, F is biproper and its value at infinity is equal to I_m. The identity

$$F(\lambda) + F(\bar{\lambda})^* = 2\big(I_m + W(\bar{\lambda})^*\big)^{-1}\big(I_m - W(\bar{\lambda})^*W(\lambda)\big)\big(I_m + W(\lambda)\big)^{-1}$$

shows that F has a nonnegative real part on $\mathbb{R}$ if and only if W is contractive on $\mathbb{R}$. Moreover, F has a positive definite real part on $\mathbb{R}$ if and only if W is strictly contractive on $\mathbb{R}$.

Assume now that W is given by the realization $W(\lambda) = C(\lambda I_n - A)^{-1}B$. Since $F(\lambda) = \big(2I_m - (I_m + W(\lambda))\big)\big(I_m + W(\lambda)\big)^{-1} = 2\big(I_m + W(\lambda)\big)^{-1} - I_m$, we see that F admits the realization

$$F(\lambda) = I_m - 2C(\lambda I_n - A^\times)^{-1}B, \tag{16.17}$$

where, as usual, $A^\times = A - BC$. Now apply Theorem 15.1 to F using the realization (16.17). For this case equation (15.3), with X replaced by Y, has the form

$$-i(A^\times)^*Y + iYA^\times - \frac{1}{2}(YB + i2C^*)(B^*Y - 2iC) = 0. \tag{16.18}$$

Using $A^\times = A - BC$ and setting $Y = -2X$, a straightforward computation shows that (16.18) is equivalent to

$$iXA - iA^*X + XBB^*X + C^*C = 0, \tag{16.19}$$

and the latter equation is precisely (16.11). By applying Theorem 15.1 to F and using the equivalence between (16.18) and (16.19) we obtain the following result.

Proposition 16.11. *Let $W(\lambda) = C(\lambda I_n - A)^{-1}B$, and assume that the pair (A, B) is controllable. Then W is contractive on the real line if and only if the equation* (16.19) *has a Hermitian solution. Moreover, if the given realization is minimal, then any Hermitian solution of* (16.19) *is invertible.*

The above proposition provides an alternative proof of Theorem 16.1(ii) for the case when W is square and $D = 0$. The details involve a transformation $\lambda \mapsto i\lambda$ (cf., the beginning of the proof of Theorem 16.9).

Notes

The state space characterizations of contractive and strictly contractive rational matrix functions given in Theorems 16.1, 16.3 and 16.4 are versions of what is known as the bounded real lemma in mathematical systems theory. These results play an important role in robust and optimal control theory, see, e.g., the text books [77] and [150]. The bounded real lemma may also be found in [4] in another form. The application to spectral factorization (Section 16.3) is classical and can be found in Chapter 7 in [4]. The result that a function of the form identity plus a strict contraction admits canonical factorization (Section 16.4) is well-known; see e.g., [29] and the references given there. A surprising fact is that this property actually characterizes the circle or the line; for this see [109]. The state space results given in Sections 16.4 and 16.5 are based on [74].

Chapter 17

J-unitary rational matrix functions

In this chapter realizations are used to study rational matrix functions of which the values on the imaginary axis are J-unitary matrices. Solutions to various factorization problems are given. Special attention is paid to factorization of J-unitary rational matrix functions into J-unitary factors. We also discuss the problem of embedding a contractive rational matrix function as the $(1, 2)$ block in a unitary rational matrix function. The latter problem is related to the Darlington synthesis problem from network theory.

This chapter consists of eight sections. Realization and minimal factorization of J-unitary rational matrix functions are the main topics of Sections 17.1 and 17.2. In Section 17.3 the factorization results are specified further for unitary rational matrix functions. The Redheffer transform, which allows one to relate J-unitary rational matrix functions to certain classes of unitary rational matrix functions, is introduced in Section 17.4. This transform is used in Section 17.5 in the study of J-inner rational matrix functions. A state space analysis of inner-outer factorization is the main topic of Section 17.6. The final two sections deal with completion problems. Section 17.7 presents state space formulas for unitary completions of minimal degree, and Section 17.8 presents such formulas for bi-inner completions of non-square inner rational matrix functions.

17.1 Realizations of J-unitary rational matrix functions

Throughout this section, J stands for an $m \times m$ signature matrix, that is, J is an invertible Hermitian matrix such that $J = J^{-1}$. An $m \times m$ matrix M is said to be *J-unitary* if $M^*JM = J$. Since all matrices in the latter identity are square and J is invertible, it follows that a J-unitary matrix M is invertible and $M^{-1} = JM^*J$.

If M is a J-unitary matrix, then M^* and M^{-1} are both J-unitary too. Indeed,

$$MJM^* = (M^*J)^{-1}M^* = J^{-1}(M^*)^{-1}M^* = J^{-1} = J,$$

$$M^{-1})^*JM^{-1} = (JM^*J)^*J(JM^*J) = J(MJM^*)J = J.$$

In this chapter we deal with rational matrix functions of which the values on the imaginary axis are J-unitary matrices. A rational $m \times m$ matrix function W is called *J-unitary on the imaginary axis* if it takes J-unitary values on the imaginary axis. In other words, W is J-unitary with respect to the imaginary axis whenever

$$W(\lambda)^*JW(\lambda) = J, \qquad \lambda \in i\mathbb{R},\ \lambda \text{ not a pole of } W. \tag{17.1}$$

Equivalently, W is J-unitary with respect to the imaginary axis if and only if

$$W(-\bar{\lambda})^*JW(\lambda) = J, \qquad \lambda, -\bar{\lambda} \text{ not a pole of } W. \tag{17.2}$$

In the sequel we shall only consider matrix functions that are J-unitary with respect to the imaginary axis and not with respect to other contours. Therefore we shall feel free to omit the phrase "with respect to the imaginary axis."

Observe that if W is J-unitary, then both the functions $W(\lambda)^{-1}$ and $W(-\bar{\lambda})^*$ are J-unitary as well. Furthermore, if W_1 and W_2 are two J-unitary rational matrix functions, their product W_1W_2 will also be J-unitary.

First we shall characterize the property of being a J-unitary rational matrix function in terms of realizations. We shall assume throughout that the rational matrix functions are proper.

Theorem 17.1. *Let $W(\lambda) = D + C(\lambda I_n - A)^{-1}B$ be a minimal realization of a proper rational $m \times m$ matrix function. The following statements are equivalent:*

(i) *W is J-unitary;*

(ii) *D is J-unitary and there exists an $n \times n$ matrix H such that*

$$AH + HA^* = BJB^*, \qquad CH = DJB^*, \qquad H = H^*; \tag{17.3}$$

(iii) *D is J-unitary and there exists an $n \times n$ matrix G such that*

$$GA + A^*G = C^*JC, \qquad GB = C^*JD, \qquad G = G^*. \tag{17.4}$$

In this case the matrices H and G are uniquely determined by the given realization, they are invertible and $G = H^{-1}$.

Proof. Assume that W is J-unitary. Taking the limit in (17.1) for $\lambda \to \infty$ we see that $D^*JD = J$. Thus D is a J-unitary matrix. In particular, D is invertible, and hence W is biproper. By (17.2) we have $J(W(-\bar{\lambda})^*)^{-1}J = W(\lambda)$ for all λ for which λ is not a pole of W and $-\bar{\lambda}$ is not a zero of W. Now one computes that

$$JW(-\bar{\lambda})^{-*}J = JD^{-*}J + JD^{-*}B^*\big(\lambda I_n - (-A^* + C^*D^{-*}B^*)\big)^{-1}C^*D^{-*}J.$$

The fact that the realization is minimal yields, by the state space similarity theorem, the existence of a unique (invertible) $n \times n$ matrix H such that

$$AH = -HA^* + HC^*D^{-*}B^*, \qquad B = HC^*D^{-*}J, \qquad JD^{-*}B^* = CH. \quad (17.5)$$

Next, take adjoints and use $D^*JD = J$ to see that (17.5) also holds with H^* in place of H. By uniqueness it follows that $H = H^*$. Hence (17.3) holds, and so (i) implies (ii), even with the additional condition that H is invertible.

Next assume D is J-unitary and there exists an $n \times n$ matrix H such that (17.3) holds. A straightforward computation gives

$$\begin{aligned} W(\lambda)JW(-\bar{\lambda})^* &= \big(I + C(\lambda - A)^{-1}BD^{-1}\big)DJD^*\big(I - D^{-*}B^*(\lambda + A^*)^{-1}C^*\big) \\ &= J + C(\lambda - A)^{-1}BJD^* - DJB^*(\lambda + A^*)^{-1}C^* \\ &\qquad -C(\lambda - A)^{-1}BJB^*(\lambda + A^*)^{-1}C^* \\ &= J + C(\lambda - A)^{-1}HC^* - CH(\lambda + A^*)^{-1}C^* \\ &\qquad -C(\lambda - A)^{-1}(H(\lambda + A^*) - (\lambda - A)H)(\lambda + A^*)^{-1}C^* \\ &= J + C(\lambda - A)^{-1}HC^* - CH(\lambda + A^*)^{-1}C^* \\ &\qquad -C(\lambda - A)^{-1}HC^* + CH(\lambda + A^*)^{-1}C^* = J. \end{aligned}$$

Thus the function $W(\lambda)^*$ is J-unitary. But then so is W.

We have now proved that (i) and (ii) are equivalent. The equivalence of (i) and (iii) can be established in the same way. Actually the implication (iii) $\Rightarrow$ (i) can be obtained directly from (17.4) without having to take recourse to the function $W(\lambda)^*$. As above, (i) implies the stronger version of (iii) with the extra requirement that G is invertible.

The uniqueness and invertibility of H and G follow from the minimality. The invertibility can also be proved directly, and in fact from slightly weaker conditions. Assume (17.3) holds and that the pair (A, B) is controllable. Then H is invertible. Indeed, assume $Hx = 0$. Then $DJB^*x = CHx = 0$, so $B^*x = 0$. Hence $(AH + HA^*)x = 0$ too. With $Hx = 0$, this gives $HA^*x = -AHx = 0$. So $\operatorname{Ker} H \subset \operatorname{Ker} B^*$ and $A^*[\operatorname{Ker} H] \subset \operatorname{Ker} H$. Thus $\operatorname{Ker} H \subset \operatorname{Ker}(B^*|A^*) = \{0\}$. So H is invertible. Likewise, one shows that if (17.4) is satisfied and the pair (C, A) is observable, then G is invertible.

Finally, let H be as in (17.3), then (17.4) holds with H^{-1} in place of G. By uniqueness it follows that $G = H^{-1}$. □

In the argument for the implication (ii) $\Rightarrow$ (i) given above, the minimality of the given realization does not play a role. Similarly the minimality condition is irrelevant for the implication (iii) $\Rightarrow$ (i). This is also reflected by the following proposition.

Proposition 17.2. *Let $W(\lambda) = D + C(\lambda I_n - A)^{-1}B$ be a realization of a rational $m \times m$ matrix function. Assume D is J-unitary, and let H and G be given Hermitian $n \times n$ matrices. Consider the following four statements:*

(i) $AH + HA^* = BJB^*, \qquad CH = DJB^*$;

(ii) $AH + HA^* = HC^*JCH, \qquad CH = DJB^*$;

(iii) $GA + A^*G = C^*JC, \qquad GB = C^*JD$;

(iv) $GA + A^*G = GBJB^*G, \qquad GB = C^*JD$.

Then (i) *and* (ii) *are equivalent, and so are* (iii) *and* (iv). *Each of* (i)–(iv) *implies that W is J-unitary. Moreover, if (A, B) is controllable and* (i) *holds, then all four statements are equivalent and the realization is minimal. Likewise, if (C, A) is observable and* (iii) *holds, then again all four statements are equivalent and the realization is minimal.*

Proof. To see the equivalence of (i) and (ii), use $D^*JD = J$ to see that $BJB^* = HC^*JCH$. In an analogous manner one sees that (iii) and (iv) are equivalent. For the case when the realization is minimal the fact that (i) and (iii) imply that W is J-unitary is covered by Theorem 17.1. The general case is proved using the type of arguments occurring the proof of Theorem 17.1. Now suppose that (A, B) is controllable, and that (i) holds. In the proof of Theorem 17.1 we have already shown that this implies that H is invertible. Taking $G = H^{-1}$ it follows that (iii) is satisfied, and hence also (iv). Next, we show that in this case (C, A) is observable. Indeed, by induction one shows that $H^{-1}\mathrm{Ker}\,(C|A) \subset \mathrm{Ker}\,(B^*|A^*) = \{0\}$. Hence the realization is minimal. The equivalence of all four statements now follows from Theorem 17.1. The reasoning for final statement of the theorem is similar. □

The next proposition shows that under certain additional conditions the first identity in (i) of Proposition 17.2 implies the second identity in (i), and analogously for (i) replaced by (iii).

Proposition 17.3. *Let $W(\lambda) = D + C(\lambda I_n - A)^{-1}B$ be a realization of a J-unitary rational $m \times m$ matrix function, and let H and G be $n \times n$ Hermitian matrices. The following two statements are true:*

(i) *If the pair (A, B) is controllable and $GA + A^*G = C^*JC$, then $GB = C^*JD$.*

(ii) *If the pair (C, A) is observable and $AH + HA^* = BJB^*$, then $CH = DJB^*$.*

Proof. We only prove the first part of the proposition, the second part can be established analogously. Assume that W is J-unitary. Computing $W(-\bar{\lambda})^*JW(\lambda)$ one sees that this is equivalent to

$$\begin{bmatrix} -B^* & D^*JC \end{bmatrix} \left(\lambda - \begin{bmatrix} -A^* & C^*JC \\ 0 & A \end{bmatrix}\right)^{-1} \begin{bmatrix} C^*JD \\ B \end{bmatrix} = 0. \qquad (17.6)$$

Now assume that $GA + A^*G = C^*JC$. Using

$$S = \begin{bmatrix} I & G \\ 0 & I \end{bmatrix}$$

as a similarity transformation in the realization (17.6), we see that (17.6) is equivalent to

$$\begin{bmatrix} -B^* & D^*JC - B^*G \end{bmatrix} \left(\lambda - \begin{bmatrix} -A^* & 0 \\ 0 & A \end{bmatrix}\right)^{-1} \begin{bmatrix} C^*JD - GB \\ B \end{bmatrix} = 0.$$

But this identity, in turn, is equivalent to $(D^*JC - B^*G)(\lambda - A)^{-1}B = 0,\ \lambda \in \rho(A)$. The fact that (A, B) is controllable now implies that $GB = C^*JD$. □

The Hermitian matrix H in Theorem 17.1(ii), which is uniquely determined by the conditions stated there, will be called the *Hermitian matrix associated with the minimal realization* $W(\lambda) = D + C(\lambda I_n - A)^{-1}B$. Our next concern is how the associated Hermitian matrix behaves under similarity transformation on the realization.

Proposition 17.4. *Let $W(\lambda) = D + C(\lambda I_n - A)^{-1}B$ be a minimal realization of a J-unitary rational $m \times m$ matrix function. Write H for the Hermitian matrix associated with this realization, and let S be an invertible $n \times n$ matrix. Then the Hermitian matrix associated with the minimal realization*

$$W(\lambda) = D + CS^{-1}(\lambda I_n - SAS^{-1})^{-1}SB \tag{17.7}$$

is given by SHS^*.

Proof. For the (minimal) realization (17.7), the matrix SHS^* satisfies the requirements of condition (ii) in Theorem 17.1. □

As a consequence of the above proposition the number of positive and the number of negative eigenvalues of the matrix H do not depend on the particular choice of the minimal realization of the function W. The number of positive eigenvalues of H will be denoted by $\pi_+(W)$. At the end of this section, in Proposition 17.10, it will be seen how to express $\pi_+(W)$ completely in terms of W itself rather than in terms of the associated Hermitian matrix H.

The next two propositions describe how the associated Hermitian matrix behaves under the operations of inversion, taking adjoints, and multiplication.

Proposition 17.5. *Let $W(\lambda) = D + C(\lambda I_n - A)^{-1}B$ be a minimal realization of a J-unitary rational $m \times m$ matrix function. Write H for the Hermitian matrix associated with this realization and, as usual, $A^\times$ for the matrix $A - BD^{-1}C$. Then the Hermitian matrices associated with the minimal realizations*

$$\begin{aligned} W(\lambda)^{-1} &= D^{-1} - D^{-1}C(\lambda I_n - A^\times)^{-1}BD^{-1}, \\ W(-\bar{\lambda})^* &= D^* - B^*(\lambda I_n + A^*)^{-1}C^*, \end{aligned}$$

are $-H$ and $-H^{-1}$, respectively.

Proof. For the first realization, use (17.3) and observe that

$$\begin{aligned} A^\times(-H)+(-H)(A^\times)^* &= -AH - HA^* + BD^{-1}CH + HC^*D^{-*}B^* \\ &= -BJB^* + BD^{-1}DJB^* + BJD^*D^{-*}B^* \\ &= BJB^* = (BD^{-1})J(D^{-*}B^*), \end{aligned}$$

and $-D^{-1}C(-H) = D^{-1}CH = JB^* = D^{-1}J(BD^{-1})^*$. The claim for the second realization is straightforward from the fact that in Theorem 17.1, the matrix G is the inverse of H. □

Proposition 17.6. *For $j = 1, 2$, let $W_j(\lambda) = D_j + C_j(\lambda I_{n_j} - A_j)^{-1}B_j$ be a minimal realization of a J-unitary rational $m\times m$ matrix function W_j having as the Hermitian matrix associated to it H_j. Suppose $W = W_1, W_2$ is a minimal factorization. Then W is a J-unitary rational matrix function,*

$$W(\lambda) = D_1D_2 + \begin{bmatrix} C_1 & D_1C_2 \end{bmatrix}\left(\lambda I_{n_1+n_2} - \begin{bmatrix} A_1 & B_1C_2 \\ 0 & A_2 \end{bmatrix}\right)^{-1}\begin{bmatrix} B_1D_2 \\ B_2 \end{bmatrix}$$

is a minimal realization of W, and the associated Hermitian matrix is the block diagonal matrix $\operatorname{diag}(H_1, H_2)$.

Proof. Applying (17.3) to both realizations, using also $D_2JD_2^* = J$, one sees that

$$\begin{aligned} &\begin{bmatrix} H_1 & 0 \\ 0 & H_2 \end{bmatrix}\begin{bmatrix} A_1^* & 0 \\ C_2^*B_1^* & A_2^* \end{bmatrix} + \begin{bmatrix} A_1 & B_1C_2 \\ 0 & A_2 \end{bmatrix}\begin{bmatrix} H_1 & 0 \\ 0 & H_2 \end{bmatrix} \\ &\qquad = \begin{bmatrix} B_1JB_1^* & B_1C_2H_2 \\ H_2C_2^*B_1^* & B_2JB_2^* \end{bmatrix} = \begin{bmatrix} B_1D_2 \\ B_2 \end{bmatrix} J \begin{bmatrix} D_2^*B_1^* & B_2^* \end{bmatrix}. \end{aligned}$$

So the first equality in (17.3) is satisfied for the product realization. Also,

$$\begin{aligned} &\begin{bmatrix} C_1 & D_1C_2 \end{bmatrix}\begin{bmatrix} H_1 & 0 \\ 0 & H_2 \end{bmatrix} = \begin{bmatrix} C_1H_1 & D_1C_2H_2 \end{bmatrix} \\ &\qquad = \begin{bmatrix} D_1JB_1^* & D_1D_2JB_2^* \end{bmatrix} = (D_1D_2)J\begin{bmatrix} D_2^*B_1^* & B_2^* \end{bmatrix}, \end{aligned}$$

and this proves the second equality of (17.3) for the product realization. □

Next, we present a few examples. As before, J stands for an $m \times m$ signature matrix.

Example 17.7. Let R be an $m \times m$ matrix such that $R^*JR = JR$, and let $\omega \notin i\mathbb{R}$. Then the rational $m \times m$ matrix function W given by

$$W(\lambda) = I_m - R + \frac{\lambda - \omega}{\lambda + \bar{\omega}} R$$

is J-unitary. To be more specific, let u be a vector in $\mathbb{C}^m$ such that $u^*Ju = \langle Ju, u\rangle \neq 0$, and take for R the rank 1 matrix

$$R = \frac{1}{u^*Ju} Juu^*.$$

Then $R^*JR = JR = R^*J = (u^*Ju)^{-1}uu^*$. (Note here that uu^* is a rank 1 matrix, while u^*Ju is just a scalar.) A minimal realization for W for this particular choice of R may be obtained by setting

$$A = -\bar{\omega}, \qquad B = u^*, \qquad C = -\frac{(\omega + \bar{\omega})}{u^*Ju}.$$

The associated Hermitian matrix satisfies $AH + HA^* = BJB^*$, which in this case becomes $-(\omega + \bar{\omega})H = u^*Ju$. So $H = -(u^*Ju)(2\Re\omega)^{-1}$.

Example 17.8. Let $\alpha \in i\mathbb{R}$, $n \in \mathbb{N}$, and let $x \in \mathbb{C}^m$ be a J-neutral vector, i.e., $x^*Jx = 0$. Then

$$W(\lambda) = I_m + \frac{i}{(\lambda - \alpha)^{2n}} Jxx^*$$

is J-unitary. A minimal realization for W can be obtained by setting $A = J_{2n}(\alpha)$, the Jordan block of size $2n$ with eigenvalue α, and

$$C = i \begin{bmatrix} Jx & 0 & \cdots & 0 \end{bmatrix}, \qquad B = \begin{bmatrix} 0 \\ \vdots \\ 0 \\ x^* \end{bmatrix},$$

where C is an $m \times 2n$ matrix and B is a $2n \times m$ matrix. The associated Hermitian matrix can be computed to be the following matrix:

$$H = [h_{p\,q}]_{p,q=1}^{2n}, \qquad h_{p\,q} = \begin{cases} 0 & \text{if } p + q \neq 2n + 1, \\ (-1)^q i & \text{if } p + q = 2n + 1. \end{cases}$$

We conclude this section with a few remarks on matrix-valued kernel functions and their state space representations. Introduce the functions

$$\begin{aligned} K_W(\lambda, \mu) &= \frac{J - W(\lambda)JW(\mu)^*}{\lambda + \bar{\mu}}, \\ K_{*,W}(\mu, \lambda) &= \frac{J - W(\mu)^*JW(\lambda)}{\lambda + \bar{\mu}}. \end{aligned}$$

Here W is a rational $m \times m$ matrix function. Furthermore, λ and μ are complex numbers, not poles of W, $\lambda \neq -\overline{\mu}$.

Lemma 17.9. *Let $W(\lambda) = D + C(\lambda I_n - A)^{-1}B$ be a minimal realization of a J-unitary rational matrix function having H as its associated Hermitian matrix. Then the following two identities hold:*

$$K_W(\lambda,\mu) = -C(\lambda - A)^{-1}H^{-1}(\bar{\mu} - A^*)^{-1}C^*, \tag{17.8}$$

$$K_{*,W}(\mu,\lambda) = -B^*(\bar{\mu} - A^*)^{-1}H^{-1}(\lambda - A)^{-1}B. \tag{17.9}$$

Proof. We shall only prove (17.9); identity (17.8) can be obtained in an analogous fashion. First note that

$$\begin{aligned} W(\mu)^*JW(\lambda) &= (D^* + B^*(\bar{\mu} - A^*)^{-1}C^*)J(D + C(\lambda - A)^{-1}B) \\ &= D^*JD + B^*(\bar{\mu} - A^*)^{-1}C^*JD + D^*JC(\lambda - A)^{-1}B \\ &\qquad + B^*(\bar{\mu} - A^*)^{-1}C^*JC(\lambda - A)^{-1}B. \end{aligned}$$

Now use the identities $D^*JD = J$, $C^*JD = HB$ and $C^*JC = H^{-1}A + A^*H^{-1}$ which hold by Theorem 17.1. Then one sees that

$$W(\mu)^*JW(\lambda) = J + (\lambda + \bar{\mu})B^*(\bar{\mu} - A^*)^{-1}H^{-1}(\lambda - A)^{-1}B.$$

From this (17.9) is immediate. □

The kernel function $K_W(\lambda,\mu)$ is said to have κ *negative squares* if for each $r \in \mathbb{N}$ and any collection of points $\omega_1,\dots,\omega_r$ in the complex plane, not poles of W, and any collection of vectors $u_1,\dots,u_r$ in $\mathbb{C}^m$ the $r \times r$ Hermitian matrix

$$\left[u_j^* K_W(\omega_j,\omega_i)u_i\right]_{i,j=1}^{r} \tag{17.10}$$

has at most κ negative eigenvalues, and it has exactly κ negative eigenvalues for at least one choice of r, $\omega_1,\dots,\omega_r$ and $u_1,\dots,u_r$. For $K_{*,W}(\mu,\lambda)$, the definition is of course similar.

Proposition 17.10. *Let $W(\lambda) = D + C(\lambda I_n - A)^{-1}B$ be a minimal realization of a J-unitary rational $m \times m$ matrix function, and let H be the Hermitian matrix associated with this realization. Then the number of negative squares of each of the functions K_W and $K_{*,W}$ is equal to $\pi_+(W)$, the number of positive eigenvalues of the matrix H.*

This result corroborates the already established fact that the integer $\pi_+(W)$ is independent of the particular minimal realization of W (cf., the paragraph after the proof of Proposition 17.4).

Proof. It follows from the previous lemma that $K_{*,W}$ has at most $\pi_+(W)$ negative squares. Indeed, if $\omega_1,\dots,\omega_r$ is a collection of points in the complex plane, not poles of W, and $u_1,\dots,u_r$ is a collection of vectors in $\mathbb{C}^m$, then the $r \times r$ Hermitian matrix (17.10) can be written in the form $-E^*H^{-1}E$, where H is the Hermitian matrix associated with the given realization of W.

Next, consider

$$M = \operatorname{span}\{(\lambda - A)^{-1}Bu \mid u \in \mathbb{C}^m,\ \lambda \in \mathbb{C} \text{ not an eigenvalue of } A\}.$$

Clearly, for $u \in \mathbb{C}^m$ and λ not an eigenvalue of A, the vector $\lambda(\lambda - A)^{-1}Bu$ belongs to M. Since M is closed in $\mathbb{C}^n$, this implies that

$$Bu = \lim_{\lambda \to \infty} \lambda(\lambda - A)^{-1}Bu \in M, \qquad u \in \mathbb{C}^m.$$

Thus $\operatorname{Im} B \subset M$. Next, note that $A(\lambda - A)^{-1}Bu = -Bu + \lambda(\lambda - A)^{-1}Bu \in M$. Hence M is invariant under A. But then $\operatorname{Im}(A|B) \subset M$. By hypothesis, the given realization of W is minimal. This implies that $\operatorname{Im}(A|B) = \mathbb{C}^n$. We conclude that $M = \mathbb{C}^n$. The latter implies that $\mathbb{C}^n$ has a basis $x_1, \ldots, x_n$ such that for each j the vector x_j is of the form $x_j = (\lambda_i - A)^{-1}Bu_j$ for some vector $u_j \in \mathbb{C}^m$ and some $\omega_j \in \mathbb{C}$. Consider the $n \times n$ matrix $X = [\ x_1 \cdots x_n\ ]$. We obtain that for these u_j and ω_i we have

$$\left[u_j^* K_{*,W}(\omega_j, \omega_i)u_i\right]_{i,j=1}^{n} = -X^* H^{-1} X.$$

As X is invertible, this matrix has exactly $\pi_+(W)$ negative eigenvalues. This settles the matter for $K_{*,W}$; for K_W the argument is similar. □

17.2 Factorization of J-unitary rational matrix functions

In this section minimal factorizations of J-unitary rational matrix functions into a product of two J-unitary rational matrix functions will be studied. Here, as in the previous section, J is an $m \times m$ signature matrix. To state the main theorem we need to recall a notion introduced in Section 11.1. Let $H = H^*$ be an invertible $n \times n$ matrix. A subspace $M \subset \mathbb{C}^n$ is called H-nondegenerate if $M \cap [HM]^\perp = \{0\}$. For such a subspace one has $M \dotplus [HM]^\perp = \mathbb{C}^n$, as a simple dimension count shows. Also note that $(HM)^\perp = H^{-1}[M^\perp]$.

Theorem 17.11. *Let $W(\lambda) = D + C(\lambda I_n - A)^{-1}B$ be a minimal realization of a J-unitary rational $m \times m$ matrix function, and let H be the Hermitian matrix associated with this realization. Let M be an A-invariant H^{-1}-nondegenerate subspace, and denote by Π the projection of $\mathbb{C}^n$ onto $H[M^\perp]$ along M. Let $D = D_1 D_2$ be a factorization of D into two J-unitary constant matrices, and put*

$$\begin{aligned} W_1(\lambda) &= D_1 + C(\lambda I_n - A)^{-1}(I - \Pi)BD_2^{-1}, \\ W_2(\lambda) &= D_2 + D_1^{-1}C\Pi(\lambda I_n - A)^{-1}B. \end{aligned}$$

Then $W = W_1 W_2$, this factorization is minimal, and the factors W_1 and W_2 are J-unitary. Conversely, any minimal factorization $W = W_1 W_2$ with J-unitary *factors W_1 and W_2 is obtained in this way. Moreover, given a fixed factorization $D = D_1D_2$, the correspondence between minimal factorizations of W with two J-unitary factors and H-nondegenerate invariant subspaces of A is one-to-one.*

Proof. From (17.3) we know that $A^\times = -HA^*H^{-1}$, where $A^\times = A - BD^{-1}C$. It follows that $H[M^\perp[$ is $A^\times$-invariant because M is A-invariant. Since the subspace M is H^{-1}-nondegenerate, the projection Π is a supporting projection. Hence the factorization $W = W_1W_2$ is a minimal one. To complete the proof of the first part of the theorem it remains to show that the factors W_1 and W_2 are J-unitary rational matrix functions. In fact, it suffices to show that one of them is J-unitary, the J-unitarity of the other one then follows automatically. Since Π is a supporting projection we know that a minimal realization,

$$W_1(\lambda) = D_1 + C_1(\lambda - A_1)^{-1}B_1,$$

of W_1 is obtained by taking

$$\begin{aligned} A_1 &= \tau_M{}^*A\tau_M : M \to M, \\ B_1 &= \tau_M{}^*(I - \Pi)BD_2^{-1} : \mathbb{C}^m \to M, \\ C_1 &= C\tau_M : M \to \mathbb{C}^m. \end{aligned}$$

Here τ_M is the canonical embedding of M into $\mathbb{C}^n$, and hence $\tau_M{}^*\tau_M$ is the orthogonal projection of $\mathbb{C}^n$ onto M. Put $G_1 = \tau_M{}^*H^{-1}\tau_M$. Then G_1 is invertible. Indeed, suppose $G_1x = 0$ for some $x \in M$. Then $H^{-1}x \in \operatorname{Ker}\tau_M{}^* = M^\perp$, i.e., $x \in H(M^\perp)$. So $x \in M \cap H(M^\perp) = \{0\}$.

Next, we shall show that the conditions of Theorem 17.1 (iii) are satisfied. First, note that

$$\begin{aligned} (G_1A_1 + A_1^*G_1) &= \tau_M{}^*H^{-1}\tau_M\tau_M{}^*A\tau_M + \tau_M{}^*A\tau_M\tau_M{}^*H^{-1}\tau_M \\ &= \tau_M{}^*(H^{-1}A + A^*H^{-1})\tau_M \\ &= \tau_M{}^*C^*JC\tau_M{}^* = C_1^*JC_1. \end{aligned}$$

Furthermore, we have

$$G_1B_1 = \tau_M{}^*H^{-1}\tau_M\tau_M{}^*(I - \Pi)BD_2^{-1} = \tau_M{}^*H^{-1}(I - \Pi)BD_2^{-1}.$$

Now, as M is H^{-1}-nondegenerate, $\operatorname{Im} H^{-1}\Pi = M^\perp$ and $H^{-1}[M]^\perp = H[M^\perp] = \operatorname{Im}\Pi$. This yields $\langle H^{-1}\Pi x, y\rangle = \langle H^{-1}\Pi x, \Pi y\rangle = \langle \Pi x, H^{-1}\Pi y\rangle = \langle x, H^{-1}\Pi y\rangle$. Hence $\Pi^*H^{-1} = H^{-1}\Pi$, that is, the projection Π is H^{-1}-selfadjoint. Therefore $H^{-1}(I - \Pi) = (I - \Pi^*)H^{-1}$. Moreover, as $(I - \Pi)\tau_M = \tau_M$ we have the identity $\tau_M{}^*(I - \Pi^*) = \tau_M{}^*$. Thus

$$\begin{aligned} G_1B_1 &= \tau_M{}^*H^{-1}(I - \Pi)BD_2^{-1} \\ &= \tau_M{}^*(I - \Pi^*)H^{-1}BD_2^{-1} \\ &= \tau_M{}^*H^{-1}BD_2^{-1} = \tau_M{}^*C^*JD_1 = C_1^*JD_1. \end{aligned}$$

Hence the conditions of Theorem 17.1 (iii) are satisfied, and thus W_1 is J-unitary.

The converse statement is a direct consequence of Proposition 17.6 and Theorem 8.5. □

As a special case of the preceding theorem we state the following proposition concerning the case where one of the factors is of degree 1.

Proposition 17.12. *Let $W(\lambda) = D + C(\lambda I_n - A)^{-1}B$ be a minimal realization of a J-unitary rational $m \times m$ matrix function, and let H be the Hermitian matrix associated with this realization. Suppose x is an eigenvector of A corresponding to the eigenvalue ω of A, and assume $\langle H^{-1}x, x\rangle \neq 0$. Then W admits a minimal factorization $W = W_1 W_2$ into two J-unitary factors where the factor W_1 is given by*

$$W_1(\lambda) = I_m + \frac{1}{(\lambda - \omega)\langle H^{-1}x, x\rangle} Cxx^*C^*J\,. \tag{17.11}$$

Furthermore, in case $\omega \notin i\mathbb{R}$ the scalar x^*C^*JCx is non-zero and

$$W_1(\lambda) = I_m - \frac{1}{x^*C^*JCx}\left(1 - \frac{\lambda + \bar{\omega}}{\lambda - \omega}\right) Cxx^*C^*J. \tag{17.12}$$

Observe that the factor W_1 is of the form as given in Example 17.7

Proof. As $\langle H^{-1}x, x\rangle \neq 0$, the subspace $M = \text{span}\,\{x\}$ is H^{-1}-nondegenerate. Therefore we can apply the previous theorem. The projection $I - \Pi$ is given by

$$(I - \Pi)v = \frac{\langle H^{-1}v, x\rangle}{\langle H^{-1}x, x\rangle} x = \frac{x^*H^{-1}v}{x^*H^{-1}x} x.$$

Taking $D_1 = I$ and $D_2 = D$ one obtains

$$\begin{aligned} W_1(\lambda) &= I + C(\lambda - A|_M)^{-1}(I - \Pi)BD^{-1} \\ &= I + \frac{Cxx^*H^{-1}BD^{-1}}{(\lambda - \omega)\langle H^{-1}x, x\rangle} \\ &= I + \frac{Cxx^*C^*J}{(\lambda - \omega)\langle H^{-1}x, x\rangle}. \end{aligned}$$

This proves (17.11).

Next we apply (17.4) in the present setting. Recall that $G = H^{-1}$. It follows that $x^*C^*JCx = (\omega + \bar{\omega})x^*H^{-1}x$. Thus, when $\omega \notin i\mathbb{R}$ or, equivalently, $\omega + \bar{\omega} \neq 0$,

$$x^*C^*JCx \neq 0, \qquad \langle H^{-1}x, x\rangle = \frac{x^*C^*JCx}{\omega + \bar{\omega}}.$$

Employing this in (17.11) immediately yields (17.12). □

17.3 Factorization of unitary rational matrix functions

In this section we shall consider the special case of rational matrix functions that are unitary on the imaginary axis, that is, we continue the theme of the previous section with $J = I$. For simplicity, we call such functions *unitary rational matrix functions* and omit the additional qualifier "on the imaginary axis."

Let W be a unitary rational matrix function. Then W is bounded by 1 on the imaginary axis, and hence W cannot have pure imaginary poles. Since W^{-1} is also a unitary rational matrix function, W cannot have pure imaginary zeros either. Replacing λ by λ^{-1} one also sees that W has to be biproper.

Lemma 17.13. *Let $W(\lambda) = D + C(\lambda - A)^{-1}B$ be a minimal realization of a unitary rational $m \times m$ matrix function, and let H be the Hermitian matrix associated with this realization. Then A has no pure imaginary eigenvalues. Let P be the spectral projection of A corresponding to the part of $\sigma(A)$ lying in the open right half plane. Then* $\operatorname{Im} P$ *is maximal H^{-1}-positive and* $\operatorname{Ker} P$ *is maximal H^{-1}-negative.*

Proof. Since the realization is minimal and W has no poles on the imaginary axis, the matrix A has no pure imaginary eigenvalues. By Theorem 17.1 with $G = H^{-1}$ we have $GA + A^*G = C^*C$. Because of the minimality of the realization we also know that the pair (C, A) is observable. Let us denote by $\nu(G)$ the number of negative eigenvalues of G, and by $\pi(G)$ the number of positive eigenvalues of G. By a well-known inertia theorem (see Theorem 13.1.4 in [107]) we have $\nu(G) = \dim \operatorname{Ker} P$ and $\pi(G) = \dim \operatorname{Im} P$.

Now put $M = \operatorname{Im} P$, let τ_M be the canonical embedding of M into $\mathbb{C}^n$, and introduce $A_M = \tau_M{}^* A \tau_M$, $G_M = \tau_M{}^* G \tau_M$ and $C_M = C\tau_M$. Then G_M is Hermitian, and (using the fact that M is invariant under A) we have

$$
\begin{aligned}
G_M A_M + A_M^* G_M &= \tau_M{}^* G \tau_M \tau_M{}^* A \tau_M + \tau_M{}^* A^* \tau_M \tau_M{}^* G \tau_M \\
&= \tau_M{}^* (GA + A^*G) \tau_M = \tau_M{}^* C^* C \tau_M = C_M^* C_M .
\end{aligned}
$$

The invariance of M under A also implies that $\operatorname{Ker}(C_M|A_M) \subset \operatorname{Ker}(C|A)$, and hence (C_M, A_M) is an observable pair too. Moreover, A_M has only eigenvalues in the open right half plane. The inertia theorem referred to above then gives that G_M is positive definite. But this is equivalent to saying that $\operatorname{Im} P$ is H^{-1}-positive. As $\pi(H^{-1}) = \dim \operatorname{Im} P$, it is actually maximal H^{-1}-positive. The other part of the proposition is proved in a similar way. □

Observe that an H^{-1}-positive subspace is in particular H^{-1}-nondegenerate . Likewise, an H^{-1}-negative subspace is H^{-1}-nondegenerate. So we are in a position to apply Theorem 17.11. This yields the following two results of which we shall only prove the second.

Theorem 17.14. *Let $W(\lambda) = D + C(\lambda I_n - A)^{-1}B$ be a minimal realization of a unitary rational $m \times m$ matrix function (so, in particular, D is invertible), and* *let $A^\times = A - BD^{-1}C$ be the associate main operator. Then W admits a minimal factorization $W = W_1W_2$ having the following additional properties:*

(i) *W_1 has its poles in the left half plane and its zeros in the right half plane,*

(ii) *W_2 has its poles in the right half plane and its zeros in the left half plane,*

(iii) *$\delta(W_1) = n - \pi_+(W)$ and $\delta(W_2) = \pi_+(W)$.*

Such a factorization can be obtained as follows. Let P denote the spectral projection corresponding to the part of $\sigma(A)$ lying in the open left half plane, and write $P^\times$ for the spectral projection of $A^\times$ corresponding to the part of $\sigma(A^\times)$ lying in the open right half plane. Then $\mathbb{C}^n = \operatorname{Im} P \dotplus \operatorname{Ker} P^\times$ and the functions

$$\begin{aligned} W_1(\lambda) &= I_m + C(\lambda I_n - A)^{-1}(I_n - \Pi)BD^{-1}, \\ W_2(\lambda) &= D + C\Pi(\lambda I_n - A)^{-1}B, \end{aligned} \tag{17.13}$$

meet the requirements. Here Π is the projection of $\mathbb{C}^n$ along $\operatorname{Im} P$ onto $\operatorname{Ker} P^\times$.

Theorem 17.15. *Let $W(\lambda) = D + C(\lambda I_n - A)^{-1}B$ be a minimal realization of a unitary rational $m \times m$ matrix function (so, in particular, D is invertible), and let $A^\times = A - BD^{-1}C$ be the associate main operator. Then W admits a minimal factorization $W = W_1W_2$ having the following additional properties:*

(i) *W_1 has its poles in the right half plane and its zeros in the left half plane,*

(ii) *W_2 has its poles in the left half plane and its zeros in the right half plane,*

(iii) *$\delta(W_1) = \pi_+(W)$ and $\delta(W_2) = n - \pi_+(W)$.*

Such a factorization can be obtained as follows. Let P denote the spectral projection corresponding to the part of $\sigma(A)$ lying in the open right half plane, and write $P^\times$ for the spectral projection of $A^\times$ corresponding to the part of $\sigma(A^\times)$ lying in the open left half plane. Then $\mathbb{C}^n = \operatorname{Im} P \dotplus \operatorname{Ker} P^\times$ and the functions

$$\begin{aligned} W_1(\lambda) &= I_m + C(\lambda I_n - A)^{-1}(I_n - \Pi)BD^{-1}, \\ W_2(\lambda) &= D + C\Pi(\lambda I_n - A)^{-1}B, \end{aligned} \tag{17.14}$$

meet the requirements. Here Π is the projection of $\mathbb{C}^n$ along $\operatorname{Im} P$ onto $\operatorname{Ker} P^\times$.

Proof. With Lemma 17.13 in mind, the idea is to apply Theorem 17.11 taking $M = \operatorname{Im} P$. We need to find $H[\operatorname{Im} P]^\perp = \operatorname{Ker}(P^*H^{-1})$.

From (17.3) we know that $A^\times = -HA^*H^{-1}$, hence $P^\times = -HP^*H^{-1}$. It follows that $H[\operatorname{Im} P]^\perp = \operatorname{Im} P^\times$. Let Π be the projection along $\operatorname{Im} P$ onto $\operatorname{Im} P^\times$. Then, by Theorem 17.11, the function W admits the factorization $W = W_1W_2$, where W_1 and W_2 are given by (17.14), and these factors are unitary. Moreover, the factorization is minimal. Finally, the poles of W_1 are the eigenvalues of $A|_{\operatorname{Im} P}$ (counting multiplicities), its zeros are the eigenvalues of $A^\times|_{\operatorname{Im} P^\times}$ (counting multiplicities too). Similarly, the poles of W_2 are the eigenvalues of $A|_{\operatorname{Ker} P}$, while the zeros of W_2 are the eigenvalues of $A^{\times}|_{\operatorname{Ker} P^{\times}}$. So the position of poles and zeros of W_1 and W_2 is as required. It also follows that

$$\delta(W_1) = \dim \operatorname{Ker} \Pi = \dim \operatorname{Im} P = \pi_+(W),$$

and hence by minimality also $\delta(W_2) = n - \pi_+(W)$. □

Our next theorem is on complete factorization of a unitary rational matrix function into unitary factors (cf., Part III in [20]).

Theorem 17.16. *Let W be a unitary rational $m \times m$ matrix function of McMillan degree n. Then W admits a minimal factorization into n factors of McMillan degree 1. Moreover, each of these factors can be taken to be unitary.*

In order to prove this theorem we first show that a unitary rational matrix function allows for a realization with very special properties.

Lemma 17.17. *Let W be a unitary rational $m \times m$ matrix function with $W(\infty) = I_m$. Then W admits a minimal realization $W(\lambda) = I_m + C(\lambda I_n - A)^{-1}B$ such that*

$$A = \begin{bmatrix} A_{11} & A_{12} \\ 0 & A_{22} \end{bmatrix}, \tag{17.15}$$

where A_{11} and A_{22} are upper triangular, A_{11} has all its eigenvalues in the open right half plane, A_{22} has all its eigenvalues in the open left half plane, and the Hermitian matrix associated with the realization is given by

$$H = \begin{bmatrix} I_n & 0 \\ 0 & -I_n \end{bmatrix}. \tag{17.16}$$

Proof. Take an arbitrary minimal realization $W(\lambda) = I + C(\lambda - A)^{-1}B$. By Schur's theorem there is an orthogonal change of basis such that A is upper triangular. In fact, we may take the eigenvalues of A on the diagonal in any order we like. This is known as the ordered Schur form of A. We apply this to construct a similarity transformation such that A is of the form

$$A = \begin{bmatrix} A_{11} & A_{12} \\ 0 & A_{22} \end{bmatrix}$$

where A_{11} is upper triangular having all its eigenvalues in the open right half plane, and A_{22} is upper triangular having all its eigenvalues in the open left half plane. The spectral projection of A corresponding to its eigenvalues in the open right half plane is given by

$$P = \begin{bmatrix} I & 0 \\ 0 & 0 \end{bmatrix}.$$

Let H be the Hermitian matrix associated with this realization, and let G be its inverse. Decompose G in the same way as A, and write

$$G = \begin{bmatrix} G_{11} & G_{12} \\ G_{12}^* & G_{22} \end{bmatrix}.$$

Because of Lemma 17.13 we have that $\operatorname{Im} P$ is maximal G-positive, and so G_{11} is positive definite. Likewise, since $\operatorname{Ker} P$ is maximal G-negative, G_{22} is negative definite.

Next, we employ the Schur complement of G_{11} in G. So we factorize G as

$$G = \begin{bmatrix} I & 0 \\ G_{12}^* & I \end{bmatrix} \begin{bmatrix} G_{11} & 0 \\ 0 & G_{22} - G_{12}^* G_{11}^{-1} G_{12} \end{bmatrix} \begin{bmatrix} I & G_{11}^{-1} G_{12} \\ 0 & I \end{bmatrix}.$$

Since G_{11} is positive definite and G_{22} is negative definite, the Schur complement $G_{22} - G_{12}^* G_{11}^{-1} G_{12}$ is negative definite too.

Now take the Cholesky decomposition of G_{11}, that is, write $G_{11} = C_{11}^* C_{11}$ with C_{11} upper triangular. Likewise, take the Cholesky decomposition of the Schur complement. Thus $G_{22} - G_{12}^* G_{11}^{-1} G_{12} = -C_{22}^* C_{22}$ with C_{22} upper triangular. Put

$$S = \begin{bmatrix} C_{11}^{-1} & -G_{11}^{-1} G_{12} C_{22}^{-1} \\ 0 & C_{22}^{-1} \end{bmatrix}.$$

Then, using Proposition 17.4, one checks that the realization

$$W(\lambda) = I + CS(\lambda - S^{-1}AS)^{-1} S^{-1} B$$

has all the desired properties. □

Proof of Theorem 17.16. Without loss of generality we may assume that W has the value I_m at infinity. Let $W(\lambda) = I_m + C(\lambda I_n - A)^{-1} B$ be a minimal realization as in the previous lemma, and let H be the Hermitian matrix associated with this realization. In particular, A is upper triangular. For this realization we have by (17.3) that $A^\times = -HA^*H^{-1}$. This is clearly a lower triangular matrix. Now let $e_1, \ldots, e_n$ be the standard basis of $\mathbb{C}^n$. For $k = 1, \ldots, n$, define Π_k to be the orthogonal projection of $\mathbb{C}^n$ onto span $\{e_k\}$. Then for $j = 1, \ldots, n-1$ the projection $\Pi_{j+1} + \cdots + \Pi_n$ is a supporting projection for the minimal realization $W(\lambda) = I_m + C(\lambda I_n - A)^{-1} B$. It then follows from Theorem 10.5 in [20] that W admits a factorization into n factors of degree 1.

It remains to prove that each of the factors is unitary. Clearly, for each integer $j = 1, \ldots, n-1$ the image and kernel of $\Pi_{j+1} + \cdots + \Pi_n$ are both H^{-1}-nondegenerate and are each other's H-orthogonal complements. From Theorem 17.11 it then follows that for each j the products $W_1 \cdots W_j$ and $W_{j+1} \cdots W_n$ are unitary. From this one concludes that each W_j separately is unitary. □

17.4 Intermezzo on the Redheffer transformation

In this section we study the Redheffer transform of a J-unitary rational matrix function. This will allow us to relate J-unitary rational matrix functions to certain classes of unitary rational matrix functions. The results obtained will be used in the next section. All the time, J will be a signature matrix.

The starting point of our considerations is a 2×2 block matrix

$$M = \begin{bmatrix} M_{11} & M_{12} \\ M_{21} & M_{22} \end{bmatrix}, \tag{17.17}$$

with M_{11} a $p \times p$ matrix and M_{22} a $q \times q$ matrix. When M_{22} is an invertible matrix, the *Redheffer transform* Λ of M is defined as follows:

$$\Lambda = \begin{bmatrix} \Lambda_{11} & \Lambda_{12} \\ \Lambda_{21} & \Lambda_{22} \end{bmatrix} = \begin{bmatrix} M_{11} - M_{12}M_{22}^{-1}M_{21} & M_{12}M_{22}^{-1} \\ -M_{22}^{-1}M_{21} & M_{22}^{-1} \end{bmatrix}. \tag{17.18}$$

We refer to the map $M \mapsto \Lambda$ as the *Redheffer transformation.*

Let $J = \operatorname{diag}(I_p, -I_q)$. The matrix M in (17.17) is said to be *J-contractive* if $M^*JM \leq J$. The next lemma shows that for such a matrix the requirement that M_{22} is invertible is automatically fulfilled. Hence the Redheffer transform of a J-contractive matrix M with $J = \operatorname{diag}(I_p, -I_q)$ is well-defined.

Lemma 17.18. *Let $J = \operatorname{diag}(I_p, -I_q)$. If the matrix M in (17.17) is J-contractive, then M_{22} is invertible, the (well-defined) Redheffer transform Λ of M is a contraction, and $\|M_{22}^{-1}M_{21}\| < 1$. Conversely, if M_{22} is invertible and the Redheffer transform Λ of M is a contraction, then M is J-contractive.*

Proof. Assume that the matrix M is J-contractive. By considering the $(2,2)$-entry of M^*JM and using $M^*JM \leq J$, we see that

$$M_{22}^*M_{22} \geq I_q + M_{12}^*M_{12}. \tag{17.19}$$

Thus $M_{22}^*M_{22}$ is positive definite, and hence, because M_{22} is square, the matrix M_{22} is invertible. Multiplying the inequality (17.19) from the left by M_{22}^{-*} and from the right by M_{22}^{-1}, we get $I_q - M_{22}^{-*}M_{12}^*M_{12}M_{22}^{-1} \geq M_{22}^{-*}M_{22}^{-1}$. Since $M_{22}^{-*}M_{22}^{-1}$ is positive definite, we may conclude that so is $I_q - M_{22}^{-*}M_{12}^*M_{12}M_{22}^{-1}$. But this is equivalent to $\|M_{22}^{-1}M_{21}\| < 1$.

Next assume that M_{22} is invertible and consider the equations

$$\begin{bmatrix} M_{11} & M_{12} \\ M_{21} & M_{22} \end{bmatrix} \begin{bmatrix} x \\ y \end{bmatrix} = \begin{bmatrix} u \\ v \end{bmatrix}. \tag{17.20}$$

Then, as M_{22} is invertible, these equations are equivalent to

$$\begin{bmatrix} \Lambda_{11} & \Lambda_{12} \\ \Lambda_{21} & \Lambda_{22} \end{bmatrix} \begin{bmatrix} x \\ v \end{bmatrix} = \begin{bmatrix} u \\ y \end{bmatrix}. \tag{17.21}$$

Indeed, rewrite (17.20) as $M_{11}x + M_{12}y = u$ and $M_{21}x + M_{22}y = v$. Solving for y in the second of these equations, one gets

$$y = -M_{22}^{-1}M_{21}x + M_{22}^{-1}v. \tag{17.22}$$

Inserting this in the first of the two equations above, we obtain

$$u = (M_{11} - M_{12}M_{22}^{-1}M_{21})x + M_{12}M_{22}^{-1}v. \tag{17.23}$$

Together, (17.22) and (17.23) prove the desired equivalence between (17.20) and (17.21).

Notice that the condition that the matrix M is J-contractive is equivalent to the inequality $\|u\|^2 - \|v\|^2 \le \|x\|^2 - \|y\|^2$. Indeed, $M^*JM \le J$ is equivalent to

$$\|u\|^2 - \|v\|^2 = \langle J\begin{bmatrix} u \\ v \end{bmatrix}, \begin{bmatrix} u \\ v \end{bmatrix}\rangle = \langle JM\begin{bmatrix} x \\ y \end{bmatrix}, M\begin{bmatrix} x \\ y \end{bmatrix}\rangle \tag{17.24}$$

$$= \langle M^*JM\begin{bmatrix} x \\ y \end{bmatrix}, \begin{bmatrix} x \\ y \end{bmatrix}\rangle \le \langle J\begin{bmatrix} x \\ y \end{bmatrix}, \begin{bmatrix} x \\ y \end{bmatrix}\rangle = \|x\|^2 - \|y\|^2.$$

Similarly, the condition that the Redheffer transform Λ is a contraction is equivalent to $\|u\|^2 + \|y\|^2 \le \|x\|^2 + \|v\|^2$. But

$$\|u\|^2 - \|v\|^2 \le \|x\|^2 - \|y\|^2 \iff \|u\|^2 + \|y\|^2 \le \|x\|^2 + \|v\|^2.$$

Thus, as desired, M is J-contractive amounts to the same as M_{22} is invertible and Λ is a contraction. □

Corollary 17.19. *Let $J = \operatorname{diag}(I_p, -I_q)$, and assume that the matrix M in* (17.17) *is J-contractive. Then M^* is J-contractive too.*

Proof. By Lemma 17.18, the fact that M is J-contractive implies that M_{22} is invertible and the Redheffer transform Λ of M is a contraction. Since M_{22} is invertible, so is M_{22}^*. Thus the Redheffer transform of M^* is well-defined. Moreover, the Redheffer transform of M^* is equal to Λ^*. As Λ is a contraction, the same holds true for Λ^*. But then the converse part of Lemma 17.18 shows that M^* is J-contractive too. □

Proposition 17.20. *Let $J = \operatorname{diag}(I_p, -I_q)$. The matrix M in* (17.17) *is J-unitary if and only if M_{22} is invertible and the Redheffer transform of M is unitary.*

Proof. Since a J-unitary matrix is J-contractive and a unitary matrix is a contraction, we see from Lemma 17.18 that without loss of generality we may assume that the matrix M_{22} is invertible. This allows us to use the equivalence of the equations (17.20) and (17.21).

Next, using a calculation as in (17.24), one sees that M is J-contractive if and only if the equality $\|x\|^2 - \|y\|^2 = \|u\|^2 - \|v\|^2$ holds. Furthermore, the condition that Λ is unitary is equivalent to $\|x\|^2 + \|v\|^2 = \|u\|^2 + \|y\|^2$. But

$$\|x\|^2 - \|y\|^2 = \|u\|^2 - \|v\|^2 \iff \|x\|^2 + \|v\|^2 = \|u\|^2 + \|y\|^2.$$

Hence M is J-unitary if and only if Λ is unitary. □

Next we pass from matrices to matrix functions. Consider a rational matrix function W,

$$W(\lambda) = \begin{bmatrix} W_{11}(\lambda) & W_{12}(\lambda) \\ W_{21}(\lambda) & W_{22}(\lambda) \end{bmatrix}, \tag{17.25}$$

with W_{11} a $p \times p$ rational matrix function and W_{22} a $q \times q$ rational matrix function. Assume W_{22} to be regular, i.e., $\det W_{22}(\lambda) \not\equiv 0$. Thus W_{22}^{-1} is a well-defined rational matrix function. Under these assumptions the *Redheffer transform* of W is defined to be the rational matrix function Σ given by

$$\begin{aligned} \Sigma(\lambda) &= \begin{bmatrix} \Sigma_{11}(\lambda) & \Sigma_{12}(\lambda) \\ \Sigma_{21}(\lambda) & \Sigma_{22}(\lambda) \end{bmatrix} \qquad (17.26) \\ &= \begin{bmatrix} W_{11}(\lambda) - W_{12}(\lambda)W_{22}(\lambda)^{-1}W_{21}(\lambda) & W_{12}(\lambda)W_{22}(\lambda)^{-1} \\ -W_{22}(\lambda)^{-1}W_{21}(\lambda) & W_{22}(\lambda)^{-1} \end{bmatrix}. \end{aligned}$$

As before, let $J = \operatorname{diag}(I_p, -I_q)$. If the rational matrix function W is J-unitary with respect to the imaginary axis, then we know from Proposition 17.20 that the Redheffer transform Σ is unitary. In particular, it has no pure imaginary poles and zeros (see the second paragraph of Section 17.3).

The following theorem is the main result of this section.

Theorem 17.21. *Let W be a rational matrix function, and let*

$$W(\lambda) = \begin{bmatrix} D_1 & 0 \\ 0 & D_2 \end{bmatrix} + \begin{bmatrix} C_1 \\ C_2 \end{bmatrix} (\lambda I_n - A)^{-1} \begin{bmatrix} B_1 & B_2 \end{bmatrix} \tag{17.27}$$

be a realization of W. Assume D_2 is invertible, and put $A_2^\times = A - B_2 D_2^{-1} C_2$. Then the Redheffer transform Σ of W has the realization

$$\Sigma(\lambda) = \begin{bmatrix} D_1 & 0 \\ 0 & D_2^{-1} \end{bmatrix} + \begin{bmatrix} C_1 \\ -D_2^{-1}C_2 \end{bmatrix} (\lambda I_n - A_2^\times)^{-1} \begin{bmatrix} B_1 & B_2 D_2^{-1} \end{bmatrix}, \tag{17.28}$$

and this realization is minimal if and only if so is the realization (17.28). *Moreover, assuming both realizations* (17.27) *and* (17.28) *to be minimal, the following holds. Let $J = \operatorname{diag}(I_p, -I_q)$ and suppose W is J-unitary on the imaginary axis. If H_W and H_Σ denote the Hermitian matrices associated with the realizations* (17.27) *and* (17.28), *respectively, then $H_W = H_\Sigma$.*

Proof. Write W in the form (17.25). From Theorem 2.4 we have

$$W_{22}(\lambda)^{-1} = D_2^{-1} - D_2^{-1}C_2(\lambda - A_2^\times)^{-1}B_2D_2^{-1},$$

and with the help of this expression one computes

$$\begin{aligned} W_{12}(\lambda)W_{22}(\lambda)^{-1} &= C_1(\lambda-A)^{-1}B_2\big(D_2^{-1}-D_2^{-1}C_2(\lambda-A_2^\times)^{-1}B_2D_2^{-1}\big) \\ &= C_1(\lambda-A_2^\times)^{-1}B_2D_2^{-1}, \end{aligned}$$

$$\begin{aligned} W_{22}(\lambda)^{-1}W_{21}(\lambda) &= \big(D_2^{-1}-D_2^{-1}C_2(\lambda-A_2^\times)^{-1}B_2D_2^{-1}\big)C_2(\lambda-A)^{-1}B_1 \\ &= D_2^{-1}C_2(\lambda-A_2^\times)^{-1}B_1. \end{aligned}$$

Now $W_{12}(\lambda)W_{22}(\lambda)^{-1}W_{21}(\lambda) = \big(C_1(\lambda-A)^{-1}B_2\big)W_{22}(\lambda)^{-1}W_{21}(\lambda)$, and hence

$$\begin{aligned} &W_{11}(\lambda)-W_{12}(\lambda)W_{22}(\lambda)^{-1}W_{21}(\lambda) \\ &\quad = D_1+C_1(\lambda-A)^{-1}B_1-\big(C_1(\lambda-A)^{-1}B_2\big)\big(D_2^{-1}C_2(\lambda-A_2^\times)^{-1}B_1\big) \\ &\quad = D_1+C_1(\lambda-A)^{-1}B_1-C_1(\lambda-A)^{-1}(A-A_2^\times)(\lambda-A^\times)^{-1}B_1 \\ &\quad = D_1+C_1(\lambda-A_2^\times)^{-1}B_1. \end{aligned}$$

This proves (17.28).

Next we deal with minimality. Assume the realization (17.27) is minimal. To prove the minimality of the realization (17.28), assume the realization (17.28) is not observable. Then

$$\operatorname{Ker}\left(\begin{bmatrix} C_1 \\ -D_2^{-1}C_2\end{bmatrix}, A_2^\times\right) \neq \{0\}.$$

Observe that the subspace on the left-hand side is invariant under $A_2^\times$. Hence there exists an eigenvalue λ_0 of $A_2^\times$ and there is a non-zero vector x such that $A_2^\times x = \lambda_0 x$, and $C_1x = 0$, $-D_2^{-1}C_2x = 0$. By the definition of $A_2^\times$ this implies that $Ax = A_2^\times x - B_2D_2^{-1}C_2x = A_2^\times x = \lambda_0 x$. So

$$\operatorname{Ker}\left(\begin{bmatrix} C_1 \\ C_2\end{bmatrix}, A\right) \neq \{0\}.$$

Hence the realization (17.27) is not observable, which is a contradiction. It follows that the realization (17.28) is observable. A similar argument proves that the realization (17.28) is controllable. The reverse implication, minimality of (17.28) implies minimality of (17.27), is proved in an analogous way.

Now assume both realizations are minimal. It remains to prove the equality of the corresponding Hermitian matrices. This is seen as follows. According to Theorem 17.1 the matrix H_W is uniquely determined by the four expressions $D_1^*D_1 = I_p$, $D_2^*D_2 = I_q$ and

$$AH_W + H_WA^* = B_1B_1^* - B_2B_2^*, \qquad \begin{bmatrix} C_1 \\ C_2 \end{bmatrix} H_W = \begin{bmatrix} D_1B_1^* \\ -D_2B_2^* \end{bmatrix}.$$

Next, using the same theorem with I_{p+q} as the signature matrix, we know that H_Σ is uniquely determined by the identities $D_1^*D_1 = I_p$, $D_2^*D_2 = I_q$ and

$$A_2^\times H_\Sigma + H_\Sigma(A_2^\times)^* = B_1B_1^* + B_2D_2^{-*}D_2^{-1}B_2^*, \tag{17.29}$$

$$\begin{bmatrix} C_1 \\ -D_2^{-1}C_2 \end{bmatrix} H_\Sigma = \begin{bmatrix} D_1B_1^* \\ D_2^{-1}D_2^{-*}B_2^* \end{bmatrix}.$$

Since $D_2^*D_2 = I_q$ and $A_2^\times = A - B_2C_2 = A + B_2B_2^*H_W^{-1}$, we obtain that the formulas for H_Σ are satisfied by H_W. Uniqueness of the associated Hermitian matrix proves then that $H_W = H_\Sigma$. □

We finish this section by returning to the examples of Section 17.1. Consider, for $J = \operatorname{diag}(I_p, -I_q)$, the function W of Example 17.7. So, taking $u = \begin{bmatrix} u_1^* & u_2^* \end{bmatrix}^*$,

$$W(\lambda) = I_{p+q} + \begin{bmatrix} -u_1 \\ u_2 \end{bmatrix} (\lambda + \bar{\omega})^{-1} \begin{bmatrix} u_1^* & u_2^* \end{bmatrix} \frac{2\Re\omega}{u^*Ju}.$$

Using Theorem 17.21 one finds, for Redheffer transform Σ of W,

$$\Sigma(\lambda) = I_{p+q} - \frac{1}{\lambda - \alpha}\frac{2\Re\omega}{u^*Ju} uu^*,$$

where

$$\alpha = -\bar{\omega} - \frac{2\Re\omega}{u^*Ju}\|u_2\|^2 = \frac{-\bar{\omega}\|u_1\|^2 - \omega\|u_2\|^2}{u^*Ju}.$$

For the Example 17.8, things are somewhat more complicated. We use the realization presented there, writing $x = \begin{bmatrix} x_1^* & x_2^* \end{bmatrix}^*$. The Redheffer transform of

$$W(\lambda) = I_{p+q} + \frac{i}{(\lambda - \alpha)^{2n}} \begin{bmatrix} x_1 \\ -x_2 \end{bmatrix} \begin{bmatrix} x_1^* & x_2^* \end{bmatrix}$$

then becomes

$$\Sigma(\lambda) = I_{p+q} + i \begin{bmatrix} x & 0 & \cdots & 0 \end{bmatrix} (\lambda - A_2^\times)^{-1} \begin{bmatrix} 0 \\ \vdots \\ 0 \\ x^* \end{bmatrix},$$

where

$$A_2^\times = J_{2n}(\alpha) + \begin{bmatrix} 0 & 0 & \cdots & 0 \\ \vdots & & & \vdots \\ 0 & & & 0 \\ i\|x_2\|^2 & 0 & \cdots & 0 \end{bmatrix}.$$

Since Σ only involves the entry of $(\lambda - A_2^\times)^{-1}$ in the upper right corner, this can be computed further. The entry in question is just 1 over the characteristic polynomial of $A_2^\times$, and so

$$\Sigma(\lambda) = I_{p+q} + \frac{i}{(\lambda-\alpha)^{2n} - i\|x_2\|^2}\, xx^*.$$

17.5 J-inner rational matrix functions

A matrix M is called a *J-contraction* if $M^*JM \leq J$. A rational matrix function W is called *J-inner* if W is J-unitary on the imaginary axis and, in addition, $W(\lambda)$ is a J-contraction for λ in the open right half plane, λ not a pole of W. Note that we restrict the attention here to functions that are J-inner relative to the imaginary axis.

If W is J-inner with $J = I$, then W is called *bi-inner* or *two-sided inner* (cf., Section 17.6 below). Clearly, if W is bi-inner it cannot have poles in the right open half plane. Also, if a unitary rational matrix W is analytic on the right half plane, then by the maximum modulus theorem $\|W(\lambda)\| \leq 1$ for $\Re\lambda > 0$, i.e., W is bi-inner. Thus a unitary rational matrix function W is bi-inner if and only if it is analytic on the right half plane. Recall from the second paragraph of Section 17.3 that a unitary rational matrix function has no pure imaginary poles or zeros, and that it is biproper.

The next theorem characterizes the property of being J-inner in terms of a minimal realization.

Theorem 17.22. *Let $W(\lambda) = D + C(\lambda I_n - A)^{-1}B$ be a minimal realization of a J-unitary rational $m \times m$ matrix function, and let H be the Hermitian matrix associated with this realization. Then W is J-inner if and only if H is negative definite.*

First we state a result that is of independent interest, and which proves one direction of Theorem 17.22.

Proposition 17.23. *If W is a J-inner rational matrix function, where the signature matrix J has the form $J = \operatorname{diag}(I_p, -I_q)$, then its Redheffer transform Σ is bi-inner. If, in addition, W is given by the minimal realization (17.27), then $A_2^\times = A - B_2D_2^{-1}C_2$ has all its eigenvalues in the open left half plane, and the Hermitian matrix H associated with (17.27) is negative definite.*

Proof. The first part of the proposition can be derived from Proposition 17.20 and Lemma 17.18. For the second part, consider a minimal realization of W written in the form (17.27) with the partitioning induced by $J = \operatorname{diag}(I_p, -I_q)$. Then we also have a minimal realization (17.28) of Σ. Since Σ is bi-inner, it is analytic in the right half plane, and by minimality of the realization this shows that $A_2^\times$ has all its eigenvalues in the left half plane.

It follows from the fact that H satisfies the Lyapunov equation (17.29) and from minimality that H is negative definite (see Corollary 1 in Section 13.1 in [107]). □

Proof of Theorem 17.22. Assume H is negative definite. For $\Re\lambda > 0$ we then have

$$\begin{aligned} J - W(\lambda)^* J W(\lambda) &= J - \big(D^* + B^*(\bar\lambda - A^*)^{-1}C^*\big)J\big(D + C(\lambda - A)^{-1}B\big) \\ &= J - D^*JD - B^*(\bar\lambda - A^*)^{-1}C^*JD - D^*JC(\lambda - A)^{-1}B \\ &\qquad -B^*(\bar\lambda - A^*)^{-1}C^*JC(\lambda - A)^{-1}B. \end{aligned}$$

Using the identities $D^*JD = J$, $C^*JD = HB$ and $C^*JC = H^{-1}A + A^*H^{-1}$, which hold by Theorem 17.1, one sees that

$$J - W(\lambda)^*JW(\lambda) = -2(\mathrm{Re}\,\lambda)B^*(\bar\lambda - A^*)^{-1}H^{-1}(\lambda - A)^{-1}B \geq 0.$$

Hence W is J-inner.

Conversely, if W is J-inner, where $J = \mathrm{diag}\,(I_p\,, -I_q)$, then H is negative definite by Proposition 17.23. So, it remains to show that we can reduce the general case to the situation where J is of the form $J = \mathrm{diag}\,(I_p\,, -I_q)$. To this end, let T be an invertible matrix such that $T^*JT = J_1 = \mathrm{diag}\,(I_p\,, -I_q)$ for some nonnegative integers p and q. Such a T does exist. Observe that $J = T^{-*}J_1T^{-1}$, and since $J = J^{-1}$, we obtain that $J = TJ_1T^*$. Consider the matrix function $W_1 = T^{-1}WT$. Then W_1 is J_1-inner, and has a minimal realization

$$W_1(\lambda) = T^{-1}DT + T^{-1}C(\lambda - A)^{-1}BT.$$

We claim that H is the Hermitian matrix associated with this minimal realization. Indeed, using $J = TJ_1T^*$ we have

$$AH + HA^* = BJB^* = BTJ_1T^*B^*,$$

$$T^{-1}CH = T^{-1}DJB^* = (T^{-1}DT)J_1T^*B^*.$$

By Theorem 17.1, the matrix H is the Hermitian matrix associated with the given minimal realization of W_1. So we can apply Proposition 17.23 to W_1 in order to conclude that H is negative definite. □

In the next theorem we analyze J-inner functions in terms of a realization which is not necessarily minimal. As always in this chapter, J stands for a signature matrix.

Theorem 17.24. *Let $W(\lambda) = D + C(\lambda I_n - A)^{-1}B$ be a (possibly non-minimal) realization of a rational $m \times m$ matrix function. Suppose $D^*JD = J$, and assume there exists a Hermitian matrix X such that*

$$XA + A^*X = C^*JC, \qquad XB = C^*JD, \qquad \mathrm{Ker}\,(C|A) \subset \mathrm{Ker}\,X.$$

Then W is J-unitary. In that case W is J-inner if and only if X is nonpositive.

Proof. With respect to the orthogonal decomposition $\mathbb{C}^n = \operatorname{Im} X \oplus \operatorname{Ker} X$ write

$$X = \begin{bmatrix} G & 0 \\ 0 & 0 \end{bmatrix}.$$

Note that G is invertible and Hermitian. Also, with respect to the decomposition $\mathbb{C}^n = \operatorname{Im} X \oplus \operatorname{Ker} X$, write

$$A = \begin{bmatrix} A_{11} & A_{12} \\ A_{21} & A_{22} \end{bmatrix}, \qquad B = \begin{bmatrix} B_1 \\ B_2 \end{bmatrix}, \qquad C = \begin{bmatrix} C_1 & C_2 \end{bmatrix}.$$

Then $XB = C^*JD$ yields

$$XB = \begin{bmatrix} GB_1 \\ 0 \end{bmatrix} = \begin{bmatrix} C_1^* \\ C_2^* \end{bmatrix} JD.$$

Since $D^*JD = J$, we know that D is invertible. Hence JD is invertible, and so $C_2 = 0$. Now $XA + A^*X = C^*JC$ gives

$$XA + A^*X = \begin{bmatrix} GA_{11} + A_{11}^*G & GA_{12} \\ A_{12}^*G & 0 \end{bmatrix} = \begin{bmatrix} C_1^*JC_1 & 0 \\ 0 & 0 \end{bmatrix}.$$

As G is invertible, one obtains $A_{12} = 0$. Therefore

$$W(\lambda) = D + C_1(\lambda - A_{11})^{-1}B_1, \tag{17.30}$$

and for this realization of W we have $GA_{11} + A_{11}^*G = C_1^*JC_1$ and $GB_1 = C_1^*JD$.

It is now sufficient to show that (17.30) is minimal. Indeed, the proof can then be completed by applying Theorems 17.1 and 17.22.

One checks that

$$\operatorname{Ker} CA^j = \operatorname{Ker} C_1A_{11}^j \oplus \operatorname{Ker} X, \quad j = 0, 1, 2, \ldots.$$

As $\operatorname{Ker}(C|A) \subset \operatorname{Ker} X$ by assumption, we obtain $\operatorname{Ker}(C_1|A_{11}) = \{0\}$. Thus (C_1, A_{11}) is an observable pair. It remains to show that (A_{11}, B_1) is controllable. For this it suffices to prove that $(A_{11}^\times, B_1)$ is a controllable pair, where $A_{11}^\times = A_{11} - B_1D^{-1}C_1$. Now $A_{11}^\times = -G^{-1}A_{11}^*G$, while $B_1 = G^{-1}C_1^*JD$. So it is enough to show that $(-A_{11}^*, C_1^*JD)$ is a controllable pair. But this is equivalent to $(D^*JC_1, -A_{11})$ being an observable pair. Now D^*J is invertible, and hence

$$\operatorname{Ker}(D^*JC_1| - A_{11}) = \operatorname{Ker}(C_1|A_{11}) = \{0\},$$

which completes the proof. □

We finish this section with a theorem on the multiplicative structure of J-inner rational matrix functions. It states that a J-inner rational matrix function admits a complete factorization into J-inner factors of McMillan degree 1.

Theorem 17.25. *Let W be a J-inner rational matrix function of McMillan degree n. Then there are J-inner rational matrix functions $W_1, \ldots, W_n$ of McMillan degree 1 such that $W = W_1 \cdots W_n$.*

Proof. Employing a similar argument as in the proof of Lemma 17.17, taking into account Theorem 17.22, one can prove that the J-inner rational matrix function W admits a realization with upper triangular main matrix and having $-I$ as its associated Hermitian matrix. Following the line of argument of the proof of Theorem 17.16 one then proves that a J-inner rational matrix function admits a minimal factorization into n factors of degree 1, and that these factors can be taken to be J-unitary.

It remains to show that the factors are actually J-inner. Let us consider for each of the factors a minimal realization of the form

$$W_j(\lambda) = D_j + \frac{1}{\lambda - a_j} D_j J B_j^* h_j^{-1} B_j.$$

The Hermitian matrix associated with this realization is denoted by h_j; it is just a real number in this case (compare Example 17.7). Consider the minimal realization for W resulting from taking the product realization of the above minimal realizations of the W_i's. According to Proposition 17.6, the Hermitian matrix H associated with this product realization is the diagonal matrix with the numbers $h_1, \ldots, h_n$ on the diagonal. According to Proposition 17.4 and the state space similarity theorem, there is an invertible matrix S such that SHS^* is the Hermitian matrix associated with the minimal realization of W mentioned in the first paragraph of this proof. That is, $SHS^* = -I$. But this is only possible if all numbers h_i are negative. Then we can apply Theorem 17.22 to conclude that each of the factors is J-inner. □

17.6 Inner-outer factorization

In this section we consider inner-outer factorization of a possibly non-square $p \times q$ rational matrix function L. First we introduce the necessary terminology.

A $p \times q$ rational matrix function V is called *inner* if V is analytic on the closed right half plane (including the imaginary axis and infinity) and the values of V on the imaginary axis are isometries. The latter means that $V(\lambda)^* V(\lambda) = I_p$ for each $\lambda \in i\mathbb{R}$. Since V is assumed to be proper, this identity also holds at infinity. By the maximum modulus principle, an inner function V satisfies

$$\|V(\lambda)\| \leq 1, \qquad \Re\lambda \geq 0.$$

Note that for V to be inner, we must have $q \leq p$. If $q = p$, then V is inner if and only if V is bi-inner (cf., the first two paragraphs of Section 17.5).

A rational square matrix-valued function X is said to be an *invertible outer function* if X is analytic on the closed right half plane (infinity included) and

$\det X(\lambda) \neq 0$ for $\Re\lambda \geq 0$ (again with infinity included). Finally, given a $p \times q$ rational matrix function L, we say that a factorization

$$L(\lambda) = V(\lambda)X(\lambda),$$

is an *inner-outer factorization* if V is a $p \times q$ inner rational function and X is a $q \times q$ invertible outer rational matrix function.[1] Clearly, for such a factorization to exist L must be analytic in the closed right half plane (infinity included) and the values of L on $i\mathbb{R} \cup \{\infty\}$ have to be left invertible matrices. As we shall see (Theorem 17.26 below), these two conditions are not only necessary for L to have an inner-outer factorization but also sufficient.

Put $\Phi(\lambda) = L(-\bar{\lambda})^* L(\lambda)$. Obviously, if L has an inner-outer factorization $L = VX$ (suppressing the variable λ), then, since V takes isometric values on the imaginary axis and at infinity, we have

$$\Phi(\lambda) = X(-\bar{\lambda})^* X(\lambda),$$

and this factorization is a left spectral factorization (with respect to $i\mathbb{R}$) of the rational $q \times q$ matrix function Φ. This gives a hint about how to construct an inner-outer factorization.

Indeed, assume L is analytic in the closed right half plane, infinity included, and let $\Phi(\lambda) = X(-\bar{\lambda})^* X(\lambda)$ be a left spectral factorization of Φ with respect to $i\mathbb{R}$. Put $V(\lambda) = L(\lambda)X(\lambda)^{-1}$. Then V is analytic in the closed right half plane (infinity included) because both L and X^{-1} are analytic there. In addition, V takes isometric values on the imaginary axis. Hence $L = VX$ is an inner-outer factorization. This leads to the following theorem.

Theorem 17.26. *Let $L(\lambda) = D + C(\lambda I_n - A)^{-1}B$ be a realization of a $p \times q$ rational matrix function. Assume A has all its eigenvalues in the open left half plane, L takes left invertible values on the imaginary axis, and $D^*D = I_q$. Then L admits an inner-outer factorization $L(\lambda) = V(\lambda)X(\lambda)$ with the inner factor V and the invertible outer factor X given by*

$$V(\lambda) = D + \big((I - DD^*)C + DB^*P\big)\big(\lambda I_n - (A - BD^*C + BB^*P)\big)^{-1}B,$$

$$X(\lambda) = I_q + (D^*C - B^*P)(\lambda I_n - A)^{-1}B.$$

Here P is the (unique) Hermitian $i\mathbb{R}$-stabilizing solution of

$$PBB^*P + P(A - BD^*C) + (A^* - C^*DB^*)P - C^*(I - DD^*)C = 0,$$

that is, the solution $P = P^*$ for which $A - BD^*C + BB^*P$ has all its eigenvalues in the open left half plane.

[1]Note that in our definition of inner-outer factorization, the outer factor is required to be invertible outer. This restricted version of inner-outer factorization is used throughout the book.

Proof. Put $\Phi(\lambda) = L(-\bar{\lambda})^* L(\lambda)$. Using $D^*D = I$ and the given realization for L we compute that Φ is given by the realization $\Phi(\lambda) = I + \widehat{C}(\lambda - \widehat{A})^{-1}\widehat{B}$, where

$$\widehat{A} = \begin{bmatrix} -A^* & C^*C \\ 0 & A \end{bmatrix}, \qquad \widehat{B} = \begin{bmatrix} C^*D \\ B \end{bmatrix}, \qquad \widehat{C} = \begin{bmatrix} -B^* & D^*C \end{bmatrix}. \quad (17.31)$$

Since L has left invertible values on the imaginary axis (that is, has full column rank there), Φ takes positive definite values on the imaginary axis. Thus we know from Section 9.2 that Φ admits a left spectral factorization with respect to $i\mathbb{R}$. It follows that an inner-outer factorization does exist under the assumptions of the theorem.

To find the spectral factorization in concrete form, we proceed as in the proof of Theorem 13.1. In other words we apply Theorem 12.5 with the data given by (17.31). The same argument as in the proof of Theorem 13.1 gives that the Riccati equation featured in the theorem has a Hermitian stabilizing solution P. Now use P to define $X(\lambda)$ by the expression given in the theorem which is the analogue of the expression for $L_-(\lambda)$ in Theorem 13.1. With the function X obtained this way, we have the left spectral factorization $\Phi(\lambda) = X(-\bar{\lambda})^* X(\lambda)$.

It remains to compute $V(\lambda) = L(\lambda)X(\lambda)^{-1}$. Note that

$$X^{-1}(\lambda) = I - (D^*C - B^*P)\big(\lambda - (A - BD^*C + BB^*P)\big)^{-1}B.$$

From $A - BD^*C + BB^*P = A - B(D^*C - B^*P)$, we now obtain

$$(\lambda - A)^{-1}BX(\lambda)^{-1} = \big(\lambda - (A - BD^*C + BB^*P)\big)^{-1}B.$$

Using the latter identity it is straightforward to deduce the formula for V given in the theorem. □

The following corollary will be useful in the final chapter of the book.

Corollary 17.27. *Let $L(\lambda) = D + C(\lambda I_n - A)^{-1}B$ be a realization of a $p \times q$ rational matrix function. Assume A has all its eigenvalues in the open left half plane, L takes left invertible values on the imaginary axis, and $D^*D = I_q$. Then there is a $q \times p$ rational matrix function $L^\sharp(\lambda)$ which has no poles on the imaginary line including infinity, such that $L^\sharp(i\omega)L(i\omega) = I_q$, $\omega \in \mathbb{R}$.*

Proof. Let $L(\lambda) = V(\lambda)X(\lambda)$ be an inner-outer factorization of L and take $L^\sharp(\lambda) = X(\lambda)^{-1}V(-\bar{\lambda})^*$. □

Next we consider the dual problem of outer-co-inner factorization. A possibly non-square rational matrix function V is called *co-inner* if V is analytic on the closed right half plane (including infinity), and takes co-isometric values on the imaginary axis. In other words, V is co-inner if $\widetilde{V}$ is inner, where $\widetilde{V}(\lambda) = V(\bar{\lambda})^*$. Note that for V to be co-inner, we must have $p \leq q$.

A factorization
$$L(\lambda) = X(\lambda)V(\lambda),$$
where X is invertible outer and V is co-inner is called an *outer-co-inner factorization.*[2] Obviously, in that case $\Phi(\lambda) = L(\lambda)L(-\bar{\lambda})^* = X(\lambda)X(-\bar{\lambda})^*$ is a right spectral factorization with respect to $i\mathbb{R}$, and conversely. Using a duality argument we obtain the following counterpart to Theorem 17.26.

Theorem 17.28. *Let $L(\lambda) = D + C(\lambda - A)^{-1}B$ be a realization of a $p \times q$ rational matrix function. Assume A has all its eigenvalues in the open left half plane, $L(i\omega)$ is right invertible for each $\omega \in \mathbb{R}$, and $DD^* = I$. Then L admits an outer-co-inner factorization*
$$L(\lambda) = X(\lambda)V(\lambda),$$
with the co-inner factor and the invertible outer factor being given by
$$V(\lambda) = D + C\big(\lambda I_n - (A - BD^*C + QC^*C)\big)^{-1}\big(B(I - D^*D) + QC^*D\big),$$
$$X(\lambda) = I_p + C(\lambda I_n - A)^{-1}(BD^* - QC^*).$$
Here Q is the (unique) Hermitian $i\mathbb{R}$-stabilizing solution
$$QC^*CQ + (A - BD^*C)Q + Q(A^* - C^*DB^*) - B(I - D^*D)B^* = 0,$$
that is, the solution $Q = Q^*$ for which $A - BD^*C + QC^*C$ has all its eigenvalues in the open left half plane.

Proof. Let $\widetilde{L}(\lambda) = D^* + B^*(\lambda - A^*)^{-1}C^*$, and apply Theorem 17.26 to $\widetilde{L}$. Note that A^* also has all its eigenvalues in the open left half plane. So, applying Theorem 17.26 to $\widetilde{L}$ yields a factorization $\widetilde{L}(\lambda) = \widetilde{V}(\lambda)\widetilde{X}(\lambda)$, where $\widetilde{V}$ is inner and $\widetilde{X}$ is invertible outer. Then $V(\lambda) = \widetilde{V}(\bar{\lambda})^*$ is co-inner and $X(\lambda) = \widetilde{X}(\bar{\lambda})^*$ is invertible outer. So $L(\lambda) = X(\lambda)V(\lambda)$ is an outer-co-inner factorization of L. Theorem 17.26 also gives formulas for the factors $\widetilde{V}$ and $\widetilde{X}$. Those for V and X are now obtained from the expressions $V(\lambda) = \widetilde{V}(\bar{\lambda})^*$ and $X(\lambda) = \widetilde{X}(\bar{\lambda})^*$. □

17.7 Unitary completions of minimal degree

In this section we deal with the following completion problem. Given a strictly proper rational $m \times p$ matrix function W, having contractive values on the imaginary axis, find an $(m + p) \times (m + p)$ rational matrix function U having unitary values on the imaginary axis, such that
$$U(\lambda) = \begin{bmatrix} U_{11}(\lambda) & W(\lambda) \\ U_{21}(\lambda) & U_{22}(\lambda) \end{bmatrix}. \tag{17.32}$$

[2]Note that in our definition of outer-co-inner factorization, the outer factor is required to be invertible outer (cf., footnote 1).

In other words, we want to find a unitary rational matrix function U such that W is embedded as a (right upper) corner in U. Moreover, we wish to find such a U which has the same McMillan degree as W. We shall normalize U so that $U(\infty) = I_{m+p}$.

This problem can be treated for the more general case of a proper W (see [75]). However, for sake of simplicity we shall confine ourselves to the strictly proper case. The following theorem describes all possible solutions.

Theorem 17.29. *Let $W(\lambda) = C(\lambda I_n - A)^{-1}B$ be a minimal realization of an $m \times p$ strictly proper rational matrix function W which is contractive on the imaginary axis. Then the set of all unitary rational $(m+p) \times (m+p)$ matrix functions U of the form (17.32) with $U(\infty) = I_{m+p}$ and $\delta(U) = \delta(W)$ is in one-to-one correspondence with the set of Hermitian solutions of the algebraic Riccati equation*

$$XC^*CX - AX - XA^* + BB^* = 0. \tag{17.33}$$

Moreover, these Hermitian solutions X are invertible, and the one-to-one correspondence referred to above is given by

$$U(\lambda) = \begin{bmatrix} I_m & 0 \\ 0 & I_p \end{bmatrix} + \begin{bmatrix} C \\ B^*X^{-1} \end{bmatrix} (\lambda I_n - A)^{-1} \begin{bmatrix} XC^* & B \end{bmatrix}. \tag{17.34}$$

Proof. Suppose U is a unitary rational matrix function with W as its right upper corner block entry, $U(\infty) = I_{m+p}$ and $\delta(U) = \delta(W)$. The McMillan degree of W is n, the size of the main matrix in the given minimal realization $W(\lambda) = C(\lambda I_n - A)^{-1}B$ of W. Hence $\delta(U) = n$, and U has a realization of the type

$$U(\lambda) = \begin{bmatrix} I_m & 0 \\ 0 & I_p \end{bmatrix} + \begin{bmatrix} \widehat{C}_1 \\ \widehat{C}_2 \end{bmatrix} (\lambda I_n - \widehat{A})^{-1} \begin{bmatrix} \widehat{B}_1 & \widehat{B}_2 \end{bmatrix}.$$

Clearly $W(\lambda) = \widehat{C}_1(\lambda I_n - \widehat{A})^{-1}\widehat{B}_2$ is realization of W. Comparing this realization with the given one, and using the state space similarity theorem for minimal realizations, we see that there exists an invertible $n \times n$ matrix with $\widehat{A} = S^{-1}AS$, $\widehat{B}_2 = S^{-1}B$ and $\widehat{C}_1 = CS$. Introducing $C_2 = \widehat{C}_2S^{-1}$ and $B_1 = SB_1$, we get

$$U(\lambda) = \begin{bmatrix} I_m & 0 \\ 0 & I_p \end{bmatrix} + \begin{bmatrix} C \\ C_2 \end{bmatrix} (\lambda I_n - A)^{-1} \begin{bmatrix} B_1 & B \end{bmatrix}, \tag{17.35}$$

and this realization of U is a minimal one.

Since U is unitary, there is a Hermitian X such that

$$AX + XA^* = \begin{bmatrix} B_1 & B \end{bmatrix} \begin{bmatrix} B_1^* \\ B^* \end{bmatrix}, \qquad \begin{bmatrix} C \\ C_2 \end{bmatrix} X = \begin{bmatrix} B_1^* \\ B^* \end{bmatrix}. \tag{17.36}$$

In particular, we have $B_1^* = CX$. Inserting $B_1^* = CX$ into the first part of (17.36) we obtain (17.33). Moreover, X is invertible by minimality of the realization of U (see Theorem 17.1), and so $C_2 = B^*X^{-1}$, which yields (17.34).

Conversely, suppose that X is a Hermitian solution of (17.33). By minimality of the realization of W we have that X is invertible. The argument is as follows. Suppose $Xx = 0$. Then (17.33) gives $x^*BB^*x = 0$, hence $B^*x = 0$. Again using (17.33) we get $XA^*x = 0$, and we see that $\operatorname{Ker} X \subset \operatorname{Ker}(B^*|A^*) = \{0\}$. Let U be given by (17.34). Then, by Theorem 17.1, the rational matrix function U is unitary. Obviously, W is the right upper corner block entry of U and $\delta(U) = \delta(W)$, and $U(\infty) = I_{m+p}$.

To show that the correspondence between Hermitian solutions X of (17.33) and the set of all unitary rational matrix functions U of the form (17.32) with $U(\infty) = I$ and $\delta(U) = \delta(W)$ is one-to-one we argue as follows. We have seen in the previous part of the theorem that any such U is necessarily of the form (17.34) for some Hermitian solution of (17.33). Assume that for two solutions X_1 and X_2 the functions U_1 and U_2 given by (17.34) with these solutions in place of X coincide. Then, from (17.36) it is seen that

$$A(X_1 - X_2) + (X_1 - X_2)A^* = 0, \qquad \begin{bmatrix} C \\ C_2 \end{bmatrix} (X_1 - X_2) = 0.$$

Hence $\operatorname{Im}(X_1 - X_2)$ is A-invariant, and it is also contained in $\operatorname{Ker} C$. This implies that $\operatorname{Im}(X_1 - X_2) \subset \operatorname{Ker}(C|A) = \{0\}$. Thus $X_1 = X_2$. □

17.8 Bi-inner completions of inner functions

Our aim in this section is to complete a possibly non-square inner function to a (square) bi-inner one. It is convenient to begin with two propositions. With the notation used in the first proposition we anticipate Theorem 17.32 below.

Proposition 17.30. *Let $V(\lambda) = \widetilde{D} + C(\lambda I_n - A)^{-1}\widetilde{B}$ be a realization of a $p \times q$ rational matrix function, and assume*

$$\widetilde{D}^*\widetilde{D} = I_q, \qquad \sigma(A) \subset \mathbb{C}_{\mathrm{left}}, \qquad Y\widetilde{B} = C^*\widetilde{D}, \tag{17.37}$$

where Y is the unique (Hermitian) solution of the Lyapunov equation

$$YA + A^*Y = C^*C. \tag{17.38}$$

Then V is inner. Conversely, if V is inner, the given realization of V is minimal, and Y is the unique (Hermitian) solution of the Lyapunov equation (17.38), *then* (17.37) *is satisfied.*

Since A has all its eigenvalues in the open left half plane, equation (17.38) has a unique solution Y, and this solution is given by

$$Y = -\int_0^\infty e^{tA^*}C^*Ce^{tA}\,dt. \tag{17.39}$$

From this representation one sees that the matrix Y is generally negative semidefinite, and that it has the stronger property of being negative definite when the realization $V(\lambda) = \widetilde{D} + C(\lambda I_n - A)^{-1}\widetilde{B}$ is minimal (or even just observable). Thus the above result can be viewed as a special case of Theorem 17.24. It is illustrative to give a direct proof.

Proof. Assume that (17.37) holds with Y as indicated in the theorem. Then $\widetilde{D}$ is an isometry by the first condition in (17.37). Thus $p \geq q$. For pure imaginary λ, a straightforward computation, using (17.37) and (17.38), gives

$$\begin{aligned} V(\lambda)^*V(\lambda) &= \big(\widetilde{D}^* + \widetilde{B}^*(\bar{\lambda} - A^*)^{-1}C^*\big)\big(\widetilde{D} + C(\lambda - A)^{-1}\widetilde{B}\big) \\ &= I_q - \widetilde{B}^*(\lambda + A^*)^{-1}Y\widetilde{B} + \widetilde{B}^*Y(\lambda - A^{-1}\widetilde{B} \\ &\qquad -\widetilde{B}^*(\lambda + A^*)^{-1}(YA + A^*Y)(\lambda - A)^{-1}\widetilde{B} \\ &= I_q - \widetilde{B}^*(\lambda + A^*)^{-1}Y\widetilde{B} + \widetilde{B}^*Y(\lambda - A)^{-1}\widetilde{B} \\ &\qquad -\widetilde{B}^*(\lambda + A^*)^{-1}\big(Y(A - \lambda) + (A^* + \lambda)Y\big)(\lambda - A)^{-1}\widetilde{B} = I_q. \end{aligned}$$

Hence V has isometric values on $i\mathbb{R}$. Since V is analytic in the open right half plane by the second condition in (17.37), we may conclude that V is inner.

Next, let V be inner and let the realization $V(\lambda) = \widetilde{D} + C(\lambda I_n - A)^{-1}\widetilde{B}$ be minimal. Clearly, since V is inner, the first two conditions in (17.37) are satisfied. Let Y be the unique solution of (17.38). It remains to show that $Y\widetilde{B} = C^*\widetilde{D}$. This is done by using the same arguments as used in the proof of Proposition 17.3. □

Proposition 17.31. *Let $U(\lambda) = D + C(\lambda I_n - A)^{-1}B$ be a realization of a $p \times p$ rational matrix function. Assume*

$$D^*D = I_q, \qquad \sigma(A) \subset \mathbb{C}_{\text{left}}, \qquad YB = C^*D, \tag{17.40}$$

where Y is the unique (Hermitian) solution of the Lyapunov equation

$$YA + A^*Y = C^*C. \tag{17.41}$$

Then U is bi-inner and the McMillan degree of U is equal to the rank of Y which, in turn, is equal to $\dim \operatorname{Ker}(C|A)^\perp$.

Proof. The fact that U is bi-inner follows from Proposition 17.30. Since $\sigma(A)$ is contained in $\mathbb{C}_{\text{left}}$, the unique solution Y of (17.41) is given by the integral representation (17.39), from which we easily obtain $\operatorname{Ker} Y = \operatorname{Ker}(C|A)$. Now consider the decomposition $\mathbb{C}^n = X_1 \oplus X_2$, where $X_1 = \operatorname{Ker}(C|A)$ and X_2 is the orthogonal compliment of X_1 in $\mathbb{C}^n$. Thus $X_1 = \operatorname{Ker} Y$ and $X_2 = \operatorname{Im} Y$. In particular $\operatorname{rank} Y = \dim X_2$. Write A, B, C and Y as block matrices according to the decomposition $\mathbb{C}^n = X_1 \oplus X_2$. Then

$$A = \begin{bmatrix} A_1 & \star \\ 0 & A_2 \end{bmatrix}, \qquad B = \begin{bmatrix} B_1 \\ B_2 \end{bmatrix}, \qquad C = \begin{bmatrix} 0 & C_2 \end{bmatrix}, \qquad Y = \begin{bmatrix} 0 & 0 \\ 0 & Y_2 \end{bmatrix}, \tag{17.42}$$

and $U(\lambda) = D + C_2(\lambda I_n - A_2)^{-1}B_2$. Since $\operatorname{rank} Y = \operatorname{rank} Y_2 = \dim X_2$, it suffices to prove that this second realization of U is minimal. From (17.41), the third identity in (17.40) and the partitioning of A, B, C and Y in (17.42), we see that

$$Y_2A_2 + A_2^*Y_2 = C_2^*C_2, \qquad Y_2B_2 = C_2^*D.$$

For $A_2^\times = A_2 - B_2D^{-1}C_2$, the associate main matrix of the realization $U(\lambda) = D + C_2(\lambda I_n - A_2)^{-1}B_2$, this gives

$$Y_2A_2^\times = Y_2A_2 - Y_2B_2D^{-1}C_2 = -A_2^*Y_2 + C_2^*C_2 - C_2^*DD^{-1}C_2 = -A_2^*Y_2.$$

Now Y_2 is invertible. Thus $A_2^\times$ and $-A_2^*$ are similar. From the second part of (17.40) and the partitioning of A in (17.42), we see that $\sigma(A_2) \subset \mathbb{C}_{\text{left}}$. Taking into account the similarity of $A_2^\times$ and $-A_2^*$, it follows that $\sigma(A_2^\times) \subset \mathbb{C}_{\text{right}}$. In particular, $\sigma(A_2)$ and $\sigma(A_2^\times)$ are disjoint. But then, by a remark made after the proof of Theorem 7.6 in [20], the realization $U(\lambda) = D + C_2(\lambda I_n - A_2)^{-1}B_2$ is minimal. □

Let V be as in Proposition 17.30, so in particular V is inner. Returning to the aim of this section, we shall now complete V to a $p \times p$ bi-inner rational matrix function. Before turning to the theorem in question, we make some preparations.

According to the first condition in (17.37) the matrix $\widetilde{D}$ is an isometry. Thus $p \geq q$. When $p = q$, there is nothing to do. Therefore in what follows we take $p > q$. The fact that $\widetilde{D}$ is an isometry, implies that $I_p - \widetilde{D}\widetilde{D}^*$ is an orthogonal projection of rank $p - q$. Thus we can choose a $p \times (p - q)$ isometry E such that $I_p - \widetilde{D}\widetilde{D}^* = EE^*$. Now note that there exists an $n \times (p - q)$ matrix $B^\sharp$ such that

$$YB^\sharp = C^*E. \tag{17.43}$$

Since Y is Hermitian, to prove that equation (17.43) has a solution of the desired form, it suffices to show that $\operatorname{Ker} Y \subset \operatorname{Ker} E^*C$. In fact, we have $\operatorname{Ker} Y \subset \operatorname{Ker} C$. Indeed, assume that $Yx = 0$, then we see from (17.38) that $x^*C^*Cx = 0$, which is equivalent to $Cx = 0$.

Theorem 17.32. *Let $V(\lambda) = \widetilde{D} + C(\lambda I_n - A)^{-1}\widetilde{B}$ be a realization of a $p \times q$ rational matrix function satisfying the conditions* (17.37)*, where Y is the unique (Hermitian) solution of the Lyapunov equation* (17.38)*. Let E be a $p \times (p - q)$ isometry such that $I_p - \widetilde{D}\widetilde{D}^* = EE^*$, and let $B^\sharp$ be an $n \times (p - q)$ matrix solution of* (17.43)*. Put $U(\lambda) = D + C(\lambda I_n - A)^{-1}B$, where B and D are the $p \times p$ matrices given by $B = \begin{bmatrix} B^\sharp & \widetilde{B} \end{bmatrix}$ and $D = \begin{bmatrix} E & \widetilde{D} \end{bmatrix}$. Then U is a $p \times p$ bi-inner completion of V, that is, U is a bi-inner rational $p \times p$ matrix function of the form $\begin{bmatrix} V^\sharp(\lambda) & V(\lambda) \end{bmatrix}$, and the McMillan degree of U is equal to the rank of Y.*

The rational $p \times (p - q)$ matrix function $V^\sharp$ can be described explicitly; it is actually given by the realization $V^\sharp(\lambda) = E + C(\lambda I_n - A)^{-1}B^\sharp$.

Proof. To prove that U is bi-inner, apply Proposition 17.30 to U with its given realization. Since (17.38) holds, it suffices to show that $YB = C^*D$ and $D^*D = I_p$. These facts follow from the third identity in (17.37) and the definitions of E and $B^\sharp$. Indeed, we have

$$YB = \begin{bmatrix} YB^\sharp & Y\widetilde{B} \end{bmatrix} = \begin{bmatrix} C^*E & C^*\widetilde{D} \end{bmatrix} = C^*D,$$

$$DD^* = \begin{bmatrix} E & \widetilde{D} \end{bmatrix} \begin{bmatrix} E^* \\ \widetilde{D}^* \end{bmatrix} = EE^* + \widetilde{D}\widetilde{D}^* = I_p,$$

and, since D is a square matrix, $DD^* = I_p$ amounts to the same as $D^*D = I_p$. The final statement is an immediate corollary of Proposition 17.31. □

Next, we return to the inner-outer factorization discussed in Section 17.6. The point we focus on here is the completion of the inner factor to a bi-inner function.

Let $L(\lambda) = D + C(\lambda I_n - A)^{-1}B$ be a realization of a $p \times q$ rational matrix function. Assume A has all its eigenvalues in the open left half plane, $L(i\omega)$ is left invertible for each $\omega \in \mathbb{R}$, and $D^*D = I_q$. Let $L(\lambda) = V(\lambda)X(\lambda)$ be the inner-outer factorization constructed in Theorem 17.26, in particular,

$$V(\lambda) = D + \big((I - DD^*)C + DB^*Y\big)\big(\lambda I_n - (A - BD^*C + BB^*Y)\big)^{-1}B,$$

where $Y = Y^*$ satisfies the algebraic Riccati equation

$$YBB^*Y + Y(A - BD^*C) + (A^* - C^*DB^*)Y - C^*(I - DD^*)C = 0,$$

and $A - BD^*C + BB^*Y$ has all its eigenvalues in the open left half plane. Choose a $p \times (p-q)$ isometry E such that $I - DD^* = EE^*$, and let $B^\sharp$ be any $n \times (p-q)$ matrix such that $YB^\sharp = C^*E$.

Corollary 17.33. *In the situation described in the previous paragraph, introduce*

$$U(\lambda) = \begin{bmatrix} V^\sharp(\lambda) & V(\lambda) \end{bmatrix},$$

where the rational $p \times (p-q)$ matrix function $V^\sharp$ is given by

$$V^\sharp(\lambda) = E + \big((I - DD^*)C + DB^*Y\big)\big(\lambda I_n - (A - BD^*C + BB^*Y)\big)^{-1}B^\sharp.$$

Then U is bi-inner.

Proof. All we need to show is that Theorem 17.32 may be applied with the matrices $A - BD^*C + BB^*Y$ and $(I - DD^*)C + DB^*Y$ in place of A and C, respectively. For this we need to verify the identities $\big((I - DD^*)C + DB^*Y\big)^*D = YB$ and

$$\begin{aligned} Y(A - BD^*C + BB^*Y) &+ (A - BD^*C + BB^*Y)^*Y \\ &= \big((I - DD^*)C + DB^*Y\big)^*\big((I - DD^*)C + DB^*Y\big). \end{aligned}$$

This involves nothing more than a routine computation using that $D^*D = I$ and that $Y = Y^*$ is a solution of the Riccati equation featured in the paragraph preceding the corollary. □

Notes

The first three sections are largely based on [3]. The Redheffer transformation of Section 17.4, which is a standard tool in the analysis of 2×2 block matrix functions, originates from [130]. Theorem 17.22 in Section 17.5 also implies that if W is a J-inner rational matrix function, then the function $K_{*,W}(\mu, \lambda)$ has no negative squares, that is, it is a positive definite kernel, see also Theorem 2.5 in [39]. Theorem 17.25 in Section 17.5 is a simple case of a more far-reaching theory concerning the multiplicative structure of general matrix-valued J-inner functions, which originates from [118]; see also Chapter 4 in [39]. Factorizations in degree 1 factors, of which Theorem 17.25 provides an example, are the main topic of Part III in [20]. Section 17.6 originates from Section 7.4 in [43]; for the corresponding state space formulas, see [146]. Section 17.7 is related to the problem of Darlington synthesis. The latter problem can be found in [4]. The presentation given here is based on [75]. For further results in this direction, including Darlington embedding for time-variant systems, see [36] and Chapter 6 in [117]. The result presented in Section 17.8 may be found in, e.g., Chapter 12 (page 249) in [149].

Part VII
Applications of J-spectral factorizations

In this part, the state space theory of J-spectral factorization, developed in the preceding two parts, is used to solve H_∞-problems. There are three chapters. The first chapter (Chapter 18) presents the solution of the Nehari interpolation problem for rational matrix functions. The second chapter (Chapter 19) reviews elements from control and mathematical systems theory that play an essential role in the final chapter. The third and final chapter (Chapter 20) treats H_∞-control problems. Here we use the J-spectral factorization theory to obtain the solutions of some of the main problems in this area, namely the standard problem, the one-sided problem, and the full model matching problem.

Chapter 18

Application to the rational Nehari problem

In this chapter the rational matrix version of the Nehari problem (relative to the imaginary axis) is solved using a J-spectral factorization approach. The data of the problem are given in realized form. This together with the state space results on J-spectral factorization derived in Chapter 14 allows us to solve the problem and to obtain an explicit linear fractional representation of all its solutions, again in realized form. The main attention is given to the so-called suboptimal case. The more general Nehari-Takagi problem is also solved using the J-spectral factorization method.

This chapter consists of six sections. Section 18.1 presents the problem statement and the main theorem. Section 18.2 deals with the theory of linear fractional maps. Such maps will play an important role in this and the final chapter. In Section 18.3 the rational matrix Nehari problem is reduced to a J-spectral factorization of a special kind, and all solutions are described in terms of the coefficients of the J-spectral factor. This result is used in Section 18.4 to prove the main theorem of Section 18.1. Section 18.5 deals with the Nehari problem for the non-stable case, when the given function does not necessarily have all its poles in the open left half plane. Section 18.6, the final section of the chapter, gives the solution of the rational matrix Nehari-Takagi problem.

18.1 Problem statement and main result

Let R be a rational $p \times q$ matrix function which does not have a pole on the imaginary axis and at infinity. In particular, R is proper. In this section we study the problem of finding all proper rational $p \times q$ matrix functions K such that K

has all its poles in the open right half plane and

$$\|K - R\|_\infty = \sup_{s \in i\mathbb{R}} \|K(s) - R(s)\| < \gamma, \tag{18.1}$$

where γ is a pre-specified positive number. Note that both R and K are proper and have no pole on the imaginary axis, and hence the so-called infinity norm $\|K - R\|_\infty$ is well-defined. We shall refer to this problem as the (*suboptimal*) *rational Nehari problem for R relative to the imaginary axis with tolerance* γ. The latter qualifier will be omitted when $\gamma = 1$. The word "suboptimal" refers to the fact that we use in (18.1) a strict inequality.

We first deal with the case when R is stable. A rational matrix function is called $i\mathbb{R}$-*stable*, or simply *stable* when no confusion is possible (as will be the case in this chapter), if all its poles are in the open left half plane. Note that such a function is proper and has no pole on $i\mathbb{R}$. We shall assume additionally that R is strictly proper.

To state the main result we start with a realization of R. Since R is stable and strictly proper, we can choose a realization of R of the form

$$R(\lambda) = C(\lambda I_n - A)^{-1}B, \tag{18.2}$$

with the property that A has all its eigenvalues in the open left half plane. Let P and Q be the unique solutions of the Lyapunov equations

$$AP + PA^* = -BB^*, \qquad A^*Q + QA = -C^*C, \tag{18.3}$$

respectively. Note that P and Q are given by

$$P = \int_0^\infty e^{\tau A}BB^*e^{\tau A^*}\,d\tau, \qquad Q = \int_0^\infty e^{\tau A^*}C^*Ce^{\tau A}\,d\tau.$$

Hence P and Q are nonnegative Hermitian matrices. One usually refers to P as the *controllability gramian*, and to Q as the *observability gramian*, corresponding to the realization (18.2). We shall prove the following theorem.

Theorem 18.1. *Let $R(\lambda) = C(\lambda I_n - A)^{-1}B$ be a realization of the $p \times q$ rational matrix function R, and assume that A has all its eigenvalues in the open left half plane. Then the rational Nehari problem for R relative to the imaginary axis with tolerance γ is solvable if and only if the matrix $\gamma^2 I_n - P^{1/2}QP^{1/2}$ is positive definite. In that case all solutions of the Nehari problem for R can be obtained in the following way. Introduce the rational matrix functions*

$$X_{11}(\lambda) = I_p + CP(\lambda I_n + A^*)^{-1}Z^{-1}C^*, \tag{18.4}$$

$$X_{12}(\lambda) = CP(\lambda I_n + A^*)^{-1}Z^{-1}QB, \tag{18.5}$$

$$X_{21}(\lambda) = -B^*(\lambda I_n + A^*)^{-1}Z^{-1}C^*, \tag{18.6}$$

$$X_{22}(\lambda) = I_q - B^*(\lambda I_n + A^*)^{-1}Z^{-1}QB, \tag{18.7}$$

where $Z = \gamma^2 I_n - QP$. Then all solutions K of the rational Nehari problem for R relative to the imaginary axis are given by

$$K(\lambda) = -\big(X_{11}(\lambda)H(\lambda) + X_{12}(\lambda)\big)\big(X_{21}(\lambda)H(\lambda) + X_{22}(\lambda)\big)^{-1}, \qquad (18.8)$$

where H is any rational $p \times q$ matrix function which has all its poles in the open right half plane and satisfies $\|H\|_\infty < \gamma$. Moreover, there is a one-to-one correspondence between the solution K and the free parameter H.

Before we prove the above theorem (in Section 18.4 below) it will be convenient first to make some preparations. The following lemma restates the necessary and sufficient condition appearing in Theorem 18.1 in operator language.

Lemma 18.2. *Let $R(\lambda) = C(\lambda I_n - A)^{-1}B$ be a realization of the $p \times q$ rational matrix function R, and assume that A has all its eigenvalues in the open left half plane. Consider the Hankel operator H_R generated by R, that is the finite rank integral operator from $L_2^q[0,\infty)$ into $L_2^p[0,\infty)$ given by*

$$(H_R f)(t) = \int_0^\infty Ce^{A(t+\tau)}Bf(\tau)\,d\tau.$$

Then $\|H_R\| < \gamma$ if and only if the matrix $\gamma^2 I_n - P^{1/2}QP^{1/2}$ is positive definite.

Proof. We need the controllability operator Ξ and the observability operator Ω associated with the realization (18.2). Thus

$$\Xi : L_2^q[0,\infty) \to \mathbb{C}^q, \qquad \Xi f = \int_0^\infty e^{\tau A}Bf(\tau)\,d\tau,$$

$$\Omega : \mathbb{C}^n \to L_2^p[0,\infty), \qquad (\Omega x)(t) = Ce^{tA}x, \quad t > 0.$$

Clearly $P = \Xi\Xi^*$, $Q = \Omega^*\Omega$ and $H_R = \Omega\Xi$. Now let $\lambda_1(X)$ denote the largest eigenvalue of an operator X all of whose non-zero spectrum consists of positive eigenvalues. Then

$$\begin{aligned} \|H_R\|^2 &= \lambda_1(H_R^*H_R) = \lambda_1(\Xi^*\Omega^*\Omega\Xi) \\ &= \lambda_1(\Xi\Xi^*\Omega^*\Omega) = \lambda_1(PQ) = \lambda_1(P^{1/2}QP^{1/2}). \end{aligned}$$

Hence $\|H_R\| < \gamma$ if and only if all the eigenvalues of $P^{1/2}QP^{1/2}$ are strictly less than γ^2. Thus $\|H_R\| < \gamma$ if and only if $\gamma^2 I - P^{1/2}QP^{1/2}$ is positive definite. □

We close the section by showing that, without loss of generality, we may assume that in Theorem 18.1 the tolerance $\gamma = 1$. Indeed, consider for the original problem $\widetilde{R}(\lambda) = \gamma^{-1}R(\lambda)$, and $\widetilde{K}(\lambda) = \gamma^{-1}K(\lambda)$. Then we have $\|R - K\|_\infty < \gamma$ if and only if $\|\widetilde{R} - \widetilde{K}\|_\infty < 1$. Moreover, if R is given by the realization (18.2), then $\widetilde{R}$ admits the realization $\widetilde{R} = \widetilde{C}(\lambda - A)^{-1}B$, where $\widetilde{C} = \gamma^{-1}C$. One easily sees that, for solutions $\widetilde{P}$ and $\widetilde{Q}$ of the corresponding Lyapunov equations (18.3), one has $\widetilde{P} = P$, $\widetilde{Q} = \gamma^{-2}Q$. Hence $\widetilde{Z} = I - \widetilde{P}\widetilde{Q} = \gamma^{-2}Z$. For the functions $X_{ij}(\lambda)$ appearing in Theorem 18.1 we have the following:

$$\begin{aligned}
\widetilde{X}_{11}(\lambda) &= I_p + \widetilde{C}P(\lambda + A^*)^{-1}\widetilde{Z}^{-1}\widetilde{C}^* = X_{11}(\lambda),\\
\widetilde{X}_{12}(\lambda) &= \widetilde{C}P(\lambda + A^*)^{-1}\widetilde{Z}^{-1}\widetilde{Q}B = \gamma^{-1}X_{21}(\lambda),\\
\widetilde{X}_{21}(\lambda) &= -B^*(\lambda + A^*)^{-1}\widetilde{Z}^{-1}\widetilde{C}^* = \gamma X_{21}(\lambda),\\
\widetilde{X}_{22}(\lambda) &= I_q - B^*(\lambda + A^*)^{-1}\widetilde{Z}^{-1}\widetilde{Q}B = X_{22}(\lambda).
\end{aligned}$$

Suppose that $\widetilde{K}(\lambda)$ is a solution to the problem with $\gamma = 1$, given by

$$\widetilde{K}(\lambda) = -(\widetilde{X}_{11}(\lambda)\widetilde{H}(\lambda) + \widetilde{X}_{12}(\lambda))(\widetilde{X}_{21}(\lambda)\widetilde{H}(\lambda) + \widetilde{X}_{22}(\lambda))^{-1},$$

for some $\widetilde{H}$ satisfying $\|\widetilde{H}\|_\infty < 1$. Now taking $H(\lambda) = \gamma\widetilde{H}(\lambda)$ we have $\|H\|_\infty < \gamma$, and with $K(\lambda) = \gamma\widetilde{K}(\lambda)$, we obtain that (18.8) holds.

18.2 Intermezzo about linear fractional maps

The expression (18.8), which assigns to the rational matrix function H a rational matrix function K, is usually called a *linear fractional map.* Such maps will play an important role in this and the final chapter. Therefore, we review some of the main properties of linear fractional maps in this section.

It will be convenient first to introduce some notation and terminology. Given a $p \times q$ rational matrix function F, we write F^* for the adjoint of F relative to the imaginary axis, that is, $F^*(\lambda) = F(-\bar{\lambda})^*$. (In engineering literature, including [76], [43]), this function is often denoted by $F^\sim$.) By RAT we shall denote the set of all rational matrix functions that are proper and have no pole on the imaginary axis $i\mathbb{R}$, and $\text{RAT}^{p\times q}$ will stand for the set of all F in RAT that are of size $p \times q$. If F belongs to $\text{RAT}^{p\times q}$, then F^* belongs to $\text{RAT}^{q\times p}$. Note that $\text{RAT}^{p\times q}$ is closed under the usual addition of matrix functions as well as under scalar multiplication. Also for $F \in \text{RAT}^{p\times q}$ and $G \in \text{RAT}^{q\times r}$, we have $FG \in \text{RAT}^{p\times r}$. In particular $\text{RAT}^{p\times p}$ is an algebra. The unit element in this algebra is E_p, the $p \times p$ matrix function which is identically equal to the $p \times p$ identity matrix I_p.

A function $F \in \text{RAT}^{p\times p}$ is said to be *invertible in* $\text{RAT}^{p\times p}$ if F has an inverse G in $\text{RAT}^{p\times p}$, that is, $G \in \text{RAT}^{p\times p}$ and $FG = GF = E_p$. For a rational $p \times p$ matrix function F such that $\det F(\lambda) \not\equiv 0$, the pointwise inverse F^{-1}, defined by $F^{-1}(\lambda) = F(\lambda)^{-1}$, is again a rational matrix function. If $F \in \text{RAT}^{p\times p}$ and $\det F(\lambda) \not\equiv 0$, then F^{-1} need not be an element of $\text{RAT}^{p\times p}$. Indeed, F^{-1} might have a pole on the imaginary axis or fail to be proper. In fact, $F^{-1} \in \text{RAT}^{p\times p}$ if and only if F is biproper and $\det F(\lambda)$ has no zero on $i\mathbb{R}$, and in that case F^{-1} is the inverse of F in the algebra $\text{RAT}^{p\times p}$.

A function F in $\mathrm{RAT}^{p\times q}$ is analytic on the imaginary axis and at infinity. Hence we can consider the norm

$$\|F\|_\infty = \sup_{s\,\in\, i\mathbb{R}} \|F(s)\|. \tag{18.9}$$

This is the usual L_∞-norm for bounded matrix functions on $i\mathbb{R}$ which we already used in (18.1). We write $F \in \mathrm{RAT}_{\mathbb{B}}^{p\times q}$, whenever F belongs to $\mathrm{RAT}^{p\times q}$ and its infinity-norm $\|F\|_\infty$ is strictly less than 1. Thus $\mathrm{RAT}_{\mathbb{B}}^{p\times q}$ is the open unit ball in $\mathrm{RAT}^{p\times q}$ with respect to the norm defined by (18.9). Note that $\|F\|_\infty < 1$ is equivalent to $I_p - F(-\bar\lambda)^*F(\lambda)$ being positive definite on $i\mathbb{R}\cup\{\infty\}$. For the latter property we use the notation $E_p - F^*F > 0$.

Now let $\Theta \in \mathrm{RAT}^{(p+q)\times(p+q)}$, and let us partition Θ as a 2×2 block matrix function in the following way:

$$\Theta(\lambda) = \begin{bmatrix} \Theta_{11}(\lambda) & \Theta_{12}(\lambda) \\ \Theta_{21}(\lambda) & \Theta_{22}(\lambda) \end{bmatrix} \tag{18.10}$$

with $\Theta_{11}(\lambda)$ a $p\times p$ matrix and $\Theta_{22}(\lambda)$ a $q\times q$ matrix. With this partitioning of Θ we associate the linear fractional map

$$(\mathcal{F}_\Theta H)(\lambda) = \big(\Theta_{11}(\lambda)H(\lambda) + \Theta_{12}(\lambda)\big)\big(\Theta_{21}(\lambda)H(\lambda) + \Theta_{22}(\lambda)\big)^{-1}. \tag{18.11}$$

Here H is assumed to be in $\mathrm{RAT}^{p\times q}$. In general, it is not clear for which H the map is well-defined. However for a J-unitary Θ, with $J = \mathrm{diag}\,(I_p\,,\, -I_q)$, we have the following result.

Theorem 18.3. *Let $\Theta \in \mathrm{RAT}^{(p+q)\times(p+q)}$ be J-unitary with $J = \mathrm{diag}\,(I_p\,,\,-I_q)$. Then Θ is invertible in $\mathrm{RAT}^{(p+q)\times(p+q)}$, the maps $\mathcal{F}_\Theta$ and $\mathcal{F}_{\Theta^{-1}}$ are well-defined on $\mathrm{RAT}_{\mathbb{B}}^{p\times q}$ and map $\mathrm{RAT}_{\mathbb{B}}^{p\times q}$ into itself. Moreover*

$$H = \mathcal{F}_{\Theta^{-1}}\mathcal{F}_\Theta H = \mathcal{F}_\Theta\mathcal{F}_{\Theta^{-1}}H, \qquad H \in \mathrm{RAT}_{\mathbb{B}}^{p\times q}. \tag{18.12}$$

Proof. We divide the proof into three parts. In the first part it is shown that Θ^{-1} is in $\mathrm{RAT}^{(p+q)\times(p+q)}$ and is J-unitary, and also that the maps $\mathcal{F}_\Theta$ and $\mathcal{F}_{\Theta^{-1}}$ are well-defined on $\mathrm{RAT}_{\mathbb{B}}^{p\times q}$. In the second part we prove that $\mathcal{F}_\Theta$ maps $\mathrm{RAT}_{\mathbb{B}}^{p\times q}$ into itself. In the final part the identities in (18.12) will be established.

Part 1. Since Θ is proper and has no pole on $i\mathbb{R}$, the fact that Θ is J-unitary implies that for each $\lambda \in i\mathbb{R}\cup\{\infty\}$ the matrix $\Theta(\lambda)$ is J-unitary and hence invertible. It follows that Θ is invertible in $\mathrm{RAT}^{(p+q)\times(p+q)}$ and that Θ^{-1} is J-unitary.

The fact that the matrix $\Theta(\lambda)$ is J-unitary for $\lambda \in i\mathbb{R}\cup\{\infty\}$ implies that $\Theta_{22}(\lambda)$ is invertible and $\|\Theta_{22}(\lambda)^{-1}\Theta_{21}(\lambda)\| < 1$ for $\lambda \in i\mathbb{R}\cup\{\infty\}$. It follows that Θ_{22} is invertible in $\mathrm{RAT}^{q\times q}$ and that

$$\|\Theta_{22}^{-1}\Theta_{21}\|_\infty = \sup_{\lambda\,\in\,\mathbb{R}} \|\Theta_{22}(\lambda)^{-1}\Theta_{21}(\lambda)\| = \max_{\lambda\,\in\, i\mathbb{R}\cup\{\infty\}} \|\Theta_{22}(\lambda)^{-1}\Theta_{21}(\lambda)\| < 1.$$

Next, take $H \in \mathrm{RAT}_{\mathbb{B}}^{p\times q}$. Then $\|\Theta_{22}^{-1}\Theta_{21}H\|_\infty \le \|\Theta_{22}^{-1}\Theta_{21}\|_\infty\|H\|_\infty < 1$. Thus $\Theta_{21}H + \Theta_{22} = \Theta_{22}\left(\Theta_{22}^{-1}\Theta_{21}H + E_q\right)$ is invertible in $\mathrm{RAT}^{q\times q}$. It follows that $\mathcal{F}_\Theta H$ is well-defined for $H \in \mathrm{RAT}_{\mathbb{B}}^{p\times q}$. Since Θ^{-1} is also J-unitary, $\mathcal{F}_{\Theta^{-1}}$ is well-defined on $\mathrm{RAT}_{\mathbb{B}}^{p\times q}$ too.

Part 2. In this part we show that $\mathcal{F}_\Theta$ maps $\mathrm{RAT}_{\mathbb{B}}^{p\times q}$ into itself. Take H in $\mathrm{RAT}_{\mathbb{B}}^{p\times q}$, and write $F = \mathcal{F}_\Theta H$. First note that

$$\begin{bmatrix} F \\ E_q \end{bmatrix} = \begin{bmatrix} (\Theta_{11}H + \Theta_{12})(\Theta_{21}H + \Theta_{22})^{-1} \\ (\Theta_{21}H + \Theta_{22})(\Theta_{21}H + \Theta_{22})^{-1} \end{bmatrix} = \Theta \begin{bmatrix} H \\ E_q \end{bmatrix} X^{-1}, \tag{18.13}$$

where $X = \Theta_{21}H + \Theta_{22}$. The fact that Θ is J-unitary, with $J = \operatorname{diag}(I_p, -I_q)$ is equivalent to the identity

$$\Theta^* \begin{bmatrix} E_p & 0 \\ 0 & -E_q \end{bmatrix} \Theta = \begin{bmatrix} E_p & 0 \\ 0 & -E_q \end{bmatrix}. \tag{18.14}$$

Hence, using (18.13), we obtain

$$\begin{aligned} E_q - F^*F &= -\begin{bmatrix} F^* & E_q \end{bmatrix} \begin{bmatrix} E_p & 0 \\ 0 & -E_q \end{bmatrix} \begin{bmatrix} F \\ E_q \end{bmatrix} \\ &= -X^{-*}\begin{bmatrix} H^* & E_q \end{bmatrix} \Theta^* \begin{bmatrix} E_p & 0 \\ 0 & -E_q \end{bmatrix} \Theta \begin{bmatrix} H \\ E_q \end{bmatrix} X^{-1} \\ &= -X^{-*}\begin{bmatrix} H^* & E_q \end{bmatrix} \begin{bmatrix} E_p & 0 \\ 0 & -E_q \end{bmatrix} \begin{bmatrix} H \\ E_q \end{bmatrix} X^{-1} \\ &= X^{-*}(E_q - H^*H)X^{-1}. \end{aligned}$$

It follows that $I_q - F(-\bar\lambda)^*F(\lambda) = X(-\bar\lambda)^{-*}\big(I_q - H(-\bar\lambda)^*H(\lambda)\big)X(\lambda)^{-1}$. Now $\|H\|_\infty < 1$. This means that $I_p - H(-\bar\lambda)^*H(\lambda)$ is positive definite on $i\mathbb{R} \cup \{\infty\}$. But then $I_q - F(-\bar\lambda)^*F(\lambda)$ is also positive definite on $i\mathbb{R} \cup \{\infty\}$. The latter is equivalent to $\|F\|_\infty < 1$. Thus $F \in \mathrm{RAT}_{\mathbb{B}}^{p\times q}$, as desired.

From what has been proved so far, we conclude that the result of the previous steps also hold with Θ^{-1} instead of Θ. Thus $\mathcal{F}_{\Theta^{-1}}$ maps $\mathrm{RAT}_{\mathbb{B}}^{p\times q}$ into itself. Therefore, to complete the proof, it remains to prove the identities in (18.12). In fact, by interchanging the roles of Θ and Θ^{-1}, it suffices to prove the first identity in (18.12). This will be done in the next part.

Part 3. Take $H \in \mathrm{RAT}_{\mathbb{B}}^{p\times q}$, and put $F = \mathcal{F}_\Theta H$, $G = \mathcal{F}_{\Theta^{-1}}F$. From (18.14) we see that

$$\Theta^{-1} = \begin{bmatrix} E_p & 0 \\ 0 & -E_q \end{bmatrix} \Theta^* \begin{bmatrix} E_p & 0 \\ 0 & -E_q \end{bmatrix} = \begin{bmatrix} \Theta_{11}^* & -\Theta_{21}^* \\ -\Theta_{12}^* & \Theta_{22}^* \end{bmatrix}. \tag{18.15}$$

By using (18.13) for Θ as well as for Θ^{-1}, we have

$$\begin{bmatrix} F \\ E_q \end{bmatrix} = \Theta \begin{bmatrix} H \\ E_q \end{bmatrix} (\Theta_{21}H + \Theta_{22})^{-1},$$

$$\begin{bmatrix} G \\ E_q \end{bmatrix} = \Theta^{-1} \begin{bmatrix} F \\ E_q \end{bmatrix} (-\Theta_{12}^* F + \Theta_{22}^*)^{-1}.$$

Now observe that

$$\begin{aligned} -\Theta_{12}^* F + \Theta_{22}^* &= \begin{bmatrix} 0 & E_q \end{bmatrix} \begin{bmatrix} \Theta_{11}^* F - \Theta_{21}^* \\ -\Theta_{12}^* F + \Theta_{22}^* \end{bmatrix} = \begin{bmatrix} 0 & E_q \end{bmatrix} \Theta^{-1} \begin{bmatrix} F \\ E_q \end{bmatrix} \\ &= \begin{bmatrix} 0 & E_q \end{bmatrix} \Theta^{-1}\Theta \begin{bmatrix} H \\ E_q \end{bmatrix} (\Theta_{21}H + \Theta_{22})^{-1} \\ &= \begin{bmatrix} 0 & E_q \end{bmatrix} \begin{bmatrix} H \\ E_q \end{bmatrix} (\Theta_{21}H + \Theta_{22})^{-1} = (\Theta_{21}H + \Theta_{22})^{-1}. \end{aligned}$$

In particular, $(\Theta_{21}H + \Theta_{22})^{-1}(-\Theta_{12}^* F + \Theta_{22}^*)^{-1} = E_q$. But then

$$\begin{aligned} G &= \begin{bmatrix} E_p & 0 \end{bmatrix} \begin{bmatrix} G \\ E_q \end{bmatrix} = \begin{bmatrix} E_p & 0 \end{bmatrix} \Theta^{-1} \begin{bmatrix} F \\ E_q \end{bmatrix} (-\Theta_{12}^* F + \Theta_{22}^*)^{-1} \\ &= \begin{bmatrix} E_p & 0 \end{bmatrix} \Theta^{-1}\Theta \begin{bmatrix} H \\ E_q \end{bmatrix} (\Theta_{21}H + \Theta_{22})^{-1}(-\Theta_{12}^* F + \Theta_{22}^*)^{-1} \\ &= \begin{bmatrix} E_p & 0 \end{bmatrix} \begin{bmatrix} H \\ E_q \end{bmatrix} = H, \end{aligned}$$

which proves the first identity in (18.12). □

We are particulary interested in proper rational $p \times q$ matrix functions that are analytic on the closed left half plane with infinity included. The class of these functions will be denoted by $\mathrm{RAT}_+^{p\times q}$. Since the functions in $\mathrm{RAT}_+^{p\times q}$ have no pole on $i\mathbb{R}$ and are proper, $\mathrm{RAT}_+^{p\times q}$ is a linear subspace of $\mathrm{RAT}^{p\times q}$. We write $\mathrm{RAT}_{+,\mathbb{B}}^{p\times q}$ for the set of all F $\mathrm{RAT}_+^{p\times q}$ such that (18.9) holds. Thus

$$\mathrm{RAT}_{+,\mathbb{B}}^{p\times q} = \mathrm{RAT}_+^{p\times q} \cap \mathrm{RAT}_{\mathbb{B}}^{p\times q}.$$

Now, as in Theorem 18.3, let $\Theta \in \text{RAT}^{(p+q)\times(p+q)}$ be J-unitary with $J = \text{diag}\,(I_p\,, -I_q)$. Fix $R \in \text{RAT}^{p\times q}$, and consider

$$V = \begin{bmatrix} V_{11} & V_{12} \\ V_{21} & V_{22} \end{bmatrix} = \begin{bmatrix} \Theta_{11} - R\Theta_{21} & \Theta_{12} - R\Theta_{22} \\ \Theta_{21} & \Theta_{22} \end{bmatrix} = \begin{bmatrix} E_p & -R \\ 0 & E_q \end{bmatrix} \Theta. \tag{18.16}$$

Since Θ is invertible in $\text{RAT}^{(p+q)\times(p+q)}$ by Theorem 18.3, it follows that the same holds true for V.

Let $\mathcal{F}_V$ be the linear fractional map defined by V. Since $V_{21} = \Theta_{21}$ and $V_{22} = \Theta_{22}$, we know from Theorem 18.3 that for each function H in $\text{RAT}_{\mathbb{B}}^{p\times q}$ the function $V_{21}H + V_{22}$ is invertible in $\text{RAT}^{q\times q}$. Thus $\mathcal{F}_V$ is well-defined on $\text{RAT}_{\mathbb{B}}^{p\times q}$. Moreover, since

$$\begin{aligned} V_{11}H + V_{12} &= (\Theta_{11} - R\Theta_{21})H + (\Theta_{12} - R\Theta_{22}) \\ &= (\Theta_{11}H + \Theta_{12}) - R(\Theta_{21}H + \Theta_{22}), \end{aligned}$$

we see that

$$\mathcal{F}_V H = \mathcal{F}_\Theta H - R, \qquad H \in \text{RAT}_{\mathbb{B}}^{p\times q}. \tag{18.17}$$

The fact $V_{22} = \Theta_{22}$ implies that V_{22} is invertible in $\text{RAT}^{q\times q}$. The following theorem is the second main result of this section.

Theorem 18.4. *Let $\Theta \in \text{RAT}^{(p+q)\times(p+q)}$ be J-unitary with $J = \text{diag}\,(I_p\,, -I_q)$, and let V be given by (18.16), where $R \in \text{RAT}^{p\times q}$. Then V is invertible in $\text{RAT}^{(p+q)\times(p+q)}$, and V_{22} is invertible in $\text{RAT}^{q\times q}$. Assume additionally that*

(a) *V and V^{-1} belong to $\text{RAT}_+^{(p+q)\times(p+q)}$,*

(b) *V_{22} and V_{22}^{-1} belong to $\text{RAT}_+^{q\times q}$.*

Then $\mathcal{F}_V$ is well-defined and one-to-one on $\text{RAT}_{+,\mathbb{B}}^{p\times q}$. Also

$$\mathcal{F}_V\big[\text{RAT}_{+,\mathbb{B}}^{p\times q}\big] = \big\{K \in \text{RAT}_+^{p\times q} \mid \|R + K\|_\infty < 1\big\}. \tag{18.18}$$

Note that conditions (a) and (b) in the above theorem are not independent. Indeed, the property that V_{22} belongs to $\text{RAT}_+^{q\times q}$ follows from the fact that V belongs to $\text{RAT}_+^{(p+q)\times(p+q)}$.

Proof. The fact that V is invertible in $\text{RAT}^{(p+q)\times(p+q)}$ and V_{22} in $\text{RAT}^{q\times q}$ has already been proved in the two paragraphs preceding Theorem 18.4. From Theorem 18.3 we know that $\mathcal{F}_\Theta$ is well-defined and one-to-one on $\text{RAT}_{\mathbb{B}}^{p\times q}$. But then we see from (18.17) that the same holds true for $\mathcal{F}_V$. Now recall that $\text{RAT}_{+,\mathbb{B}}^{p\times q} \subset \text{RAT}_{\mathbb{B}}^{p\times q}$. This allows us to conclude that $\mathcal{F}_V$ is well-defined and one-to-one on $\text{RAT}_{+,\mathbb{B}}^{p\times q}$. It remains to show the identity (18.18). This will be done in two parts. The first part covers the inclusion

$$\mathcal{F}_V\big[\text{RAT}_{+,\mathbb{B}}^{p\times q}\big] \subset \big\{K \in \text{RAT}_+^{p\times q} \mid \|R + K\|_\infty < 1\big\}. \tag{18.19}$$

The reverse inclusion is proved in the second step.

Part 1. Take H in $\text{RAT}_{+,\mathbb{B}}^{p\times q}$. We first show that $\mathcal{F}_V H$ belongs to $\text{RAT}_{+}^{p\times q}$. From condition (a) we know that V is analytic on the closed left half plane. Hence the same holds true for the entries $V_{ij},\ i,j=1,2$. Now $V_{22}=\Theta_{22}$ is invertible in $\text{RAT}_{+}^{q\times q}$, and so $V_{22}^{-1}V_{21}H$ is analytic on the closed left half plane. Moreover,

$$\|V_{22}^{-1}V_{21}H\|_\infty \le \|V_{22}^{-1}V_{21}\|_\infty\,\|H\|_\infty \le \|\Theta_{22}^{-1}\Theta_{21}\|_\infty\,\|H\|_\infty < 1.$$

By the maximum modulus principle, this gives $\|V_{22}^{-1}(\lambda)V_{21}(\lambda)H(\lambda)\|<1$ for λ in the closure of $\mathbb{C}_{\text{left}}$. It follows that $V_{22}^{-1}(\lambda)V_{21}(\lambda)H(\lambda)+I_q$ is invertible for each λ in the closed left half plane, and that the function $\big(V_{22}^{-1}(\lambda)V_{21}(\lambda)H(\lambda)+I_q\big)^{-1}$ is again analytic on the closed left half plane. Thus $V_{22}^{-1}V_{21}H+E_q$ is invertible in $\text{RAT}_{+}^{q\times q}$. By assumption, V_{22} is invertible in $\text{RAT}_{+}^{q\times q}$ too. Combining these facts we obtain that $V_{21}H+V_{22}$ belongs to $\text{RAT}_{+}^{q\times q}$ and is invertible in $\text{RAT}_{+}^{q\times q}$. But then $\mathcal{F}_V H$ belongs to $\text{RAT}_{+}^{p\times q}$, as desired.

Next, consider $K=\mathcal{F}_V H$. Using (18.17), we see that $R+K=\mathcal{F}_\Theta H$. Since $\text{RAT}_{+,\mathbb{B}}^{p\times q}$ is a subset of $\text{RAT}_{\mathbb{B}}^{p\times q}$ and $\mathcal{F}_\Theta$ maps $\text{RAT}_{\mathbb{B}}^{p\times q}$ into itself (by Theorem 18.3), we know that $R+K$ belongs to $\text{RAT}_{\mathbb{B}}^{p\times q}$, that is, $\|R+K\|_\infty<1$. Thus (18.19) is proved

Part 2. Take $K\in\text{RAT}_{+}^{p\times q}$, and suppose $\|R+K\|_\infty<1$. Since $R+K$ belongs to $\text{RAT}_{\mathbb{B}}^{p\times q}$, we know from Theorem 18.3 that there exists a unique H in $\text{RAT}_{\mathbb{B}}^{p\times q}$ such that $\mathcal{F}_\Theta H=R+K$. In fact, by (18.12), the function in question is $H=\mathcal{F}_{\Theta^{-1}}(R+K)$. Furthermore, according to (18.17), the equality $\mathcal{F}_\Theta H=R+K$ yields $\mathcal{F}_V H=K$. Note that H has no poles on $i\mathbb{R}\cup\{\infty\}$. The main difficulty is to show that H is analytic on the open left half plane $\mathbb{C}_{\text{left}}$.

From (18.15) and $H=\mathcal{F}_{\Theta^{-1}}(R+K)$ we know that

$$H=\mathcal{F}_{\Theta^{-1}}(R+K)=\big(\Theta_{11}^*(R+K)-\Theta_{21}^*\big)\big(-\Theta_{12}^*(R+K)+\Theta_{22}^*\big)^{-1}.$$

Put

$$H_1=\Theta_{11}^*(R+K)-\Theta_{21}^*,\qquad H_2=-\Theta_{12}^*(R+K)+\Theta_{22}^*.$$

Then H_2 is invertible in $\text{RAT}^{q\times q}$ and $H=H_1H_2^{-1}$. Moreover,

$$\begin{bmatrix} H_1 \\ H_2 \end{bmatrix}=\Theta^{-1}\begin{bmatrix} R+K \\ E_q \end{bmatrix}=V^{-1}\begin{bmatrix} K \\ E_q \end{bmatrix}. \tag{18.20}$$

Since V^{-1} and K belong to $\text{RAT}_{+}^{(p+q)\times(p+q)}$ and $\text{RAT}_{+}^{p\times q}$, respectively, we see from the second equality in (18.20) that H_1 belongs to $\text{RAT}_{+}^{p\times q}$ and H_2 belongs to $\text{RAT}_{+}^{q\times q}$. In other words, H_1 and H_2 are analytic in the open left half plane. Hence, in order to prove that H is analytic on the open left half plane $\mathbb{C}_{\text{left}}$, it remains to show H_2^{-1} is analytic in $\mathbb{C}_{\text{left}}$.

Multiplying (18.20) from the left by V we get $V_{21}H_1 + V_{22}H_2 = E_q$, hence

$$V_{22}^{-1} = V_{22}^{-1}(V_{21}H_1 + V_{22}H_2) = V_{22}^{-1}(V_{21}H + V_{22})H_2 = (V_{22}^{-1}V_{21}H + E_q)H_2.$$

Now introduce the scalar rational functions

$$\begin{aligned} f(\lambda) &= \det V_{22}(\lambda)^{-1}, \\ g(\lambda) &= \det \big(V_{22}(\lambda)^{-1}V_{21}(\lambda)H(\lambda) + I_q\big), \\ h(\lambda) &= \det H_2(\lambda). \end{aligned}$$

Then $f = gh$. Also f, g and h have no poles or zeros on $i\mathbb{R} \cup \{\infty\}$. This allows us to use winding number arguments (see Section IV.5 in [32]; also [53], pages 143 and 152). For simplicity we write $\mathrm{wn}_\circ(f)$ for the winding number around the origin of f, and we use the analogous notation for g and h. Note that $\mathrm{wn}_\circ(f)$ is just equal to the difference of the number of zeros and number of poles (multiplicities taken into account) of f in $\mathbb{C}_{\mathrm{left}}$, and similarly for $\mathrm{wn}_\circ(g)$ and $\mathrm{wn}_\circ(h)$. First observe that, by condition (b) in our theorem, both V_{22} and V_{22}^{-1} are analytic in the closed left half plane. Thus f has no zeros or poles in the closed left half plane, which implies that $\mathrm{wn}_\circ(f) = 0$. Since

$$\|V_{22}^{-1}V_{21}H\|_\infty \leq \|V_{22}^{-1}V_{21}\|_\infty \|H\|_\infty < 1,$$

it follows that g is analytic on the closed left half plane and has no zeros in the closed left half plane. Thus $\mathrm{wn}_\circ(g)$ is also zero. The fact that $f = gh$ implies that $\mathrm{wn}_\circ(f)$ is the sum of $\mathrm{wn}_\circ(g)$ and $\mathrm{wn}_\circ(h)$. Hence $\mathrm{wn}_\circ(h) = 0$. We already know that h is analytic on the closed left half plane. Thus $\mathrm{wn}_\circ(h) = 0$ tells us that h has no zeros on the closed left half plane. This implies H_2 is analytic on $\mathbb{C}_{\mathrm{left}}$, and hence the same holds true for H. □

Next we present a more general version of Theorem 18.4. In this more general version $K \in \mathrm{RAT}^{p\times q}$ is not supposed to be analytic on the open left half plane $\mathbb{C}_{\mathrm{left}}$ but K is required to have a prescribed number of poles in $\mathbb{C}_{\mathrm{left}}$. To state the result we need the following terminology. Let $F \in \mathrm{RAT}^{p\times q}$. By *the number of poles of F in the open left half plane, multiplicities taken into account*, we mean the nonnegative integer

$$\sum_{\lambda \in \mathbb{C}_{\mathrm{left}}} \delta(F;\lambda). \tag{18.21}$$

Here $\delta(F;\lambda)$ is the local degree of F at λ defined in the one but last paragraph of Section 8.2. Since $\delta(F;\lambda)$ is non-zero if and only if λ is a pole of F, the sum in (18.21) is finite.

Theorem 18.5. *Let $\Theta \in \mathrm{RAT}^{(p+q)\times(p+q)}$ be J-unitary with $J = \mathrm{diag}\,(I_p\,, -I_q)$, and let V be given by* (18.16)*, where $R \in \mathrm{RAT}^{p\times q}$. Then V is invertible in $\mathrm{RAT}^{(p+q)\times(p+q)}$, and V_{22} is invertible in $\mathrm{RAT}^{q\times q}$. Assume additionally that*

(α) *V and V^{-1} belong to* $\mathrm{RAT}_+^{(p+q)\times(p+q)}$,

(β) *V_{22} belongs to* $\mathrm{RAT}_+^{q\times q}$ *and V_{22}^{-1} has precisely κ poles, multiplicities taken into account, in* $\mathbb{C}_{\mathrm{left}}$.

Then $\mathcal{F}_V$ is well-defined and one-to-one on $\mathrm{RAT}_{+,\mathbb{B}}^{p\times q}$. *Also*

$$\mathcal{F}_V\left[\mathrm{RAT}_{+,\mathbb{B}}^{p\times q}\right] = \big\{K \in \mathrm{RAT}^{p\times q} \;\mid\; \|R+K\|_\infty < 1 \text{ and } K \text{ has } \kappa \text{ poles in } \mathbb{C}_{\mathrm{left}}, \text{ multiplicities taken into account}\big\}. \tag{18.22}$$

For $\kappa = 0$ the above theorem is just Theorem 18.4. To prove Theorem 18.5 one can use the same line of reasoning as in the proof of Theorem 18.4 above. However, the winding number argument employed in the final paragraph of the proof of Theorem 18.4 has to be used in a more sophisticated way. For the details we refer to the literature; see, e.g., [86] and the references therein.

18.3 The J-spectral factorization approach

In this section we shall exhibit the connection between the rational Nehari problem and J-spectral factorization. From the final paragraph of Section 18.1 we know that without loss of generality the tolerance γ can be assumed to be equal to 1. Therefore, in what follows we take $\gamma = 1$.

Let R be a stable rational $p \times q$ matrix function. With R we associate the $(p+q)\times(p+q)$ matrix function W given by

$$W(\lambda) = G(-\bar{\lambda})^* J G(\lambda), \tag{18.23}$$

where

$$J = \begin{bmatrix} I_p & 0 \\ 0 & -I_q \end{bmatrix}, \qquad G(\lambda) = \begin{bmatrix} I_p & R(\lambda) \\ 0 & I_q \end{bmatrix}. \tag{18.24}$$

Note that J is a $(p+q)\times(p+q)$ signature matrix.

The fact that R is stable implies that G and G^{-1} are analytic on the closed right half plane (infinity included), and hence the right-hand side of (18.23) is a left J-spectral factorization of W relative to $i\mathbb{R}$. In this section we shall show that the rational Nehari problem for R relative to the imaginary axis is solvable if and only if W admits a right J-spectral factorization of W relative to $i\mathbb{R}$ with an additional condition on the inverse of the spectral factor.

The first step is given by the next proposition. This proposition, which does not involve realizations and does not require R to be stable, will also provide one of the main steps in the proof of Theorem 18.1 which will be given in the next section.

Proposition 18.6. *Let R be a proper rational $p \times q$ matrix function, and consider the factorization $W(\lambda) = G(-\bar{\lambda})^* J G(\lambda)$, where J and G are defined by* (18.24).

Assume that W admits a right J-spectral factorization with respect to the imaginary axis, $W(\lambda) = L_+(-\bar{\lambda})^* J L_+(\lambda)$, with the additional property that the rational $q \times q$ matrix function in the right lower corner of $L_+^{-1}(\lambda)$ is biproper and its inverse is analytic on the closed left half plane. Then the rational Nehari problem for R relative to the imaginary axis is solvable. Moreover, all solutions can be obtained in the following way. Partition $L_+^{-1}(\lambda)$ as a 2×2 block matrix function,

$$L_+^{-1}(\lambda) = \begin{bmatrix} Y_{11}(\lambda) & Y_{12}(\lambda) \\ Y_{21}(\lambda) & Y_{22}(\lambda) \end{bmatrix}, \tag{18.25}$$

where $Y_{22}(\lambda)$ has size $q \times q$. Then all solutions K of the rational Nehari problem for R relative to the imaginary axis are given by

$$K(\lambda) = -\big(Y_{11}(\lambda)H(\lambda) + Y_{12}(\lambda)\big)\big(Y_{21}(\lambda)H(\lambda) + Y_{22}(\lambda)\big)^{-1}, \tag{18.26}$$

where H is any rational $p \times q$ matrix function which has all its poles in the open right half plane and satisfies $\|H\|_\infty < 1$. Finally, there is a one-to-one correspondence between the solution K and the free parameter H.

Proof. We shall apply the results of the previous section. Put

$$\Theta(\lambda) = \begin{bmatrix} I_p & R(\lambda) \\ 0 & I_q \end{bmatrix} L(\lambda)^{-1}.$$

Then $\Theta \in \mathrm{RAT}^{(p+q)\times(p+q)}$ and Θ is J-unitary on the imaginary axis. Introduce $V(\lambda) = L_+^{-1}(\lambda)$. Then

$$V = \begin{bmatrix} E_p & -R \\ 0 & E_q \end{bmatrix} \Theta,$$

and thus (18.16) is satisfied. From $V = L_+^{-1}$ and the properties of L_+ and L_+^{-1} we see that V satisfies all conditions necessary to apply Theorem 18.4. Thus

$$\mathcal{F}_V\big[\mathrm{RAT}^{p\times q}_{+,\mathbb{B}}\big] = \big\{ -K \in \mathrm{RAT}^{p\times q}_{+} \mid \|R-K\|_\infty < 1\big\}.$$

This proves that (18.26) indeed describes the set of all solutions of the rational Nehari problem for R relative to the imaginary axis. Since $\mathcal{F}_V$ is one-to-one on $\mathrm{RAT}^{p\times q}_{+,\mathbb{B}}$, by Theorem 18.3, we also obtain the one-to-one correspondence between the solutions K and the free parameter H. □

In Proposition 18.6 we have that W admits a J-spectral factorization $W(\lambda) = L_+(-\bar{\lambda})^* J L_+(\lambda)$ with the additional property that the $q \times q$ matrix function in the right lower corner of L_+^{-1} is biproper and has an analytic inverse on the closed left half plane. This property, which involves an inverse of a block of the inverse of L_+, can be replaced by the following more simple condition: the $p \times p$ matrix function in the left upper corner of L_+ is biproper and its inverse is analytic in the closed left half plane. To see this, write

$$L_+(\lambda) = \begin{bmatrix} L_{11}(\lambda) & L_{12}(\lambda) \\ L_{21}(\lambda) & L_{22}(\lambda) \end{bmatrix}, \qquad L_+^{-1}(\lambda) = \begin{bmatrix} X_{11}(\lambda) & X_{12}(\lambda) \\ X_{21}(\lambda) & X_{22}(\lambda) \end{bmatrix}.$$

A straightforward Schur complement argument gives that L_{11}^{-1} is analytic in the closed left half plane if and only if X_{22}^{-1} is analytic in the closed left half plane. Indeed, from Section 2.2 in [20] we have that

$$\begin{aligned} X_{22}^{-1}(\lambda) &= L_{22}(\lambda) - L_{21}(\lambda)L_{11}^{-1}(\lambda)L_{12}(\lambda), \\ L_{11}^{-1}(\lambda) &= X_{11}(\lambda) - X_{12}(\lambda)X_{22}^{-1}(\lambda)X_{21}(\lambda). \end{aligned}$$

This observation will be used in the final chapter to smoothen the phrasing of several theorems.

18.4 Proof of the main result

Proof of Theorem 18.1. We split the proof into five parts. Throughout this section we take $\gamma = 1$. As has been explained in the final paragraph of Section 18.1, this can be done without loss of generality. Furthermore, in what follows R is the strictly proper $p \times q$ rational matrix function given by formula (18.2).

Part 1. Let K be a solution of the rational Nehari problem for R relative to the imaginary axis. Define F to be the $p \times q$ rational matrix function on $i\mathbb{R}$ given by $F(i\lambda) = K(i\lambda) - R(i\lambda)$. Note that F is continuous on the imaginary axis, $\lim_{\lambda\in\mathbb{R},\,|\lambda|\to\infty} F(i\lambda)$ exists and is equal to a $p \times q$ matrix D, say. Furthermore,

$$\|F\|_\infty = \sup_{\lambda\in\mathbb{R}} \|F(i\lambda)\| < 1.$$

Now, since K is analytic, the Hankel operator generated by F is equal to the Hankel operator generated by $-R$, that is, $H_F = H_{K-R} = H_R$ and $\|H_F\| < 1$ (see, e.g., Section XII.2 in [51]). So $\|H_R\| < 1$, and hence, by Lemma 18.2, the matrix $I - P^{1/2}QP^{1/2}$ is positive definite.

In the remaining Parts 2–5 of the proof it is assumed that $I - P^{1/2}QP^{1/2}$ is positive definite. We show that under this condition the Nehari problem is solvable and we derive all its solutions. The main work is done in Parts 3 and 4. Part 2 has a preliminary character, and in Part 5 we finish the proof by applying Proposition 18.6.

Part 2. As a first step we show that $I - P^{1/2}QP^{1/2}$ is positive definite implies that $I - Q^{1/2}PQ^{1/2}$ is positive definite too. To see this, we argue as follows. Introduce $T = Q^{1/2}P^{1/2}$. Clearly $I - T^*T$ is positive definite, and hence T is a strict contraction (i.e., $\|T\| < 1$). But then so is $T^* = P^{1/2}Q^{1/2}$. Thus, as desired, $I - Q^{1/2}PQ^{1/2}$ is positive definite.

Next, put $K = Z^{-1}Q$, where $Z = I - QP$ while Q and P are the unique solutions to the Lyapunov equations (18.3). Note that Z is invertible, because the matrix $I - P^{1/2}QP^{1/2}$ is positive definite. We claim that K is nonnegative and that the following identity holds:

$$KA + A^*K = KBB^*K - Z^{-1}C^*CZ^{-*}. \tag{18.27}$$

To prove that K is nonnegative , we use

$$ZQ^{1/2} = (I - QP)Q^{1/2} = Q^{1/2}(I - Q^{1/2}PQ^{1/2}).$$

This yields $Z^{-1}Q^{1/2} = Q^{1/2}(I - Q^{1/2}PQ^{1/2})^{-1}$, and hence

$$K = Z^{-1}Q = Q^{1/2}(I - Q^{1/2}PQ^{1/2})^{-1}Q^{1/2} \geq 0. \tag{18.28}$$

To prove (18.27) we first multiply the second identity in (18.3) from the left by Z^{-1} and from the right by Z^{-*}. Using $K = Z^{-1}Q = QZ^{-*}$, this yields

$$KAZ^{-*} + Z^{-1}A^*K = -Z^{-1}C^*CZ^{-*}.$$

Now observe that

$$\begin{aligned} KAZ^{-*} &= KA(I - PQ)^{-1} = KA\Big(I + P(I - QP)^{-1}Q\Big) \\ &= KA + KAPZ^{-1}Q = KA + KAPK. \end{aligned}$$

But then, taking advantage of the first identity in (18.3) , we obtain

$$\begin{aligned} KAZ^{-*} + Z^{-1}A^*K &= KA + A^*K + K(AP + A^*P)K \\ &= KA + A^*K - KBB^*K. \end{aligned}$$

Thus

$$\begin{aligned} KA + A^*K &= KAZ^{-*} + Z^{-1}A^*K + KBB^*K \\ &= KBB^*K - Z^{-1}C^*CZ^{-*}, \end{aligned}$$

which proves (18.27).

Part 3. Put $W(\lambda) = G(-\bar{\lambda})^*JG(\lambda)$, where J and G are defined by (18.24). It was already observed that this factorization is a left J-spectral factorization with respect to $i\mathbb{R}$. In this part we prove that W also admits a right J-spectral factorization with respect to $i\mathbb{R}$. To do this we use that $I - P^{1/2}QP^{1/2}$ is positive definite and apply Theorem 14.14 with $L_-(\lambda) = G(\lambda)$.

Employing the realization (18.2) of R, one gets

$$L_-(\lambda) = \begin{bmatrix} I_p & R(\lambda) \\ 0 & I_q \end{bmatrix} = \begin{bmatrix} I_p & 0 \\ 0 & I_q \end{bmatrix} + \begin{bmatrix} C \\ 0 \end{bmatrix} (\lambda - A)^{-1} \begin{bmatrix} 0 & B \end{bmatrix}.$$

So, with

$$\widehat{A} = A, \qquad \widehat{B} = \begin{bmatrix} 0 & B \end{bmatrix}, \qquad \widehat{C} = \begin{bmatrix} C \\ 0 \end{bmatrix},$$

we have $L_-(\lambda) = I_{p+q} + \widehat{C}(\lambda - \widehat{A})^{-1}\widehat{B}$, and the associate main matrix of this realization $\widehat{A}^\times = \widehat{A} - \widehat{B}\widehat{C}$ obviously coincides with the main matrix $\widehat{A} = A$.

For the realization considered here we denote by $\widehat{P}$ and $\widehat{Q}$ the solutions of the equations (14.53) and (14.52), respectively. In other words, $\widehat{P}$ and $\widehat{Q}$ are the unique solutions of

$$A\widehat{P} + \widehat{P}A^* = BB^*, \qquad A^*\widehat{Q} + \widehat{Q}A = C^*C.$$

So $\widehat{P} = -P$ and $\widehat{Q} = -Q$. It follows that $I - \widehat{P}\widehat{Q} = I - PQ$, and therefore (ii) implies that $I - \widehat{P}\widehat{Q} = (I - PQ) = P^{1/2}(I - P^{1/2}QP^{1/2})P^{-1/2}$ is invertible. Hence $I - \widehat{Q}\widehat{P}$ is invertible too.

Thus by Theorem 14.14 the rational $(p+q) \times (p+q)$ matrix function W admits a right J-spectral factorization, $W(\lambda) = L_+(-\bar{\lambda})^* J L_+(\lambda)$, with respect to $i\mathbb{R}$. In fact, for L_+ one can take

$$L_+(\lambda) = \begin{bmatrix} I_p & 0 \\ 0 & I_q \end{bmatrix} + \begin{bmatrix} -CP \\ B^* \end{bmatrix} Z^{-1}(\lambda + A^*)^{-1} \begin{bmatrix} C^* & QB \end{bmatrix}, \tag{18.29}$$

where $Z = I - QP$. Theorem 14.14 also tells us that for this choice of the right J-spectral factor L_+ we have

$$L_+^{-1}(\lambda) = \begin{bmatrix} I_p & 0 \\ 0 & I_q \end{bmatrix} - \begin{bmatrix} -CP \\ B^* \end{bmatrix} (\lambda + A^*)^{-1} Z^{-1} \begin{bmatrix} C^* & QB \end{bmatrix}, \tag{18.30}$$

where, as before, $Z = I - QP$.

Now partition $L_+^{-1}(\lambda)$ as

$$V(\lambda) = L_+^{-1}(\lambda) = \begin{bmatrix} X_{11}(\lambda) & X_{12}(\lambda) \\ X_{21}(\lambda) & X_{22}(\lambda) \end{bmatrix}, \tag{18.31}$$

where the block in the right lower corner has size $q \times q$. Comparing (18.30) and (18.31) we see that the rational matrix functions X_{ij}, $i, j = 1, 2$, are precisely the functions given by (18.4)–(18.7).

Part 4. In this part, again assuming $I - P^{1/2}QP^{1/2}$ to be positive definite, we show that the $q \times q$ rational matrix function $X_{22}(\lambda)$ in the right lower corner of the block matrix in (18.31) has precisely the properties which will allow us to apply Proposition 18.6.

Obviously, X_{22} is biproper. Since the eigenvalues of A are in the open left half plane, those of $-A^*$ are in the open right half plane as well, and hence X_{22} is analytic on the closed left half plane. It remains to show that X_{22}^{-1} is also analytic on the closed left half plane. From the expression for $X_{22}(\lambda)$ we see that

$$X_{22}^{-1}(\lambda) = I + B^*(\lambda - A_0)^{-1}Z^{-1}QB,$$

where $A_0 = -A^* + Z^{-1}QBB^* = -A^* + KBB^*$, with K as in Part 2 of the present proof. Thus, in order to show that X_{22}^{-1} is analytic on the closed left half plane, it suffices to show that A_0 has all its eigenvalues in the open right half plane.

To determine the location of the eigenvalues of A_0 we first prove that

$$A_0K + KA_0^* = KBB^*K + Z^{-1}C^*CZ^{-*}. \tag{18.32}$$

This identity follows from (18.27). Indeed, using the definition of A_0, we have

$$A_0K = (-A^* + KBB^*)K = -A^*K + KBB^*K.$$

But then, using (18.27), we see that

$$A_0K + KA_0^* = -A^*K - KA + 2KBB^*K = KBB^*K + Z^{-1}C^*CZ^{-*},$$

which proves (18.32).

The identity (18.32) implies that A_0 does not have pure imaginary eigenvalues. Indeed, suppose A_0 has a pure imaginary eigenvalue. Then the same holds true for A_0^*, that is, there is a pure imaginary λ_0 and a non-zero vector x such that $A_0^*x = \lambda_0 x$. This implies $x^*A_0 = -\lambda_0 x^*$, and hence $x^*(A_0K + KA_0^*)x = 0$. From (18.32) it then follows that $x^*KBB^*Kx = 0$. In other words, $x^*KB = 0$. Using the definition of A_0, we see that

$$-\lambda_0 x^* = x^*A_0 = -x^*A^* + x^*KBB^* = -x^*A^*.$$

We conclude that A^* has a pure imaginary eigenvalue which is impossible because by assumption A (and hence A^* too) has all its eigenvalues in the open left half plane. Thus a contradiction has been obtained, and we conclude that A_0 has no pure imaginary eigenvalue.

It remains to show that A_0 has no eigenvalues in the open left half plane. If K would be invertible, then K would be positive definite, and the statement that A_0 has no eigenvalues in the open left half plane would now follow immediately from $A_0K + KA_0^* \geq 0$ and the classical Carlson-Schneider inertia theorem (see Theorem 13.1.3 in [107]). However since K may not be invertible an additional argument is required, which will be presented in the next two paragraphs.

Let n be the order of the square matrix A. Note that K, Q, and Z are also square matrices of order n. Put $\mathcal{X}_1 = \operatorname{Im} K$ and $\mathcal{X}_2 = \operatorname{Ker} K$. Since K is selfadjoint, we have the orthogonal direct sum decomposition $\mathbb{C}^n = \mathcal{X}_1 \oplus \mathcal{X}_2$. The identity $K = Z^{-1}Q$ implies that $\operatorname{Ker} Q = \operatorname{Ker} K$. Hence, by selfadjointness, $\operatorname{Im} Q = \operatorname{Im} K$. It follows that relative to the decomposition $\mathbb{C}^n = \mathcal{X}_1 \oplus \mathcal{X}_2$ the matrices K and Q admit the following 2×2 block matrix representation:

$$K = \begin{bmatrix} K_1 & 0 \\ 0 & 0 \end{bmatrix}, \qquad Q = \begin{bmatrix} Q_1 & 0 \\ 0 & 0 \end{bmatrix},$$

where both K_1 and Q_1 are positive definite. Next, we partition A, B, and C relative to the decomposition $\mathbb{C}^n = \mathcal{X}_1 \oplus \mathcal{X}_2$. This yields

$$A = \begin{bmatrix} A_{11} & 0 \\ A_{21} & A_{22} \end{bmatrix}, \qquad B = \begin{bmatrix} B_1 \\ B_2 \end{bmatrix}, \qquad C = \begin{bmatrix} C_1 & 0 \end{bmatrix}.$$

Here we used that $\mathcal{X}_2 = \operatorname{Ker} Q = \operatorname{Ker}(C|A)$, which implies that $\mathcal{X}_2$ is A-invariant and that C is zero on $\mathcal{X}_2$. From $ZK = Q$, we see that $Z[\operatorname{Im} K] = \operatorname{Im} Q$, and hence $Z[\mathcal{X}_1] = \mathcal{X}_1$. Thus. relative to $\mathbb{C}^n = \mathcal{X}_1 \oplus \mathcal{X}_2$, the matrix Z partitions as

$$Z = \begin{bmatrix} Z_{11} & Z_{12} \\ 0 & Z_{22} \end{bmatrix},$$

where both Z_{11} and Z_{22} are invertible. Employing the block matrix representations for K, A and B we compute A_0. We have

$$A_0 = -\begin{bmatrix} A_{11}^* & A_{21}^* \\ 0 & A_{22}^* \end{bmatrix} + \begin{bmatrix} K_1 & 0 \\ 0 & 0 \end{bmatrix} \begin{bmatrix} B_1B_1^* & B_1B_2^* \\ B_2B_1^* & B_2B_2^* \end{bmatrix}.$$

Thus A_0 has the form

$$A_0 = \begin{bmatrix} A_{0,11} & \star \\ 0 & A_{0,22} \end{bmatrix},$$

where $A_{0,11} = -A_{11}^* + K_1B_1B_1^*$ and $A_{0,22} = A_{22}^*$. Since A has all its eigenvalues in the open left half plane, the same holds true for A_{22}. Hence $A_{0,22}$ has all its eigenvalues in the open right half plane. Thus, in order to prove that A_0 has all its eigenvalues in the open right half plane, it suffices to show that $A_{0,11}$ has this property. This will be done in the next paragraph.

Since A_0 has no pure imaginary eigenvalue, the same holds true for $A_{0,11}$. From (18.32), using the block matrices in the previous paragraph, we see that $A_{0,11}K_1 + K_1A_{0,11}^* \geq 0$. As K_1 is positive definite we can now apply the Carlson-Schneider inertia theorem (i.e., Theorem 13.1.3 in [107]) to show that the inertia of $A_{0,11}$ is equal to the inertia of K_1. Using again that K_1 is positive definite, it follows that all the eigenvalues of $A_{0,11}$ are in the open right half plane, as desired.

Part 5. We are now ready to complete the proof. Assume $I - P^{1/2}QP^{1/2}$ is positive definite. By the previous two parts of the proof, the rational matrix function

$W(\lambda) = G(-\bar{\lambda})^* J G(\lambda)$ admits a right J-spectral factorization with respect to the imaginary axis, written $W(\lambda) = L_+(-\bar{\lambda})^* J L_+(\lambda)$, with the additional property that the $q \times q$ matrix function in the right lower corner of $L_+^{-1}(\lambda)$ is biproper and its inverse is analytic on the closed left half plane. It was also shown that $L_+^{-1}(\lambda)$ partitions as

$$L_+^{-1}(\lambda) = \begin{bmatrix} X_{11}(\lambda) & X_{12}(\lambda) \\ X_{21}(\lambda) & X_{22}(\lambda) \end{bmatrix},$$

where the rational matrix functions X_{ij}, $i, j = 1, 2$, are precisely the functions given by (18.4)–(18.7). But then we can apply Proposition 18.6 to get the desired description of all solutions. □

18.5 The case of a non-stable given function

In this section we return to the general case, where the rational $p \times q$ matrix function R is not necessarily stable, i.e., does not necessarily have all its poles in the open left half plane. Throughout we assume R to be proper and to have no poles on the imaginary axis. Write $R = R_- + R_+$, where R_- is a stable rational $p \times q$ matrix function which is strictly proper, and R_+ is a proper rational $p \times q$ matrix function which has all its poles in the open right half plane. The required location of the poles determines R_- and R_+ uniquely. Recall that we seek proper rational $p \times q$ matrix functions K such that K has all its poles in the open right half plane and

$$\|R - K\|_\infty = \|R_- - (K - R_+)\|_\infty < \gamma. \tag{18.33}$$

The second term in (18.33) gives us a hint of how to solve the Nehari problem for R. In fact, from (18.33) we see that K is a solution to the Nehari problem with tolerance γ for R if and only if $K - R_+$ is a solution to the Nehari problem with tolerance γ for R_-. This remark allows us to extend Theorem 18.1 to the case when the given function R is non-stable.

To describe the resulting theorem, we shall assume that R_- and R_+ are given in the form

$$R_-(\lambda) = C_-(\lambda I_n - A_-)^{-1} B_-, \qquad R_+(\lambda) = D + C_+(\lambda I_n - A_+)^{-1} B_+, \tag{18.34}$$

where A_- has all its eigenvalues in the open left half plane, and A_+ has all its eigenvalues in the open right half plane. In the situation where the realizations in (18.34) are minimal, these conditions on the location of the spectra of A_- and A_+ are automatically fulfilled. Put

$$P_- = \int_0^\infty e^{\tau A_-} B_- B_-^* e^{\tau A_-^*}\, d\tau, \qquad Q_- = \int_0^\infty e^{\tau A_-^*} C_-^* C_- e^{\tau A_-}\, d\tau. \tag{18.35}$$

Note that P_- and Q_- are well-defined because all the eigenvalues of A_- are in the open left half plane. The following theorem is the main result of this section.

Theorem 18.7. *Let $R = R_- + R_+$ with R_- and R_+ being given by (18.34). Assume A_- and A_+ have all their eigenvalues in the open left and open right half plane, respectively, and let P_- and Q_- be given by (18.35). Then the rational Nehari problem for R relative to the imaginary axis with tolerance γ is solvable if and only if the matrix $\gamma^2 I_n - P_-^{1/2} Q_- P_-^{1/2}$ is positive definite. In this case the matrix $Z_- = \gamma^2 I_n - P_- Q_-$ is invertible and all solutions of the Nehari problem under consideration can be obtained in the following way. Introduce rational matrix functions Y_{ij}, $i, j = 1, 2$, by setting*

$$\begin{bmatrix} Y_{11}(\lambda) & Y_{12}(\lambda) \\ Y_{21}(\lambda) & Y_{22}(\lambda) \end{bmatrix} = \begin{bmatrix} I_p & -D \\ 0 & I_q \end{bmatrix} + \begin{bmatrix} -C_+ & C_-P_- + DB_-^* \\ 0 & -B_-^* \end{bmatrix} \tag{18.36}$$
$$\cdot \left(\lambda I_{2n} - \begin{bmatrix} A_+ & -B_+B_-^* \\ 0 & -A_-^* \end{bmatrix} \right)^{-1} \begin{bmatrix} 0 & B_+ \\ Z_-^{-1}C_-^* & Z_-^{-1}Q_-B_- \end{bmatrix}.$$

Then all solutions K to the rational Nehari problem for R relative to the imaginary axis with tolerance γ are given by

$$K(\lambda) = -\big(Y_{11}(\lambda)H(\lambda) + Y_{12}(\lambda)\big)\big(Y_{21}(\lambda)H(\lambda) + Y_{22}(\lambda)\big)^{-1}, \tag{18.37}$$

where H is any rational $p \times q$ matrix function which has all its poles in the open right half plane and satisfies $\|H\|_\infty < \gamma$. Moreover, there is a one-to-one correspondence between the solution K and the free parameter H.

Proof. From Theorem 18.1 we know that the Nehari problem with tolerance γ for R_- is solvable if and only if the matrix $\gamma^2 I - P_-^{1/2} Q_- P_-^{1/2}$ is positive definite. On the other hand, we also know (see the second paragraph of this section) that the Nehari problem with tolerance γ for R is solvable if and only if the Nehari problem with tolerance γ for R_+ is solvable. These two "if and only if" statements together yield the first part of the theorem.

Next, assume that the matrix $\gamma^2 I - P_-^{1/2} Q_- P_-^{1/2}$ is positive definite. As we have already seen, K is a solution to the Nehari problem with tolerance γ for R if and only if K is of the form $\widetilde{K} + R_+$, where $\widetilde{K}$ is an arbitrary solution to the Nehari problem with tolerance γ for R_-. By Theorem 18.1, applied to R_- in place of R, the latter solutions are given by

$$\widetilde{K}(\lambda) = -\big(X_{11}(\lambda)H(\lambda) + X_{12}(\lambda)\big)\big(X_{21}(\lambda)H(\lambda) + X_{22}(\lambda)\big)^{-1},$$

with the coefficients in this linear fractional representation given by

$$\begin{aligned} X_{11}(\lambda) &= I_p + C_-P_-(\lambda + A_-^*)^{-1}Z_-^{-1}C_-^*, \\ X_{12}(\lambda) &= C_-P_-(\lambda + A_-^*)^{-1}Z_-^{-1}Q_-B_-, \\ X_{21}(\lambda) &= -B_-^*(\lambda + A_-^*)^{-1}Z_-^{-1}C_-^*, \\ X_{22}(\lambda) &= I_q - B_-^*(\lambda + A_-^*)^{-1}Z_-^{-1}Q_-B_-, \end{aligned}$$

where $Z_- = \gamma^2 I - P_- Q_-$, which is invertible. It follows that

$$\begin{aligned}
K(\lambda) &= R_+(\lambda) + \widetilde{K}(\lambda) \\
&= R_+(\lambda) - \big(X_{11}(\lambda)H(\lambda) + X_{12}(\lambda)\big)\big(X_{21}(\lambda)H(\lambda) + X_{22}(\lambda)\big)^{-1} \\
&= R_+(\lambda)\big(X_{21}(\lambda)H(\lambda) + X_{22}(\lambda)\big)\big(X_{21}(\lambda)H(\lambda) + X_{22}(\lambda)\big)^{-1} \\
&\qquad - \big(X_{11}(\lambda)H(\lambda) + X_{12}(\lambda)\big)\big(X_{21}(\lambda)H(\lambda) + X_{22}(\lambda)\big)^{-1} \\
&= -\Big(\big(X_{11}(\lambda) - R_+(\lambda)X_{21}(\lambda)\big)H(\lambda) + \big(X_{12}(\lambda) - R_+(\lambda)X_{22}(\lambda)\big)\Big) \\
&\qquad .\big(X_{21}(\lambda)H(\lambda) + X_{22}(\lambda)\big)^{-1} \\
&= -\big(Y_{11}(\lambda)H(\lambda) + Y_{12}(\lambda)\big)\big(Y_{21}(\lambda)H(\lambda) + Y_{22}(\lambda)\big)^{-1},
\end{aligned}$$

where H is any rational $p \times q$ matrix function having all its poles in the open right half plane and satisfies $\|H\|_\infty < \gamma$. Moreover, the coefficient matrix

$$Y(\lambda) = \begin{bmatrix} Y_{11}(\lambda) & Y_{12}(\lambda) \\ Y_{21}(\lambda) & Y_{22}(\lambda) \end{bmatrix}$$

is given by

$$Y(\lambda) = \begin{bmatrix} I_p & -R_+(\lambda) \\ 0 & I_q \end{bmatrix} \begin{bmatrix} X_{11}(\lambda) & X_{12}(\lambda) \\ X_{21}(\lambda) & X_{22}(\lambda) \end{bmatrix}.$$

Now, using the formulas for X_{ij}, $i, j = 1, 2$, one gets

$$\begin{bmatrix} X_{11}(\lambda) & X_{12}(\lambda) \\ X_{21}(\lambda) & X_{22}(\lambda) \end{bmatrix} = \begin{bmatrix} I_p & 0 \\ 0 & I_q \end{bmatrix} + \begin{bmatrix} C_- P_- \\ -B_-^* \end{bmatrix} (\lambda + A_-^*)^{-1} \begin{bmatrix} Z_-^{-1} C_-^* & Z_-^{-1} Q_- B_- \end{bmatrix}.$$

Furthermore, employing the realization of R_+,

$$\begin{bmatrix} I_p & -R_+(\lambda) \\ 0 & I_q \end{bmatrix} = \begin{bmatrix} I_p & -D \\ 0 & I_q \end{bmatrix} + \begin{bmatrix} -C_+ \\ 0 \end{bmatrix} (\lambda - A_+)^{-1} \begin{bmatrix} 0 & B_+ \end{bmatrix}.$$

Taking the product of these realizations (see Theorem 2.5) we reach the conclusion that the coefficient matrix $Y(\lambda)$ admits the desired realization (18.36).

The fact that there is one-to-one correspondence between the solution K and the free parameter H in (18.37) follows directly from the corresponding result in Theorem 18.1. □

18.6 The Nehari-Takagi problem

In the Nehari-Takagi problem the given function R is the same as in the Nehari problem. However the solutions K are allowed to come from a wider class. To be more more specific, let the rational $p \times q$ matrix function R be as in the first paragraph of Section 18.1. Thus R is proper and does not have a pole on the imaginary axis. Let κ be a non-negative integer. Then the (*rational*) *Nehari-Takagi problem* (*relative to the imaginary axis*) is the problem of finding all proper rational $p \times q$ matrix functions K such that K has no pole on the imaginary axis and at most κ poles in the open left half plane (multiplicities taken into account), and

$$\|K - R\|_\infty = \sup_{s \in i\mathbb{R}} \|K(s) - R(s)\| < \gamma, \tag{18.38}$$

where γ is a pre-specified positive number. When $\kappa = 0$, the conditions on K reduce to the requirement that K has all its poles in the open right half plane. Thus with $\kappa = 0$ the Nehari-Takagi problem is just the Nehari problem considered in the preceding sections.

In this section we take $\gamma = 1$, which can be done without loss of generality (cf., the last paragraph of Section 18.1), and we assume that R is strictly proper and stable. Thus R admits a realization $R(\lambda) = C(\lambda I_n - A)^{-1}B$ where A has all its eigenvalues in the open left half plane. The following result is the analogue of Theorem 18.1 for the Nehari-Takagi problem.

Theorem 18.8. *Let* $(\lambda) = C(\lambda I_n - A)^{-1}B$ *be a realization of the rational* $p \times q$ *matrix function* R*, assume* A *has all its eigenvalues in the open left half plane, and let*

$$P = \int_0^\infty e^{sA}BB^*e^{sA^*}\,ds, \qquad Q = \int_0^\infty e^{sA^*}C^*Ce^{sA}\,ds$$

(*i.e.,* P *and* Q *are the controllability and observability gramians corresponding to the given realization*)*. Suppose* $I_n - PQ$ *is invertible. Then the rational Nehari-Takagi problem for* R *relative to the imaginary axis with* $\gamma = 1$ *is solvable if and only if the matrix* PQ *has at most* κ *eigenvalues* (*multiplicities taken into account*) *larger than* 1*. Moreover, if* κ_0 *is the number of eigenvalues of* PQ *larger than* 1*, then all solutions* K *of the Nehari-Takagi problem for* R *relative to the imaginary axis with* $\gamma = 1$ *such that* K *has precisely* κ_0 *poles in the open left half plane are given by the linear fractional formula*

$$K(\lambda) = -\big(\Theta_{11}(\lambda)G(\lambda) + \Theta_{12}(\lambda)\big)\big(\Theta_{21}(\lambda)G(\lambda) + \Theta_{22}(\lambda)\big)^{-1}. \tag{18.39}$$

Here the free parameter G *is an arbitrary rational* $p \times q$ *matrix function which has all its poles in the open right half plane and* $\|G\|_\infty < 1$*. Furthermore, the coefficients* Θ_{ij}*,* $i, j = 1, 2$*, are given by*

$$\begin{aligned}
\Theta_{11}(\lambda) &= I_p + CP(\lambda I_n + A^*)^{-1}(I_n - QP)^{-1}C^*, \\
\Theta_{12}(\lambda) &= CP(\lambda I_n + A^*)^{-1}(I_n - QP)^{-1}QB, \\
\Theta_{21}(\lambda) &= -B^*(\lambda I_n + A^*)^{-1}(I_n - QP)^{-1}C^*, \\
\Theta_{22}(\lambda) &= I_q - B^*(\lambda I_n + A^*)^{-1}(I_n - QP)^{-1}QB.
\end{aligned}$$

To prove the above theorem one can follow the same line of reasoning as used in this chapter to prove Theorem 18.1. The role of Theorem 18.4 has to be taken over by Theorem 18.5. For further details we refer to the literature; see for example [86] and the references therein.

Notes

The Nehari problem has its roots in the classical papers of Nehari [114] and Adamjan-Arov-Krein [1], [2]. The rational matrix version played an important role in the early development of H-infinity control theory; see, e.g., the lecture notes [43]. Here one already finds the J-spectral factorization approach. For an overview of the various methods to deal with the matrix Nehari problem we refer to the notes to Chapter 20 in [7]. The Takagi version of the Nehari problem has its roots in [142]. The result with a full proof can also be found in Section 20.5 of [7]. For an abstract approach to the Nehari-Takagi problem, covering applications to time-invariant infinite-dimensional systems and time-varying finite-dimensional linear systems, we refer to [86].

Chapter 19

Review of some control theory for linear systems

In this chapter a brief survey is given of a number of basic elements of control and mathematical systems theory. The main aim is to give the reader some understanding for the type of problems that will be treated in the final chapter.

The chapter consists of two sections. Section 19.1 introduces the concepts of stability of systems and the method of feedback to stabilize a system. Section 19.2 deals with the notion of internal stability of a closed loop system. In particular the Youla-Jabr-Bongiorno parametrization of all stabilizing compensators is presented.

19.1 Stability and feedback

In this section we consider a causal input-output system Σ as in the figure below:

As usual (cf., Section 2.1) the symbol u denotes the input and y the output. Mathematically input and output are vector-valued functions of a (time) parameter t. Such an input-output system is called *externally stable* or *bounded-input bounded-output stable* (BIBO-stable) if a bounded input u produces a bounded output y, that is, $\sup_{t\geq 0} \|u(t)\| < \infty$ implies $\sup_{t\geq 0} \|y(t)\| < \infty$.

Now let us assume that Σ is a causal linear time invariant system given by the following finite dimensional state space representation:

$$\begin{cases} x'(t) &= Ax(t) + Bu(t), \\ y(t) &= Cx(t) + Du(t), \quad t \geq 0. \end{cases} \tag{19.1}$$

Here A, B, C, D are matrices of appropriate sizes, and A is a square matrix. We refer to (19.1) as a *realization* of the system. The realization (19.1) is called *stable* if for any initial value $x(0)$, with zero input u, the state $x(t)$ will go to zero if $t \to \infty$. It is easily seen that stability of the realization (19.1) is equivalent to the requirement that the matrix A has all its eigenvalues in the open left half plane. If the latter holds, A is said to be a *stable matrix*.

Given (19.1) the effect of inputs on outputs can be described in the time domain by a lower triangular integral operator

$$y(t) = Ce^{tA}x(0) + \int_0^t k(t-s)u(s)\,ds + Du(t), \tag{19.2}$$

where $k(t)$ is the so-called impulse response function. As we have already seen in Section 2.1, in the frequency domain with $x(0) = 0$ the connection between input and output is given by $\widehat{y}(\lambda) = W(\lambda)\widehat{u}(\lambda)$, where W is the transfer function of the system, and $\widehat{u}$ and $\widehat{y}$ denote the Laplace transforms of the input u and the output y, respectively. In terms of (19.1) we have

$$k(t) = Ce^{tA}B, \qquad W(\lambda) = D + C(\lambda - A)^{-1}B. \tag{19.3}$$

From (19.2) and the first identity in (19.3) it is clear that stability of the realization (19.1) implies external stability of the corresponding system. The converse is also true when the realization is minimal, that is, when the pair (A, B) is controllable and the pair (C, A) is observable. We summarize this and related results in the following theorems.

Theorem 19.1. *Let (19.1) be a minimal realization, then the corresponding system is externally stable if and only if the realization is stable.*

Theorem 19.2. *Let k be the impulse response function and let W be the transfer function of the linear time invariant system given by (19.1). The following statements are equivalent:*

1. *The system given by (19.1) is externally stable;*
2. $\int_0^\infty \|k(t)\|\,dt < \infty$;
3. *The rational matrix function W is $i\mathbb{R}$-stable, that is, W has all its poles in the open left half plane.*

An important issue is stabilizing an unstable system. The simplest method is that of static state feedback. To explain this method consider the system given by the state space representation:

$$\begin{cases} x'(t) &= Ax(t) + Bu(t), \\ y(t) &= x(t), \qquad t \geq 0. \end{cases}$$

Note that the output is equal to the state. This case is sometimes referred to as the full information case. The problem is to find a static feedback control law $u(t) = Fx(t) + v(t)$ that will make the system sending v to x stable. That is, to find a matrix F of appropriate size such that

$$x'(t) = (A + BF)x(t) + Bv(t)$$

is stable. This amounts to requiring that the matrix $A + BF$ is stable, i.e., all its eigenvalues are in the open left half plane. For such a matrix F to exist the pair (A, B) should be stabilizable in the sense of Section 13.2. Two questions appear: first, when is a pair of matrices (A, B) stabilizable, and second, how to construct a stabilizing matrix F?

We start with an observation concerning the so-called single input case. In that situation, the matrix B is an $n \times 1$ vector, and one may assume without loss of generality that A and B have the form

$$A = \begin{bmatrix} 0 & 1 & & \\ & & \ddots & \\ & & & 1 \\ -a_n & \cdots & \cdots & -a_1 \end{bmatrix}, \qquad B = \begin{bmatrix} 0 \\ \vdots \\ 0 \\ 1 \end{bmatrix}.$$

Consider $F = [\, f_n \ \cdots \ f_1 \,]$. Then

$$A + BF = \begin{bmatrix} 0 & 1 & & \\ & & \ddots & \\ & & & 1 \\ f_n - a_n & \cdots & \cdots & f_1 - a_1 \end{bmatrix}.$$

So, in this case, *any* polynomial can be obtained as the characteristic polynomial of $A + BF$ by an appropriate choice of F.

Next we make a second observation. Let A be an $n \times n$ matrix, let B be an $n \times m$ matrix, and write $\mathbb{C}^n = \operatorname{Im}(A|B) \dotplus X_0$. With respect to this direct sum decomposition, the matrices A and B can be written as

$$A = \begin{bmatrix} A_{11} & A_{10} \\ 0 & A_{00} \end{bmatrix}, \qquad B = \begin{bmatrix} B_1 \\ 0 \end{bmatrix},$$

with (A_{11}, B_1) controllable. Thus, for any $m \times n$ matrix $F = \begin{bmatrix} F_1 & F_0 \end{bmatrix}$, one has

$$A + BF = \begin{bmatrix} A_{11} + B_1F_1 & A_{10} + B_1F_0 \\ 0 & A_{00} \end{bmatrix},$$

and hence $\sigma(A+BF) = \sigma(A_{11}+B_1F_1) \cup \sigma(A_{00})$. Note that $\sigma(A_{00})$, the second part in the right-hand side of the preceding identity, is independent of the particular choice of X_0 and also of the choice of F. Therefore the eigenvalues of A_{00} are called the *uncontrollable eigenvalues of A relative to the matrix B*. Clearly, A has no uncontrollable eigenvalues relative to B if and only if the pair (A, B) is controllable.

From the discussion in the previous paragraph we conclude that, in order for (A, B) to be stabilizable, it is necessary that the uncontrollable eigenvalues of A relative to B are in the open left half plane. The converse of this observation would follow if any controllable pair is stabilizable. This is the case for single input as we have already seen. That it is true in general appears from the next result which is actually quite a bit stronger, and is known as the *pole placement theorem*.

Theorem 19.3. *Let A be an $n \times n$ matrix, and let B be an $n \times m$ matrix. The following two statements are equivalent:*

(i) *The pair (A, B) is controllable;*

(ii) *For any scalar polynomial $p(\lambda) = \lambda^n + p_1\lambda^{n-1} + \cdots + p_{n-1}\lambda + p_n$, there is an $m \times n$ matrix F such that the characteristic polynomial of $A + BF$ coincides with p.*

Corollary 19.4. *Let A be an $n \times n$ matrix and let B be an $n \times m$ matrix. The pair (A, B) is stabilizable if and only if the uncontrollable eigenvalues of A relative to the matrix B are in the open left half plane.*

Let A be an $n \times n$ matrix and let C be an $m \times n$ matrix. The pair (C, A) is called *detectable* when there exists an $n \times m$ matrix R such that $A - RC$ is stable. In other words the pair (C, A) is detectable if and only if the pair (A^*, C^*) is stabilizable. By definition the *unobservable eigenvalues of A relative to C* are the uncontrollable eigenvalues of A^* relative to C^*. It is also possible to give a direct definition of the latter notion, involving a decomposition of the type $\mathbb{C}^n = \operatorname{Ker}(C|A) \dotplus X_0$. From the above definitions and Corollary 19.4 it is clear that the pair (C, A) is detectable if and only if the unobservable eigenvalues of A relative to the matrix C are in the open left half plane.

19.2 Parametrization of internally stabilizing compensators

In this section G is the transfer function of a system Σ with two inputs u and w, and two outputs y and z. Here u is the control input, w a disturbance, y is the output which can be measured and z is the output to be controlled. Throughout, we shall assume that the system Σ is given in state space form as follows:

$$\left\{\begin{array}{rcl} x'(t) & = & Ax(t) + B_1 w(t) + B_2 u(t), \\ z(t) & = & C_1 x(t) + D_1 u(t), \\ y(t) & = & C_2 x(t) + D_2 w(t), \qquad t \geq 0. \end{array}\right. \tag{19.4}$$

It will be convenient to rewrite the realization (19.4) in the form

$$\left\{\begin{array}{rcl} x'(t) & = & Ax(t) + \begin{bmatrix} B_1 & B_2 \end{bmatrix} \begin{bmatrix} w(t) \\ u(t) \end{bmatrix}, \\ \begin{bmatrix} z(t) \\ y(t) \end{bmatrix} & = & \begin{bmatrix} C_1 \\ C_2 \end{bmatrix} x(t) + \begin{bmatrix} 0 & D_1 \\ D_2 & 0 \end{bmatrix} \begin{bmatrix} w(t) \\ u(t) \end{bmatrix}. \end{array}\right.$$

From the latter representation we see that the transfer function of (19.4) is given by

$$G(\lambda) = \begin{bmatrix} G_{11}(\lambda) & G_{12}(\lambda) \\ G_{21}(\lambda) & G_{22}(\lambda) \end{bmatrix} = \begin{bmatrix} 0 & D_1 \\ D_2 & 0 \end{bmatrix} + \begin{bmatrix} C_1 \\ C_2 \end{bmatrix} (\lambda - A)^{-1} \begin{bmatrix} B_1 & B_2 \end{bmatrix}.$$

In particular, the transfer function G_{22} is strictly proper.

Let C be a causal finite dimensional linear time invariant system of the type considered in the previous section, and let K be its transfer function. Thus K is a proper rational matrix function To define what it means that C is an internally stabilizing compensator for Σ we introduce two additional inputs v_1 and v_2 as in the following figure:

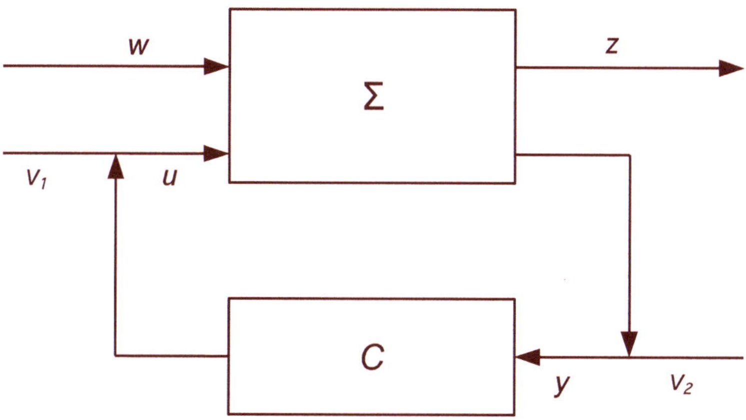

These two additional inputs are regarded as disturbances: v_1 is a disturbance on the control input u, while v_2 is a disturbance on the measured output. Then the system

C with transfer function K is said to be an *internally stabilizing compensator for the system* Σ if the nine transfer functions from the disturbances w, v_1, v_2 to z, u and y are all stable rational matrix functions. In this case, by slight abuse of terminology, we shall also say that K is an internally stabilizing compensator for the transfer function G of Σ.

After Laplace transform, the nine transfer functions from the disturbances w, v_1, v_2 to z, u and y are given by

$$\begin{bmatrix} \hat{z} \\ \hat{u} \\ \hat{y} \end{bmatrix} = \begin{bmatrix} I & -G_{12} & 0 \\ 0 & I & -K \\ 0 & -G_{22} & I \end{bmatrix}^{-1} \begin{bmatrix} G_{11} & 0 & 0 \\ 0 & I & 0 \\ G_{21} & 0 & I \end{bmatrix} \begin{bmatrix} \hat{w} \\ \hat{v}_1 \\ \hat{v}_2 \end{bmatrix}. \tag{19.5}$$

Now G_{22} is strictly proper and K is proper. Hence the rational matrix functions $I - G_{22}(\lambda)K(\lambda)$ and $I - K(\lambda)G_{22}(\lambda)$ are biproper with the value I at infinity. It follows that the inverses $I - G_{22}K$ and $I - KG_{22}$ are well-defined. Using these facts, the product of the first two matrices in the right-hand side of the identity (19.5)can be computed as

$$\begin{bmatrix} G_{11} + G_{12}K(I - G_{22}K)^{-1}G_{21} & G_{12}(I - KG_{22})^{-1} & G_{12}K(I - G_{22}K)^{-1} \\ K(I - G_{22}K)^{-1}G_{21} & (I - KG_{22})^{-1} & K(I - G_{22}K)^{-1} \\ (I - G_{22}K)^{-1}G_{21} & G_{22}(I - KG_{22})^{-1} & (I - G_{22}K)^{-1} \end{bmatrix}.$$

Theorem 19.5. *Let G be the transfer function of the system Σ given by* (19.4), *and let C be a causal finite dimensional linear time invariant system whose transfer function K is a proper rational matrix function. Then C is an internally stabilizing compensator for Σ if and only if K stabilizes G_{22} in the sense that the transfer functions from v_1 and v_2 to u and y are stable rational matrix functions, that is, the four functions*

$$(I - KG_{22})^{-1}, \quad K(I - G_{22}K)^{-1}, \quad G_{22}(I - KG_{22})^{-1}, \quad (I - G_{22}K)^{-1},$$

are stable.

There is a beautiful parametrization of all internally stabilizing compensators, known as the Youla-Jabr-Bongiorno parametrization. In order to state the parametrization we need a *doubly coprime factorization* of G_{22}, that is, a factorization

$$G_{22}(\lambda) = N(\lambda)M(\lambda)^{-1} = \widetilde{M}(\lambda)^{-1}\widetilde{N}(\lambda), \tag{19.6}$$

where $N, M, \widetilde{N}$ and $\widetilde{M}$ are $i\mathbb{R}$-stable rational matrix functions of appropriate sizes, with the additional property that there exist $i\mathbb{R}$-stable rational matrix functions

$X, Y, \widetilde{X}$ and $\widetilde{Y}$ such that

$$\begin{bmatrix} \widetilde{X}(\lambda) & -\widetilde{Y}(\lambda) \\ -\widetilde{N}(\lambda) & \widetilde{M}(\lambda) \end{bmatrix} \begin{bmatrix} M(\lambda) & Y(\lambda) \\ N(\lambda) & X(\lambda) \end{bmatrix} = \begin{bmatrix} M(\lambda) & Y(\lambda) \\ N(\lambda) & X(\lambda) \end{bmatrix} \begin{bmatrix} \widetilde{X}(\lambda) & -\widetilde{Y}(\lambda) \\ -\widetilde{N}(\lambda) & \widetilde{M}(\lambda) \end{bmatrix} = I. \tag{19.7}$$

Such a factorization always exists, in fact we can readily give formulas for all matrix functions involved in terms of the realization of G_{22}. To do this we assume that the realization

$$G_{22}(\lambda) = C_2(\lambda I - A)^{-1}B_2,$$

has two additional properties, namely (C_2, A) is detectable, and (A, B_2) is stabilizable. That is, there exist matrices F and H such that the matrices $A_F = A + B_2F$ and $A_H = A + HC_2$ are both stable. Then, one choice of a doubly coprime factorization is given by the functions

$$\begin{cases} M(\lambda) = I + F(\lambda - A_F)^{-1}B_2, & N(\lambda) = C_2(\lambda - A_F)^{-1}B_2, \\ \widetilde{M}(\lambda) = I + C_2(\lambda - A_H)^{-1}H, & \widetilde{N}(\lambda) = C_2(\lambda - A_H)^{-1}B_2, \\ X(\lambda) = I - C_2(\lambda - A_F)^{-1}H, & Y(\lambda) = -F(\lambda - A_F)^{-1}H, \\ \widetilde{X}(\lambda) = I - F(\lambda - A_H)^{-1}B_2, & \widetilde{Y}(\lambda) = -F(\lambda - A_H)^{-1}H. \end{cases} \tag{19.8}$$

Next, we give the Youla-Jabr-Bongiorno parametrization, which describes all internally stabilizing compensators of Σ in terms of $i\mathbb{R}$- stable, proper rational matrix functions in a one-to-one way.

Theorem 19.6. *Let G be the transfer function of the system Σ given by* (19.4), *and let M, N, X, Y be the $i\mathbb{R}$-stable rational matrix functions related to the doubly coprime factorization of G_{22}. Let C be a causal finite dimensional linear time invariant system whose transfer function K is a proper rational matrix function. Then C is an internally stabilizing compensator of Σ if and only if K has the form*

$$K(\lambda) = \big(Y(\lambda) - M(\lambda)Q(\lambda)\big)\big(X(\lambda) - N(\lambda)Q(\lambda)\big)^{-1}, \tag{19.9}$$

where Q is an $i\mathbb{R}$-stable rational matrix function. Moreover, the map from Q to K is one-to-one.

Replacing M, N, X, Y by $\widetilde{M}$, $\widetilde{N}$, $\widetilde{X}$, $\widetilde{Y}$ we have the following alternative expression for the transfer function K of the compensator:

$$K(\lambda) = \big(\widetilde{X}(\lambda) - Q(\lambda)\widetilde{N}(\lambda)\big)^{-1}\big(\widetilde{Y}(\lambda) - Q(\lambda)\widetilde{M}(\lambda)\big).$$

Notes

The results of the first section are standard results in mathematical systems theory, see, e.g., [94] or the more recent [33], [84]. For analogous results in the discrete time case we refer to [94], Chapter 21 of [150], and to [85]. A proof of Theorem 19.5 can be found in Chapter 4 of [43]. The formulas (19.8) giving the doubly coprime factorization in state space terms were derived in [115], see also Section 4.5 in [43]. Theorem 19.6 presents a result of [148].

Chapter 20

H-infinity control applications

The focus of the chapter is on a part of control theory called H-infinity control. The problem involved is the general H-infinity control problem, the so-called standard problem. It concerns the construction of a stabilizing controller with additional constraints on the maximum of the norm of the closed loop transfer function, taken over the values of the argument on the imaginary line. In its simplest form the problem is equivalent to the rational matrix Nehari problem considered in Chapter 18. The label H-infinity is related to the fact that a proper rational matrix function is stable if and only if it is analytic and uniformly bounded in the open right half plane. A function with the latter properties is usually referred to as an H_∞-function (on the right half plane).

The chapter consists of four sections. Section 20.1 introduces the standard problem mentioned above, and shows how this problem can be reduced to a model matching problem. In the next two sections we discuss a one-sided model matching problem (Section 20.2) and the two-sided model matching problem (Section 20.3). In particular, it will be shown how these two problems reduce to J-spectral factorization problems involving certain rational matrix functions. All of this will be done in general terms, without any state space formulas as yet. In the final section (Section 20.4) we use results from Chapter 14 and present the solution to the model matching problem in state space terms. This leads to the solution of the standard problem in these terms too.

In this chapter, as in Section 18.2, we use the following notation: if R is a rational matrix function, then R^* denotes the rational matrix function given by $R^*(\lambda) = R(-\bar{\lambda})^*$. (In engineering literature, including [76] and [43], this function is often denoted by $R^\sim$.) Recall also from Section 18.2 that RAT *denotes the set of all proper rational matrix functions that are analytic on the imaginary axis. Furthermore,* $\mathrm{RAT}^{p\times q}$ *stands for the set of all F in* RAT *that are of size $p \times q$, and* $\mathrm{RAT}_+^{p\times q}$ *denotes the set of all F in* $\mathrm{RAT}^{p\times q}$ *that are analytic on the closed left half plane. In the present chapter we shall also use the notation* $\mathrm{RAT}_-^{p\times q}$ (RAT_-) *which will* *denote the set of all F in* $\mathrm{RAT}^{p\times q}$ *(in* RAT*) that are analytic in the closed right half plane. In other words, F belongs to* $\mathrm{RAT}_-^{p\times q}$ *if and only if F is an $i\mathbb{R}$ stable $p \times q$ rational matrix function. Note also that* $F \in \mathrm{RAT}_-^{p\times q}$ *if and only if* $F^* \in \mathrm{RAT}_+^{q\times p}$.

20.1 The standard problem and model matching

Throughout this chapter G is the transfer function of a system Σ with two inputs u and w, and two outputs y and z. The input u is the control input, w is a disturbance, y is the output we can measure, and z is the output to be controlled. As in Section 19.2 we assume that the system is given by the state space representation

$$\begin{cases} x'(t) &= Ax(t) + B_1 w(t) + B_2 u(t), \\ z(t) &= C_1 x(t) + D_1 u(t), \\ y(t) &= C_2 x(t) + D_2 w(t), \qquad t \geq 0. \end{cases} \tag{20.1}$$

In particular, the function G is of the form

$$\begin{aligned} G(\lambda) &= \begin{bmatrix} G_{11}(\lambda) & G_{12}(\lambda) \\ G_{21}(\lambda) & G_{22}(\lambda) \end{bmatrix} \\ &= \begin{bmatrix} 0 & D_1 \\ D_2 & 0 \end{bmatrix} + \begin{bmatrix} C_1 \\ C_2 \end{bmatrix} (\lambda - A)^{-1} \begin{bmatrix} B_1 & B_2 \end{bmatrix}. \end{aligned} \tag{20.2}$$

Taking Laplace transforms and assuming the system to be at rest at $t = 0$ we have

$$\begin{bmatrix} \widehat{z}(\lambda) \\ \widehat{y}(\lambda) \end{bmatrix} = \begin{bmatrix} G_{11}(\lambda) & G_{12}(\lambda) \\ G_{21}(\lambda) & G_{22}(\lambda) \end{bmatrix} \begin{bmatrix} \widehat{w}(\lambda) \\ \widehat{u}(\lambda) \end{bmatrix}. \tag{20.3}$$

Our goal is to find a proper rational matrix function K such that: (1) K is the transfer function of an internally stabilizing compensator C of Σ (see Section 19.2), and (2) the influence of w on z is kept small in a sense we shall explain presently.

Inserting $\widehat{u}(\lambda) = K(\lambda)\widehat{y}(\lambda)$ into (20.3), one sees that

$$\begin{cases} \widehat{z}(\lambda) = G_{11}(\lambda)\widehat{w}(\lambda) + G_{12}(\lambda)K(\lambda)\widehat{y}(\lambda), \\ \widehat{y}(\lambda) = G_{21}(\lambda)\widehat{w}(\lambda) + G_{22}(\lambda)K(\lambda)\widehat{y}(\lambda). \end{cases} \tag{20.4}$$

Since G_{22} is strictly proper, so is $G_{22}K$, and hence the determinant of the matrix $I - G_{22}(\lambda)K(\lambda)$ does not vanish identically. By the second equation in (20.4) we have $\widehat{y}(\lambda) = \big(I - G_{22}(\lambda)K(\lambda)\big)^{-1} G_{21}(\lambda)\widehat{w}(\lambda)$. Inserting this into the first equation of (20.4), we obtain that the closed loop transfer function from $\widehat{w}$ to $\widehat{z}$ is given by the Redheffer representation

$$\begin{aligned}\widehat{z}(\lambda) &= (\mathcal{R}_G(K)(\lambda))\widehat{w}(\lambda) \qquad (20.5)\\ &= \left(G_{11}(\lambda) + G_{12}(\lambda)K(\lambda)\big(I - G_{22}(\lambda)K(\lambda)\big)^{-1}G_{21}(\lambda)\right)\widehat{w}(\lambda).\end{aligned}$$

The second requirement on K is that, given a tolerance γ, we want $\mathcal{R}_G(K)$ to be in RAT and to satisfy the bound

$$\|\mathcal{R}_G(K)\|_\infty = \max_{\lambda \in i\mathbb{R}} \|\mathcal{R}_G(K)(\lambda)\| < \gamma. \qquad (20.6)$$

This problem is known in control theory as the *standard problem of H-infinity control*.

The approach to solving this problem using J-spectral factorization techniques starts from the Youla parametrization of internally stabilizing compensators which we reviewed in Section 19.2. This leads, as we shall see in the final two paragraphs of this section, to an equivalent and easier to handle problem. Indeed, from the given rational matrix function G one constructs three rational matrix functions, T_1, T_2 and T_3 such that internally stabilizing compensators for which (20.6) holds are in one-to-one correspondence with $i\mathbb{R}$-stable rational matrix functions Q for which

$$\|T_1 - T_2QT_3\|_\infty < \gamma. \qquad (20.7)$$

The latter problem is called the *model matching problem* . It turns out that under mild assumptions (see Section 20.4 below) the rational matrix functions T_1, T_2 and T_3 are $i\mathbb{R}$ stable. In particular, these functions have no poles on the imaginary axis and at infinity, and hence they are all in RAT. Furthermore, we shall see that T_2 has a left inverse in RAT and T_3 has a right inverse in RAT.

A particular case (see the next section) of the model matching problem, when T_2 is square and $T_3 = I$, is a variation on the Nehari problem as discussed in Chapter 18.

Next, we present the reduction of the standard problem to a model matching problem. All necessary calculations take place in RAT, i.e., in the set of rational matrix functions that are analytic on $i\mathbb{R}$ and at infinity. As before, we partition the transfer function G as in the first part of (20.3). Also we shall employ the same notation as in Section 19.2 insofar as it concerns the doubly coprime factorization of G_{22} in (19.6) and the parametrization of the transfer functions of the internally stabilizing compensators of the system Σ in Theorem 19.6. We can then introduce three new functions, namely

$$\begin{aligned}T_1(\lambda) &= G_{11}(\lambda) + G_{12}(\lambda)M(\lambda)\widetilde{Y}(\lambda)G_{21}(\lambda), \qquad (20.8)\\ T_2(\lambda) &= G_{12}(\lambda)M(\lambda), \qquad (20.9)\\ T_3(\lambda) &= \widetilde{M}(\lambda)G_{21}(\lambda). \qquad (20.10)\end{aligned}$$

Recall that the problem we wish to solve is to find, if possible, internally stabilizing compensators C of the system Σ with a proper transfer function K such that $\mathcal{R}_G(K)$ belongs to RAT and (20.6) is satisfied, i.e.,

$$\|\mathcal{R}_G(K)\|_\infty = \max_{\lambda\in i\mathbb{R}} \|\mathcal{R}_G(K)(\lambda)\| < \gamma.$$

Here $\mathcal{R}_G(K)$ is given by

$$\mathcal{R}_G(K)(\lambda) = G_{11}(\lambda) + G_{12}(\lambda)K(\lambda)\big(I - G_{22}(\lambda)K(\lambda)\big)^{-1}G_{21}(\lambda); \qquad (20.11)$$

see (20.5). In case K is given by (19.9) involving the function Q featured there, we can rewrite $\mathcal{R}_G(K)$ as follows.

Theorem 20.1. *With K as in* (19.9), *the closed loop transfer function is given by*

$$\mathcal{R}_G(K)(\lambda) = T_1(\lambda) - T_2(\lambda)Q(\lambda)T_3(\lambda),$$

where T_1, T_2 and T_3 are given by (20.8), (20.9) *and* (20.10), *respectively*

Proof. Inserting $G_{22}(\lambda) = \widetilde{M}(\lambda)^{-1}N(\lambda)$ and (19.9) into $\big(I - G_{22}(\lambda)K(\lambda)\big)^{-1}$, and suppressing the variable λ for notational convenience, we get

$$\begin{aligned}(I - G_{22}K)^{-1} &= (X - NQ)\big(\widetilde{M}(X - NQ) - \widetilde{N}(Y - MQ)\big)^{-1}\widetilde{M}\\ &= (X - NQ)\widetilde{M}.\end{aligned}$$

In the actual derivation of these identities, the doubly coprime factorization in (19.6) and the defining properties given by (19.7) are employed. Again using (19.9), we arrive at $K(I - G_{22}K)^{-1} = (Y - MQ)\widetilde{M}$. Substituting this in the formula for the closed loop transfer function (20.11) yields

$$\begin{aligned}\mathcal{R}_G(K) &= (G_{11} + G_{12}Y\widetilde{M}G_{21}) - G_{12}MQ\widetilde{M}G_{21}\\ &= (G_{11} + G_{12}Y\widetilde{M}G_{21}) - T_2QT_3.\end{aligned}$$

Now from the defining properties of a doubly coprime factorization (19.6) one sees that $M\widetilde{Y} = Y\widetilde{M}$. Inserting this in the formula above we obtain that $T_1 = G_{11} + G_{12}Y\widetilde{M}G_{21}$. This completes the proof. □

20.2 The one-sided model matching problem

In this section we consider the model matching problem (20.7) with $T_1 \in \mathrm{RAT}^{l\times p}$, $T_2 \in \mathrm{RAT}_-^{l\times q}$ and $T_3 = I_p$. Furthermore, we assume that T_2 has a left inverse in $\mathrm{RAT}^{q\times l}$. In particular, T_1 is analytic on the imaginary axis (with infinity included) and T_2 is $i\mathbb{R}$-stable. Note that the left invertibility of T_2 implies that $l \geq q$, that is, T_2 is a "tall" matrix.

Given T_1 and T_2 as in the previous paragraph, the problem is to find necessary and sufficient conditions for the existence of an $i\mathbb{R}$-stable rational $q \times p$ matrix function Q, i.e., $Q \in \text{RAT}_-^{q\times p}$, such that $\|T_1 - T_2Q\|_\infty < \gamma$, and to give a full parametrization of all such Q. We refer to this problem as the *one-sided model matching problem corresponding to T_1 and T_2.*

We shall explain how this problem reduces to the Nehari problem, and we shall present a necessary and sufficient condition for its solution in terms of a J-spectral factorization. The following theorem is the main result of this section.

Theorem 20.2. *Let $T_1 \in \text{RAT}^{l\times p}$ and $T_2 \in \text{RAT}_-^{l\times q}$ be given, and assume T_2 has a left inverse in* RAT. *Let $\gamma > 0$, and put*

$$\Upsilon(\lambda) = \begin{bmatrix} T_2^*(\lambda) & 0 \\ T_1^*(\lambda) & I_p \end{bmatrix} \begin{bmatrix} I_l & 0 \\ 0 & -\gamma^2 I_p \end{bmatrix} \begin{bmatrix} T_2(\lambda) & T_1(\lambda) \\ 0 & I_p \end{bmatrix},$$

$$J = \begin{bmatrix} I_q & 0 \\ 0 & -I_p \end{bmatrix}.$$

Then there exists $Q \in \text{RAT}_-^{q\times p}$ such the norm constraint $\|T_1 - T_2Q\|_\infty < \gamma$ is satisfied if and only if Υ admits a left J-spectral factorization

$$\Upsilon(\lambda) = W^*(\lambda)JW(\lambda), \tag{20.12}$$

with respect to the imaginary axis having the additional property that the $q\times q$ block in the left upper corner of $W(\lambda)$ has an inverse in $\text{RAT}_-^{q\times q}$. Moreover, writing $W^{-1}(\lambda) = \begin{bmatrix} \omega_{ij}(\lambda) \end{bmatrix}_{i,j=1}^2$, where $\omega_{11}(\lambda)$ and $\omega_{22}(\lambda)$ are of sizes $q\times q$ and $p \times p$, respectively, all solutions Q of the one-sided model matching problem corresponding to T_1 and T_2 are given by

$$Q(\lambda) = -\big(\omega_{11}(\lambda)U(\lambda) + \omega_{12}(\lambda)\big)\big(\omega_{21}(\lambda)U(\lambda) + \omega_{22}(\lambda)\big)^{-1}, \tag{20.13}$$

where U is a rational matrix function in $\text{RAT}_-^{q\times p}$ with $\|U\|_\infty < 1$.

Proof. Since T_2 belongs to $\text{RAT}_-^{l\times q}$ and has a left inverse in RAT, we know from Theorem 17.26 that T_2 admits an inner-outer factorization with an invertible outer factor. Thus $T_2 = VX$, where V is inner, and both X and X^{-1} are analytic in the closed right half plane. If T_2 happened to be square, the reduction to the Nehari problem would now be easy. Indeed, in that case V is bi-inner, and hence

$$\|T_1 - T_2Q\|_\infty = \|T_1 - VXQ\|_\infty = \|V^*T_1 - XQ\|_\infty = \|R - \widehat{Q}\|_\infty,$$

where $R = V^*T_1$ and $\widehat{Q} = XQ$. Actually, since both X and Q are in RAT_-, also $\widehat{Q}$ is in RAT_-. Thus, this is not quite the Nehari problem as presented in Chapter 18, but applying the results of Section 18.3 to R^* yields $\widehat{Q}^*$. Also note that R^* is not stable, but it is just in RAT. At this point we use the fact that Proposition 18.6, when applied to R^*, does not require R^* to be stable. Recall that this point was made explicitly in the paragraph preceding the statement of Proposition 18.6.

However, in general, T_2 is only left invertible and not square, in which case V is only inner and not bi-inner. To deal with this more general case, we proceed as follows (see Section 17.8): take $V^\sharp$ such that $\widetilde{U} = \begin{bmatrix} V & V^\sharp \end{bmatrix}$ is bi-inner. We choose $V^\sharp$ such that $\widetilde{U}$ has the same McMillan degree as V, that is, in the way outlined in Section 17.8. Then

$$\|T_1 - T_2 Q\|_\infty = \| \begin{bmatrix} V^* \\ (V^\sharp)^* \end{bmatrix} T_1 - \begin{bmatrix} XQ \\ 0 \end{bmatrix} \|_\infty .$$

It follows that $\|T_1 - T_2 Q\|_\infty < \gamma$ if and only if for each for $\lambda \in i\mathbb{R} \cup \{\infty\}$ the following two conditions hold:

(a) $\Phi(\lambda) = \gamma^2 I_p - T_1^*(\lambda)V^\sharp(\lambda)(V^\sharp)^*(\lambda)T_1(\lambda) > 0,$

(b) $\gamma^2 I_p - T_1^*(\lambda)V^\sharp(\lambda)(V^\sharp)^*(\lambda)T_1(\lambda)$
$\qquad -(V^*(\lambda)T_1(\lambda) - X(\lambda)Q(\lambda))^*(V^*(\lambda)T_1(\lambda) - X(\lambda)Q(\lambda)) > 0.$

Using (a), the inequality (b) can be reduced to $\Phi - (V^*T_1 - XQ)^*(V^*T_1 - XQ) > 0$ where, for notational convenience, the variable λ being suppressed.

Now, let $\Phi(\lambda) = N^*(\lambda)N(\lambda)$ be a left canonical factorization of Φ relative to the imaginary axis. Then condition (b) above is equivalent to

$$I_p - N^{-*}(V^*T_1 - XQ)^*(V^*T_1 - XQ)N^{-1} > 0,$$

i.e., to $\|V^*T_1N^{-1} - XQN^{-1}\|_\infty < 1$. Observe that this, in turn, is precisely an instance of Nehari's problem, with $R = V^*T_1N^{-1}$ and $\widehat{Q} = XQN^{-1}$.

We apply the Nehari problem to R. Applying the result of Section 18.3, in particular Proposition 18.6 (which we apply with left half plane and right half plane interchanged) one sees that this Nehari problem is solvable if and only if the function $\Psi(\lambda)$, defined by

$$\Psi(\lambda) = \begin{bmatrix} I_q & 0 \\ N^{-*}(\lambda)T_1^*(\lambda)V(\lambda) & I_p \end{bmatrix} \begin{bmatrix} I_q & 0 \\ 0 & -I_p \end{bmatrix} \begin{bmatrix} I_q & V^*(\lambda)T_1(\lambda)N^{-1}(\lambda) \\ 0 & I_p \end{bmatrix},$$

has a left J-spectral factorization of the form

$$\Psi(\lambda) = L_-^*(\lambda) \begin{bmatrix} I_q & 0 \\ 0 & -I_p \end{bmatrix} L_-(\lambda), \tag{20.14}$$

with the additional property that the $p \times p$ block entry in the right lower corner of L_-^{-1} has an inverse in $\mathrm{RAT}_-^{p\times p}$. Moreover, in that case, if we partition $L_-^{-1}(\lambda)$

as $L_-^{-1}(\lambda) = \left[\ L_{ij}(\lambda)\ \right]_{i,j=1}^{2}$, with L_{11} a $q \times q$ rational matrix function, then all solutions to this Nehari problem are given by

$$\widehat{Q}(\lambda) = -(L_{11}(\lambda)U(\lambda) + L_{12}(\lambda))(L_{21}(\lambda)U(\lambda) + L_{22}(\lambda))^{-1},$$

where U runs over all functions in $\text{RAT}_-^{q\times p}$ for which $\|U\|_\infty < 1$. Finally, recall (see the final paragraph of Section 18.3) that the additional property of the $p \times p$ block entry in the right lower corner of L_-^{-1} is equivalent to the $q \times q$ block entry in the left upper corner of L_- having an inverse in $\text{RAT}_-^{p\times p}$.

Put $Q(\lambda) = X^{-1}(\lambda)\widehat{Q}(\lambda)N(\lambda)$. From the results of the previous paragraph, we get that all solutions to the one-sided model matching problem are given by

$$Q(\lambda) = -\big(X^{-1}(\lambda)L_{11}(\lambda)U(\lambda) + X^{-1}(\lambda)L_{12}(\lambda)\big)$$
$$\cdot\big(N^{-1}(\lambda)L_{21}(\lambda)U(\lambda) + N^{-1}(\lambda)L_{22}(\lambda)\big)^{-1},$$

where U runs over all functions in $\text{RAT}_-^{q\times p}$ for which $\|U\|_\infty < 1$.

Next, introduce

$$W(\lambda) = L_-(\lambda)\begin{bmatrix} X(\lambda) & 0 \\ 0 & N(\lambda) \end{bmatrix}. \tag{20.15}$$

Note that the $q \times q$ block entry in the left upper corner of W has an inverse in $\text{RAT}_-^{q\times q}$. Furthermore,

$$W^{-1}(\lambda) = \begin{bmatrix} X^{-1}(\lambda)L_{11}(\lambda) & X^{-1}(\lambda)L_{12}(\lambda) \\ N^{-1}(\lambda)L_{21}(\lambda) & N^{-1}(\lambda)L_{22}(\lambda) \end{bmatrix}.$$

So all solutions are parametrized by the function W^{-1}.

It remains to establish the identity (20.12), that is, once more suppressing the variable λ,

$$\Upsilon = W^*JW.$$

Let us denote the right side of the previous identity by Ξ. Thus $\Xi = W^*JW$. Using the definition of W in (20.15) together with formula (20.14), we see that

$$\Xi = \begin{bmatrix} X^* & 0 \\ 0 & N^* \end{bmatrix}\Psi\begin{bmatrix} X & 0 \\ 0 & N \end{bmatrix}.$$

It follows that

$$\begin{aligned}
\Xi &= \begin{bmatrix} X^* & 0 \\ T_1^* V & N^* \end{bmatrix} \begin{bmatrix} I & 0 \\ 0 & -I \end{bmatrix} \begin{bmatrix} X & V^* T_1 \\ 0 & N \end{bmatrix} \\
&= \begin{bmatrix} X^* & 0 \\ T_1^* V & N^* \end{bmatrix} \begin{bmatrix} V^* V & 0 \\ 0 & -I \end{bmatrix} \begin{bmatrix} X & V^* T_1 \\ 0 & N \end{bmatrix} \\
&= \begin{bmatrix} T_2^* & 0 \\ T_1^* V V^* & I \end{bmatrix} \begin{bmatrix} I & 0 \\ 0 & -N^* N \end{bmatrix} \begin{bmatrix} T_2 & V V^* T_1 \\ 0 & I \end{bmatrix} \\
&= \begin{bmatrix} T_2^* & 0 \\ T_1^* V V^* & I \end{bmatrix} \begin{bmatrix} I & 0 \\ 0 & -\gamma^2 I + T_1^* V^\sharp (V^\sharp)^* T_1 \end{bmatrix} \begin{bmatrix} T_2 & V V^* T_1 \\ 0 & I \end{bmatrix} \\
&= \begin{bmatrix} T_2^* T_2 & T_2^* V V^* T_1 \\ T_1^* V V^* T_2 & -\gamma^2 I + T_1^* (V V^* + V^\sharp (V^\sharp)^*) T_1 \end{bmatrix} \\
&= \begin{bmatrix} T_2^* T_2 & X^* V^* T_1 \\ T_1^* V X & -\gamma^2 I + T_1^* T_1 \end{bmatrix} \\
&= \begin{bmatrix} T_2^* T_2 & T_2^* T_1 \\ T_1^* T_2 & -\gamma^2 I + T_1^* T_1 \end{bmatrix} \\
&= \begin{bmatrix} T_2^* & 0 \\ T_1^* & I \end{bmatrix} \begin{bmatrix} I & 0 \\ 0 & -\gamma^2 I \end{bmatrix} \begin{bmatrix} T_2 & T_1 \\ 0 & I \end{bmatrix} = \Upsilon.
\end{aligned}$$

Thus we conclude that we may obtain W from a J-spectral factorization of a function that is easily described in terms of T_1 and T_2, as desired. Note also that the positivity of $\gamma^2 - T_1^* V^\sharp (V^\sharp)^* T_1$ on $i\mathbb{R} \cup \{\infty\}$ is implied by the J-spectral factorization. □

20.3 The two-sided model matching problem

In this section we extend the analysis of the previous section to the two-sided model matching problem. It will turn out that in this case we need two J-spectral factorizations.

Theorem 20.3. *Let $T_1 \in \mathrm{RAT}_-^{l\times p}$, $T_2 \in \mathrm{RAT}_-^{l\times q}$ and $T_3 \in \mathrm{RAT}_-^{m\times p}$. Assume that T_2 has a left inverse in* RAT, *and T_3 has a right inverse in* RAT. *Let $\gamma > 0$, and* *put*

$$\Omega(\lambda) = \begin{bmatrix} T_3(\lambda) & 0 \\ T_1(\lambda) & I_l \end{bmatrix} \begin{bmatrix} I_p & 0 \\ 0 & -\gamma^2 I_l \end{bmatrix} \begin{bmatrix} T_3^*(\lambda) & T_1^*(\lambda) \\ 0 & I_l \end{bmatrix}. \tag{20.16}$$

Then there exists $Q \in \mathrm{RAT}_-^{q\times m}$ *such that* $\|T_1 - T_2QT_3\| < \gamma$ *if and only if two conditions* (i) *and* (ii) *hold. The first condition* (i) *is as follows:*

(i) *With respect to the imaginary axis,* Ω *admits a right* J*-spectral factorization*

$$\Omega(\lambda) = V(\lambda)JV(-\bar{\lambda})^*, \text{ where } J = \begin{bmatrix} I_m & 0 \\ 0 & -I_l \end{bmatrix}, \tag{20.17}$$

having the additional property that the $m \times m$ *block in the upper left-hand corner of* V *has an inverse in* $\mathrm{RAT}_-^{m\times m}$.

With V *as in* (20.17), *define*

$$\widetilde{\Omega}(\lambda) = \begin{bmatrix} 0 & -T_2^*(\lambda) \\ I & 0 \end{bmatrix} V^{-*}(\lambda) \begin{bmatrix} -I_m & 0 \\ 0 & I_l \end{bmatrix} V^{-1}(\lambda) \begin{bmatrix} 0 & I \\ -T_2(\lambda) & 0 \end{bmatrix}. \tag{20.18}$$

Then the second condition (ii) *is:*

(ii) *With respect to the imaginary axis,* $\widetilde{\Omega}$ *admits a left* J*-spectral factorization of the form*

$$\widetilde{\Omega}(\lambda) = W(-\bar{\lambda})^* JW(\lambda), \text{ where } J = \begin{bmatrix} I_q & 0 \\ 0 & -I_m \end{bmatrix}, \tag{20.19}$$

having the additional property that the $q \times q$ *block in the upper left-hand corner of* W *has an inverse in* $\mathrm{RAT}_-^{q\times q}$.

Moreover, when (i) *and* (ii) *are satisfied,* (*all*) *the solutions* Q *to the two-sided model matching problem corresponding to* T_1, T_2 *and* T_3 *can be obtained as follows. Partition* $W^{-1} = \left[\ X_{ij}\ \right]_{i,j=1}^2$, *with* X_{11} *a* $q \times q$ *rational matrix function. Then*

$$Q = -(X_{11}U + X_{12})(X_{21}U + X_{22})^{-1}, \tag{20.20}$$

where U *is an* $i\mathbb{R}$*-stable rational* $q \times m$ *matrix function with* $\|U\|_\infty < 1$.

Proof. The idea of the proof is to reduce the two-sided model matching problem to the one-sided model matching problem discussed in the previous section. The proof is divided into several steps.

Part 1. We first show that condition (i) in the theorem is a necessary condition. To this end, introduce $T_1^\diamond(\lambda) = T_1(\bar{\lambda})^*$ and $T_3^\diamond(\lambda) = T_3(\bar{\lambda})^*$. Note the crucial difference with the functions T_1^* and T_3^*: the functions $T_1^\diamond$ and $T_3^\diamond$ are analytic in the closed right half plane, infinity included. With the help of these functions, rewrite $\|T_1 - T_2QT_3\|_\infty < \gamma$ in the following way: $\|T_1^\diamond - T_3^\diamond \widehat{Q}\|_\infty < \gamma$, where $\widehat{Q} = Q^\diamond T_2^\diamond$ (with the obvious interpretations for these functions). Taking into account Theorem 20.2, this gives that the first condition is necessary. Indeed, with

$$L = \begin{bmatrix} T_3^\diamond & T_1^\diamond \\ 0 & I \end{bmatrix}$$

and $V = W^\diamond$, we obtain

$$L^* \begin{bmatrix} I & 0 \\ 0 & -\gamma^2 I \end{bmatrix} L = W^* \begin{bmatrix} I & 0 \\ 0 & -\gamma^2 I \end{bmatrix} W.$$

Part 2. The next step is to rewrite the two-sided model matching problem in an equivalent way.

Use Theorem 17.28 to write $T_3(\lambda) = Y(\lambda)V_1(\lambda)$ where Y is an $m \times m$ invertible outer function and V_1 is an $m \times p$ co-inner function. Let $V_1^\sharp$ be such that $\widetilde{V} = \begin{bmatrix} V_1^* & (V_1^\sharp)^* \end{bmatrix}$ is bi-inner (see Corollary 17.33). Write $R = T_1\widetilde{V} = \begin{bmatrix} T_1V_1^* & T_1(V_1^\sharp)^* \end{bmatrix} = \begin{bmatrix} R_1 & R_2 \end{bmatrix}$, where R_1 is an $l \times m$ and R_2 is an $l \times (p-m)$ matrix function. As $\widetilde{V}$ is bi-inner, $T_3\widetilde{V} = \begin{bmatrix} Y & 0 \end{bmatrix}$. Thus we have

$$\|T_1 - T_2QT_3\|_\infty < \gamma$$

if and only if

$$\| \begin{bmatrix} R_1 & R_2 \end{bmatrix} - \begin{bmatrix} T_2QY & 0 \end{bmatrix} \|_\infty < \gamma.$$

In turn, this can be rewritten as

$$\gamma^2 I_l > \big(R_1(\lambda) - T_2(\lambda)Q(\lambda)Y(\lambda)\big)\big(R_1(\lambda) - T_2(\lambda)Q(\lambda)Y(\lambda)\big)^* + R_2(\lambda)R_2^*(\lambda),$$

for all $\lambda \in i\mathbb{R} \cup \{\infty\}$, or equivalently, suppressing the variable λ again, as

$$\gamma^2 I_l - R_2R_2^* > (R_1 - T_2QY)(R_1 - T_2QY)^*.$$

This implies that $\gamma^2 I_l - R_2R_2^* > 0$, and if we write $\gamma^2 I_l - R_2R_2^* = MM^*$ with M and M^{-1} in $\text{RAT}_-^{l\times l}$, then we can rewrite the inequality above as

$$I_l > M^{-1}(R_1 - T_2QY)(R_1 - T_2QY)^* M^{-*}.$$

Thus $\|T_1 - T_2QT_3\|_\infty < \gamma$ if and only if the following two conditions hold:

$$\gamma^2 I_l - R_2R_2^* > 0, \qquad \|M^{-1}R_1 - M^{-1}T_2QY\|_\infty < 1. \tag{20.21}$$

Note that the last of these two conditions is a one-sided model matching problem for QY, as both Y and Y^{-1} are in $\text{RAT}_-^{m\times m}$. Observe also that $M^{-1}R_1 = M^{-1}T_1V_1^*$ is in $\text{RAT}^{l\times m}$, because V_1^* is inner and hence analytic in the closed left half plane, infinity included. Also $M^{-1}T_2$ is in $\text{RAT}_-^{l\times q}$. Although we do not know that $M^{-1}R_1$ is in $\mathrm{RAT}_-^{l\times m}$ (that is, we do not know that it is analytic in the closed right half plane), still all conditions of Theorem 20.2 are met. Thus we may apply Theorem 20.2, to see that solvability of the one-sided model matching problem, which is the second condition in (20.21), is equivalent to a J-spectral factorization problem in the following way.

Put

$$K = \begin{bmatrix} M^{-1}T_2 & M^{-1}R_1 \\ 0 & I_m \end{bmatrix}.$$

Then, by Theorem 20.2, solvability of the one-sided model matching problem, which (as just noted) is the second part of (20.21), is equivalent to existence of a matrix function P such that P and P^{-1} are in $\mathrm{RAT}_-^{(m+q)\times(m+q)}$,

$$K^* \begin{bmatrix} I_l & 0 \\ 0 & -I_m \end{bmatrix} K = P^* \begin{bmatrix} I_q & 0 \\ 0 & -I_m \end{bmatrix} P,$$

and, in addition, the $q \times q$-block of P in the upper left corner has an inverse in $\mathrm{RAT}_-^{m\times m}$. Recall that the last condition is equivalent to the requirement that the $m \times m$-block in the right lower corner of P^{-1} is in $\mathrm{RAT}_-^{m\times m}$. Moreover, (all) the solutions Q to the one-sided model matching problem corresponding to $M^{-1}T_2$ and $M^{-1}R_1$ are generated by P^{-1} as follows: if $P^{-1} = \left[P_{ij}\right]_{i,j=1}^{2}$, with P_{11} of size $q \times q$, then

$$\begin{aligned} Q &= -(P_{11}U + P_{12})(P_{21}U + P_{22})^{-1}Y^{-1} \\ &= -(P_{11}U + P_{12})(YP_{21}U + YP_{22})^{-1}. \end{aligned}$$

Introduce

$$W = P \begin{bmatrix} I_q & 0 \\ 0 & Y^{-1} \end{bmatrix}.$$

Then W and W^{-1} are analytic in the right half plane and the $m \times m$ block in the right lower corner of W^{-1} is equal to YP_{22}, which is also in $\mathrm{RAT}_-^{m\times m}$. Finally, W generates all solutions Q.

Let

$$\widetilde{K} = K \begin{bmatrix} I_q & 0 \\ 0 & Y^{-1} \end{bmatrix}.$$

We conclude that solvability of the one-sided model matching problem, which is the second part of (20.21), is equivalent to existence of a J-spectral factorization of the form

$$W^* \begin{bmatrix} I_q & 0 \\ 0 & -I_m \end{bmatrix} W = \widetilde{K}^* \begin{bmatrix} I_l & 0 \\ 0 & -I_m \end{bmatrix} \widetilde{K}, \tag{20.22}$$

with the additional property that the $m \times m$ block in the right lower corner of W^{-1} is in $\mathrm{RAT}_-^{m\times m}$.

Part 3. Continuing with the considerations above, we compute

$$\begin{aligned}
\widetilde{K} &= \begin{bmatrix} M^{-1}T_2 & M^{-1}R_1 \\ 0 & I_m \end{bmatrix} \begin{bmatrix} I_q & 0 \\ 0 & Y^{-1} \end{bmatrix} \\
&= \begin{bmatrix} -M^{-1} & M^{-1}R_1Y^{-1} \\ 0 & Y^{-1} \end{bmatrix} \begin{bmatrix} -T_2 & 0 \\ 0 & I_m \end{bmatrix} \\
&= \begin{bmatrix} -M & R_1 \\ 0 & Y \end{bmatrix}^{-1} \begin{bmatrix} -T_2 & 0 \\ 0 & I_m \end{bmatrix} \\
&= \begin{bmatrix} 0 & Y \\ -M & R_1 \end{bmatrix}^{-1} \begin{bmatrix} 0 & I_m \\ -T_2 & 0 \end{bmatrix}.
\end{aligned}$$

It follows that

$$\widetilde{K}^* \begin{bmatrix} I_l & 0 \\ 0 & -I_m \end{bmatrix} \widetilde{K}$$

is equal to

$$\begin{bmatrix} 0 & -T_2^* \\ I_m & 0 \end{bmatrix} \begin{bmatrix} 0 & -M^* \\ Y^* & R_1^* \end{bmatrix}^{-1} \begin{bmatrix} I_l & 0 \\ 0 & -I_m \end{bmatrix} \begin{bmatrix} 0 & Y \\ -M & R_1 \end{bmatrix}^{-1} \begin{bmatrix} 0 & I_m \\ -T_2 & 0 \end{bmatrix},$$

which, in turn, can be written as,

$$\begin{bmatrix} 0 & -T_2^* \\ I_m & 0 \end{bmatrix} \left(\begin{bmatrix} 0 & Y \\ -M & R_1 \end{bmatrix} \begin{bmatrix} I_l & 0 \\ 0 & -I_m \end{bmatrix} \begin{bmatrix} 0 & -M^* \\ Y^* & R_1^* \end{bmatrix} \right)^{-1} \begin{bmatrix} 0 & I_m \\ -T_2 & 0 \end{bmatrix}.$$

Now the product of the middle three terms is easily seen to be equal to

$$\begin{bmatrix} YY^* & YR_1^* \\ R_1Y & R_1R_1^* - MM^* \end{bmatrix}.$$

Observe also that $YY^* = T_3T_3^*$ and $YR_1^* = YV_1T_1^* = T_3T_1^*$. Furthermore,

$$\begin{aligned}
R_1R_1^* - MM^* &= R_1R_1^* - \gamma^2 I_l + R_2R_2^* \\
&= -\gamma^2 I_l + \begin{bmatrix} R_1 & R_2 \end{bmatrix} \begin{bmatrix} R_1^* \\ R_2^* \end{bmatrix} \\
&= -\gamma^2 I_l + T_1\widetilde{V}^*\widetilde{V}T_1^* = -\gamma^2 I_l + T_1T_1^*.
\end{aligned}$$

Hence

$$\begin{bmatrix} YY^* & YR_1^* \\ R_1Y & R_1R_1^* - MM^* \end{bmatrix} = \begin{bmatrix} T_3T_3^* & T_3T_1^* \\ T_1T_3^* & -\gamma^2 I_l + T_1T_1^* \end{bmatrix}$$

$$= \begin{bmatrix} T_3 & 0 \\ T_1 & I \end{bmatrix} \begin{bmatrix} I_p & 0 \\ 0 & -\gamma^2 I_L \end{bmatrix} \begin{bmatrix} T_3^* & T_1^* \\ 0 & I \end{bmatrix} = \Omega.$$

Part 4. After these preliminaries we can now complete the proof in one direction. Indeed, to show that both the conditions (i) and (ii) need to be satisfied, note that we already saw at the beginning of the proof that (i) is necessary. Assuming that (i) holds, we continue the computation above, with V as in (20.17), and see that

$$\widetilde{K}^* \begin{bmatrix} I_l & 0 \\ 0 & -I_m \end{bmatrix} \widetilde{K} = \begin{bmatrix} 0 & -T_2^* \\ I_m & 0 \end{bmatrix} \left(V \begin{bmatrix} I_m & 0 \\ 0 & -I_l \end{bmatrix} V^* \right)^{-1} \begin{bmatrix} 0 & I_m \\ -T_2 & 0 \end{bmatrix}$$

$$= \begin{bmatrix} 0 & -T_2^* \\ I_m & 0 \end{bmatrix} V^{-*} \begin{bmatrix} I_m & 0 \\ 0 & -I_l \end{bmatrix} V^{-1} \begin{bmatrix} 0 & I_m \\ -T_2 & 0 \end{bmatrix}.$$

Thus, by (20.22), the second condition (ii) is necessary as well.

Part 5. For the converse, assume that both (i) and (ii) are satisfied. As in the proof of Theorem 20.2, applied to $T_1^\diamond$ and $T_3^\diamond$, in place of T_1 and T_2, we see that (i) implies that the first condition in (20.21) holds. Now follow the arguments in Parts 3 and 4 backwards to see that also the second condition in (20.21) is met. As we have already seen that these two conditions taken together are equivalent to the two-sided model matching problem, the proof is complete. □

Note that for the factorization (20.17) we need the analogue of Theorem 14.7 for right J-spectral factorization, applied to the function Ω given by (20.16). This analogue can be obtained by applying the left factorization result of Theorem 14.7 to the function $\Omega(-\lambda)$; cf., the paragraphs immediately following Theorem 14.8. In addition, the analogue of Theorem 14.7 for right J-spectral factorization provides us with a formula for the right J-spectral factor $\widetilde{V}$, satisfying

$$\Omega(\lambda) = \widetilde{V}(-\bar{\lambda})^* \begin{bmatrix} I_m & 0 \\ 0 & -I_l \end{bmatrix} \widetilde{V}(\lambda).$$

The function we need will then be $V(\lambda) = \widetilde{V}(-\bar{\lambda})^*$. We state the result of carrying out all this in state space form as a lemma, which will be useful in the next section.

Lemma 20.4. *Let* $H(\lambda) = D + C(\lambda I_n - A)^{-1}B$ *be a realization of an* $(m+l)\times(p+l)$ *rational matrix function* H. *Write* $J' = \operatorname{diag}(I_p, -I_l)$, $J = \operatorname{diag}(I_m, -I_l)$, *and* *assume that* $DJ'D^* = J$. *Also assume that* A *has all its eigenvalues in the open left half plane. Put* $\Omega(\lambda) = H(\lambda)J'H(-\bar{\lambda})^*$. *Then* Ω *admits a right* J*-spectral factorization with respect to the imaginary axis if and only if the algebraic Riccati equation*

$$XC^*JCX + X(A^* - C^*J^{-1}DJ'B^*) + (A - BJ'D^*J^{-1}C)X$$
$$+BJ'D^*JDJ'B^* - BJB^* = 0$$

has a Hermitian solution X *such that* $A^* - C^*J^{-1}(DJ'B^* - CX)$ *has its eigenvalues in the open left half plane. If* X *is such a solution (necessarily unique), and*

$$V(\lambda) = I_{m+l} + C(\lambda I_n - A)^{-1}(BJ'D^* - XC^*)J^{-1},$$

then $\Omega(\lambda) = V(\lambda)JV(-\bar{\lambda})^*$ *is a right* J*-spectral factorization of* Ω *with respect to the imaginary axis.*

20.4 State space solution of the standard problem

In this section we return to the standard problem. We recall the basic facts about the problem. The starting point is a system in state space form

$$\begin{cases} x'(t) &= Ax(t) + B_1w(t) + B_2u(t), \\ z(t) &= C_1x(t) + D_1u(t), \\ y(t) &= C_2x(t) + D_2w(t), \qquad t \geq 0. \end{cases} \tag{20.23}$$

The input vector $u(t)$ belongs to $\mathbb{C}^q$, the noise vector $w(t)$ belongs to $\mathbb{C}^p$, the state vector $x(t)$ belongs to $\mathbb{C}^n$, the measured output $y(t)$ belongs to $\mathbb{C}^m$, and finally, the output $z(t)$ to be controlled belongs to $\mathbb{C}^l$. Thus the sizes of the matrices featured in (20.23) are as follows: A is $n \times n$, B_1 is $n \times p$, B_2 is $n \times q$, C_1 is $l \times n$, C_2 is $m \times n$, D_1 is $l \times q$, and D_2 is $m \times p$.

Throughout the section we assume that the following simplifying assumptions hold:

A1. (A, B_1) is controllable and (C_1, A) is observable,

A2. (A, B_2) is stabilizable and (C_2, A) is detectable, that is, there are matrices F and H so that both $A + B_2F$ and $A + HC_2$ have all their eigenvalues in the open left half plane.

A3. $D_1^*C_1 = 0,\ D_1^*D_1 = I_q,\ D_2B_1^* = 0,\ D_2D_2^* = I_m$.

Given is also $\gamma > 0$. The problem we consider is to find an internally stabilizing compensator K from $\widehat{y}$ to $\widehat{u}$ such that (20.6) holds.

As we have explained in Section 20.1 this problem can be transformed into a model matching problem, using the rational matrix functions T_1, T_2, and T_3

appearing in (20.8)–(20.10). First we shall use (20.23) to derive state space realizations for T_1, T_2, and T_3. For this purpose we fix matrices H and F such that $A_F = A + B_2F$ and $A_H = A + HC_2$ are stable matrices. Recall that assumption A2 guarantees the existence of matrices H and F with these properties. It is a matter of straightforward calculations to check that the following proposition holds.

Proposition 20.5. *Write $G_{22}(\lambda) = C_2(\lambda I_n - A)^{-1}B_2$, and assume assumption* A2 *is satisfied. Let F and H be matrices such that $A_F = A + B_2F$ and $A_H = A + HC_2$ are stable matrices. Suppose a doubly coprime factorization of $G_{22}(\lambda)$ is given by the functions in* (19.8). *Then*

$$T_1(\lambda) = \begin{bmatrix} C_1 + D_1F & -D_1F \end{bmatrix} \left(\lambda I_{2n} - \begin{bmatrix} A_F & -B_2F \\ 0 & A_H \end{bmatrix}\right)^{-1} \begin{bmatrix} B_1 \\ B_1 + HD_2 \end{bmatrix},$$

$$T_2(\lambda) = D_1 + (C_1 + D_1F)(\lambda I_n - A_F)^{-1}B_2,$$

$$T_3(\lambda) = D_2 + C_2(\lambda I_n - A_H)^{-1}(B_1 + HD_2).$$

Observe that T_1, T_2 and T_3 are in RAT_-. Next, we show that T_2 has a left inverse, while T_3 has a right inverse, both in RAT.

Lemma 20.6. *Under the assumptions* A1, A2, A3, *the matrix function T_2 has a left inverse in* RAT *and T_3 has a right inverse in* RAT.

Proof. By Corollary 17.27, it suffices to show that $T_2(\lambda)$ is left invertible for all $\lambda \in i\mathbb{R}$ and that $T_3(\lambda)$ is right invertible for all $\lambda \in i\mathbb{R}$. First we show that

$$\begin{bmatrix} A - \lambda I_n & B_2 \\ C_1 & D_1 \end{bmatrix} \tag{20.24}$$

is left invertible for all $\lambda \in i\mathbb{R}$ if and only if $T_2(\lambda)$ is left invertible for all $\lambda \in i\mathbb{R}$.

To see that this is the case, we first establish that $T_2(\lambda)$ is left invertible for all $\lambda \in i\mathbb{R}$ if and only if

$$\begin{bmatrix} A_F - \lambda I_n & B_2 \\ C_1 + D_1F & D_1 \end{bmatrix} \tag{20.25}$$

is left invertible for all $\lambda \in i\mathbb{R}$. Indeed, assume that $T_2(\lambda)$ is left invertible for all pure imaginary λ, and that for some $\lambda_0 \in i\mathbb{R}$ and some vectors u and x we have

$$\begin{bmatrix} A_F - \lambda_0 I_n & B_2 \\ C_1 + D_1F & D_1 \end{bmatrix} \begin{bmatrix} x \\ u \end{bmatrix} = \begin{bmatrix} 0 \\ 0 \end{bmatrix}. \tag{20.26}$$

Then, since λ_0 is not an eigenvalue of A_F, it follows that $x = (\lambda_0 - A_F)^{-1}B_2u$. Inserting this in $(C_1 + D_1F)x + D_1u = 0$, gives $T_2(\lambda_0)u = 0$. Since $T_2(\lambda_0)$ is left invertible $u = 0$, and hence also $x = 0$.

Conversely, assume $T_2(\lambda_0)u = 0$ for some u and some pure imaginary λ_0. Suppose that (20.25) is left invertible for all $\lambda \in i\mathbb{R}$. Put $x = (\lambda_0 - A_F)^{-1}B_2u$, then (20.26) holds, hence $x = 0$ and $u = 0$.

Now (20.25) can be written as

$$\begin{bmatrix} A_F - \lambda_0 I_n & B_2 \\ C_1 + D_1 F & D_1 \end{bmatrix} = \begin{bmatrix} A - \lambda I_n & B_2 \\ C_1 & D_1 \end{bmatrix} \begin{bmatrix} I_n & 0 \\ F & I_q \end{bmatrix}.$$

Thus we see that (20.25) is left invertible if and only if (20.24) is left invertible.

Next we show that

$$\begin{bmatrix} A - \lambda I_n & B_2 \\ C_1 & D_1 \end{bmatrix}$$

is left invertible for all $\lambda \in i\mathbb{R}$. Indeed, assume that for some $\lambda_0 \in i\mathbb{R}$ and some vectors u and x we have

$$\begin{bmatrix} A - \lambda_0 I_n & B_2 \\ C_1 & D_1 \end{bmatrix} \begin{bmatrix} x \\ u \end{bmatrix} = \begin{bmatrix} 0 \\ 0 \end{bmatrix}.$$

Then, in particular, $C_1x + D_1u = 0$. Using $D_1^*D_1 = I_q$ and $D_1^*C_1 = 0$, this implies that $u = 0$. But then $(A - \lambda_0 I_n)x = 0$ and $C_1x = 0$. Since the pair (C_1, A_1) is observable by assumption, it follows that $x = 0$. □

For sake of convenience, and without loss of generality, we shall assume from now on that $\gamma = 1$. The first main result in this section is the following theorem.

Theorem 20.7. *Suppose the system* (20.23) *satisfies the assumptions* A1, A2 *and* A3, *and let* $\gamma = 1$. *Then there is an internally stabilizing compensator* K *for the system* (20.23) *satisfying* (20.6) *if and only if the following two conditions hold:*

(i) *there is a Hermitian solution* Y *of the Riccati equation*

$$Y(C_1^*C_1 - C_2^*C_2)Y + AY + YA^* + B_1B_1^* = 0 \qquad (20.27)$$

with the additional properties that $A^* + (C_1^*C_1 - C_2^*C_2)Y$ *is stable and* $Y > 0$,

(ii) *with the unique* Y *from* (i) *there is a Hermitian solution* Z *of the Riccati equation*

$$Z(YC_2^*C_2Y - B_2B_2^*)Z + Z(A + YC_1^*C_1) + (A^* + C_1^*C_1Y)Z + C_1^*C_1 = 0 \qquad (20.28)$$

with the additional properties that $A + YC_1^*C_1 - B_2B_2^*Z + YC_2^*C_2YZ$ *is stable and* $Z > 0$.

Moreover, when (i) *and* (ii) *are satisfied,* (*all*) *the internally stabilizing compensators can be obtained as follows. Introduce*

$$\Psi(\lambda) = \begin{bmatrix} \Psi_{11}(\lambda) & \Psi_{12}(\lambda) \\ \Psi_{21}(\lambda) & \Psi_{22}(\lambda) \end{bmatrix} = \begin{bmatrix} I_q & 0 \\ 0 & I_m \end{bmatrix} + \begin{bmatrix} -B_2^*Z \\ C_2(I_n + YZ) \end{bmatrix} (\lambda I_n - \widetilde{A})^{-1} \begin{bmatrix} B_2 & YC_2^* \end{bmatrix}, \quad (20.29)$$

where $\widetilde{A} = A + YC_1^*C_1 - B_2B_2^*Z + YC_2^*C_2YZ$. *Then* (*all*) *the internally stabilizing compensators satisfying* (20.6) *are given by*

$$K(\lambda) = \big(\Psi_{11}(\lambda)U(\lambda) + \Psi_{12}(\lambda)\big)\big(\Psi_{21}(\lambda)U(\lambda) + \Psi_{22}(\lambda)\big)^{-1},$$

where U *is an* $i\mathbb{R}$*-stable rational* $q \times m$ *matrix function with* $\|U\|_\infty < 1$.

Note that condition (i) requires the Riccati equation (20.27) to have a positive definite $i\mathbb{R}$-stabilizing solution. From Theorem 13.3 we know that the $i\mathbb{R}$-stabilizing solution is unique. Similarly, condition (ii) requires (20.28) to have a positive definite $i\mathbb{R}$-stabilizing solution, which is unique for the same reason.

It will be convenient to split the proof in a number of lemmas.

Lemma 20.8. *The existence of a right* J*-spectral factorization* (20.17) *in condition* (i) *of Theorem* 20.3 *is equivalent to the existence of an* $i\mathbb{R}$*-stabilizing Hermitian solution* Y *to the Riccati equation* (20.27). *Moreover, the additional property that the* $m \times m$ *block in the left upper corner of* V *has an inverse in* $\mathrm{RAT}_-^{m\times m}$ *is equivalent to* $Y > 0$.

Proof. We split the proof in two parts.

Part 1. Starting from the formulas for T_1 and T_3 given in Proposition 20.5 we form

$$L = \begin{bmatrix} T_3 & 0 \\ T_1 & I_l \end{bmatrix}.$$

This matrix function has the realization $L(\lambda) = D + \widetilde{C}(\lambda I_n - \widetilde{A})^{-1}\widetilde{B}$, where

$$\widetilde{A} = \begin{bmatrix} A_F & -B_2F \\ 0 & A_H \end{bmatrix}, \qquad \widetilde{B} = \begin{bmatrix} B_1 & 0 \\ B_1 + HD_2 & 0 \end{bmatrix},$$

$$\widetilde{C} = \begin{bmatrix} 0 & C_2 \\ C_1 + D_1F & -D_1F \end{bmatrix}, \qquad D = \begin{bmatrix} D_2 & 0 \\ 0 & I_l \end{bmatrix}.$$

It will be more convenient however to work with a similar realization. Put

$$S = \begin{bmatrix} I_n & I_n \\ 0 & I_n \end{bmatrix}. \quad (20.30)$$

Note that

$$\widetilde{A} = S \begin{bmatrix} A_F & -HC_2 \\ 0 & A_H \end{bmatrix} S^{-1}, \quad \widetilde{B} = S \begin{bmatrix} -HD_2 & 0 \\ B_1 + HD_2 & 0 \end{bmatrix},$$

$$\widetilde{C} = \begin{bmatrix} 0 & C_2 \\ C_1 + D_1 F & -C_1 \end{bmatrix} S^{-1}.$$

Also put $J' = \operatorname{diag}(I_p, -I_l)$ and $J = \operatorname{diag}(I_m, -I_l)$.

Using the factorization principle from Section 2.6 one sees that L can be factored as $L(\lambda) = L_1(\lambda)L_2(\lambda)$, where

$$L_1(\lambda) = \begin{bmatrix} I_m & 0 \\ 0 & I_l \end{bmatrix} + \begin{bmatrix} 0 \\ C_1 + D_1 F \end{bmatrix} (\lambda - A_F)^{-1} \begin{bmatrix} -H & 0 \end{bmatrix},$$

$$L_2(\lambda) = \begin{bmatrix} D_2 & 0 \\ 0 & I_l \end{bmatrix} + \begin{bmatrix} C_2 \\ C_1 \end{bmatrix} (\lambda - A_H)^{-1} \begin{bmatrix} B_1 + HD_2 & 0 \end{bmatrix}.$$

Because L_1 is of the form

$$L_1(\lambda) = \begin{bmatrix} I_m & 0 \\ \Xi(\lambda) & I_l \end{bmatrix},$$

where

$$\Xi(\lambda) = -(C_1 + D_1 F)(\lambda I_n - A_H)^{-1} H,$$

we have that L_1 and its inverse are in RAT_-. Thus Ω admits a right J-spectral factorization if and only if $\Omega_2 = L_2 J' L_2^*$ admits a right J-spectral factorization. Moreover, $\Omega = VJV^*$ with V and its inverse in RAT_- if and only if $\Omega_2 = V_2 J V_2^*$, where $V_2 = L_1^{-1} V$, and V_2 and its inverse are in RAT_-.

Now applying Lemma 20.4 to Ω_2, and using that $D_2 D_2^* = I_m$ and $D_2 B_1^* = 0$, we obtain that a right J-spectral factorization of Ω_2 exists if and only if the algebraic Riccati equation

$$X(C_2^* C_2 - C_1^* C_1)X + XA^* + AX - B_1 B_1^* = 0$$

has a Hermitian solution X for which $A^* + (C_2^* C_2 - C_1^* C_1)X$ has all its eigenvalues in the open left half plane. Comparing with (20.27) we see that this is equivalent to taking $X = -Y$. Observe also that this solution Y is unique since X is unique.

In addition $V(\lambda) = L_1(\lambda)^{-1} V_2(\lambda)$, where

$$\begin{aligned} V_2(\lambda) &= \begin{bmatrix} I_m & 0 \\ 0 & I_l \end{bmatrix} + \begin{bmatrix} C_2 \\ C_1 \end{bmatrix} (\lambda I_n - A_H)^{-1} \begin{bmatrix} H - XC_2^* & XC_1^* \end{bmatrix} \\ &= \begin{bmatrix} I_m & 0 \\ 0 & I_l \end{bmatrix} + \begin{bmatrix} C_2 \\ C_1 \end{bmatrix} (\lambda I_n - A_H)^{-1} \begin{bmatrix} H + YC_2^* & -YC_1^* \end{bmatrix}. \end{aligned}$$

Part 2. Next, we show that the property that the $m \times m$ block in the left upper corner of V has an inverse in RAT_-, is equivalent to Y being positive definite.

Because of the special form of H_1, we have that the $m \times m$ block in the left upper corner of V is equal to the $m \times m$ block in the left upper corner of V_2. Let us denote this block by V_{11}. Then

$$V_{11}(\lambda)^{-1} = I_m - C_2\big(\lambda I_n - (A - YC_2^*C_2)\big)^{-1}(H + YC_2^*).$$

Now using (20.27) we have that

$$\begin{aligned}(A - YC_2^*C_2)Y &+ Y(A^* - C_2^*C_2Y)\\ &= -B_1B_1^* - Y(C_1^*C_1 + C_2^*C_2)Y \le -B_1B_1^* \le 0. \qquad (20.31)\end{aligned}$$

Since the pair (A, B_1) is controllable it follows from standard arguments concerning Lyapunov equations (see, e.g., Theorem 4 in Section 13.1 in [107]) that $A - YC_2^*C_2$ has its spectrum in the open left half plane if and only if Y is positive definite. □

This finishes the first part of the proof of Theorem 20.7. Next we consider the second condition in Theorem 20.3 and its equivalence to the remaining parts of Theorem 20.7.

Lemma 20.9. *The existence of a left J-spectral factorization as in* (20.19) *in condition* (ii) *of Theorem* 20.3 *is equivalent to the existence of an $i\mathbb{R}$-stabilizing solution Z of* (20.28). *Moreover, the additional property that the $q \times q$ block in the upper left corner of W is in $\mathrm{RAT}_-^{q\times q}$ is equivalent to Z being positive definite.*

Proof. Again we shall split the argument into several parts.

Part 1. For the first step we start by computing the function from condition (ii) of Theorem 20.3 as follows. Using the notation of the proof of Lemma 20.8, define

$$\widetilde{L}(\lambda) = V(\lambda)^{-1}\begin{bmatrix} 0 & I_m \\ -T_2(\lambda) & 0\end{bmatrix} = V_2(\lambda)^{-1}L_1(\lambda)^{-1}\begin{bmatrix} 0 & I_m \\ -T_2(\lambda) & 0\end{bmatrix}.$$

Observe that the function $\widetilde{\Omega}$ in condition (ii) of Theorem 20.3 is given by

$$\widetilde{\Omega}(\lambda) = \widetilde{L}^*(\lambda)J'\widetilde{L}(\lambda), \quad \text{where } J' = \begin{bmatrix} -I_m & 0 \\ 0 & I_l\end{bmatrix}.$$

First we show that the existence of a left J-spectral factorization

$$\widetilde{\Omega} = W^*JW, \text{ where } J = \begin{bmatrix} I_q & 0 \\ 0 & -I_l\end{bmatrix}$$

amounts to the existence of a left J-spectral factorization of the matrix function $\widetilde{L}_1^*J'\widetilde{L}_1$, where $\widetilde{L}_1$ arises from a certain factorization of $\widetilde{L}$. In fact, the argument will be similar to the one used in the proof of the previous lemma.

Using the product rule and then simplifying, we get

$$L_1(\lambda)^{-1}\begin{bmatrix}0 & I_m\\ -T_2(\lambda) & 0\end{bmatrix} = \left(I_{m+l} + \begin{bmatrix}0\\ C_1+D_1F\end{bmatrix}(\lambda I_n - A_F)^{-1}\begin{bmatrix}H & 0\end{bmatrix}\right)$$

$$\cdot\left(\begin{bmatrix}0 & I_m\\ -D_1 & 0\end{bmatrix} + \begin{bmatrix}0\\ C_1+D_1F\end{bmatrix}(\lambda I_n - A_F)^{-1}\begin{bmatrix}-B_2 & 0\end{bmatrix}\right)$$

$$= \left(\begin{bmatrix}0 & I_m\\ -D_1 & 0\end{bmatrix} + \begin{bmatrix}0\\ C_1+D_1F\end{bmatrix}(\lambda I_n - A_F)^{-1}\begin{bmatrix}-B_2 & H\end{bmatrix}\right).$$

Thus, again applying the multiplication rule, we obtain a formula for $\widetilde{L}(\lambda)$, by pre-multiplying the above expression with $V_2(\lambda)^{-1}$. Using also $C_1^*D_1 = 0$, this yields $\widetilde{L}(\lambda) = \widetilde{D} + \widetilde{C}(\lambda - \widetilde{A})^{-1}\widetilde{B}$, where

$$\widetilde{A} = \begin{bmatrix}A - YC_2^*C_2 + YC_1^*C_1 & -YC_1^*C_1\\ 0 & A_F\end{bmatrix}, \qquad \widetilde{B} = \begin{bmatrix}0 & H + YC_2^*\\ -B_2 & H\end{bmatrix},$$

$$\widetilde{D} = \begin{bmatrix}0 & I_m\\ -D_1 & 0\end{bmatrix}, \qquad \widetilde{C} = \begin{bmatrix}-C_2 & 0\\ -C_1 & C_1 + D_1F\end{bmatrix}.$$

It is convenient to consider another realization. With S as in (20.30) and writing $A_Y = A - YC_2^*C_2 + YC_1^*C_1$, we have

$$\widetilde{A} = S\begin{bmatrix}A_Y & -B_2F - YC_2^*C_2\\ 0 & A_F\end{bmatrix}S^{-1}, \quad \widetilde{B} = S\begin{bmatrix}B_2 & YC_2^*\\ -B_2 & H\end{bmatrix},$$

$$\widetilde{C} = \begin{bmatrix}-C_2 & -C_2\\ -C_1 & D_1F\end{bmatrix}S^{-1}.$$

It is now easily checked that $\widetilde{L} = \widetilde{L}_1\widetilde{L}_2$, where

$$\widetilde{L}_1(\lambda) = \begin{bmatrix}0 & I_m\\ -D_1 & 0\end{bmatrix} + \begin{bmatrix}-C_2\\ -C_1\end{bmatrix}(\lambda I_n - A_Y)^{-1}\begin{bmatrix}B_2 & YC_2^*\end{bmatrix},$$

$$\widetilde{L}_2(\lambda) = \begin{bmatrix}I_q & 0\\ 0 & I_m\end{bmatrix} + \begin{bmatrix}-F\\ -C_2\end{bmatrix}(\lambda I_n - A_F)^{-1}\begin{bmatrix}-B_2 & H\end{bmatrix}.$$

Since A_F is stable, $\widetilde{L}_2$ is in $\mathrm{R{\small AT}}_-$, and as

$$A_F - \begin{bmatrix}-B_2 & H\end{bmatrix}\begin{bmatrix}-F\\ -C_2\end{bmatrix} = A_F - B_2F + HC_2 = A_H$$

has all its eigenvalues in the open left half plane, $\widetilde{L}_2^{-1}$ is in RAT_- too.

From the considerations in the previous paragraph it follows that the rational matrix function $\widetilde{\Omega} = \widetilde{L}^* J' \widetilde{L}$ admits a left J-spectral factorization if and only if the function $\widetilde{L}_1 J' \widetilde{L}_1$ admits a left J-spectral factorization.. In that case, if W_1 is a left J-spectral factor of $\widetilde{L}_1^* J' \widetilde{L}_1$, then $W = W_1 \widetilde{L}_2$ is a J-spectral factor of $\widetilde{L}^* J' \widetilde{L}$.

Part 2. In this part we continue to use the notation of the previous part. We now apply Theorem 14.7 to $\widetilde{L}_1$. This yields that there exists a left J-spectral factorization of $\widetilde{\Omega} = \widetilde{L}^* J' \widetilde{L}$ if and only if there is a Hermitian solution X of the algebraic Riccati equation

$$X(B_2B_2^* - YC_2^*C_2Y)X + X(A + YC_1^*C_1) + (A^* + C_1^*C_1Y)X - C_1^*C_1 = 0$$

having the additional property $\sigma(A + YC_1^*C_1 + B_2B_2^*X - YC_2^*C_2YX) \subset \mathbb{C}_{\text{left}}$. This solution X is unique.

Taking $Z = -X$ we see that Z satisfies the algebraic Riccati equation (20.28) and is the $i\mathbb{R}$-stabilizing solution of that equation. Thus the left J-spectral factor W_1 of $\widetilde{L}_1^* J' \widetilde{L}_1$ is given by

$$W_1(\lambda) = I_{q+m} + \begin{bmatrix} B_2^*Z \\ -C_2 - C_2YZ \end{bmatrix} (\lambda I_n - A_Y)^{-1} \begin{bmatrix} B_2 & YC_2^* \end{bmatrix}, \tag{20.32}$$

and the product $W(\lambda) = W_1(\lambda)\widetilde{L}_2(\lambda)$ becomes

$$I_{q+m} + \begin{bmatrix} B_2^*Z & -F \\ -C_2 - C_2YZ & -C_2 \end{bmatrix} \left(\lambda - \begin{bmatrix} A_Y & -B_2F - YC_2^*C_2 \\ 0 & A_F \end{bmatrix} \right)^{-1} \begin{bmatrix} B_2 & YC_2^* \\ -B_2 & H \end{bmatrix}.$$

Part 3. We now consider the additional property that the $q \times q$ block in the upper left corner of W has an inverse in $\mathrm{RAT}_-^{q\times q}$, and prove that this is equivalent to Z being positive definite. Let us denote the $q \times q$ block in the upper left corner of W by W_{11}. Then

$$W_{11}(\lambda) = I_q + \begin{bmatrix} B_2^*Z & -F \end{bmatrix} \left(\lambda I_{2n} - \begin{bmatrix} A_Y & -B_2F - YC_2^*C_2 \\ 0 & A_F \end{bmatrix} \right)^{-1} \begin{bmatrix} B_2 \\ -B_2 \end{bmatrix}.$$

Thus the main operator in the realization of W_{11}^{-1} is

$$\begin{aligned} \widetilde{A} &= \begin{bmatrix} A_Y & -B_2F - YC_2^*C_2 \\ 0 & A_F \end{bmatrix} - \begin{bmatrix} B_2 \\ -B_2 \end{bmatrix} \begin{bmatrix} B_2^*Z & -F \end{bmatrix} \\ &= \begin{bmatrix} A_Y - B_2B_2^*Z & -YC_2^*C_2 \\ B_2B_2^*Z & A \end{bmatrix}. \end{aligned}$$

We have to show that this matrix has all its eigenvalues in the open left half plane if and only if Z is positive definite.

In order to do this, it is helpful to consider a similar matrix. Take

$$S = \begin{bmatrix} I_n & 0 \\ -I_n & I_n \end{bmatrix},$$

and put

$$\widehat{A} = S^{-1}\widetilde{A}S = \begin{bmatrix} A + YC_1C_1^* - B_2B_2^*Z & -YC_2^*C_2 \\ YC_1^*C_1 & A - YC_2^*C_2 \end{bmatrix}.$$

We shall show that $Z > 0$ if and only if $\widehat{A}$ has all its eigenvalues in the left half plane. To this end, consider $\begin{bmatrix} Z & 0 \\ 0 & Y^{-1} \end{bmatrix} \widehat{A} + \widehat{A}^* \begin{bmatrix} Z & 0 \\ 0 & Y^{-1} \end{bmatrix}$

$$= \begin{bmatrix} -C_1^*C_1 - ZB_2B_2^* - ZYC_2^*C_2YZ & C_1^*C_1 - ZYC_2^*C_2 \\ C_1^*C_1 - C_2^*C_2YZ & \Lambda \end{bmatrix},$$

where, because of (20.31),

$$\begin{aligned} \Lambda &= Y^{-1}(A - YC_2^*C_2) + (A^* - C_2^*C_2Y)Y^{-1} \\ &= -Y^{-1}B_1B_1^*Y^{-1} - C_1^*C_1 - C_2^*C_2. \end{aligned}$$

Substituting the latter expression for Λ in the right lower corner of the matrix above, we obtain

$$\begin{aligned} &\begin{bmatrix} Z & 0 \\ 0 & Y^{-1} \end{bmatrix} \widehat{A} + \widehat{A}^* \begin{bmatrix} Z & 0 \\ 0 & Y^{-1} \end{bmatrix} \\ &= \begin{bmatrix} -C_1^*C_1 - ZB_2B_2^* - ZYC_2^*C_2YZ & C_1^*C_1 - ZYC_2^*C_2 \\ C_1^*C_1 - C_2^*C_2YZ & -Y^{-1}B_1B_1^*Y^{-1} - C_1^*C_1 - C_2^*C_2 \end{bmatrix} \\ &= -\begin{bmatrix} C_1^* & ZB_2 & 0 & ZYC_2^* \\ -C_1^* & 0 & Y^{-1}B_1 & C_2^* \end{bmatrix} \begin{bmatrix} C_1 & -C_1 \\ B_2^*Z & 0 \\ 0 & B_1^*Y^{-1} \\ C_2YZ & C_2 \end{bmatrix}. \end{aligned}$$

With the notation $\widehat{C}$ as shorthand for the latter factor, this reduces to

$$\begin{bmatrix} Z & 0 \\ 0 & Y^{-1} \end{bmatrix} \widehat{A} + \widehat{A}^* \begin{bmatrix} Z & 0 \\ 0 & Y^{-1} \end{bmatrix} = -\widehat{C}^*\widehat{C}. \tag{20.33}$$

Next, we show that the pair $\left(\widehat{C}, \widehat{A}\right)$ is observable. Suppose

$$\widehat{A}\begin{bmatrix} x \\ y \end{bmatrix} = \lambda_0 \begin{bmatrix} x \\ y \end{bmatrix}, \qquad \widehat{C}\begin{bmatrix} x \\ y \end{bmatrix} = 0,$$

or, which comes down to the same,

$$\begin{aligned} (A + YC_1^*C_1 - B_2B_2^*Z)x - YC_2C_2^*y &= \lambda_0 x, \\ YC_1^*C_1x + (A - YC_2^*C_2)y &= \lambda_0 y, \end{aligned}$$

and

$$C_1x = C_1y, \qquad B_2^*Zx = 0, \qquad B_1^*Y^{-1}y = 0, \qquad C_2YZx = -C_2y.$$

Using $C_1x = C_1y$, it follows that $(A - YC_2^*C_2 + YC_1^*C_1)y = \lambda_0 y$. Combining this with $B_1^*Y^{-1}y = 0$, and putting $w = Y^{-1}y$, we obtain

$$(AY - YC_2^*C_2Y + YC_1C_1^*Y + B_1B_1^*)w = \lambda_0 Yw, \qquad B_1^*w = 0.$$

Now use (20.27) to see that this implies $YA^*w = \lambda_0 Yw$. As Y is invertible we have $A^*w = \lambda_0 w$ and $B_1^*w = 0$. Since (A, B_1) is controllable, it follows that $w = 0$. Hence $y = 0$ too. From

$$(A + YC_1^*C_1 - B_2B_2^*Z)x - YC_2C_2^*y = \lambda_0 x,$$

combined with $y = 0$, $C_1x = C_1y = 0$ and $B_2^*Zx = 0$ we then have $Ax = \lambda_0 x$. The observability of the pair (C_1, A) finally gives $x = 0$.

We finish by applying the result of Theorem 4 in Section 13.1 in [107] to the equation (20.33). Combined with the fact that $Y > 0$, this gives that $Z > 0$ if and only if $\widehat{A}$ has all its eigenvalues in the open left half plane. □

This concludes the proof of the equivalence of (i) and (ii) in Theorem 20.7. We bring the argument to a close as follows.

Proof of Theorem 20.7. In view of the two preceding lemmas, it remains to prove the formulas for the parametrization of the internally stabilizing compensators satisfying (20.6). Recall from Theorem 19.6, in particular from formula (19.9), that $K = (Y - MQ)(X - NQ)^{-1}$. Also we have formula (20.20), that is the expression $Q = -(X_{11}U + X_{12})(X_{21}U + X_{22})^{-1}$, where

$$W^{-1} = \begin{bmatrix} X_{11} & X_{12} \\ X_{21} & X_{22} \end{bmatrix}.$$

Here W is obtained from Part 1 of the proof of Lemma 20.9. Combining the expressions, we see that

$$\begin{aligned} K &= ((YX_{21}+MX_{11})U+(YX_{22}+MX_{12})) \\ &\qquad \cdot((XX_{21}+NX_{11})U+(XX_{22}+NX_{12}))^{-1} \\ &= (\Psi_{11}U+\Psi_{12})(\Psi_{21}U+\Psi_{22})^{-1}, \end{aligned}$$

with Ψ given by

$$\Psi = \begin{bmatrix} \Psi_{11} & \Psi_{12} \\ \Psi_{21} & \Psi_{22} \end{bmatrix} = \begin{bmatrix} M & Y \\ N & X \end{bmatrix} \begin{bmatrix} X_{11} & X_{12} \\ X_{21} & X_{22} \end{bmatrix}.$$

The formulas in (19.8) now give

$$\begin{bmatrix} M(\lambda) & Y(\lambda) \\ N(\lambda) & X(\lambda) \end{bmatrix} = \begin{bmatrix} I_q & 0 \\ 0 & I_m \end{bmatrix} + \begin{bmatrix} F \\ C_2 \end{bmatrix} (\lambda I_n - A_F)^{-1} \begin{bmatrix} B_2 & -H \end{bmatrix}.$$

Fortuitously, this is equal to the function $\widetilde{L}_2(\lambda)$ from Part 1 of the proof of the previous lemma. Since $W = W_1\widetilde{L}_2$, we get $\Psi = \widetilde{L}_2W^{-1} = W_1^{-1}$, where W_1 is given by (20.32). Hence Ψ is given by (20.29), as desired. □

We conclude with the second main result of this chapter.

Theorem 20.10. *Suppose the system* (20.23) *satisfies the assumptions* A1, A2 *and* A3, *and let* γ *be an arbitrary positive number. Then there is an internally stabilizing compensator* K *for the system* (20.23) *satisfying* (20.6) *if and only if the following three conditions hold:*

(i) *there is a positive definite* $i\mathbb{R}$*-stabilizing solution* X *of the Riccati equation*

$$X(\gamma^{-2}B_1B_1^* - B_2B_2^*)X + A^*X + XA + C_1^*C_1 = 0, \tag{20.34}$$

(ii) *there is a positive definite* $i\mathbb{R}$*-stabilizing solution* Y *of the Riccati equation*

$$Y(\gamma^{-2}C_1^*C_1 - C_2^*C_2)Y + AY + YA^* + B_1B_1^* = 0, \tag{20.35}$$

(iii) $X < \gamma^{-2}Y^{-1}$ *or, equivalently, all eigenvalues of* XY *are in the open disc* $\{z \mid |z| < \gamma^{-2}\}$.

In that case (all) the internally stabilizing compensators K *can be obtained as follows. Introduce*

$$\begin{aligned} \Phi(\lambda) &= \begin{bmatrix} \Phi_{11}(\lambda) & \Phi_{12}(\lambda) \\ \Phi_{21}(\lambda) & \Phi_{22}(\lambda) \end{bmatrix} \\ &= \begin{bmatrix} 0 & I_q \\ I_m & 0 \end{bmatrix} - \begin{bmatrix} B_2^*X \\ C_2 \end{bmatrix} (I-\gamma^{-2}YX)^{-1}(\lambda - \widehat{A})^{-1} \begin{bmatrix} YC_2^* & B_2 \end{bmatrix}, \end{aligned}$$

where $\widehat{A} = A - Y(C_2^*C_2 - \gamma^{-2}C_1^*C_1) - B_2B_2^*X(I_n - \gamma^{-2}YX)^{-1}$. *Then (all) the internally stabilizing compensators satisfying* (20.6) *are given by*

$$K(\lambda) = \Phi_{11}(\lambda) + \Phi_{12}(\lambda)U(\lambda)(I_m - \Phi_{22}(\lambda)U(\lambda))^{-1}\Phi_{21}(\lambda),$$

where U *is an* $i\mathbb{R}$*-stable rational* $q \times m$ *matrix function satisfying* $\|U\|_\infty < \gamma$.

Proof. The theorem may be derived from the previous one upon giving the connections between X and Z. Again we assume $\gamma = 1$ without loss of generality.

Under this assumption, condition (i) in Theorem 20.7 is exactly the same as the second condition in the present theorem. Henceforth we suppose it is satisfied. Thus, throughout the proof, Y will be a positive definite $i\mathbb{R}$-stabilizing solution of (20.27), or, equivalently, of (20.35) with $\gamma = 1$. The argument below is divided into four parts.

Part 1. Introduce the block matrices

$$H = \begin{bmatrix} -A^* & -C_1^*C_1 \\ B_1B_1^* - B_2B_2^* & A \end{bmatrix}, \qquad \widetilde{H} = \begin{bmatrix} -A^* - C_1^*C_1Y & -C_1^*C_1 \\ YC_2^*C_2Y - B_2B_2^* & A + YC_1^*C_1 \end{bmatrix}.$$

In the terminology of Section 12.1 the matrix H is the Hamiltonian of the Riccati equation (20.34) with $\gamma = 1$, while $\widetilde{H}$ is the Hamiltonian of the Riccati equation (20.28). Introduce also

$$S = \begin{bmatrix} I_n & 0 \\ Y & I_n \end{bmatrix}.$$

Since Y is a solution of the Riccati equation (20.35) and $\gamma = 1$, a direct computation gives $S^{-1}HS = \widetilde{H}$.

Part 2. Here we assume that Z is the (unique) Hermitian $i\mathbb{R}$-stabilizing solution of equation (20.28), and in addition that Z is positive definite. That is, it is assumed that condition (ii) in Theorem 20.7 is met. Since Z is $i\mathbb{R}$-stabilizing, the space $\operatorname{Im} \begin{bmatrix} Z^* & I_n \end{bmatrix}^*$ is the spectral subspace of $\widetilde{H}$ corresponding to the open left half plane. It follows that

$$S\operatorname{Im} \begin{bmatrix} Z \\ I_n \end{bmatrix} = \operatorname{Im} \begin{bmatrix} Z \\ I_n + YZ \end{bmatrix}$$

is the spectral subspace of H corresponding to the open left half plane.

Our next concern is the invertibility of $I_n + YZ$. Since Z is positive definite, $I_n + YZ = Z^{-1/2}(I_n + Z^{1/2}YZ^{1/2})Z^{1/2}$ is similar to a positive definite matrix. Consequently, $I_n + YZ$ is invertible.

Next, put $X = Z(I_n + YZ)^{-1}$. We shall show that X is positive definite, X is the $i\mathbb{R}$-stabilizing solution of (20.34) (with $\gamma = 1$), and that $X < Y^{-1}$. For this, note that $X = Z(I_n + YZ)^{-1} = (Z^{-1} + Y)^{-1}$, so that X is positive definite. Furthermore,

$$\operatorname{Im} \begin{bmatrix} Z \\ I_n + YZ \end{bmatrix} = \operatorname{Im} \begin{bmatrix} X \\ I_n \end{bmatrix}.$$

Hence X is the Hermitian $i\mathbb{R}$-stabilizing solution of (20.34). In addition, since Z is positive definite also $X^{-1} > Y$, and as both X and Y are positive definite this yields $X < Y^{-1}$. We conclude that all conditions of Theorem 20.10 are satisfied.

Part 3. This part deals with the reverse implication. So, we start with the positive definite $i\mathbb{R}$-stabilizing solution X of (20.34) with $\gamma = 1$ such that $X < Y^{-1}$. We show that $Z = (I_n - YX)^{-1}$ is well-defined and positive definite, and that Z is the $i\mathbb{R}$-stabilizing solution of (20.28). Since $X < Y^{-1}$, the matrix $I - YX$ is invertible, hence Z is well-defined. In addition, $Z = X(I_n - YX)^{-1} = (X^{-1} - Y)^{-1}$ is positive definite because $X < Y^{-1}$.

Recall that $\operatorname{Im}\begin{bmatrix} X^* & I_n \end{bmatrix}^*$ is the spectral subspace of H corresponding to the open left half plane. It follows that

$$S^{-1}\operatorname{Im}\begin{bmatrix} X \\ I_n \end{bmatrix} = \operatorname{Im}\begin{bmatrix} X \\ I_n - YX \end{bmatrix} = \operatorname{Im}\begin{bmatrix} Z \\ I_n \end{bmatrix}$$

is the spectral subspace of $\widetilde{H}$ corresponding to the open left half plane. Thus Z is the Hermitian $i\mathbb{R}$-stabilizing solution of (20.28) and, in addition, Z is positive definite. So all conditions of Theorem 20.7 are met.

Part 4. We have shown that the conditions in Theorem 20.7 are equivalent to the conditions in Theorem 20.10. It remains to show that the parametrizations in both theorems are equivalent. The parametrization in Theorem 20.10 is obtained by applying the Redheffer transformation to the function Ψ of Theorem 20.7 in order to arrive at a formula for Φ. Indeed,

$$(\Psi_{11}U + \Psi_{12})(\Psi_{21}U + \Psi_{22})^{-1} = \Phi_{11} + \Phi_{12}U(I_m - \Phi_{22}U)^{-1}\Phi_{21}$$

if the functions Ψ and Φ are connected via

$$\Phi = \begin{bmatrix} \Psi_{12}\Psi_{22}^{-1} & \Psi_{11} - \Psi_{12}\Psi_{22}^{-1}\Psi_{21} \\ \Psi_{22}^{-1} & -\Psi_{22}^{-1}\Psi_{21} \end{bmatrix},$$

and this, up to an interchange of the columns, is the Redheffer transform. The desired expression for Φ is now obtained by applying Theorem 17.21 to Ψ. □

Notes

The approach to H-infinity control using factorization presented in this chapter follows closely the lines of [76], see also [77]. A precursor of this approach is [43]. Theorem 20.10 originates from [38]; the proof given there is based on arguments from optimal control theory, rather than on a factorization approach, see also [84], [150]. An interpolation approach to the problems considerd in this chapter can be found in Part V of [7]. The present chapter discusses the H-infinity control problem for systems in continuous time. The H-infinity control problem for systems in discrete time was first considered in [138], see also [139], or [42] which employs commutant lifting techniques.

Bibliography

[1] V.M. Adamjan, D.Z. Arov, M.G. Krein, Infinite Hankel matrices and generalized problems of Carathéodory-Fejér. *Functional Anal. i Prilozen* **2** (1968), 1–19 [Russian].

[2] V.M. Adamjan, D.Z. Arov, M.G. Krein, Infinite block Hankel matrices and their connection with the interpolation problem. *Amer. Math. Soc. Transl.* (2) **111** (1978), 133–156 [Russian original 1971].

[3] D. Alpay, I. Gohberg, Unitary rational matrix functions. In: *Topics in Interpolation Theory of Rational Matrix-valued Functions.* OT **33**, Birkhäuser Verlag, Basel, 1988, pp. 175–222.

[4] B.D.O. Anderson, S. Vongpantilerd, *Network analysis and synthesis.* Prentice Hall, Englewood Cliffs, N.J., 1973.

[5] D.Z. Arov, H. Dym, *J-contractive matrix valued functions and related topics.* Encyclopedia of Mathematics and its Applications **116**, Cambridge University Press, 2008.

[6] T.Ya. Azizov, I.S. Iohvidov, *Linear operators in spaces with an indefinite metric.* John Wiley and Sons, Chichester, 1989 [Russian original 1979].

[7] J.A. Ball, I. Gohberg, L. Rodman, *Interpolation of rational matrix functions.* OT **45**, Birkhäuser Verlag, Basel, 1986.

[8] J.A. Ball, A.C.M. Ran, Left versus right canonical Wiener-Hopf factorization. In: *Constructive methods of Wiener-Hopf factorization.* OT **21**, Birkhäuser Verlag, Basel, 1986, pp. 9–38.

[9] J.A. Ball, A.C.M. Ran, Hankel norm approximation of rational matrix functions in terms of their realization. In: *Modelling, Identification and Robust Control* (eds. C. Byrnes, A. Lindquist), North Holland, Amsterdam, 1986, pp. 285–296.

[10] S. Barnett, *Introduction to mathematical control theory.* Oxford, Clarendon Press, 1975.

[11] H. Bart, I. Gohberg, M.A. Kaashoek, *Minimal factorization of matrix and operator functions.* OT **1**, Birkhäuser Verlag, Basel, 1979.

[12] H. Bart, I. Gohberg, M.A. Kaashoek, Wiener-Hopf integral equations. In: *Toeplitz Centennial.* OT **4**, Birkhäuser Verlag, Basel, 1982, pp. 85–135.

[13] H. Bart, I. Gohberg, M.A. Kaashoek, The coupling method for solving integral equations. In: *Topics in operator theory and networks, the Rehovot workshop.* OT **12**, Birkhäuser Verlag, Basel, 1984, pp. 39–74.

[14] H. Bart, I. Gohberg, M.A. Kaashoek, Fredholm theory of Wiener-Hopf equations in terms of realization of their symbols. *Integral Equations and Operator Theory* **8** (1985), 590–613.

[15] H. Bart, I. Gohberg, M.A. Kaashoek, Wiener-Hopf equations with symbols analytic in a strip. In: *Constructive methods of Wiener-Hopf factorization.* OT **21**, Birkhäuser Verlag, Basel, 1986, 39–74.

[16] H. Bart, I. Gohberg, M.A. Kaashoek, Wiener-Hopf factorization, inverse Fourier transforms and exponentially dichotomous operators. *J. Funct. Anal.* **8** (1986), 1–42.

[17] H. Bart, I. Gohberg, M.A. Kaashoek, Explicit Wiener-Hopf factorization and realization. In: *Constructive methods of Wiener-Hopf factorization.* OT **21**, Birkhäuser Verlag, Basel, 1986, pp. 235–316.

[18] H. Bart, I. Gohberg, M.A. Kaashoek, Invariants for Wiener-Hopf equivalence of analytic operator functions. In: *Constructive methods of Wiener-Hopf factorization.* OT **21**, Birkhäuser Verlag, Basel, 1986, 317–355.

[19] H. Bart, I. Gohberg, M.A. Kaashoek, A.C.M. Ran, Schur complements and realizations of transfer functions. *Lin. Alg. Appl.* **399** (2005), 203–224.

[20] H. Bart, I. Gohberg, M.A. Kaashoek, A.C.M. Ran, *Factorization of matrix and operator functions: the state space method.* OT **178**, Birkhäuser Verlag, Basel, 2008.

[21] H. Bart, I. Gohberg, M.A. Kaashoek, P. Van Dooren, Factorizations of transfer functions. *SIAM J. Control Opt.* **18** (1980), 675–696.

[22] A. Ben-Artzi, I. Gohberg, M.A. Kaashoek, Invertibility and dichotomy of differential operators on a half line. *J. Dynamics and Differential Equations* **5** (1993), 1–36.

[23] V. Belevitch, *Classical Network Theory.* Holden Day, San Francisco, 1968.

[24] A. Boettcher, B. Silbermann. *Analysis of Toeplitz operators.* Springer Verlag, Berlin, 1990.

[25] J. Bognar. *Indefinite inner product spaces.* Springer Verlag, Berlin, 1974.

[26] R.W. Brockett, R.A. Skoog, A new perturbation theory for the synthesis of nonlinear networks. *Mathematical aspects of electrical network analysis (Proc. SIAM-AMS Sympos. Appl. Math., New York, 1969)*, pp. 17–33. SIAM-AMS Proc., Vol. III, *Amer. Math. Soc., Providence, R.I.*, 1971.

[27] M.S. Brodskii, *Triangular and Jordan representations of linear operators.* Transl. Math. Monogr., Vol. **32**, Amer. Math. Soc., Providence, R.I., 1970.

[28] K.M. Case, P.F. Zweifel, *Linear Transport Theory.* Addison-Wesley, Reading MA, 1967.

[29] K. Clancey, I. Gohberg, *Factorization of Matrix Functions and Singular Integral Operators.* OT **3**, Birkhäuser Verlag, Basel, 1981.

[30] D.J. Clements, Rational spectral factorization using state space methods. *Systems and Control Letters* **20** (1993), 335–343.

[31] D.J. Clements, K. Glover, Spectral factorization by Hermitian pencils. *Lin. Alg. Appl.* **122–124** (1989), 797-846.

[32] J.B. Conway, *Functions of one complex variable.* Graduate Texts in Math. **11**, Springer Verlag, Berlin, 1973.

[33] M.J. Corless, A.E. Frazho, *Linear systems and control. An operator perspective.* Marcel Dekker, Inc., New York, NY, 2003.

[34] A.N. Curilov, On the solutions of quadratic matrix equations. *Nonlinear Vibrations and Control Theory* (Udmurt State University, Izhevsk) **2** (1978), 24–33 [Russian].

[35] R.F. Curtain, H.J. Zwart, *An introduction to infinite-dimensional linear systems theory.* Texts in Applied Math. **21**, Springer Verlag, New York, 1995.

[36] P. Dewilde, Generalized Darlington synthesis. *IEEE Trans. Circ. Syst. I: Fund. Theory Appl.*, **45**, (1999), 41–58.

[37] R.G. Douglas, *Banach algebra techniques in operator theory* (Second Edition). Springer Verlag, New York, 1998.

[38] J.C. Doyle, K. Glover, P.P. Khargonekar, B.A. Francis, State-space solutions to standard H_2 and H_∞ control problems. *IEEE Trans. Automatic Control*, **34**, (1989), 831–847.

[39] H. Dym, *J-contractive matrix functions, reproducing kernel Hilbert spaces and interpolation.* CBMS Lecture Notes **71**, Amer. Math. Soc, 1989.

[40] I.A. Feldman, Wiener-Hopf operator equation and its application to the transport equation. *Mat. Issled.* **6** (1971), 115–132 [Russian]; English translation: *Integral Equations and Operator Theory* **3** (1980), 43–61.

[41] L. Finesso, G. Picci, A characterization of minimal square spectral factors. *IEEE Trans. Automatic Control*, **27** (1982), 122–127.

[42] C. Foias, A.E. Frazho, I. Gohberg, M. A. Kaashoek, *Metric constrained interpolation, commutant lifting and systems.* OT **100**, Birkhäuser Verlag, Basel, 1998.

[43] B.A. Francis, *A course in H_∞ Control Theory.* Springer Verlag, Berlin etc. 1987.

[44] A.E. Frazho, M.A. Kaashoek, A.C.M. Ran, The Non-symmetric Discrete Algebraic Riccati Equation and Canonical Factorization of Rational Matrix Functions on the Unit Circle. *Integral Equations and Operator Theory* **66** No. 2 (2010), 215–229.

[45] P.A. Fuhrmann, On symmetric rational transfer functions. *Lin. Alg. Appl.* **50**, (1983) 167–250.

[46] P. Fuhrmann, A. Gombani, On the Lyapunov equation, coinvariant subspaces and some problems related to spectral factorizations. *International Journal of Control* **73** (2000), 1129–1159.

[47] I. Gelfand, D. Raikov, G. Shilov, *Commutative normed rings.* Chelsea Publ.Inc., Bronx, NY, 1964.

[48] I.C. Gohberg, On systems of singular integral equations. *Kishinev. State. Univ. Uč. Zap.* **11** (1954), 55–60 [Russian].

[49] I.C. Gohberg, The problem of factorization of operator functions. *Izvestija Akad. Nauk SSSR*, Series Math. **28** (1964), 1055–1082 [Russian].

[50] I.C. Gohberg, I.A. Feldman, *Convolution equations and projection methods for their solution.* Transl. Math. Monogr., Vol. **41**, Amer. Math. Soc., Providence, R.I., 1974.

[51] I. Gohberg, S. Goldberg, M.A. Kaashoek, *Classes of Linear Operators*, Vol. I. OT **49**, Birkhäuser Verlag, Basel, 1990.

[52] I. Gohberg, S. Goldberg, M.A. Kaashoek, *Classes of Linear Operators*, Vol. II. OT **63**, Birkhäuser Verlag, Basel, 1993.

[53] I. Gohberg, S. Goldberg, M.A. Kaashoek, *Basic Classes of Linear Operators.* Birkhäuser Verlag, Basel, 2003.

[54] I. Gohberg, M.A. Kaashoek (Eds), *Constructive methods of Wiener-Hopf factorization.* OT **21**, Birkhäuser Verlag, Basel, 1986.

[55] I. Gohberg, M.A. Kaashoek, Block Toeplitz operators with rational symbols. In: *Contributions to Operator Theory and its Applications.* OT **35**, Birkhäuser Verlag, Basel, 1988, pp. 385–440.

[56] I. Gohberg, M.A. Kaashoek, The state space method for solving singular integral equations. In: *Mathematical System Theory. The influence of R.E. Kalman*, Springer Verlag, Berlin 1991, pp. 509–523.

[57] I. Gohberg, M.A. Kaashoek, D.C. Lay, Equivalence, linearization and decompositions of holomorphic operator functions. *J. Funct. Anal.* **28** (1978), 102–144.

[58] I. Gohberg, M.A. Kaashoek, F. van Schagen, *Partially specified matrices and operators, classification, completion, applications.* OT **79**, Birkhäuser Verlag, Basel, 1995.

[59] I. Gohberg, M.A. Kaashoek, I.M. Spitkovsky, An overview of matrix factorization theory and operator applications. In: *Factorization and integrable systems.* OT **141**, Birkhäuser Verlag, Basel, 2003, pp. 1–102.

[60] I.C. Gohberg, M.G. Kreĭn, The basic propositions on defect numbers, root numbers, and indices of linear operators. *Uspehi Mat. Nauk* **12**, no. 2 (74) (1957), 43–118 [Russian]; English translation: *Amer. Math. Soc. Transl.* (Series 2) **13** (1960), 185–265.

[61] I.C. Gohberg, M.G. Kreĭn, Systems of integral equations on a half line with kernels depending on the difference of arguments. *Uspehi Mat. Nauk* **13**, no. 2 (80) (1958), 3–72 [Russian]; English translation: *Amer. Math. Soc. Transl.* (2) **14** (1960), 217–287.

[62] I. Gohberg, N. Krupnik, *Einfüring in die Theorie der eindimensionalen singulären Integral operatoren.* Birkhäuser Verlag, Basel, 1979.

[63] I. Gohberg, N. Krupnik, *One-Dimensional Linear Singular Integral Equations. I. Introduction.* OT **53**, Birkhäuser Verlag, Basel, 1992.

[64] I. Gohberg, N. Krupnik, *One-Dimensional Linear Singular Integral Equations. II. General Theory and Applications.* OT **54**, Birkhäuser Verlag, Basel, 1992.

[65] I. Gohberg, P. Lancaster, L. Rodman, Spectral analysis of matrix polynomials. II. The resolvent form and spectral divisors. *Lin. Alg. Appl.* **21** (1978), 65–88.

[66] I. Gohberg, P. Lancaster, L. Rodman, *Spectral analysis of selfadjoint matrix polynomials.* Annals of Mathematics **112** (1980), 33–71.

[67] I. Gohberg, P. Lancaster, L. Rodman, *Matrix polynomials.* Academic Press, Inc., New York, NY, 1882.

[68] I. Gohberg, P. Lancaster, L. Rodman, *Matrices and Indefinite Scalar Products.* OT **8**, Birkhäuser Verlag, Basel, 1983.

[69] I. Gohberg, P. Lancaster, L. Rodman, *Invariant subspaces of matrices with applications.* John Wiley & Sons, Inc., New York, NY, 1986 = Classics in Appl. Math. **51**, SIAM, Philadelphia, PA, 2006.

[70] I. Gohberg, P. Lancaster, L. Rodman, *Indefinite linear algebra and applications.* Birkhäuser Verlag, Basel, 2005.

[71] I.C. Gohberg, J. Leiterer, Factorization of operator functions relative to a contour. III. Factorization and algebras. *Math. Nachr.* **55** (1973) 33–61 [Russian].

[72] I.C. Gohberg, J. Leiterer, A criterion for factorization of operator functions with respect to a contour. *Sov. Math. Dokl.* **14** (1973), 425–429.

[73] I. Gohberg, J. Leiterer, *Holomorphic Operator Functions of One Variable and Applications. Methods from Complex Analysis in Several Variables.* OT **192**, Birkhäuser Verlag, Basel, 2009.

[74] I. Gohberg, A.C.M. Ran, On pseudo-canonical factorization of rational matrix functions. *Indagationes Mathematica* N.S. **4** (1993), 51–63.

[75] I. Gohberg, S. Rubinstein, Proper contractions and their unitary minimal completions. In: *Topics in Interpolation Theory of Rational Matrix-valued Functions.* OT **33**, Birkhäuser Verlag, Basel, 1988, pp. 223–247.

[76] M. Green, K. Glover, D. Limebeer, J. Doyle, A J-spectral approach to H_∞-control. *SIAM Journal of Control and Optimization* **28** (1990), 1350–1371.

[77] M. Green, D.J.N. Limebeer, *Linear robust control.* Prentice Hall, Engelwood Cliffs, NJ, 1995.

[78] W. Greenberg, C.V.M. van der Mee, V. Protopopescu, *Boundary value problems in abstract kinetic theory.* OT **25**, Birkhäuser Verlag, Basel, 1986.

[79] G.J. Groenewald, *Wiener-Hopf factorization of rational matrix functions in terms of realizations, an alternative version.* Ph. D. Thesis, Vrije Universiteit, Amsterdam, 1993.

[80] R.J. Hangelbroek, *A functional analytic approach to the linear transport equation.* Ph. D. Thesis Universiteit Groningen, 1973; also in: *Transport Theory and Statist. Phys.* **5**, 1–85.

[81] R.J. Hangelbroek, On the derivation of some formulas in linear transport theory for media with anisotropic scattering. Dept. Math. Kath. Univ. Nijmegen, Netherlands, Report 7720 (1978).

[82] R.J. Hangelbroek, C.G. Lekkerkerker, Decompositions of a Hilbert space and factorization of a W-A determinant. *Siam J. Math. Analysis* **8** (1977), 459–472.

[83] B. Hassibi, A.H. Sayed, T. Kailath, *Indefinite-quadratic estimation and control, a unified approach to H^2 and H^∞ theories.* SIAM Studies in Applied and Numerical Mathematics, Vol. **16**, SIAM, 1999.

[84] M. Hautus, H.L. Trentelman and A.A. Stoorvogel, *Control Theory for Linear Systems.* Springer Verlag, 2001.

[85] C. Heij, A.C.M. Ran, F. van Schagen, *Introduction to Mathematical Systems Theory.* Birkhäuser Verlag, Basel, 2007.

[86] O. Iftime, M.A. Kaashoek, A. Sasane, A Grassmannian band method approach to the Nehari-Takagi problem. *J. Math. Anal. Appl.* **310** (2005), 97–115.

[87] I.S. Iohvidov, M.G. Krein, H. Langer, *Introduction to the spectral theory of operators in spaces with an indefinite metric.* Akademie-Verlag, Berlin, 1982.

[88] V.A. Jacubovič, Factorization of symmetric matrix polynomials, *Soviet Math. Doklady* **11** (1970), 1261–1264.

[89] M.A. Kaashoek, Ascent, descent, nullity and defect, a note on a paper of A.E. Taylor. *Math. Annalen* **172** (1967), 105–115.

[90] M.A. Kaashoek, C.V.M. van der Mee, A.C.M. Ran, Weighting operator patterns of Pritchard-Salamon realizations. *Integral Equations and Operator Theory* **27** (1997), 48–70; Errata in: *Integral Equations and Operator Theory* **52** (2005), 451–453.

[91] M.A. Kaashoek, C.V.M. van der Mee, A.C.M. Ran, Wiener-Hopf factorization of transfer functions of extended Pritchard-Salamon realizations. *Math. Nachr.* **196** (1998), 71–102.

[92] M.A. Kaashoek, J.N.M. Schermer, Inversion of convolution equations on a finite interval and realization triples. *Integral Equations and Operator Theory* **13** (1990), 76–103.

[93] M. A. Kaashoek, S.M. Verduyn Lunel, An integrability condition on the resolvent for hyperbolicity of the semigroup. *J. Differential Equations* **112** (1994), 374–406.

[94] T. Kailath, *Linear systems.* Prentice-Hall, Inc., Englewoord Cliffs, N.J., 1980.

[95] R.E. Kalman, M.A. Falb, M.A. Arbib, *Topics in mathematical system theory.* McGraw-Hill, New York, 1969.

[96] H.G. Kaper, C.G. Lekkerkerker, J. Hejtmanek, *Spectral methods in linear transport theory.* OT **5**, Birkhäuser Verlag, Basel, 1982.

[97] I. Karelin, L. Lerer, A.C.M. Ran, J-symmetric factorizations and algebraic Riccati equations. In: *Recent Advances in Operator Theory,* The Israel Gohberg Anniversary Volume, A. Dijksma, M.A. Kaashoek and A.C.M. Ran (Eds.). OT **124**, 2001, Birkhäuser Verlag, Basel, 1997, pp. 319–360.

[98] T. Katayama, (J, J')-spectral factorization and conjugation for discrete-time descriptor system. *Circuits Systems Signal Process.* **15** (1996), 649–669.

[99] T. Kato, *Perturbation theory for linear operators.* Springer, Berlin-Heidelberg-New York, 1966.

[100] N.M. Kostenko, Necessary and sufficient condition for factorization of rational matrix functions. *Funct. Anal. Appl.* **12** (1978), 87–88 [Russian].

[101] M.G. Kreĭn, Introduction to the geometry of indefinite J-spaces and to the theory of operators in these spaces. *Amer. Math. Soc. Transl.* (2) **93** (1970), 103–176.

[102] V. Kučera, A contribution to matrix quadratic equations. *IEEE Trans. Automatic Control.* **AC-17** (1972), no. 3, 344–347.

[103] H. Kwakernaak, M. Şebek, Polynomial J-spectral factorization. *IEEE Trans. Automatic Control* **39** (1994), 315–328.

[104] P. Lancaster, A.C.M. Ran, L. Rodman, Hermitian solutions of the discrete algebraic Riccati equation. *Int. J. Control* **44** (1986), 777–802.

[105] P. Lancaster, L. Rodman, Existence and uniqueness theorems for algebraic Riccati equations. *Int. J. Control* **32** (1980), 285–309.

[106] P. Lancaster, L. Rodman, *Algebraic Riccati equations.* Clarendon Press, Oxford, 1995.

[107] P. Lancaster, M. Tismenetsky, *The Theory of Matrices.* Academic Press, San Diego, 1985.

[108] C.G. Lekkerkerker, On eigendistributions in linear transport theory. *Proc. Royal Soc. Edinburgh* **83A** (1979), 303–326.

[109] A.S. Markus, V.I. Macaev, Two remarks on factorization of matrix functions. *Mat. Issled.* **42** (1976), 216–223.

[110] C. van der Mee, *Semigroup and factorization methods in transport theory.* Ph. D. Thesis, Vrije universteit, Amsterdam, October 21, 1981 = Mathematical Centre Tracts **146**, Mathematisch Centrum Amsterdam, 1981.

[111] C. van der Mee, *Exponentially dichotomous operators and applications.* OT **182**, Birkhäuser Verlag, Basel 2008.

[112] G.J. Meinsma, J-spectral factorization and equalizing vectors. *Systems Control Lett.* **25** (1995), 243–249.

[113] B. Mitiagin, Linearization of holomorphic operator functions. I, II. *Integral Equations and Operator Theory* **1** (1978), 114–131 and 226–249.

[114] Z. Nehari, On bounded bilinear forms. *Ann. of Math.* **65** (1957), 153–162.

[115] C.N. Nett, C.A. Jacobson, M.J. Balas, A connection between state-space and doubly coprime fractional representations. *IEEE Trans. Automatic Control,* **29** (1984), 831–832.

[116] M.A. Petersen, A.C.M. Ran, Minimal nonsquare spectral factors. *Lin. Alg. Appl.* **351-352** (2002), 553-565.

[117] D.R. Pik, *Block lower triangular operators and optimal contractive systems.* Ph. D. Thesis, Vrije Universiteit, Amsterdam, June 2, 1999.

[118] V.P. Potapov, The multiplicative structure of J-contractive matrix functions. *Trudy Moskov. Mat. Obşç.* **4** (1955), 125–236, [Russian]; English translation: *Amer. Math. Soc. Transl.* (2) **15** (1960), 131–243.

[119] A.C.M. Ran, Minimal factorization of selfadjoint rational matrix functions. *Integral Equations and Operator Theory* **5** (1982), 850–869.

[120] A.C.M. Ran, *Semidefinite invariant subspaces, stability and applications.* Ph. D. Thesis, Vrije Universiteit, Amsterdam, June 29, 1984.

[121] A.C.M. Ran, Necessary and sufficient conditions for existence of J-spectral factorization for para-Hermitian rational matrix functions. *Automatica* **39** (2003), 1935–1939.

[122] A.C.M. Ran, L. Rodman, Stability of invariant maximal semidefinite subspaces I. *Lin. Alg. Appl.* **62** (1984), 51–86.

[123] A.C.M. Ran, L. Rodman, Stability of invariant maximal semidefinite subspaces II. Applications: Self-adjoint rational matrix functions, algebraic Riccati equations. *Lin. Alg. Appl.* **63** (1984), 133–173.

[124] A.C.M. Ran, L. Rodman, A matricial boundary value problem which appears in transport theory. *J. Math. Anal. Appl.* **130** (1988), 200–222.

[125] A.C.M. Ran, L. Rodman, On symmetric factorizations of rational matrix functions. *Linear and Multilinear Algebra* **29** (1991), 243–261.

[126] A.C.M. Ran, L. Rodman, Generalized canonical factorization of matrix and operator functions with definite Hermitian part. *Integral Equations and Operator Theory* **15** (1992), 673–696.

[127] A.C.M. Ran, L. Rodman, A Class of Robustness Problems in Matrix Analysis. In: *The Harry Dym Anniversary Volume.* OT **134**, Birkhäuser Verlag, Basel 2002, pp. 337–383.

[128] A.C.M. Ran, L. Rodman, D. Temme, Stability of pseudo-spectral factorizations. In: *Operator Theory and Analysis, The M.A. Kaashoek Anniversary Volume.* OT **122**, Birkhäuser Verlag, Basel, 2001, pp. 359–383

[129] A.C.M. Ran, D. Temme, Invariant semidefinite subspaces of dissipative matrices in an indefinite inner product space, existence, construction and uniqueness. *Lin. Alg. Appl.* **212/213** (1994), 169–214.

[130] R.M. Redheffer, On a certain linear fractional transformation. *J. Math. Physics* **39** (1960), 269–286.

[131] L. Rodman, *An introduction to operator polynomials.* OT **38**, Birkhäuser Verlag, Basel, 1989.

[132] L. Roozemond, Canonical pseudo-spectral factorization and Wiener-Hopf integral equations. In: *Constructive methods of Wiener-Hopf factorization.* OT **21**, Birkhäuser Verlag, Basel, 1986, pp. 127–156.

[133] L. Roozemond, *Systems of non-normal and first kind Wiener-Hopf equations.* Ph. D. Thesis, Vrije Universiteit, Amsterdam, March 5, 1987.

[134] W. Rudin, *Real and complex analysis* (International Student Edition). McGraw-Hill, London, 1970.

[135] M.A. Shayman, Geometry of the algebraic Riccati equation I. *SIAM J. Control Optim.* **21** (1983), 375–394.

[136] M.A. Shayman, Geometry of the algebraic Riccati equation II. *SIAM J. Control Optim.* **21** (1983), 395–409.

[137] A.A. Shkalikov, Selection principles and properties of some proper and adjoined elements of operator pencils. *Moscow Univ. Math. Bull.* **43**, 1988, 16–25.

[138] A.A. Stoorvogel, The discrete time H_∞ control problem with measurement feedback. *SIAM J. Control Optim.* **30** (1992), 182–202.

[139] A.A. Stoorvogel, *The H_∞ control problem: a state space approach.* Ph. D. Thesis, Eindhoven University of Technology, Eindhoven, 1990.

[140] F. Stummel, Diskrete Konvergenz linearer Operatoren. II. *Math. Z.* **120** (1971), 231–264.

[141] B. Sz-Nagy, C. Foias, *Analyse harmonique des opérateurs de l'espace de Hilbert.* Paris, Masson and Akad. Kiado, Budapest, 1967.

[142] T. Takagi, On an algebraic problem related to an analytic theorem of Carathédory and Fejer. *Japan J. Math.* **1** (1924), 83–93.

[143] S.H. Tan, J. Vandewalle, On factorization of rational matrices. *IEEE Trans. Circuits and Systems* **35** (1988), 1179-1181.

[144] A.E. Taylor, D.C. Lay, *Introduction to Functional Analysis* (Second Edition). John Wiley and Sons, New York, 1980.

[145] H.L. Trentelman, P. Rapisarda, New algorithms for polynomial J-spectral factorization. *Math. Control Signals Systems* **12** (1999), 24–61.

[146] M. Weiss, Spectral and inner-outer factorizations through the constrained Riccati equation. *IEEE Trans. Automatic Control* **AC-39** (1994), 677–681.

[147] J.C.Willems, Least squares stationary optimal control and the algebraic Riccati equation. *IEEE Trans. Automatic Control* **AC-16** (1971), 621–634.

[148] D.C. Youla, H.A. Jabr, J.J. Bongiorno Jr., Modern Wiener-Hopf design of optimal controllers. II. The multivariable case. *IEEE Trans. Automatic Control* **AC-21** (1976), no. 3, 319–338.

[149] K. Zhou, J.C. Doyle, *Essentials of Robust Control.* Prentice Hall, Upper Saddle River, New Jersey, 1998.

[150] K. Zhou, J.C, Doyle, K. Glover, *Robust and optimal control.* Prentice Hall, Engelwood Cliffs, NJ, 1996.

[151] Y. Zucker, *Constructive factorization and partial indices of rational matrix functions.* Ph. D. Thesis, Tel-Aviv University, Tel-Aviv, 1998.

List of symbols

Symbol	Description
$\emptyset$	empty set
$\mathbb{N}$	the set consisting of the natural numbers $1, 2, 3, \ldots$
$\Re\lambda$	real part of complex number λ
$\Im\lambda$	imaginary part of complex number λ
$\bar{\lambda}$	complex conjugate of complex number λ
$\mathbb{R}$	real line
$\mathbb{R}_+$	$[0, \infty)$, set of nonnegative real numbers
$\mathbb{R}_\infty$	extended real line
$\mathbb{C}$	complex plane
$\mathbb{C}_\infty$	Riemann sphere
$\mathbb{C}_{\text{left}}$	open left half plane
$\mathbb{C}_{\text{right}}$	open right half plane
$\mathbb{C}_+$	open upper half plane
$\mathbb{C}_-$	open lower half plane
$i\mathbb{R}$	imaginary axis
$\mathbb{T}$	unit circle in the complex plane
$\mathbb{D}$	open unit disc in the complex plane
$\mathbb{D}_{\text{ext}}$	complement of the closure of the unit disc in $\mathbb{C}_\infty$
F_+	interior domain (of oriented curve in the complex plane)
F_-	exterior domain (of oriented curve in the complex plane) infinity included
$\mathbb{C}^m$	Euclidean space of complex m-vectors
x^*	complex conjugate of vector x in $\mathbb{C}^n$
f'	derivative of a differentiable or absolutely continuous function f
$h * f$	convolution product of h and f
a.e.	abbreviation of almost everywhere
$J_n(\alpha)$	Jordan block of order n and with eigenvalue α

$\operatorname{diag}(T_1, \ldots, T_k)$	block-diagonal matrix (also called block-diagonal sum) with $T_1, \ldots, T_k$ on the main block-diagonal
$\sharp V$	number of elements in (finite) set V
$\overline{V}$	closure of subset V of topological space
$\dim M$	dimension of linear manifold M
$\operatorname{codim} M$	codimension of linear manifold M
$\frac{M}{N}$, M/N	quotient space of M over N
$\perp$	symbol for orthogonality in Hilbert space
$M^\perp$	orthogonal complement of subspace M in Hilbert space
$V \perp W$	orthogonality of sets V and W
$\oplus$	orthogonal direct sum (of subspaces) of Hilbert spaces
$\dot{+}$	algebraic (possibly non-orthogonal) direct sum of linear manifolds or (sub)spaces
$\lVert \cdot \rVert$	norm on a Hilbert or Banach space
I	identity matrix or identity operator on a Hilbert or Banach space
I_m	$m \times m$ identity matrix or identity operator on $\mathbb{C}^m$
I_X	identity operator on X
$\operatorname{Ker} A$	kernel or null space of operator or matrix A
$\operatorname{Im} A$	range or image of operator or matrix A
$\det A$	determinant of matrix A
A^*	adjoint of (complex) Hilbert space operator or (complex) matrix
$A \geq 0$	nonnegative matrix or operator A
$A > 0$	positive definite matrix or operator A
$A^{1/2}$	square root of positive definite matrix or operator
A^{-1}	inverse of invertible operator or matrix
A^{-*}	stands for $(A^*)^{-1}$
$\lambda - A$	shorthand for $\lambda I - A$ (standard practice)
$\rho(A)$	resolvent set of operator or matrix
$\sigma(A)$	spectrum of operator or matrix A
$P(A;\Gamma)$	stands for $\frac{1}{2\pi i}\int_\Gamma (\lambda - A)^{-1}\, d\lambda$, the Riesz or spectral projection associated with A and Γ
AM	image of M under operator A (also denoted by $A[M]$)
$A[M]$	image of M under operator A (also denoted by AM)
$A^{-1}[M]$	inverse image of M under operator A
$A\vert_M$	restriction of operator A to subspace M
$A(X_1 \to X_2)$	(possibly) unbounded operator A with domain in X_1 and range in X_2
$\mathcal{D}(A)$	domain of (possibly) unbounded operator A

$\mathcal{L}(Y)$	Banach algebra of all bounded linear operators on Banach space Y
$\mathcal{L}(U, Y)$	Banach space of all bounded linear operators from Banach space U into Banach space Y
$C(\Gamma, U)$	Banach space of all U-valued continuous functions on Γ endowed with the supremum norm
$L_p(\Omega)$	Lebesgue space of p-integrable functions on a measurable set Ω
$L_p^m(\Omega)$	space of $\mathbb{C}^m$-valued functions of which the entries are in $L_p(\Omega)$
$L_p^{m\times r}(\Omega)$	space of $m \times r$ matrix functions of which the columns are in $L_p^m(\Omega)$
$L_{1,\omega}^m(\mathbb{R})$	a weighted L_1^m-space; see Section 5.3
$\mathbf{D}_1^m(\mathbb{R})$	a certain linear submanifold of $L_1^m(\mathbb{R})$; see Section 5.3
$\mathbf{D}_1^m[0, \infty)$	linear manifold of all functions $f \in \mathbf{D}_1^m(\mathbb{R})$ with $f(t) = 0$ for $t < 0$
$L_2(\mathbb{R}_+, \mathcal{H})$	the space of all square integrable functions on $[0, \infty)$ with values in Hilbert space $\mathcal{H}$
$\langle \cdot\, , \cdot \rangle$	standard inner product in $\mathbb{C}^m$ or $L_2[-1, 1]$
$[\cdot\, , \cdot]$	alternative inner product in $\mathbb{C}^m$ or $L_2[-1, 1]$
$A^{[\star]}$	adjoint of an operator with respect to alternative inner product (in $L_2[-1, 1]$)
$D + C(\lambda I - A)^{-1}B$	realization
$A^{\times}$	associate state space operator (or matrix), associate main operator (or matrix) corresponding to a realization
$\operatorname{Ker}(C\vert A)$	stands for $\operatorname{Ker} C \cap \operatorname{Ker} CA \cap \operatorname{Ker} CA^2 \cap \cdots$
$\operatorname{Im}(A\vert B)$	stands for $\operatorname{Im} B + \operatorname{Im} AB + \operatorname{Im} A^2B + \cdots$
$E(\cdot\,; A)$	bisemigroup generated by exponentially dichotomous operator A
e^{tS}	value at $t(< 0)$ of the left semigroup generated by S
e^{tS}	value at $t(> 0)$ of the right semigroup generated by S
P_Θ	separating projection for $-iA$ where A is the main operator of the spectral triple Θ
$\operatorname{pr}_\Pi(\Theta)$	projection of realization triple $\Theta = (A, B, C)$ associated with a projection Π
W^{-1}	pointwise inverse of rational matrix function W, defined by $W^{-1}(\lambda) = W(\lambda)^{-1}$)
$\delta(W)$	McMillan degree of a rational matrix function W
$\delta(W; \lambda_0)$	local degree of W at λ_0

$\pi_+(W)$	number of positive eigenvalues of the Hermitian matrix associated with a minimal realization of J-unitary rational matrix function W
F^*	adjoint of the rational matrix function F relative to the imaginary axis, defined by $F^*(\lambda) = F(-\bar{\lambda})^*$
RAT	the set of all rational matrix functions that are proper and have no pole at the imaginary axis
$\text{RAT}^{p\times q}$	the set of all $p \times q$ matrix functions in RAT
$\text{RAT}^{p\times q}_{\mathbb{B}}$	the set of all F in $\text{RAT}^{p\times q}$ such that $\sup_{s\in i\mathbb{R}} \|F(s)\| \leq 1$
$\text{RAT}^{p\times q}_{+}$	the set of all matrix functions in $\text{RAT}^{p\times q}$ that are analytic on the closed left half plane, infinity included
$\text{RAT}^{p\times q}_{+,\mathbb{B}}$	the set $\text{RAT}^{p\times q}_{+} \cap \text{RAT}^{p\times q}_{\mathbb{B}}$
RAT_{-}	the set of all rational matrix functions that are analytic on the closed right half plane, infinity included, that is, the set of all $i\mathbb{R}$-stable rational matrix functions
$\text{RAT}^{p\times q}_{-}$	the set of all $p \times q$ matrix functions in RAT_{-}
E_p	the unit element in the algebra $\text{RAT}^{p\times p}$

Index

Subseries *Linear Operators and Linear Systems*

BIRKHÄUSER

Edited by
Daniel Alpay, Ben Gurion University of the Negev, Beer Sheeva, Israel
Birgit Jacob, Bergische Universität Wuppertal, Germany
André M.C. Ran, Vrije Universiteit, Amsterdam, The Netherlands

The theory of linear operators has had an important influence on the development of mathematical systems theory. On the other hand, mathematical systems theory serves as a direct source of motivation and new techniques for the theory of linear operators and its applications. The subseries "Linear Operators and Linear Systems" (LOLS) is dedicated to these connections between the theory of linear operators and the mathematical theory of linear systems. LOLS will continue in the tradition of the series Operator Theory: Advances and Applications and maintain the high quality of the volumes. Books published in LOLS will be either monographs or consist of essays presenting the state of the art and new results. The volumes will be addressed to a wide range of mathematicians from beginners to experts in these fields.

OT 204: Frazho, A. / Bhosri, W.,
An Operator Perspective on Signals and Systems (2009).
ISBN 978-3-0346-0291-4

This book provides an operator perspective to solve problems arising in systems theory and signal processing. It is divided into four parts. The first part is an introduction to basic operator techniques used throughout the monograph. The second part uses finite section methods with interpolation theory to develop algorithms to compute inner-outer factorizations and solve sinusoid estimation problems. The third part develops the Riccati method to solve factorization problems arising in systems and control. We also included the Kalman filter and its connection to Wiener filtering. The final part uses isometric extensions to present an introduction to positive real and contractive interpolation theory.

OT 200: Bart, H. / Gohberg, I. / Kaashoek, M.A., Ran, A.C.M.,
A State Space Approach to Canonical Factorization with Applications (2010).
ISBN 978-3-7643-8752-5

OT 184: van der Mee, C.V.M.,
Exponentially Dichotomous Operators and Applications (2008).
ISBN 978-3-7643-8731-0

In this monograph the natural evolution operators of autonomous first-order differential equations with exponential dichotomy on an arbitrary Banach space are studied in detail. Characterizations of these so-called exponentially dichotomous operators in terms of their resolvents and additive and multiplicative perturbation results are given. The general theory of the first three chapters is then followed by applications to Wiener-Hopf factorization and Riccati equations, transport equations, diffusion equations of indefinite Sturm-Liouville type, noncausal infinite-dimensional linear continuous-time systems, and functional differential equations of mixed type.

OT 178: Bart, H. / Gohberg, I. / Kaashoek, M.A. / Ran, A.C.M.,
Factorization of Matrix and Operator Functions: The State Space Method (2007).
ISBN 978-3-7643-8267-4

OT 165: Alpay, D. / Gohberg, I. (Eds.),
Interpolation, Schur Functions and Moment Problems (2006).
ISBN 978-3-7643-7546-1

OT 161: Alpay, D. / Gohberg, I. (Eds.),
The State Space Method. Generalizations and Applications (2005).
ISBN 978-3-7643-7370-2

MIX
Papier aus verantwortungsvollen Quellen
Paper from responsible sources
FSC® C105338

If you have any concerns about our products,
you can contact us on
ProductSafety@springernature.com

In case Publisher is established outside the EU,
the EU authorized representative is:
Springer Nature Customer Service Center GmbH
Europaplatz 3, 69115 Heidelberg, Germany

Printed by Libri Plureos GmbH
in Hamburg, Germany